Microbial Nanotechnology

Microbial Nanotechnology

Guest Editor

Kamel Ahmed Abd-Elsalam

MDPI • Basel • Beijing • Wuhan • Barcelona • Belgrade • Novi Sad • Cluj • Manchester

Guest Editor
Kamel Ahmed Abd-Elsalam
Plant Pathology Research Institute
Agricultural Research Center
Giza
Egypt

Editorial Office
MDPI AG
Grosspeteranlage 5
4052 Basel, Switzerland

This is a reprint of the Special Issue, published open access by the journal *Microorganisms* (ISSN 2076-2607), freely accessible at: www.mdpi.com/journal/microorganisms/special_issues/microbial_nanotechnology.

For citation purposes, cite each article independently as indicated on the article page online and using the guide below:

Lastname, A.A.; Lastname, B.B. Article Title. *Journal Name* **Year**, *Volume Number*, Page Range.

ISBN 978-3-7258-3446-4 (Hbk)
ISBN 978-3-7258-3445-7 (PDF)
https://doi.org/10.3390/books978-3-7258-3445-7

Contents

About the Editor

Kamel Ahmed Abd-Elsalam

Kamel A. Abd-Elsalam, Ph.D., is currently a Research Professor at the Plant Pathology Research Institute, Agricultural Research Center, Giza, Egypt. Dr. Kamel's research interests include developing and improving plant biosecurity diagnostic tools, understanding fungal pathogen genomes, and creating eco-friendly hybrid nanomaterials for controlling toxigenic fungi and plant diseases, as well as applications in agroecosystems. Kamel earned his Ph.D. in Molecular Plant Pathology from Christian Albrechts University of Kiel (Germany) and Suez Canal University (Egypt). He has published 26 books related to nanobiotechnology applications in agriculture and plant protection with major publishing houses such as Springer, Taylor & Francis, and Elsevier. Since 2019, he has served as the Editor-in-Chief of the Elsevier book series "*Nanobiotechnology for Plant Protection*" and as the Series Editor for "*Applications of Genome Modified Plants and Microbes in Food and Agriculture*". Additionally, he has been an active member of the Elsevier Advisory Panel since 2020, providing feedback and suggestions to improve Elsevier's products and services. He has published more than 252 scientific research articles in international and regional specialized journals with high impact factors, and has an h-index of 44, an i-10 index of 137, and over 7421 citations. He served as a Guest Editor for the *Journal of Fungi*, *Plants* and *Microorganisms*, and as a Review Editor for Frontiers in Genomic Assay Technology, Agrochemicals, and Plant Nano Biology. He was ranked in the top 2% of the most influential scientists in the world in nanobiotechnology for 2020, 2021, and 2022 by Elsevier and Stanford University. In 2014, he was awarded the Federation of Arab Scientific Research Councils Prize for excellent scientific research in biotechnology (fungal genomics), ranking first. Additionally, he received the Egyptian State Award for Agricultural Scientific Excellence from ASRT, Egypt, in 2022 with a silver medal.

microorganisms

Editorial

Special Issue: Microbial Nanotechnology

Kamel A. Abd-Elsalam

Plant Pathology Research Institute, Agricultural Research Centre, Giza 12619, Egypt; kamelabdelsalam@gmail.com

1. Introduction

Microbial nanotechnology (MN), or microbial nanobiotechnology, is a rapidly expanding research area with the potential to transform various fields, including bioremediation, energy production, medicine, and agriculture [1]. This state-of-the-art technology involves merging nanotechnology with microorganisms to produce innovative materials with distinct properties and functions. For instance, microbial nanotechnology has been utilized to create eco-friendly methods for bioremediating industrial effluents, which can aid in reducing pollution and safeguarding the environment [2]. Moreover, microbial biosynthetic nanoparticles, consisting of a biomolecule capping layer adsorbed on the surface, have been fabricated using microbial nanotechnology. These nanoparticles have demonstrated encouraging results in drug delivery, imaging, and cancer therapy [3]. The immense potential of microbial nanotechnology makes it a stimulating research area with substantial implications for numerous fields.

Microorganisms have evolved remarkable biochemical pathways and mechanisms that enable them to interact with and transform materials at the nanoscale. They possess inherent biological machinery that can produce nanoparticles with precise sizes, shapes, and surface functionalities. These microbial nanoparticles exhibit distinct physical, chemical, and biological properties, making them highly versatile and applicable in various technological domains [4].

One of the primary areas where microbial nanotechnology has shown immense potential is in the synthesis of nanoparticles. Microorganisms can produce nanoparticles through both intracellular and extracellular processes. Intracellular synthesis involves the accumulation of nanoparticles within the microbial cells, whereas extracellular synthesis occurs when microorganisms secrete biocompatible nanoparticles into their environment. These microbial nanoparticles can be synthesized using a wide range of materials, including metals, metal oxides, semiconductors, and organic compounds. The ability to engineer and control the properties of these nanoparticles opens avenues for diverse applications in fields such as medicine, environmental remediation, energy, and electronics [5].

Furthermore, microorganisms can act as living factories to produce nanomaterials with specific functionalities. Through genetic engineering and manipulation, researchers can engineer microorganisms to express proteins or peptides that self-assemble into nanoscale structures. This approach enables the creation of tailored nanomaterials with enhanced properties, such as improved stability, biocompatibility, and targeted interaction with biological systems [6].

Microbial nanotechnology also offers promising possibilities for the development of innovative drug delivery systems. By leveraging the unique properties of microbial nanoparticles, such as their small size, high surface-to-volume ratio, and surface functionalization potential, scientists can design nanocarriers that efficiently transport therapeutic agents to targeted sites within the body. This targeted drug delivery approach enhances treatment efficacy while minimizing side effects associated with conventional drug formulations [7].

check for updates

Citation: Abd-Elsalam, K.A. Special Issue: Microbial Nanotechnology. *Microorganisms* **2024**, *12*, 352. https://doi.org/10.3390/microorganisms12020352

Received: 2 January 2024
Accepted: 6 February 2024
Published: 8 February 2024

In addition to synthesis and drug delivery, microorganisms can play a crucial role in the development of nanosensors, nanoelectronics, and nanodevices. By interfacing microorganisms with nanomaterials, researchers can exploit their inherent biological sensing capabilities to detect and respond to specific environmental or biological signals. This integration of microbial systems with nanotechnology holds promise for applications in environmental monitoring, disease diagnostics, and bioelectronics [8].

As microbial nanotechnology continues to advance, it presents both exciting opportunities and challenges. Ethical considerations, safety assessments, and regulatory frameworks need to be established to ensure the responsible development and deployment of microbial nanotechnology. Additionally, interdisciplinary collaborations among microbiologists, materials scientists, engineers, and medical professionals are essential to drive innovation and maximize the potential of this field [9].

Nanotechnology plays a significant role in multiple aspects of agriculture and food security, including crop development, plant protection, food processing, and the targeted administration of agrochemicals to create disease resistance and boost plant growth. According to research, nanotechnology shows potential for more accurate input applications in the development of these sectors [10]. In conclusion, microbial nanotechnology represents a cutting-edge and rapidly evolving field with vast potential for transforming various sectors. By leveraging the unique capabilities of microorganisms, researchers can synthesize nanomaterials, engineer functional nanodevices, and develop novel applications in areas such as medicine, energy, electronics, and environmental remediation. As we continue to explore and understand the interactions between microorganisms and nanomaterials, microbial nanotechnology is poised to revolutionize numerous fields and contribute to the advancement of science and technology.

2. An Overview of Published Articles

One study evaluated the biosynthesis and characterization of selenium nanoparticles (SeNPs) using two distinct endophytic selenobacteria for potential use as biofortifying agents and in various biotechnological applications (contribution 1).

In another contribution, the effects of food-grade titanium dioxide nanoparticles (TiO_2 NPs), a popular food ingredient, on epithelial permeability, intestinal alkaline phosphatase activity, and nutrient transport across the epithelium were studied using this model. The use of human cells, a synthetic meal, and a bacterial mock community allowed researchers to better understand the effects of dietary changes on small intestinal function, including the microbiota (contribution 2)

Research on non-*Streptomyces* species (i.e., rare actinomycetes) for AgNP production is relatively unexplored. In light of this, another study investigated the biosynthesis of AgNPs for the first time, using two rare actinomycetes, *G. nicotianae* SNPRA1 and L. aridicollis SNPRA2. Additionally, they evaluated the efficacy of these AgNPs as antifungal and anti-mycotoxin agents against mycotoxigenic fungi at non-toxic doses (contribution 3).

To assess the potential of the freshwater snail *B. alexandrina* as a bioindicator for SeONPs, another study investigated the ecotoxicity of naturally synthesized SeONPs in aquatic systems by different aquatic microorganisms. The study evaluated the impact of SeONPs on survival, alanine aminotransferase, and aspartate aminotransferase concentrations, as well as the digestive and hermaphrodite gland structure of *B. alexandrina*. Moreover, in silico molecular docking studies were conducted to analyze the interactions between nano-selenium oxide and ALT/AST. Nanotoxicity was confirmed using *D. magna*, which is a sensitive organism for detecting aquatic environmental pollution (contribution 4).

Additionally, AgNPs were biosynthesized for the first time from the waste leaf extracts of local doum palms in Tabuk, Saudi Arabia, and these demonstrated excellent antibacterial and anticandidal properties against pathogenic strains of bacteria and various yeasts (contribution 5).

TeNPs, produced as intracellular nanorods by *Gayadomonas* sp. TNPM15 isolated from mangrove sediments, were also characterized for their impact on *F. oxysporum* and *A.*

alternata. The biosynthesized TeNPs induced notable ultrastructure alterations in the cell wall, exhibiting antifungal potential through inhibitory effects on spore germination and the disruption of membrane permeability and integrity (contribution 6).

Although the use of various microorganisms for the biosynthesis of AgNPs is well documented, there is limited research available on the biosynthesis of AgNPs using actinomycetes. Given this backdrop, another contribution reported that the majority of actinomycetes that produce AgNPs are found in the Streptomyces genus (contribution 7).

Another article explored the synergistic effects of selenium nanoparticles (SeNPs) produced by *Rhizobium pusense* and ciprofloxacin (Cipro) in combating *Acanthamoeba* sp. This combination reduces the sub-lethal dose of ciprofloxacin required and inhibits cyst formation in Acanthamoeba sp. Additionally, this treatment damages the cellular integrity of the protozoa, leading to the leakage of essential proteins, enzymes, amino acids, and sugars. (contribution 8).

In a different research work, a novel approach was developed to self-assemble hybrids of *Shewanella oneidensis* and CdS nanoparticles, resulting in an improved reductive biodegradation capacity. The hybrids were then encapsulated in hydrogels to create a living material that enhanced their recyclability and long-term stability while confining them and protecting them from environmental stress. The living materials developed were found to efficiently biodegrade various organic dyes, including azo and nitroso dyes, through photocatalysis. This study highlights the potential of self-assembled nano-bacteria hybrids for bioremediation and living material development (contribution 9).

Understanding the response of bacteria to metal nanoparticles is crucial in studying the quorum sensing (QS) system of *Pseudomounas. Aeruginosa* and its regulation. Given this knowledge, the exposure of bacteria to low concentrations of bio-AgNPs was found to enhance the expression of QS regulatory genes, affecting functional genes involved in the formation of biofilms and production of virulence factors. As such, this contribution demonstrated that knowledge of bacterial virulence and pathogenicity mechanisms is essential for developing potential targets for anti-virulence therapy (contribution 10).

Furthermore, alginate-based floating microbeads were developed as a promising carrier to enhance the therapeutic index and bioavailability of clarithromycin. The inclusion of therapeutic oils in the microbeads resulted in synergistic effects with this drug, effectively eradicating infections. The synergistic activity produced was assessed through antibacterial assays, comparing the zone of inhibition of the microbeads with both the pure drug and the oils (contribution 11).

In another study, AgNPs were produced using ethyl acetate extracts of *Urtica diocia* (UD) leaves. A characterization of synthesized NPs revealed that the particles were spherical and small, and, because small NPs have desirable therapeutic activities, the NPs synthesized in this study offer promising prospects for development as pharmaceuticals. This characterization investigation revealed the existence of bioactive chemicals in the leaves and that they were incorporated into the AgNPs. (contribution 12).

Additionally, *Trichoderma harzianum* isolates have demonstrated an ability to synthesize a variety of proteins and enzymes without the need for chemical reducers and stabilizers. Biosynthesized AgNPs have shown potential in protecting cotton plants from fungal invasion caused by damping off. However, more research is required to identify the essential bioreducing and biotemplating molecules present in hyphal extracts in order to have greater control over the size and polydispersity of AgNPs (contribution 13).

Another contribution explored π-conjugated polymer nanoparticles (CPNs), given that these nanoparticles are frequently employed as probe materials for bioimaging and medication administration. Moreover, CPN use is on the rise, and they are continually being employed in an increasing number of cutting-edge biomedical sectors due to their specific photophysical and physicochemical characteristics, high compatibility, and ease of functionalization. (contribution 14).

Finally, antimicrobial peptides (AMPs) were investigated as potential alternatives to antibiotics, considering the rising demand for new antimicrobial agents. Derived from microorganisms and widely distributed in nature, AMPs exhibit a broad range of antimicrobial protection, making them suitable for the treatment of infections caused by various pathogenic microorganisms (contribution 15).

3. Conclusions

Microbial nanotechnology has several potential applications in the agroecosystem, including the following:

Soil remediation: The utilization of microbial nanotechnology presents a promising approach for the remediation of contaminated soils through the synthesis of nanoparticles that possess the capability to bind to and eliminate pollutants. For instance, microorganisms can synthesize nanoparticles of metals, such as iron, which can be employed to eliminate heavy metals and other contaminants from the soil.

Crop enhancement: Microbial nanotechnology can be used to enhance crop growth and productivity. For example, nanoparticles can be synthesized by microorganisms that can be used as fertilizers, pesticides, or herbicides. These nanoparticles can be engineered to slowly release nutrients or other substances, providing sustained benefits to crops.

Disease control: Microbial nanotechnology can be used to control human and plant diseases by synthesizing nanoparticles that can target and destroy pathogens. For example, nanoparticles of silver, copper, or zinc can be synthesized with microorganisms and used as a natural fungicide or bactericide.

Water management: The utilization of microbial nanotechnology in agriculture has the potential to enhance water management. One way this can be achieved is through the creation of nanoparticles via microorganisms, which have the ability to absorb and gradually release water. As a result, the need for irrigation can be reduced, leading to increased efficiency in water usage and greater sustainability in agricultural practices. Nevertheless, further investigation is necessary to fully comprehend the advantages and potential hazards associated with this technology when applied in agriculture.

Sustainable agriculture: Microbial nanotechnology can help create ecofriendly and sustainable farming methods. Microbial nanotechnology can contribute to the development of a more sustainable agroecosystem by encouraging environmental stewardship, maximizing resource usage, and lowering the use of chemicals.

This Special Issue delves into the use of microorganisms, including bacteria, fungi, and algae, in the production of nanomaterials. However, the wide range of microbiological systems available for use requires standardization in order to produce nanomaterials with consistent and repeatable properties. Given this background, this Special Issue covers the current state of knowledge, mechanisms of microbial synthesis, and challenges in the process. Advancements in biosynthetic pathways and genetic engineering have led to breakthroughs in microbial-based nanosynthesis, which may have potential commercial applications in the creation of sensoristic devices and diagnostic tools, as well as in the control of microbial diseases. Additionally, microbial-generated nanoparticles have shown promise in removing hazardous metals from the environment. To further explore this phenomenon, this Special Issue includes articles examining the sources of nanoparticle exposure and ecotoxicity. Overall, this Special Issue focuses on various aspects of nanomaterials, including microbial synthesis methods, characterization, applications, regulations, nanotoxicity, and challenges, and its principal aim is to present current research on the development of microbial-based nanosynthesis and its potential applications in the biomedical, environmental, and agri-food sectors.

Funding: This research received no external funding.

Conflicts of Interest: The authors declare no conflicts of interest.

List of Contributions

1. Sans-Serramitjana, E.; Gallardo-Benavente, C.; Melo, F.; Pérez-Donoso, J.M.; Rumpel, C.; Barra, P.J.; Durán, P.; Mora, M.d.L.L. A Comparative Study of the Synthesis and Characterization of Biogenic Selenium Nanoparticles by Two Contrasting Endophytic Selenobacteria. *Microorganisms* **2023**, *11*, 1600. https://doi.org/10.3390/microorganisms11061600.
2. Malik, M.; Tanzman, J.V.; Dash, S.K.; Marques, C.N.H.; Mahler, G.J. An In Vitro Small Intestine Model Incorporating a Food Matrix and Bacterial Mock Community for Intestinal Function Testing. *Microorganisms* **2023**, *11*, 1419. https://doi.org/10.3390/microorganisms11061419.
3. Abd El-Ghany, M.N.; Hamdi, S.A.; Korany, S.M.; Elbaz, R.M.; Emam, A.N.; Farahat, M.G. Biogenic Silver Nanoparticles Produced by Soil Rare Actinomycetes and Their Significant Effect on *Aspergillus*-derived mycotoxins. *Microorganisms* **2023**, *11*, 1006. https://doi.org/10.3390/microorganisms11041006.
4. Ibrahim, A.M.; Ghazy, M.; El-Sayed, H.; Abd El-Hameed, R.M.; Khalil, R.G.; Korany, S.M.; Aloufi, A.S.; Hammam, O.A.; Morad, M.Y. Histopathological, Immunohistochemical, Biochemical, and In Silico Molecular Docking Study of Fungal-Mediated Selenium Oxide Nanoparticles on *Biomphalaria alexandrina* (Ehrenberg, 1831) Snails. *Microorganisms* **2023**, *11*, 811. https://doi.org/10.3390/microorganisms11030811.
5. Alabdallah, N.M.; Kotb, E. Antimicrobial Activity of Green Synthesized Silver Nanoparticles Using Waste Leaves of *Hyphaene thebaica* (Doum Palm). *Microorganisms* **2023**, *11*, 807.
6. Abd El-Ghany, M.N.; Hamdi, S.A.; Korany, S.M.; Elbaz, R.M.; Farahat, M.G. Biosynthesis of Novel Tellurium Nanorods by *Gayadomonas* sp. TNPM15 Isolated from Mangrove Sediments and Assessment of Their Impact on Spore Germination and Ultrastructure of Phytopathogenic Fungi. *Microorganisms* **2023**, *11*, 558. https://doi.org/10.3390/microorganisms11030558.
7. AboElmaaty, S.A.; Shati, A.A.; Alfaifi, M.Y.; Elbehairi, S.E.I.; Sheraba, N.S.; Hassan, M.G.; Badawy, M.S.E.M.; Ghareeb, A.; Hamed, A.A.; Gabr, E.Z. Biofilm Inhibitory Activity of Actinomycete-Synthesized AgNPs with Low Cytotoxic Effect: Experimental and In Silico Study. *Microorganisms* **2023**, *11*, 102. https://doi.org/10.3390/microorganisms11010102.
8. Nikam, P.B.; Salunkhe, J.D.; Marathe, K.R.; Alghuthaymi, M.A.; Abd-Elsalam, K.A.; Patil, S.V. *Rhizobium pusense*-Mediated Selenium Nanoparticles–Antibiotics Combinations against *Acanthamoeba* sp. *Microorganisms* **2022**, *10*, 2502. https://doi.org/10.3390/microorganisms10122502.
9. Tao, M.; Jin, C.; Lu, H.; Jin, K.; Yu, L.; Liu, J.; Zhang, J.; Zhu, X.; Wu, Y. Living and Regenerative Material Encapsulating Self-Assembled *Shewanella oneidensis*-CdS Hybrids for Photocatalytic Biodegradation of Organic Dyes. *Microorganisms* **2022**, *10*, 2501. https://doi.org/10.3390/microorganisms10122501.
10. Saeki, E.K.; Martins, H.M.; Camargo, L.C.d.; Anversa, L.; Tavares, E.R.; Yamada-Ogatta, S.F.; Lioni, L.M.Y.; Kobayashi, R.K.T.; Nakazato, G. Effect of Biogenic Silver Nanoparticles on the Quorum-Sensing System of *Pseudomonas aeruginosa* PAO1 and PA14. *Microorganisms* **2022**, *10*, 1755. https://doi.org/10.3390/microorganisms10091755.
11. Khan, I.U.; Shoukat, M.; Asif, M.; Khalid, S.H.; Asghar, S.; Munir, M.U.; Irfan, M.; Rasul, A.; Qari, S.H.; Qumsani, A.T.; et al. Assessing the Synergistic Activity of Clarithromycin and Therapeutic Oils Encapsulated in Sodium Alginate Based Floating Microbeads. *Microorganisms* **2022**, *10*, 1171. https://doi.org/10.3390/microorganisms10061171.
12. Binsalah, M.; Devanesan, S.; AlSalhi, M.S.; Nooh, A.; Alghamdi, O.; Nooh, N. Biomimetic Synthesis of Silver Nanoparticles Using Ethyl Acetate Extract of *Urtica diocia* Leaves; Characterizations and Emerging Antimicrobial Activity. *Microorganisms* **2022**, *10*, 789. https://doi.org/10.3390/microorganisms10040789.
13. Zaki, S.A.; Ouf, S.A.; Abd-Elsalam, K.A.; Asran, A.A.; Hassan, M.M.; Kalia, A.; Albarakaty, F.M. Trichogenic Silver-Based Nanoparticles for Suppression of Fungi Involved in Damping-Off of Cotton Seedlings. *Microorganisms* **2022**, *10*, 344. https://doi.org/10.3390/microorganisms10020344.
14. Elgiddawy, N.; Elnagar, N.; Korri-Youssoufi, H.; Yassar, A. π-Conjugated Polymer Nanoparticles from Design, Synthesis to Biomedical Applications: Sensing, Imaging, and Therapy. *Microorganisms* **2023**, *11*, 2006. https://doi.org/10.3390/microorganisms11082006.
15. Sowers, A.; Wang, G.; Xing, M.; Li, B. Advances in Antimicrobial Peptide Discovery via Machine Learning and Delivery via Nanotechnology. *Microorganisms* **2023**, *11*, 1129. https://doi.org/10.3390/microorganisms11051129.

References

1. Shukla, P. Microbial nanotechnology for bioremediation of industrial wastewater. *Front. Microbiol.* **2020**, *11*, 590631.
2. Azizi, Z.L.; Daneshjou, S. Bacterial nano-factories as a tool for the biosynthesis of TiO_2 nanoparticles: Characterization and potential application in wastewater treatment. *Appl. Biochem. Biotechnol.* **2024**, *4*, 1–25. [CrossRef] [PubMed]
3. Grasso, G.; Zane, D.; Dragone, R. Microbial nanotechnology: Challenges and prospects for green biocatalytic synthesis of nanoscale materials for sensoristic and biomedical applications. *Nanomaterials* **2019**, *10*, 11. [CrossRef] [PubMed]
4. Zhang, J.; Li, F.; Liu, D.; Liu, Q.; Song, H. Engineering extracellular electron transfer pathways of electroactive microorganisms by synthetic biology for energy and chemicals production. *Chem. Soc. Rev.* **2024**, *53*, 1375–1446. [CrossRef] [PubMed]
5. de Jesus, R.A.; de Assis, G.C.; de Oliveira, R.J.; Costa, J.A.S.; da Silva, C.M.P.; Iqbal, H.M.; Ferreira, L.F.R. Metal/metal oxide nanoparticles: A revolution in the biosynthesis and medical applications. *Nano-Struct. Nano-Objects* **2024**, *37*, 101071. [CrossRef]
6. Reches, M.; Gazit, E. Molecular self-assembly of peptide nanostructures: Mechanism of association and potential uses. *Curr. Nanosci.* **2006**, *2*, 105–111. [CrossRef]
7. Rodríguez-Carmona, E.; Villaverde, A. Nanostructured bacterial materials for innovative medicines. *Trends Microbiol.* **2010**, *18*, 423–430. [CrossRef] [PubMed]
8. Kumar, H.; Kuča, K.; Bhatia, S.K.; Saini, K.; Kaushal, A.; Verma, R.; Bhalla, T.C.; Kumar, D. Applications of nanotechnology in sensor-based detection of foodborne pathogens. *Sensors* **2020**, *20*, 1966. [CrossRef] [PubMed]
9. Rabbani, M.; Hoque, M.E.; Mahbub, Z.B. Nanosensors in biomedical and environmental applications: Perspectives and prospects. In *Nanofabrication for Smart Nanosensor Applications*; Elsevier: Amsterdam, The Netherlands, 2020; pp. 163–186.
10. Kashyap, P.L.; Rai, P.; Kumar, R.; Sharma, S.; Jasrotia, P.; Srivastava, A.K.; Kumar, S. Microbial nanotechnology for climate resilient agriculture. In *Microbes for Climate Resilient Agriculture*; Wiley: Hoboken, NJ, USA, 2018; pp. 279–344.

microorganisms

MDPI

Review

π-Conjugated Polymer Nanoparticles from Design, Synthesis to Biomedical Applications: Sensing, Imaging, and Therapy

Nada Elgiddawy [1,2], Noha Elnagar [1,3], Hafsa Korri-Youssoufi [1,*] and Abderrahim Yassar [4]

1 CNRS, Institut de Chimie Moléculaire et des Matériaux d'Orsay (ICMMO), Université Paris-Saclay, ECBB, 91400 Orsay, France
2 Department of Biotechnology and Life Sciences, Faculty of Postgraduate Studies for Advanced Sciences (PSAS), Beni-Suef University, Beni-Suef 62 511, Egypt
3 Materials Science and Nanotechnology Department, Faculty of Postgraduate Studies for Advanced Sciences (PSAS), Beni-Suef University, Beni-Suef 62 511, Egypt
4 LPICM, CNRS, Ecole Polytechnique, Institut Polytechnique de Paris, Route de Saclay, 91128 Palaiseau, France; abderrahim.yassar@polytechnique.edu
* Correspondence: hafsa.korri-youssoufi@universite-paris-saclay.fr

Abstract: In the past decade, π-conjugated polymer nanoparticles (CPNs) have been considered as promising nanomaterials for biomedical applications, and are widely used as probe materials for bioimaging and drug delivery. Due to their distinctive photophysical and physicochemical characteristics, good compatibility, and ease of functionalization, CPNs are gaining popularity and being used in more and more cutting-edge biomedical sectors. Common synthetic techniques can be used to synthesize CPNs with adjustable particle size and dispersion. More importantly, the recent development of CPNs for sensing and imaging applications has rendered them as a promising device for use in healthcare. This review provides a synopsis of the preparation and functionalization of CPNs and summarizes the recent advancements of CPNs for biomedical applications. In particular, we discuss their major role in bioimaging, therapeutics, fluorescence, and electrochemical sensing. As a conclusion, we highlight the challenges and future perspectives of biomedical applications of CPNs.

Keywords: conjugated polymer nanoparticles; biomedical application; organic semiconducting; fluorescence sensing; electrochemical sensing; drug delivery; bioimaging

Citation: Elgiddawy, N.; Elnagar, N.; Korri-Youssoufi, H.; Yassar, A. π-Conjugated Polymer Nanoparticles from Design, Synthesis to Biomedical Applications: Sensing, Imaging, and Therapy. *Microorganisms* **2023**, *11*, 2006. https://doi.org/10.3390/microorganisms11082006

Academic Editors: Juan Carlos Gutiérrez and Kamel Ahmed Abd-Elsalam

Received: 15 May 2023
Revised: 23 June 2023
Accepted: 25 July 2023
Published: 3 August 2023

1. Introduction

The last decade has witnessed great progress in the use of conjugated polymer nanoparticles (CPNs), sometimes referred to in the literature as semiconducting polymer nanoparticles or polymer dots (Pdots), as an emerging class of multifunctional nanoscale materials for biomedical applications, including fluorescence and photoacoustic imaging, biosensors, and photothermal therapy for microbial infections and cancer [1–4]. CPNs exhibit outstanding properties combined with a simple preparation method, easy surface modification, and water dispersibility [5]. Their straightforward synthesis [4] in desired sizes and tunable properties makes them highly attractive materials for the aforementioned applications [6,7]. Their photophysical properties can be tailor-made by an appropriate rational molecular design of the structure of the conjugated polymers (CPs) or aggregate states. Until now, CPNs and, more generally, soft fluorescent nanoparticles (NPs) have emerged as novel biomedical analysis tools due to their exceptional photophysical properties, outstanding photochemical stability, extraordinary fluorescence efficiency, tunable spectral properties, and amplified fluorescent quenching effect with less toxicity. They have been exploited in fluorescence imaging [8], cancer cell bioimaging [9], photodynamics, photothermal therapy [10], antimicrobial therapy, and bacteria killing, as shown in Figure 1 [11]. For biomedical applications, CPNs present many advantages over small organic molecule nanoparticles. They can be obtained with accessible synthetic schemes, thus providing a much larger platform for material manipulation to meet the specific requirements for different biomedical applications. Their chemical structures, band gap energy, functionalities, topology, and morphology can easily be tuned compared to those of the organic dye nanoparticles. Furthermore, many recent reports have shown that the effectiveness of CPNs as fluorescent probes for bioimaging can surpass that of small molecules and inorganic quantum dots due to their extraordinary fluorescence brightness, excellent photochemical stability, high-fluorescence quantum yield, biocompatibility, and low toxicity to cells and tissues [12]. Among the key factors that affect their utility as biomedical tools are fluorescence emission wavelength, the brightness of the fluorescence, the size of the particle, and the Stokes shift, which should be small to capture the pertinent fluorescent signals and avoid cross-talk between the excitation light and fluorescence. When CPNs are designed as fluorescent probes, they should be as small as possible for complete elimination from the body by renal filtration, i.e., their size should be less than about 5.5 nm.

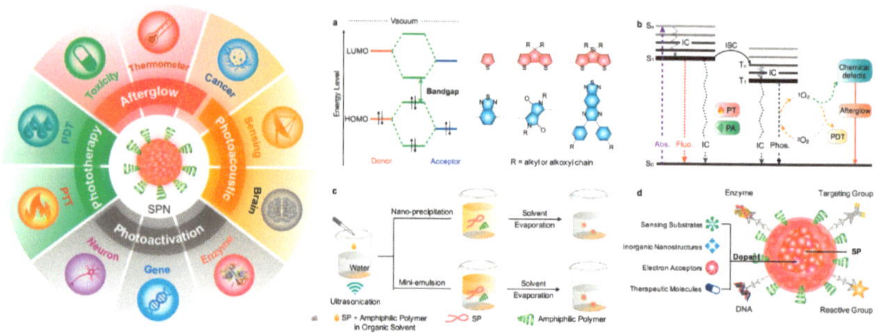

Figure 1. (**Left**): schematic illustration of CPNs for biomedical applications. (**Right**): schematic representation of donor and acceptor molecules developed for low-band-gap conjugated polymers and photophysical processes using a Jablonski diagram. (**a**) Representative chemical structures of monomers for low-band-gap polymers. (**b**) Absorption; internal conversion; fluorescence, intersystem crossing, phosphorescence; photothermal. (**c**) Cartoon presentation of the synthesis of CPNs. (**d**) Cartoon representation of surface functionalization of CPNs. Reproduced from Ref. [13] with permission from the ACS publisher.

According to Web of Science (accessed spring 2023), the number of articles published in the field of NPs and "conjugated polymers" has exponentially increased, indicating the importance and interest in such emerging fields, as shown in Figure 2. The interest in nanoparticle synthesis began in 2002, with the work of Landfester et al. on the mini-emulsion technique. Since this seminal work, a number of reviews dealing with the developments and applications of organic fluorescent nanoparticles have been published. These reviews have highlighted small organic molecules [9,14,15], conjugated polymer nanoparticles [3,16], aggregation-induced emission dots [17–19], carbon dots [20,21], and up-conversion nanoparticles [22]. Some of these reviews cover very broad topics, including the synthesis of organic and polymeric materials and the preparation of NPs, whereas others focus on applications, including sensing [23], imaging [24], and diagnostic and therapeutic potential (phototherapy). In contrast, this review showcases the recent achievements in the development of high-performance tailor-made CPs with unique optical properties, combined with their incorporation into NPs and use in biomedical applications. A particular focus is placed on material design and synthesis, structures, and topologies, as well as the functionalities and biomedical applications of CPs.

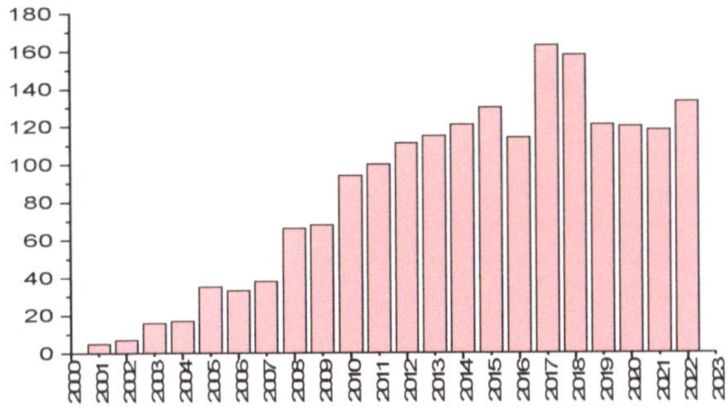

Figure 2. Histogram showing the number of publications in recent years according to Web of Science on the topic of "conjugated polymers" and nanoparticles.

2. Preparation Methods of CPNs

Two main strategies are used to produce CPNs: direct polymerization to NPs and the dispersion of preformed polymers (namely post-polymerization). Each strategy has its unique advantages and limitations. Direct polymerization does not require the use of polymers that are only soluble in organic solvents. The post-polymerization approach, however, produces NP dispersion from existing polymers using mini-emulsification, nano-precipitation, self-assembly processes, or vesicle and micelle formation methods. Therefore, this approach does not require any special equipment or expertise in organic and polymer synthesis, and it can rely on commercially available polymers.

2.1. Direct Synthesis of Conjugated Polymer NPs

In the direct synthesis of CPNs approach, the reaction of the polymerization and the NPs formation occurred simultaneously in a disperse heterophase system. In such approaches, the polymerization occurs inside the mini- or micro-emulsion formed in continuous immiscible phases. Mini- and micro-emulsion polymerization differ mainly in their stability, while the microemulsion is thermodynamically stable and forms spontaneous microdroplets under appropriate conditions. However, mini-emulsion is a metastable phase which can be achieved with the help of shearing the immiscible phase through the sonication or the homogenization of the solution. While micro-emulsion polymerization produces NPs with a diameter as small as 5 nm, these sizes are typically greater than 50 nm in mini-emulsion polymerization. Finally, both emulsion polymerizations are not only useful techniques for preparing NPs from existing polymers, but they can also serve as a template reaction medium for synthetizing new CPNs from novel monomers [2].

While the majority of the literature described the creation of NPs utilizing oil-in-water method, Mullen et al. showed that a non-aqueous emulsion (oil-in-oil) can be used to elaborate conjugated polymer NPs through direct polymerization of monomers [25]. In their work, cyclohexane and acetonitrile were used as a continuous and dispersed phase, respectively, while polyisoprene-block-poly(methyl methacrylate) (PI-*b*-PMMA) was utilized as an emulsifying agent. NPs of Poly(3,4-ethylenedioxythiophene) (PEDOT), polyacetylene and poly(thiophene-3-acetic acid) were prepared by catalytic and oxidative polymerization [26]. Additionally, the emulsifying agent (PI-*b*-PMMA) was washed with a THF solution leading to CPNs with an average diameter of 23 nm. In 2006, Huber et al. synthesized processable submicron polyacetylene NPs using catalytic emulsion polymerization [27]. The key to the success of their strategy relies on the use of catalytic complexes that are less oxophilic and very reactive towards the polymerization of the acetylene monomers. For this purpose, a small volume of an emulsion of hexane/ethanol was prepared; then, a catalytic amount of Pd^{II}, 1,3-bis(di-tert-butyl)phosphino-propane and $Pd(OAc)_2$ was added followed by the addition of an aqueous solution of surfactant, sodium dodecyl sulfate (SDS), and methane sulfonic acid. The mixture was subsequently sonicated to produce a mini-emulsion. Polymerization of acetylene under ambient pressure resulted in the formation of an intensely colored, black CPN dispersion. The diameter of the NPs was estimated to be 20 nm by using the transmission electron microscopy (TEM) [27]. The same group extended this concept to the aqueous step-growth polymerization of conjugated polymer NPs [28]. They demonstrated that the highly fluorescent CPNs could be prepared by direct polymerization in aqueous mini-emulsion, under Glaser coupling conditions as a suitable step-growth reaction. The molecular weights estimated by gel permeation chromatography are in the range of 10^4 to 10^5 g mol^{-1}, and NP sizes determined by TEM are around 30 nm.

2.2. *Conventional Synthesis Method of CNPs*

Conventional methods to obtain CNPs involve post-polymerization to produce CPNs from previously synthesized CPs. They can be generated using mini- or macro-emulsion, nanoprecipitation, or self-assembly processes.

2.2.1. Mini-Emulsion Method

The mini-emulsion approach is the oldest known and most widely used method to prepare CPNs. In their seminal work, Landfester and coworkers demonstrated the preparation of CPNs using the mini-emulsion technique for the first time. They reported the preparation of a series of water-dispersible CPs, methyl-substituted ladder-type poly(para-phenylene), polyfluorene, and polycyclopentadithiophene, with NPs sizes in the range of 75–250 nm by controlling the amount of the surfactant [29]. In such a method, the CP is solubilized in a good organic solvent, which is then injected in the water phase containing a surfactant, a bad solvent of CPs. High shear forces via ultra-sonication were then applied to obtain homogeneous and monodisperse droplets stabilized by surfactant molecules. In this way, a suspension of water dispersible NPs is formed by a slow evaporation of organic solvent. The polymer/surfactant ratio alters the size of the NPs, which typically range from 20 nm to several hundred nanometers. The mini-emulsion polymerization method has two benefits over nano-precipitation: the large choice of water immiscible organic solvents and the use of a surfactant to prevent the coalescence of emulsion droplets. The major drawback related to this method is destabilization risks due to Ostwald ripening and droplet coalescence [30]. The risk of flocculation may be avoided by an appropriate choice of surfactants used, while Ostwald ripening can be suppressed by the addition of a hydrophobic agent to the dispersed phase. The hydrophobic additive promotes the formation of an osmotic pressure within the droplets, which balances the Laplace pressure, the difference in pressure between a droplet's interior and exterior, and prevents diffusion from one droplet to the surrounding watery medium.

Recently, a cradle-to-grave study of CPN-based indaceno-dithiophene and diketopyrrolopyrrole copolymers was reported by Turner's group [31]. The study aims to achieve high-performance organic semiconductor (OSC) device based on CPNs and therefore, confirms the effect of chemical synthesis on their physico-chemical properties such as light absorption and emission.

The process utilizes a water-in-oil mini-emulsion Suzuki–Miyaura polymerization, fewer steps, lower reaction temperatures, a significant reduction in volatile organic compounds (VOCs) (>99%) and avoids halogenated solvents; Figure 3.

Suzuki–Miyaura polymerization of borylated benzothiadiazole (BT(Bpin)$_2$) and dihalogenated indacenodithiophene or diketopyrrolopyrrole-thiophene, under conventional conditions, in a water-in oil mini-emulsion, was performed to prepare a series of copolymers; Figure 4. The optimized mini-emulsion polymerization conditions led to a complete monomer consumption and the formation of copolymers with molecular weights comparable to those obtained by conventional polymerization methods. This process generates a highly concentrated NP dispersion; these NPs are surfactant coated on their surface. These dispersions were used to fabricate thin films for electronic devices. The material performance was found to be similar to that obtained using large amount of VOCs and halogenated solvents. The complete approach developed addresses all environmental concerns and enables a viable guideline for the delivery of future CP materials using only aqueous media for synthesis, purification, and thin-film processing.

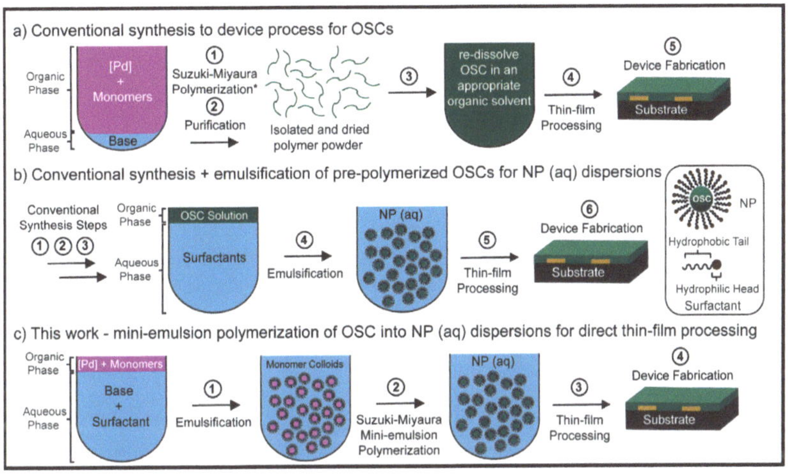

Figure 3. A cartoon overview of the synthesis-to-device process of OSCs. From top to bottom: (**a**) Conventional synthesis: (1) Polymerization based on organometallic cross-coupling reactions in a Ibiphasic system, (2) purification of OSCs using organic solvent (Soxhlet or preparatory size-exclusion chromatography); (3), (4) thin-film processing from halogenated solvents. (**b**) Process for the preparation of water-dispersible CPNs, (**c**) water in oil mini-emulsion Suzuki–Miyaura polymerization. Reproduced from [31]. Figure is available via license: Creative Commons CC BY license.

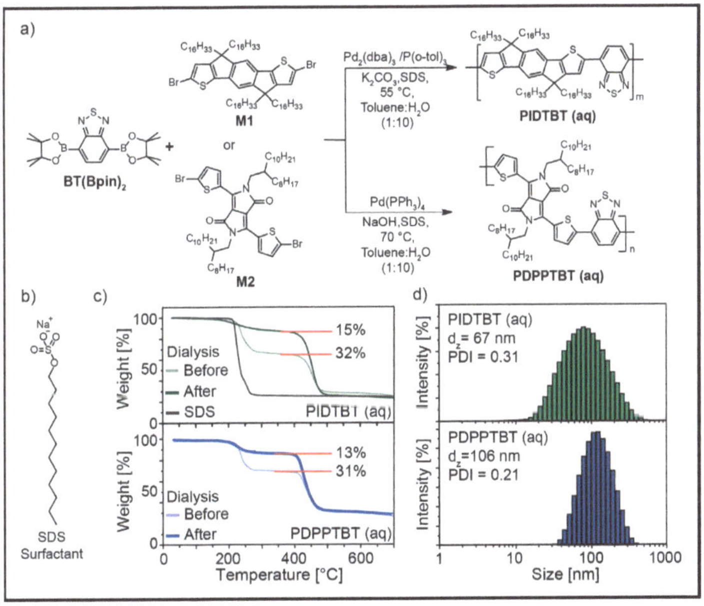

Figure 4. (**a**) Synthetic procedures for the copolymers: reagents and reaction conditions for miniemulsion Suzuki–Miyaura polymerization; (**b**) Chemical structure of SDS; NP dispersion characterization;

(**c**) TGA characterization of NPs showing the elimination of the excess of SDS via dialysis, and (**d**) DLS characterization of NPs dispersion. Reproduced from [31], figure available via license: Creative Commons CC BY license.

2.2.2. Nano-Precipitation Methods: Solvent Mediated Self-Assembly

The interaction between the CP chains and solvent molecules is a crucial parameter in the formation of NPs. In the nano-precipitation approach, a given CP is dissolved in a good organic solvent and then injected into a bad CP solvent. By contrast, this bad solvent is miscible with the solvent used to dissolve the CP, leading to fine precipitation of CPNs. To assist the formation of the NPs, the resulting mixture is vigorously sonicated. The hydrophobic effect, which is caused by the tendency of CP chains in solution to avoid contact with solvent molecules, is the main driving force for the formation of NPs. In fact, the formation of the NPs is attributed to the aggregation of the CP chains caused by the significant change in solvent polarity, and, consequently, to avoid their exposure to solvent molecules, they fold into spherical shapes. This method does not involve the use of any extra additives such as surfactants and hydrophobic agents and can be applied to a wide range of organic solvent-soluble CPs. By this method, it is possible to fine-tune the size of NPs by adjusting the polymer concentration and an appropriate choice of the molecular weight. Using this approach, one can easily control the NP size down to 40 nm by changing the polymer concentration. Furthermore, many studies indicated that the initial concentration of CP affects the size of the NPs [32,33]. The diameter of NPs is greater the higher the initial CP concentration.

2.2.3. Self-Assembly Method through Supramolecular Interaction

Comparatively, the self-assembly strategy, where the polymers assemble into predefined structures guided by specific supramolecular interactions, has been less employed. In this strategy, oppositely charged π-conjugated polyelectrolytes and co-assembling reagents are independently dissolved in a solvent (Figure 5). To create charge-neutral NPs, the resulting solutions or dispersions are successively mixed in a specific ratio while being stirred. The functionalized CPNs can be easily separated by centrifugation. Using this method, Chong and coworkers synthetized photo-activated CPNs with antitumor activity. In their process, an aqueous solution of disodium salt, 3,3′-dithiodipropionic acid (SDPA) was added under vigorous stirring to a solution of a copolymer-containing fluorene and boron-dipyrromethene repeat units in the backbone (PBF). CPNs with a diameter of 100 nm were obtained by centrifugation [34].

2.3. *Vesicles and Micelles Formation*

Alternatively, rod–rod conjugated block copolymers with a hydrophobic and a hydrophilic block have the ability to self-assemble into block copolymer micelles with a solid hydrophobic core and a hydrophilic shell. Various groups described the successful synthesis of thiophene-based conjugated diblock copolymers and their self-assembly to produce well-defined structures at the nanometer scale [35]. N. Elgidawy et al. reported a strategy of self-assembly amphiphilic all-conjugated block copolymers, poly(3-hexylthiophene)-b-poly(3-triethylene-glycol-thiophene), P3HT-b-P3TEGT, consisting of hydrophobic block P3HT and hydrophilic block P3TEGT, into core−shell NPs, as shown in Figure 6. These NPs were further surface coated with mannose to lead to an easy-to-use solution-processable NP material for fabrication of a biosensing layer [36].

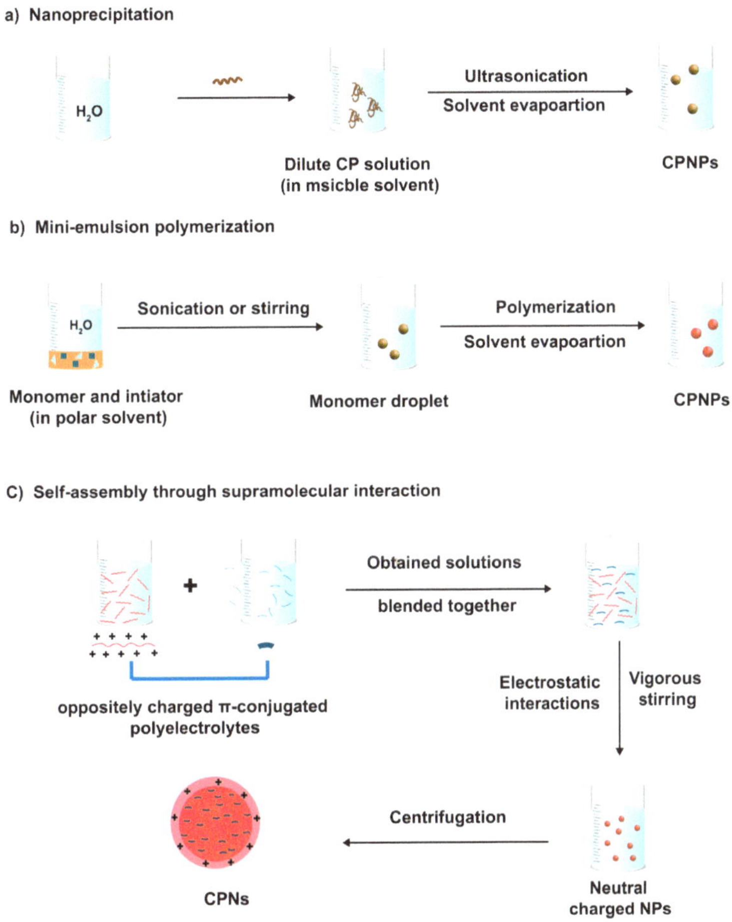

Figure 5. General schematic diagram for the preparation methods of conjugated polymer nanoparticles.

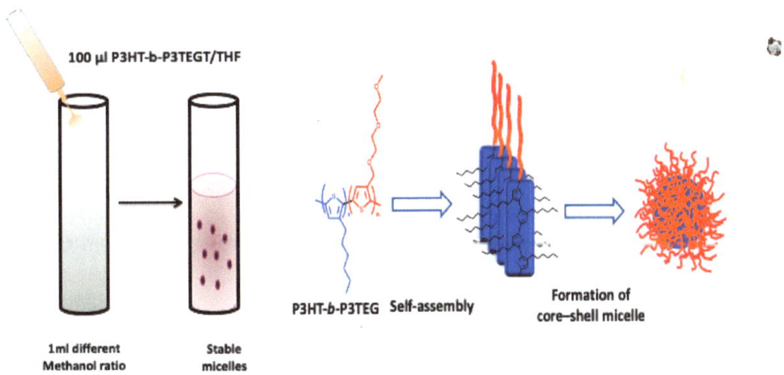

Figure 6. Cartoon overview of an oil-in-water mini-emulsion to prepare P3HT-*b*-P3TGT micelles. Illustration of the chemical structure of P3HT-*b*-P3TEGT and its self-assembly into nano-micelles. Reproduced from Ref. [35].

Another group described the formation of a CP vesicle with a crystalline core and an amorphous shell using a diblock copolymer of poly(3-hexylthiophene-b-3-(2-(2-{2-[2-(2-methoxy–ethoxy)-ethoxy]-ethoxy}-ethyl))thiophene)P(3HT-b-3EGT) with hydrophobic and hydrophilic side chains [37]. The resulting nanostructured NPs were utilized to investigate how the crystallinity and the optical properties in solution and thin films are affected by the block length. During the self-assembly process, the hydrophobic interactions, which are kinetically or thermodynamically controlled, can be regulated, leading to the creation of either small polymeric micelles or large polymeric vesicles (polymersomes), respectively, by using copolymers with a block of P3HT 26 mol %. They showed that the copolymer can self-assembly at dilute concentrations via a slow dialysis method producing highly ordered polymeric vesicles 200−250 nm in size under thermodynamic control, as shown in Figure 7. By utilizing the polystyrene, the competitive hydrophobic interactions could be used to alter the size of the micelles, while the kinetic control using fast precipitation produced micelles 5–20 nm in size.

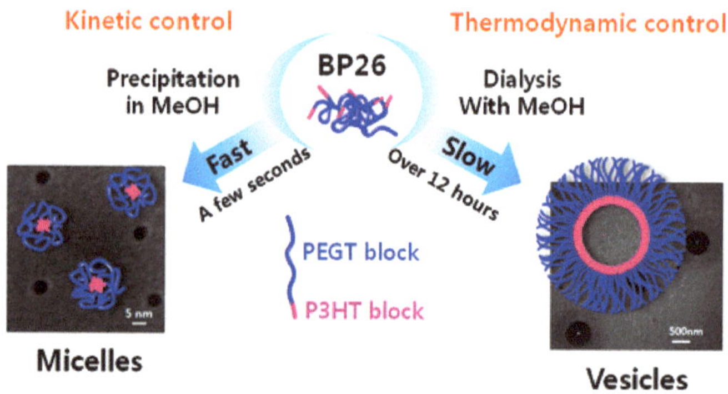

Figure 7. Schematic illustration of the self-assembly processes during precipitation (**left**) or dialysis (**right**) reproduced from Ref. [37] with permission from ACS.

2.4. *Effect of Synthesis Conditions on the Optical Properties of CNPs*

CPNs are interesting materials with morphology-dependent optoelectronic properties. In addition, their intrinsic properties (e.g., energy band gap, light absorption and emission, fluorescence quantum yield, etc.) can be tuned through chemical modification. In this section, we discuss the recent strategies to enhance their photophysical properties, fluorescence brightness, fluorescence emission and yield, as well as their fabrication from solution to solid state for future advanced applications. The fluorescence brightness of CPNs, which is proportional to $\varepsilon \times \Phi$, at the maximum absorption wavelength is an important parameter for imaging applications. However, the fluorescence properties of CNPs are generally limited by quenching phenomena caused by the inter-chain aggregation in the confined NP space. In a recent work, Braeken and coworkers proposed a strategy to enhance the fluorescence brightness of CPNs based on the introduction of branching molecules [38]. Thus, the tri- and tetra-functional branching building blocks were used to construct branched CPNs with an enhanced photoluminescence quantum yield (PLQY). Poly(p-phenylene ethynylene)s (PPEs) were selected as the CP materials not only for their simple structure, but also because of their rigid backbone structure that leads to a relatively high PLQY combined with a large molar extinction coefficient and a high photo-chemical stability. The introduction of acceptor moieties in the PPE backbone allows lowering the band gap energy and obtaining an emission in the NIR region. NPs of all samples were prepared via the combined mini-emulsion/solvent evaporation method. Sonogashira polymerization was used to produce crosslinked CPNs, following the previously published protocols. The influence of the branching molecules was studied; when 3 mol % of any of the branching molecules was used, the copolymers became more gel-like and viscous, and this effect was even more pronounced for 5 mol %. At higher concentrations, 7.5 mol % produced insoluble, likely cross-linked, polymer gel that could not be processed and characterized. The optical properties of the branched polymers in solution and in NP dispersion are investigated and then compared to those of their linear PPE counterparts. The PLQY of the CPNs increase from 5 to 11% for the samples containing 1,3,5-tribromobenzene. The one-photon fluorescence brightness doubles when 5 mol % of either branching molecule is used. The cytotoxicity of branched CPNs was assessed; they show low cytotoxicity in A549 human lung carcinoma cells up to a concentration of 100 μg/mL for 24 h. They also exhibit good NP uptake into cells and compatibility with two-photon imaging.

CPNs that generate a white-light emission when immobilized on a substrate were synthesized by Kim et al. via the Suzuki coupling reaction [39]. To achieve various emission colors with the same excitation wavelength, the phenylene units were added to the polymer's main chain structure. CPNs with emission colors of red, green, and blue were obtained from the resulting polymers, as shown in Figure 8. Electrospun nanofibers or glass slides coated with such CPNs were used to fabricate white-light-emitting materials under single-wavelength excitation. The benefit of this method is that it avoids the unnecessary energy transfer from a short- to a long-wavelength emission, which is typically known to be challenging when producing a white emission from mixed colors. White light emission was successfully achieved using spatially separated CPNs located on the nanofibers and on the glass.

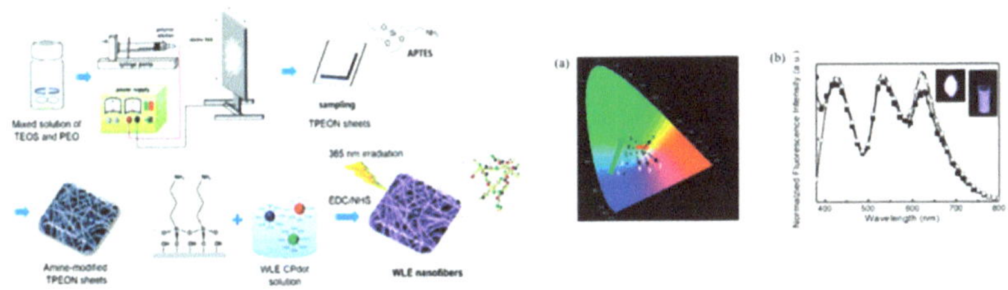

Figure 8. (**Left**) schematic presentation showing the fabrication of CPNs for white light emission. (**Right**) (**a**) CIE coordinates of CPN dispersion (point 8) (0.342, 0.340) and WLE nanofiber (point 9) (0.333, 0.331). The arrows and numbers are related to the composition of CPNs in solution. (**b**) Fluorescence emission of CPNs dispersion inset photographs: WLE CNPs dispersion (**right**) and WLE nanofiber (**left**). Reproduced with permission from RSC [39].

2.5. Effect of Synthesis Process on Morphological and Chemical Properties of CPNs

There are many recent researches focusing on the effect of CNP size and shape, as well as the role of surface chemistry of functionalized CNPs in cellular uptake [40–42]. Large NPs can obstruct the trafficking of the targeted biomolecules, limit access to congested cellular sites, and induce steric hindrance, which affects activities including targeting specificity and binding affinity of the attached biomolecules, while, small NPs that range in size from a few to several hundred nanometers enter the cells via pino- or macropinocytosis. It is noteworthy that compared with the inorganic quantum dots, CPNs exhibit less toxicity, which is confirmed by the methyl thiazolyl diphenyl tetrazolium assay [43]. Although there is no comparative study of the toxicity of CNPs and their inorganic counterparts, the low cytotoxicity of CNPs is due to the fact that the CP structures do not contain any toxic elements and they are highly biocompatible. According to Choi et al., NPs with a diameter of less than 6 nm can be eliminated through the renal system, whereas those with a diameter larger than that are primarily absorbed by the reticulo-endothelial system and end up in the liver and spleen [44]. Mendez et al. showed that the subcellular localization and toxicity of CPNs are related to their surface chemistry, the chemical structure and the side chain of the π-conjugated backbone structures. CPNs containing primary amines showed a strong Golgi localization without any toxicity. However, introducing short ethylene oxide side chains and tertiary amine side chains resulted in a decrease of Golgi localization and a highetoxicity, respectively [45]. The study of the biocompatibility of CPNs with relevant blood components was demonstrated by Khanbeigi et al. who investigated three different CPN formulations by varying the type of fluorescent CPs used, the concentration of material and the surface modifying layer. All CPN systems were compared with carboxyl-modified polystyrene beads whose negative surface charge has been proven to activate human platelet cells and influence coagulation pathways. CPNs polyethylene glycol (PEG)-coated surface does not stimulate platelet activation or aggregation, but may cause a small amount of hemolysis in the presence of free surfactant (PEG segments) and can inhibit physiological mediators of platelet aggregation, such as adenosine diphosphate [45]. The same group demonstrated the benefit of the encapsulation of PCPDTBT NPs with biocompatible poly(D,L-lactide-co-glycolide)-poly(ethylene glycol)-PLGA-PEG. When utilizing PEG2kDa-PLGA4kDa, with a content of 5% of PCPDTBT, the mean sizes of NPs were the smallest (<100 nm). The cytotoxicity (IC50 value) of the PEG2kDa-PLGA4kDa system is three times lower than that of the other two systems. No platelet aggregation or inhibition of ADP-induced platelet aggregation was seen, and hemolytic activity was less than 2.5% across all systems. The fluorescence emission shifted toward red wavelengths, and the PLQY dropped compared to the level of the THF-solution. In comparison with the high-molecular-weight polymers, the low-molecular-weight PEG2kDa-PLGA4kDa NPs had the advantages of smaller NP size, less cytotoxicity, and improved imaging performance [46]. In the field of bioimaging, obtaining accurate information about biomolecules is a challenging task [47]. As a result, both the target biomolecules and other extraneous factors such as probe concentrations, excitation power, and cellular context have an impact on the emission wavelength, emission intensity, shape, peak position, or lifetime of fluorescent probes. One approach to address these relies on the design of a single-type of probe with dual emission signals, one to reflect interactions with the target analyte and the other to serve as the internal intensity reference. An example of CPNs with a dual emission property was reported by Yao and Fukui [48].

Fluorescent π-conjugated polyelectrolyte NPs were prepared via the nano-precipitation method. The mean NP sizes were tuned by the net charge ratio between the polyelectrolyte and cation. Using the fluorescence anisotropy approach to investigate the photophysical characteristics of CPNs, including exciton migration, it was observed that a relative increase in efficiency of green-site emission brings a dual emission behavior. As the size of the NPs decreased, the intensity of the green-site emission increased. The non-radiative excitation energy transfer to an emissive trap site, low-energy defect on the conjugated backbone, or the exclusive excitation of the conjugated segments with a long effective conjugation length can all be the causes of the green-site emission. The latter mechanism is important in the small-particle-size region.

A common photophysical processes occurring in CPNs is the excitation energy transfer from the NP to the biological systems or vice versa called excitation energy transfer (EET) or Förster resonant energy transfer which consists in the simultaneous de-excitation of one of the subsystems and the excitation of the other subsystem. EET is one of the key factors that determine the biomedical and bio-imagining applications of CPNs. Due to the excellent funneling of the excitation energy to the trap sites, EET often results in a quenching of the fluorescence in CP materials at length scales of a few nanometers to tens of nanometers. Understanding the relationship between the sizes of the NPs, fluorescence brightness of CPNs, FET has been led to fabricate ultra-small (3.0–4.5 nm) CPNs with excellent photostability through a proper control of the precipitation conditions and the interchain interactions within the CNP [49].

3. Functionalization of CNPs for Biomedical Applications

For biosensing and biomedical applications, the surface of the CPNs needs to be functionalized for subsequent bioconjugation to guide the NPs toward a selective target, specific cells or organelles. Such functionalization can be achieved following three approaches: (i) direct functionalization of CPs, (ii) an affinity-driven binding of secondary capping layer strategy and (iii) the last approach, which is an extension of the affinity-driven approach.

3.1. Direct Functionalization

This approach relies on the covalent functionalization of the monomer followed by its polymerization. The post-polymerization functionalization approach is also relevant to this strategy. Direct functionalization is the most convenient method to graft functional groups on CP, enabling a direct immobilization of the probe (or targeted biomolecules). The most used organic functions are carboxylic acid, amine or N-hydroxysuccinimide (NHS) groups that allow an easy covalent grafting of targeted biomolecules through the formation of an amide bond. However, the major drawbacks of the direct functionalization strategy lie in its expensive and labor-intensive synthesis processes. For instance, using this strategy, Wang and co-workers prepared many CPNs for fluorescence imaging and clinical diagnosis, with cell imaging and transfection functions, by linking a lipid-modified cationic poly(fluorenylene phenylene) (Lipid-PFPL) and ammonium group [50]. The PFPL NPs showed excellent photostability and little cytotoxicity. These modified NPs can easily enter the cytoplasm via endocytosis (within 4 h) and have the potential to be used for cell imaging. In 2011, the fabrication of cisplatin-bonded CPNs (PT-Pt) was reported. Due to its amphiphilic nature, the copolymer forms NPs in water. Even at a high concentration (100 g mL), the polymer exhibits good photostability and has no harmful effects on human lung epithelial (A549) cells. As a result, this copolymer satisfies the essential criteria for its use as a fluorescent probe for cellular imaging. As a model, cisplatin, the cis-diamminedichloroplatinum(II) molecule widely used to treat cancer, is coupled to polythiophene to create the conjugate "PT-Pt" by coordinating interactions with the amine groups on the side chain of the polythiophene. By using fluorescence microscopy, the PT-Pt can be utilized to track the cellular distribution of cisplatin. The fluorescent imaging platform is provided by amphiphilic copolymers [51].

3.2. Functionalization through Affinity Interaction

Regarding the second approach, it involves secondary capping layers that are bound by affinity and depends on the interactions between conjugated chains and a secondary capping agent. In this strategy, the electrostatic and hydrophobic interactions are the driving force to encapsulate the naked CPs chains. The great tendency of naked CPNs to aggregate in water due to the high hydrophobicity of the conjugated backbone causes an unintentional self-quenching of fluorescence. To overcome this issue, surface modification with PEG is a desirable option which improves hydrophilicity and biocompatibility. Most importantly, the PEG chain's length can be controlled to accommodate various biological applications, which is crucial. Additionally, PEG is a useful functional polymer for biological applications due to its ease of modification with certain functional groups (amino or carboxyl groups, lipid, and biomolecules). For fluorescent imaging, multicolor regulation, gene delivery, and medication delivery and release, several PEG-functionalized CPNs have been reported. For instance, recently, we described the self-assembly of P3HT-b-P3TEGT into core–shell nanoparticles which were then mannose-decorated. The amphiphilic behavior of P3HT-b-P3TEGT helps it to form nanoparticles in water or methanol, in which hydrophobic moieties come together to form the inner core and the PEG segments on the outer surface of the NPs. Using non-covalent and hydrophilic–hydrophilic interactions, mannose was coated during the formation of NPs, as shown in Figure 9 [36].

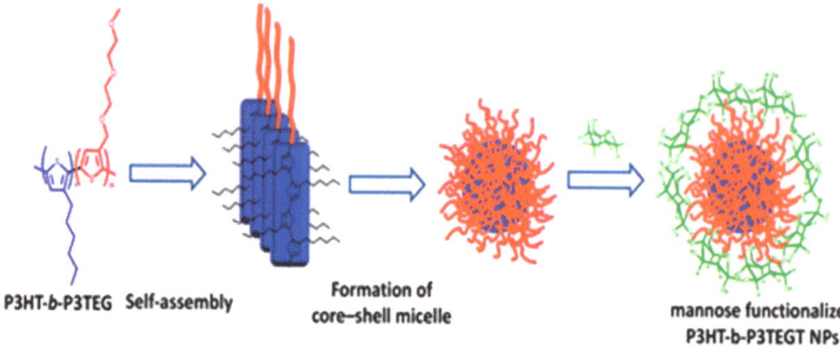

Figure 9. Cartoon overview of self-assembly of P3HT-b-P3TEGT in methanol. In methanol, $\pi-\pi$ interactions help to self-assembly the nanoparticles, in which hydrophobic moieties, blue segments, come together to form the inner core and the PEG segments, in red, on the outer surface of the NPs. Mannose was coated on P3HT-b-P3TEGT nanoparticles by noncovalent interactions between hydrophilic TEGT segments on the outer surface of the nanoparticles and hydroxyl groups of the mannose.

In order to enhance the NPs hydrophilicity and the incorporation of anchoring sites for subsequent functionalization for biosensing, imaging, and treatment applications, Wu et al. reported CPNs modified by an amphiphilic copolymer, poly(styrene-alt-maleic anhydride), that was adsorbed on the surface of NPs [52]. In the same idea, Green and co-workers reported the elaboration of multicolor CPNs by encapsulation of commercial CPs within the PEG-phospholipid layer [53]. Their protocol is simple and reproducible. The resulting NPs exhibited high colloidal stability and outstanding optical properties. Likewise, Christensen et al. combined CP based on PFBT with lipid to create CPNs with enhanced brightness and PLQY up to 46%. In addition to phospholipid or PEG encapsulation approaches, functional CPNs can also be produced by electrostatic interactions between a cationic CP and an anionic functional group [54]. By combining cationic PFO and poly(L-glutamic acid) modified anticancer drug doxorubicin (PG-Dox), bifunctional CPNs with the ability of both drug transport and imaging activities have been reported [55]. Moreover, cationic PBF polymers and negatively charged SDPA have been assembled to form multifunctional CPNs with light-activated anticancer and antibacterial activity, as well as imaging capability. The distinctive characteristics of the second approach compared to the first are its simplicity in the functionalization and the use of commercial availability CPs [34].

3.3. Multi-Functionalization Approaches

The third design strategy can be considered as an extension of the second one. In such approach, the surface of CPNs is first functionalized with a carboxyl or an amino group, as in the second approach. The functional biomolecule (peptide, sugar, protein, antibody) is then covalently bonded to the surface of the nanoparticle via carboxyl condensation or bio-orthogonal click reactions. Based on bioconjugation method, Chiu and co-workers developed functional CPNs for labeling cellular targets through the antigen–antibody interaction or the biotin–streptavidin system [56]. In their work, the authors functionalized CPNs by entrapping polymer chains into a single nanoparticle driven by hydrophobic interactions during nanoparticle formation. An amphiphilic polymer bearing functional groups is co-precipitated with a conjugated polymer to modify the nanoparticle surface for subsequent covalent functionalization with biomolecules, such as streptavidin and immunoglobulin. In this approach, the functionalization and bioconjugation models are almost applicable to all kinds of hydrophobic fluorescence CPs. Later, the same approach has been used by others to prepare a new bioimaging probe based on poly(9,9-dihexylfluorene-alt-2,1,3-benzoxadiazole) (PFBD) NP surface functionalized with PEG-mono-carboxylate.

The carboxylate-PEG functionalized NPs have several desirable structural and photophysical properties: a high colloidal stability in a broad range of pH values, an outstanding PLQ of 46%, and the presence of a carboxylic group as a surface chemical functionality. PEG-NPs were then bioconjugated with a cyclic RGDfK targeting peptide for labeling of membrane aVb3 integrin receptors on live HT-adenocarcinoma cells [57].

4. Biomedical Applications

4.1. Antimicrobial Properties of CPNs

CPs have attracted a great deal of interest as emerging materials in photodynamic therapy (PDT) due to their high light-harvesting capability and efficient energy transfer. They possess a large π-delocalized electronic system where the exciton can migrate along the polymer backbone chains or hop between chains. The ability of the exciton to diffuse throughout CP chains enables efficient energy transfer to dye acceptors such as photosensitizers (PSs) [58]. Moreover, CPs upon photo-excitation can directly sensitize oxygen molecules to produce reactive oxygen species (ROS) which can strongly damage biomacromolecules for antimicrobial and anticancer applications [59]. Whitten and co-workers reported a cationic conjugated polyelectrolyte derivate with the irradiation of visible light which can kill Gram-negative as well as Gram-positive bacteria [60]. This pioneer work opened up a new field of research of CP in the photodynamic therapy. Since then, various conjugated polyelectrolytes and hydrophobic CPNs have demonstrated their utility for PDT. Regarding the biological and biomedical applications of CPNs, the Wang group has reviewed in detail the development of CPNs for fluorescence imaging, anti-microorganism and antitumor, and gene delivery and drug delivery/release. Since then, various families of CPNs have been developed very fast and prosperously from polymer design to applications. Besides PDT, CPNs have recently emerged as a novel anti-bacterial material with a remarkable activity and are considered promising candidates to overcome the bacterial resistance with enhanced efficacy by two different modes of activity, under light illumination and in the dark.

4.1.1. Light-Activated Mode Activity

PDT is a promising method not only for cancer treatment and but also for antibacterial treatment that involves PSs and light irradiation to produce reactive toxic ROS from intracellular oxygen (O_2) to kill bacteria. Compared to antibiotics, PDT has a number of benefits, including an effective therapeutic impact, few side effects, and negligible drug resistance [61]. In a PDT process, the ROS is produced by irradiating a PS to sensitize oxygen. As a result, the ability of PS to produce ROS has an impact on the PDT's effectiveness. When the right molecular design is used, CPs could have superior light-harvesting capacity and efficient intra- and intermolecular energy transfer, which can result in a notable increase in ROS generation efficiency compared to standard PSs. The ability of conjugated polymers to sensitize oxygen to generate singlet oxygen and other ROS is due to their aptitude to strongly absorb visible light and to their outstanding fluorescence properties with a high yield of triplet state formation. Both light-activated modes of microbial killing or cancer treatment involve the generation of ROS. This is because of their exceptional fluorescence features, which include a high yield of triplet state production and, as a result, a high generation of singlet oxygen and ROS, as well as their capacity to significantly absorb visible light. The production of ROS is a necessary component of both light-activated cancer treatment and microbial eradication [34]. Luckily, CPs have good light-harvesting and energy transfer properties in comparison to other current PSs with poor solubility and small absorption cross-section [3]. In the last decade, different approaches combining common PSs with CPs to fabricate various CPNs have been developed to advance the improved PDT. By virtue of electrostatic and ligand–receptor interactions, positively charged specific elements can be associated with cells because cell membrane surfaces are negatively charged and have many specific markers. Exploiting this property, Wang and co-workers designed cationic multifunctional CPNs with light-activated antibacterial activity with imaging capability [34]. These positively charged CPNs opened up a new avenue in the design of PDT materials. Whitten and co-workers developed different CP systems to disclose and verify the biocidal mechanism of CPs by studying the effect of (i) positively charged CPs which effectively capture negatively charged microbial pathogens by electrostatic and hydrophobic interactions; (ii) singlet oxygen and/or other ROS generated upon irradiation of the CPs; (iii) ROS damaging effect of the membrane of pathogens and killing pathogens [62]. Due to easy synthesis and efficient ROS generation, CPNs are the most popular PDT material. Zhao et al. designed a biocompatible and biodegradable e-poly-Llysine (EPL)/poly (e-caprolactone) (PCL) copolymer [63]. This amphiphilic copolymer showed good ability to form monodispersed NPs, with a broad-spectrum antibacterial activity against Escherichia coli, Staphylococcus aureus and Bacillus subtilis. It is worth noting that the new NPs had a stronger antibacterial activity than the cationic peptide EPL. Several possible antibacterial pathways have been proposed in order to investigate the underlying mechanism of the biodegradable cationic NPs. It has been found that NPs can damage bacterial walls and membranes and cause an increase in ROS levels and alkaline phosphatase. More importantly, the self-assembled NPs changed bacterial osmotic pressure, resulting in cell invagination and formation of holes which cause a leakage of the cytoplasm. All these points show that the EPL–PCL NPs can be further developed to be a promising antimicrobial material for infectious disease treatment or as surfactants and emulsifiers to enhance drug encapsulation efficiency and antimicrobial activity, as shown in Figure 10.

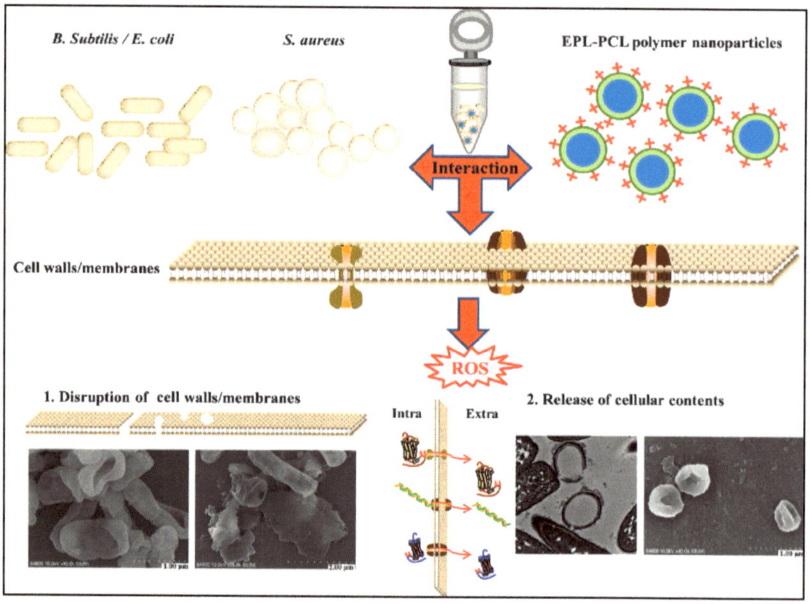

Figure 10. A cartoon scheme to illustrate the antibacterial mechanisms of the EPL–PCL NPs. Reproduced from Ref. [62] with permission from Elsevier. The scale bare in each image is 1 µzm.

Zhang et al. demonstrated dual-mode antibacterial CPNs by combining photothermal therapy and PDT. the authors explored them for efficient killing of ampicillin-resistant *Escherichia coli* (Ampr *E. coli*), as shown in Figure 11 [64].

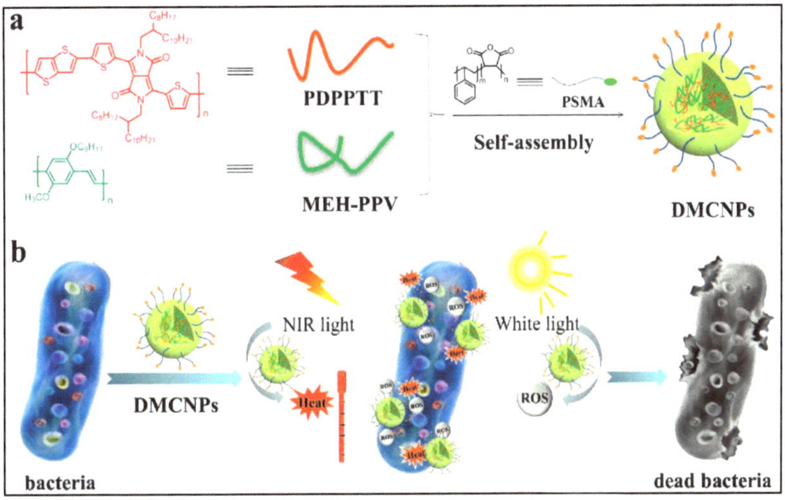

Figure 11. From top to bottom: (**a**) chemical structures of photodynamic conjugated polymers, (**b**) cartoon schematic illustration of photothermal and photodynamic therapy of CPNs for antimicrobial application. Reproduced from [64] with the permission of copyright Wiley.

The CPNs with a size of 50.4 ± 0.6 nm for dual-mode phototherapy were fabricated by co-precipitation of poly(diketopyrrolopyrrole-thienothiophene) (PDPPTT) and the PS poly [2-methoxy-5-((2-ethylhexyl)oxy)-p-phenylenevinylene] (MEHPPV) in the presence of poly(styrene-co-maleic anhydride). The latter polymer makes NPs disperse well in water via hydrophobic interactions. Thus, the dispersion of these NPs has a photothermal effect and a bacterial killing ability, simultaneously, since NPs are sensitive to the oxygen in the surrounding environment and generate ROS upon the illumination of light. With a combination of near-infrared light (550 mW cm^{-2}, 5 min), white light (65 mW cm^{-2}, 5 min) and a concentration of 9.6×10^{-4} µm of NPs, one could reach a 93% inhibition rate against Ampr *E. coli*, which is higher than the efficiency treated by PTT or PDT alone. These NPs with a dual mode offer the possibility of treating pathogenic infections brought on by resistant microorganisms in a clinical setting.

Wang et al. proposed a facile and rapid photothermal antimicrobial platform based on CPNs functionalized with the cell-penetrating peptide CPNs-Tat that have a near-infrared (NIR)-active and photothermal-responsive ability, as shown in Figure 12 [65]. Due to the positively charged Tat peptide, CPNs-Tat can enhance the interaction with bacteria cells by forming a CPNs-Tat/bacteria aggregation. Under NIR irradiation, CPNs-Tat could efficiently convert light into heat and create local hyperthermia to quickly kill bacteria. This photothermal approach provides an expedient and efficient tool for eradicating bacterial infections.

Chong et al. synthesized water-soluble cationic CPNs based on polymer containing PBF repeat units in the backbone with negatively charged SDPA in an aqueous solution through electrostatic interactions [34]. Upon photoexcitation of the NPs with white light (400–800 nm) with 90 and 45 mW cm^{-2}, they rapidly kill bacteria and cancer cells through the generated ROS. Additionally, these PBF NPs simultaneously offer optical imaging and bacteria-killing capability. PBF NPs are thus a promising multifunctional material for the simultaneous provision of optical monitoring capabilities and the treatment of bacterial infections and malignancies, as shown in Figure 13.

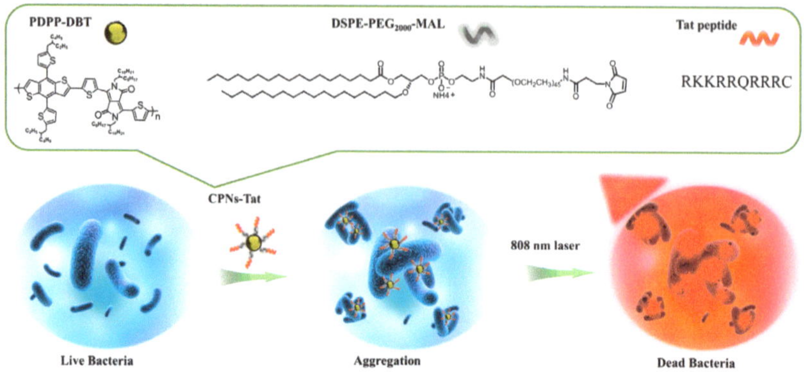

Figure 12. (**Top**) chemical structure of photothermal conjugated polymer and penetrating peptide. (**Bottom**) cartoon illustration of the photothermal killing of microbes by using CPNs-Tat as a photothermal agent upon NIR light. Reproduced from Ref. [64] with permission from ACS.

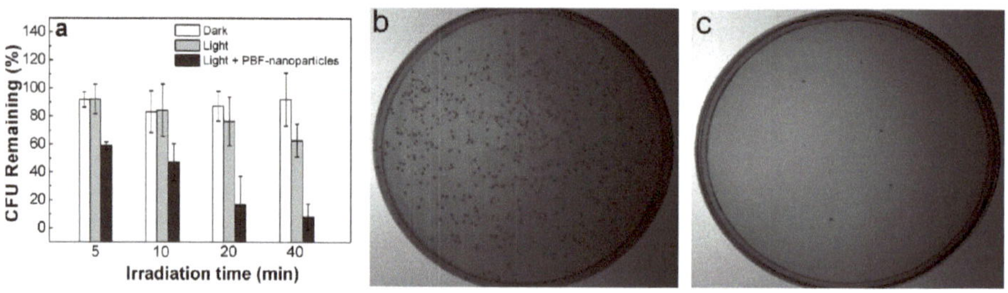

Figure 13. (**a**) Biocidal activity of PBF NPs toward Ampr *E. coli* in the dark and under white light for different irradiation times. (**b**) Plate photographs of Ampr *E. coli* incubated with PBF NPs for 40 min in the dark (**c**) and upon exposure to white light. Reproduced from Ref. [33] with permission from ACS.

4.1.2. Dark Mode Activity

The mechanism of antimicrobial activity of CPNs is still under investigation. It may involve membrane destruction by physical/chemical properties of NPs, as do many antibacterial peptides. Some recent works have demonstrated that naturally occurring antimicrobial peptides and their synthetic counterparts act only when in contact with bacterial cell walls by targeting the lipid bilayer of the bacterial membrane. The ability of these molecules to perturb the bacterial cell is highly dependent on the lipid composition of the membrane [66]. Furthermore, most antimicrobial peptides are cationic. Therefore, both the light-activated and dark biocidal activities of cationic CPNs are linked to their interactions with bacterial cell membranes. The dark killing abilities of these compounds correlate with their perturbation ability in relation to bacterial membranes. Charge density is important for the initial binding step between antimicrobial compounds and bacteria [67–69]. Recently, cationic fluorescent CPNs were shown to exhibit a broad-spectrum dark antimicrobial activity. Outstandingly positively charged cationic CPNs can physically combine with bacteria during bacterial incubation, then destroying the bacteria membrane and killing the bacteria in the dark, as shown in Figure 14 [70]. Then, NPs could diffuse through the cell wall and irreversibly disrupt the membrane structure of bacteria, which leads to the release of the cytoplasmic constituents from cells and bacteria death.

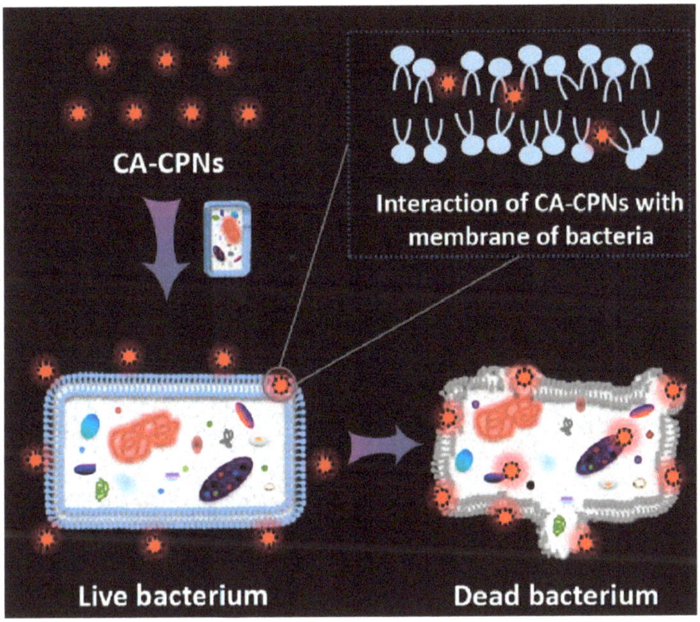

Figure 14. Schematic of dark antibacterial activity of cationic conjugated polymer NPs reproduced from [70] with permission from ACS.

In our previous work, we described a facile and cost-effective approach for preparing water-based cationic CPNs (CTAB-P3HT NPs) [71]. The combination of cationic surfactant cetyltrimethylammonium bromide (CTAB) with neutral CP (P3HT) provides positively charged NPs that not only allow electrostatic interaction with bacteria, but also synergize the antimicrobial activity of CP in the dark. The prepared cationic CPNs (CTAB-P3HT NPs) showed a broad spectrum of anti-bacterial and antifungal activity without requirement of light with a good biocompatibility toward normal human cells. This work provides a new biocompatible theragnostic NPs for bacteria detection as well as for the therapeutic treatment.

4.2. Fluorescence Sensing

In recent years, CPNs have emerged as exceptional fluorescent probes for in vitro and in vivo bioimaging and biosensing applications because of their exceptional optical properties combined with high fluorescence brightness, good photostability, and lower toxicity [1,3]. CPNs are made of conjugated polymers with π-delocalized electronic structures with distinct emission colors, a narrow bandwidth and are strongly fluorescent. This latter property, together with an easy preparation, high fluorescence brightness, low toxicity, stable photoluminescence and small particle size, makes them promising materials as fluorescent probes and fluorescent biosensors. As discussed in the previous section, CPNs are easy to prepare and their optical properties as well as their size can be easily tuned.

The most important examples of fluorescent CPs developed are compiled in Tables 1 and 2. These tables compile key parameters on physical properties of these nanoparticles. The representative family of CPs includes polyfluorene (e.g., PDHF and PFO), poly(phenylene ethynylene) (e.g., PPE), poly(phenylenevinylene) (e.g., MEH-PPVand CN-PPV), fluorene-based copolymers (e.g., PFPV, PFBT, and PFDBT), poly(thiophene), and related derivatives, which manufacture NPs with emission colors spanning the whole range of the visible spectrum.

On the other hand, fluorescent-based biosensor is a widely used approach for quick and easy medical diagnosis due to its inherent sensitivity and high selectivity [72,73]. This has stimulated enormous research and development of highly fluorescent π-CP as an active material in light-emitting devices and as a sensing layer in biosensor devices [74]. Thus, various fluorescent biosensing materials have been developed based on fluorescent π-CP for sensing metal ions, small molecules to microorganisms and cells [75–77]. The detection mechanism could be achieved by two approaches, by quenching the fluorescence of the polymers after the reaction with the target or by the emission of fluorescence.

4.2.1. Detection through Fluorescence Quenching

Wu et al. explored the size-controlled preparation of several new CPNs that show remarkable high-fluorescence quantum yield, radiative rate, photostability, high effective chromophore density, and minimal levels of aggregation-induced fluorescence quenching, resulting in fluorescence quantum yields exceeding 70%, even in the solid-state, thin films [78]. Other advantageous features of these CPNs include the lack of small dye molecules and heavy metal ions that might dissolve into a solution. Based on the CPN approach, Yuan et al. proposed a straightforward and accurate fluorescent technique for hydroquinone detection [79]. They characterized their CNPs by TEM, DLS, zeta potential, and FT-IR. The possible quenching mechanism may be caused by the electron transfer from hydroquinone to CPNs that turn off the fluorescence, as shown in Figure 15, since CPNs catalyzed the oxidation of hydroquinone to benzoquinone which could effectively quench the fluorescence of CPNs. This result was confirmed through the incubation of CPNs and the hydroquinone mixture with glutathione which induced a fluorescence recovering due to the strong antioxidant property of glutathione, so the benzoquinone was reduced into hydroquinone. The detection limit of hydroquinone down to 0.005 μM with excellent selectivity was achieved. This result of real sample analysis showed that this approach could be used in a real application. The proposed biosensor is expected to be beneficial for tracking the level of hydroquinone in the environment.

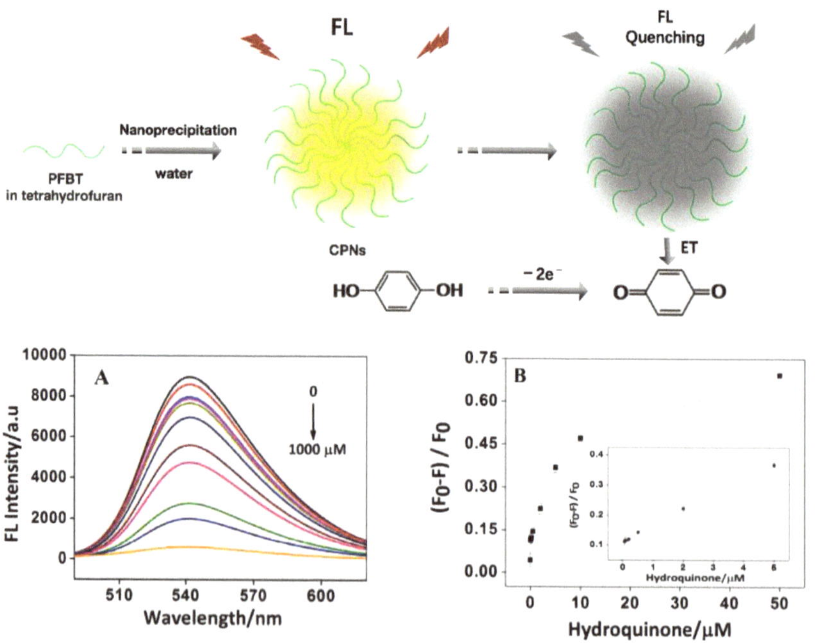

Figure 15. (**Top**): Schematic illustration of fabrication of the hydroquinone biosensor based on fluorescent CPNs. (**Bottom**): (**A**) Fluorescence spectra at different concentrations of hydroquinone

range from 0.01 mM to 1000 mM. (**B**) Plot of fluorescence peak intensities versus the hydroquinone concentration. Inset: linear relationship between the fluorescence peak intensity and the logarithm of hydroquinone concentration. Reproduced from Ref. [79]. with permission from Elsevier.

Recently, in our work, we reported a fluorescence-sensing scheme based on CTAB-P3HT NPs for the detection of the E. coli in the concentration range of (50–0.5 $\times$ 10^6 CFU/mL) with a limit of detection (LOD) less than 5 CFU/mL using a simple mix and detect strategy, as shown in Figure 16 [71]. As seen in Figure 16a, the interactions between positively charged CTAB-P3HT NPs and hole microbial particles result a decrease in fluorescence intensity. The aggregation of CTAB-P3HT NPs on E. coli, which results from the electrostatic interactions between the cationic groups of CTAB-P3HT NPs and a negative charge on the bacterial cell wall, can be used to explain the fluorescence quenching.

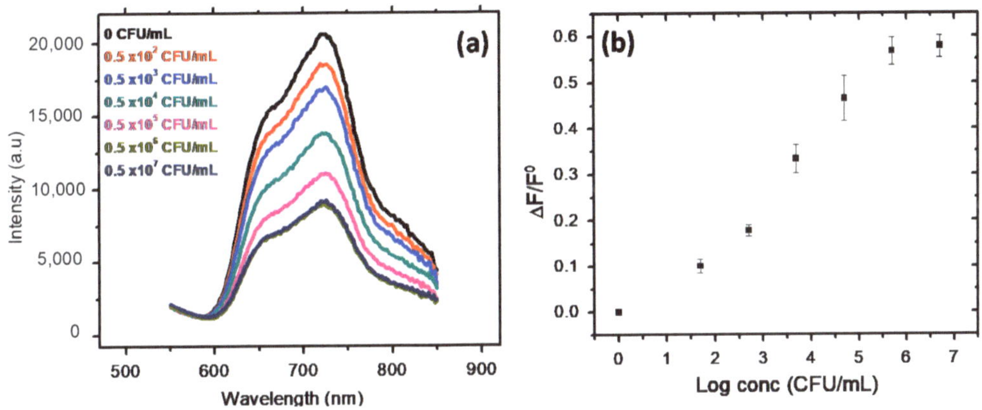

Figure 16. (**a**) Fluorescence spectra of CTAB-P3HT NPs at different concentrations of *E. coli* range from 0 CFU/mL to 0.5 $\times$ 10^7 CFU/mL, (**b**) calibration curve, a plot of fluorescence peak intensities, $\Delta F/F^0$ vs. Log concentrations of *E. coli* loads.

Li et al. reported a ratiometric fluorescent probe based on coordination CPNs for direct and fast detection of Zn^{2+} in aqueous solutions with excellent stability and selectivity [80]. The probe was prepared from aggregation-induced emission (AIE) fluorophore (4,4′-(hydrazine-1,2-diylidene bis (methanylylidene)) bis (3-hydroxybenzoic acid)) (HDBB) molecules with metal ions. The CPNs are made up of HDBB molecules and Tb3+ (referred to as Tb-HDBB-CPNs), which exhibited a matrix coordination-induced emission peak at a wavelength of 590 nm. In contrast, Zn-HDBB-CPNs exhibited a distinctive fluorescent property with a blue emission peak wavelength of 470 nm (Figure 17). Based on the cation exchange process of Tb HDBB-CPNs with Zn^{2+}, a highly selective ratiometric fluorescent probe in the range of 100 nM to 60 mM and a detection limit of 50 nM was developed for the determination of Zn^{2+} in an aqueous solution. This research demonstrates the benefits of using AIE materials as a fluorescent probe and exhibits the advantages of a direct detection procedure. Thus, it paves a way and opens up the avenue for further research in the AIE functionalized materials with ratiometric fluorescence response for chemical sensing.

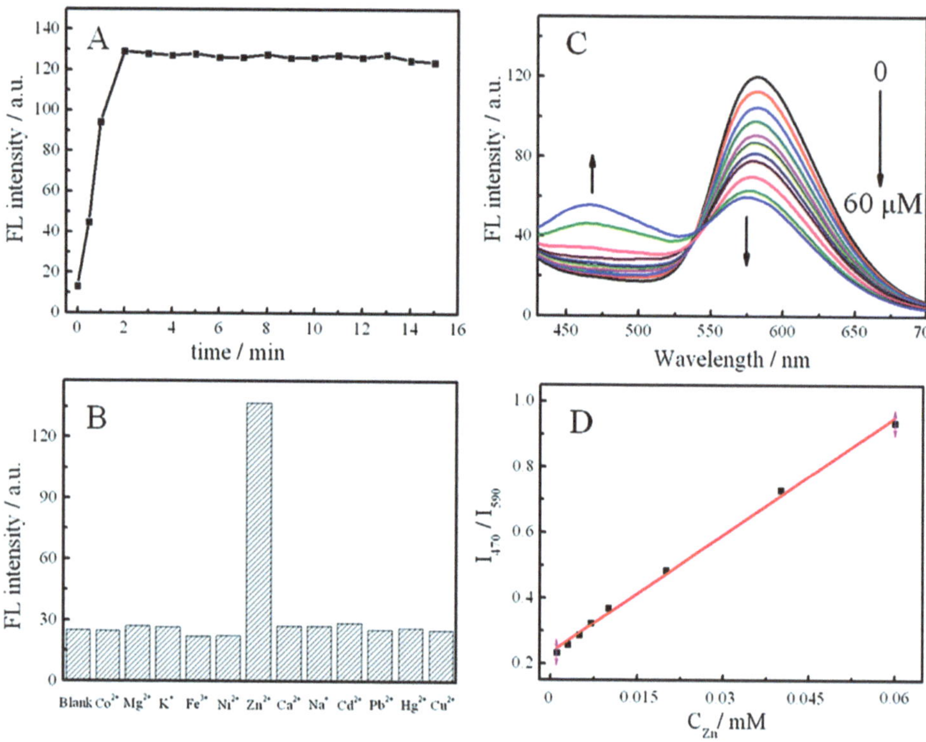

Figure 17. (**A**) Plot of fluorescence intensity versus time at 470 of 0.02 mg mL^{-1} Tb-HDBB-CPNs to 0.1 mM Zn^{2+} in pH 7.0. (**B**) A plot of selectivity of Tb-HDBB-CPN-based sensor for Zn^{2+} to other cations. (**C**) Fluorescence spectra of Tb-HDBB-CPNs probe upon addition of different concentrations of Zn^{2+}. (**D**) The linear plot of the variation of the maximum of the emission wavelength at 470/590 versus the concentration of Zn^{2+}. Reproduced from Ref. [79], copyright figures available under a Creative Commons Attribution Non-Commercial 3.0 Unported License.

4.2.2. Sensing through Fluorescence Emission

Zhang et al. demonstrated a novel and efficient melamine biosensor based on fluorescence FRET between CPNs and AuNPs, as shown in Figure 18 [81]. The system is highly selective and sensitive to the detection of melamine, with a LOD of 1.7 nmol/L. The energy transfer efficiency can reach 82.1%. The interaction mechanism occurred when the CPNs were able to close the surface of the AuNPs. Then, the fluorescence of CPNs was effectively quenched. However, the aggregation of AuNPs by the N-Au interaction after melamine addition reduced FRET of CPNs-AuNPs and promoted the recovery of CPN fluorescence. After the addition of melanin, the absorption spectrum Au-NPs overlapped significantly with the emission spectrum of the CPNs, which reduced the energy transfer efficiency. Furthermore, the proposed technique was carried out for melamine detection in real powdered infant formula with satisfactory results.

Wang et al. described a fluorescent probe for sulfide based on carboxylic functionalized CPNs [82]. The NPs were manufactured by the co-precipitation method from poly[(9,9-dioctylfluorenyl-2,7-diyl)-co-(1,4-benzo-2,1′-3-thiadiazole)] bearing carboxylic functional group and referred to as COOH-PFBT. The green fluorescence of these carboxylic CPNs is quenched due to the aggregate upon the addition of Cu(II) ions. This quenching process is fully reversible; thus, the fluorescence emission is restored when the sulfide is added to the aggregates, inducing the formation of CuS. This quenching recovery (Boff-on^) mechanism forms the basis for a new sulfide detection scheme. The fluorescence intensities vary linearly with the concentration of the sulfide in the range of 1.25 to 75.0 μM, with a LOD of 0.45 μM. This biosensor has a good selectivity for sulfur anions over other anions. When used for the determination of sulfide in spiked real water samples, this method demonstrates recoveries ranging between 98.6% and 105.7%.

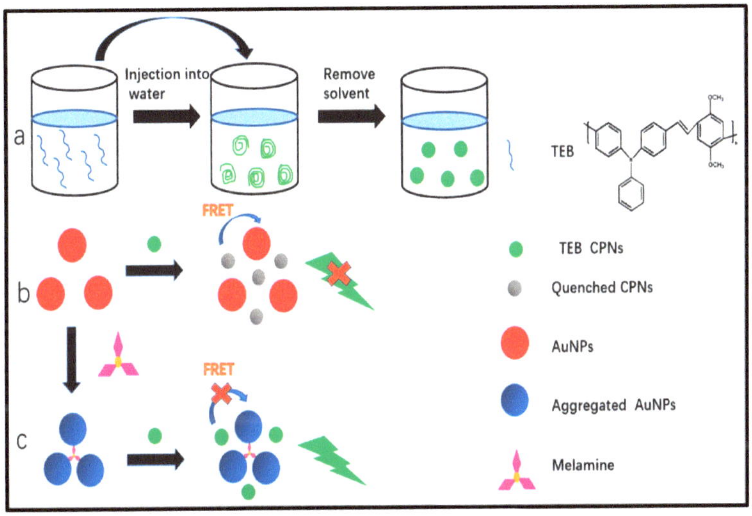

Figure 18. Schematic illustration of working principle of melamine FRET biosensor based on CPNs- and AuNPs; (**a**) mini-emulsion preparation process of CPNs; (**b**) CPNs fluorescence quenching by a FRET from AuNPs; (**c**) restoring of the fluorescence of CPNs after adding melamine. Reproduced from [81]. Copyright figure available under Creative Commons Attribution (CC BY) license.

4.2.3. Fluorescent CPNs for Bioimaging

Bioimaging is crucial in the field of medical diagnosis, biological and biomedical research to gain insights into biological processes and malfunctions in many ways, including the observation of morphology and structure, study of functions, and diagnosis and therapeutics of diseases. Some of the notable progress made in this field includes the development of various probe materials with outstanding properties for cell imaging, subcellular imaging, molecule imaging, and so on. Besides probe material development, progress in the fluorescence imaging techniques has become a prevalent choice providing effective research tools to investigate many fundamental processes in the life sciences. It is widely used in a broad variety of preclinical investigations for the visualization of biological processes and provides cellular and subcellular resolution and image enhancement in 3D and real time in a non-destructive way. As a result, fluorescence-based diagnosis of diseases and fluorescence-image-guided surgery found many real applications in human disease diagnosis [3,83]. Nevertheless, the future search for improved cost-effective, time-dependent and safe bioimaging techniques with excellent resolution is still ongoing. Fluorescent NPs have proven their excellent intrinsic properties including their resonator geometry, which allow amplified emission from a single fluorescent NP. This latter property is of interest for tracking of individual cells in larger united cell structures. Whispering-Gallery-Mode-based biosensors (WGMS) can be categorized as label-free technologies. Their basic principle relies upon the total internal reflection of light trapped inside a circular resonator. Certain circular nanoparticles satisfy the conditions of biosensing with WGMs mode lasers. Among these criteria, the NP diameter needs to be in the range of several micrometers to support the lasing modes. However, with these sizes, the NPs are less important for in vivo applications; however, there might be solutions to make a single particle laser resonator smaller by confining the fluorescence and gain to subwavelength surface plasmon polaritons.

Traditional fluorescent probe materials for imaging such as fluorescent organic dyes have excellent fluorescence properties, including high PLQYs, and a large variety of these dyes with tunable optical characteristics are readily available. However, organic dyes suffer a low photo stability and light bleaching issue. Inorganic quantum dots overcome these limitations; however, they contain heavy metals like lead, cadmium, or indium, which significantly increase their cytotoxicity. Despite the recent progress in this filed, cytotoxicity and photo-bleaching issues may significantly restrict their use in real applications [84,85]. In the last decade, CPNs have surpassed conventional organic dyes and become a new class of probe materials for bioimaging due to their excellent photostability and high fluorescence brightness. Alongside with their applications in imaging, CPNs are appealing for cell culture research because of their beneficial intrinsic properties such as low cytotoxicity, inertness to intracellular processes, and their high resistance to bleaching and easy chemical functionalization [83,86]. For potential in vivo applications, CPNs should meet certain requirements including an ideal size range, as they are small enough to allow access to smaller capillary blood vessels and large enough to not be cleared immediately, which impacts therapeutic strategies, surface charges, cell viability and the associated pharmacokinetics. Both non-specific and specific imaging can be targeted by CNPs. Non-specific imaging relies on the direct use of CPNs without functionalization with a specific targeting element. Many prior studies have mainly focused on non-specific cellular imaging of CPNs. This mechanism is driven by non-specific cellular uptake, and CPNs are mainly located in the cytoplasm. Therefore, the biocompatibility and low cytotoxicity are the main criteria for assessing the imaging performance of CPNs. Additionally, CPNs are also a new class of probe materials for theranostic agents [87]. As an example of application, CATB-P3HT-NPs that were designed by our group [71] were used to image bacteria by fluorescence confocal microscopy. The bacteria emit bright red fluorescence, indicating that NPs with positive charge are bonded to bacteria via electrostatic interactions, as shown in Figure 19.

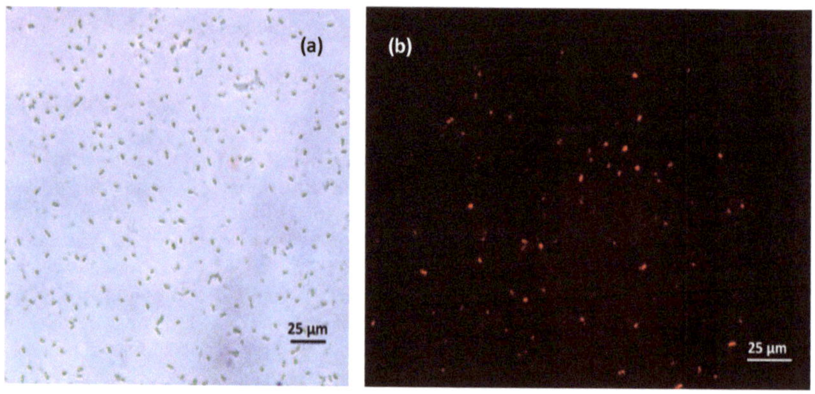

Figure 19. Fluorescence confocal microscopy images showing *E. coli* incubated with CTAB-P3HT NPs, (**a**) Phase-contrast image and (**b**) fluorescence image.

Tang et al. developed fluorescent CPNs based on the model CPs, poly (fluorene divinylenebenzothiadiazole) PFVBT, and the model quarternary ammonium salts, cationic surfactant CTAB. These CA-CPNs were successfully used in the imaging of both bacterial cells and mammalian cells with selective toxicity against bacteria, as shown in Figure 20 [70].

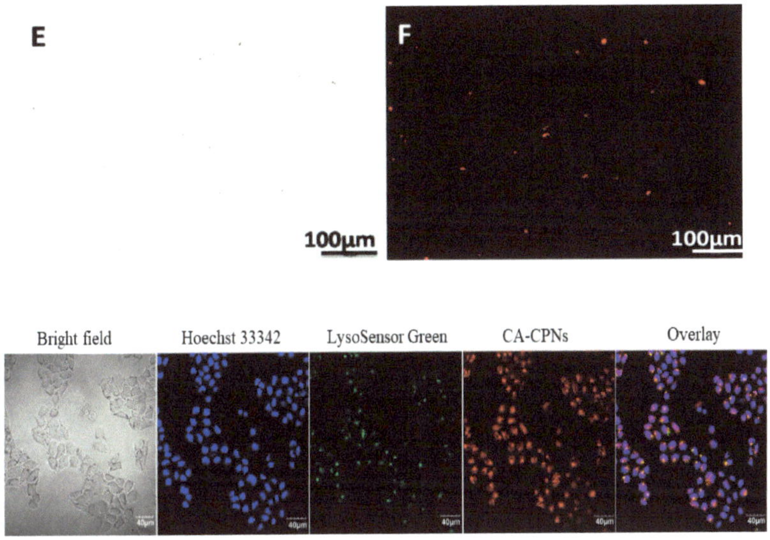

Figure 20. (**Top left**) Phase-contrast image and (**Top right**) fluorescence image of *E. coli* incubated with CA-CPNs. The excitation wavelength was 530 nm. [CA − CPNs] = 1.0 μg/mL. (**bottom**) Confocal fluorescence microscopy images of MCF-7 treated with 1.0 μg/mL of CA-CPNs for 18 h. Reproduced from Ref. [70] with permission from ACS.

Recently, other innovative CPNs have been designed and synthesized for various modal imaging techniques, as shown in Tables 1 and 2.

Table 1. Chemical names and corresponding abbreviations of conjugated polymers and nanoparticle stabilizing agents.

Abbreviation	Chemical Name	Ref.
Conjugated polymers		
DPA-CNPPV	(Poly[{2-methoxy-5-(2-ethylhexyloxy)-1,4-(1-cyanovinylenephenylene)}-co-{2,5-bis(N,N′-diphenylamino)-1,4-phenylene}]	[88]
(APNs)	Polyfluorophore Nano sensors	[89]
PBTQ4F	Poly[4-(4,8-bis(5-(2-ethylhexyl)thiophen-2-yl)benzo[1,2-b:4,5-b′]dithiophen-2-yl)-2-(2-ethylhexyl)-6,7-bis(3-((2-ethylhexyl)oxy)-4,5-difluorophenyl)-2H-[1,2,3]triazolo[4,5-g]quinoxaline]	[90]
(PTZTPA-BBT)	Triphenylamine (TPA) functionalized phenothiazine (PTZ) as the donor, strong acceptor benzothiazole (BBT).	[91]
DPP-BTzTD	Poly([2,5-bis(2-decyltetradecyl)-2,5-dihydropyrrolo[3,4-c]pyrrole-1,4-dione-3,6-dithienyl]-co-[6-(2-ethylhexyl)-[1,2,5]thiadiazolo [3,4-f]benzotriazole-4,8-diyl])	[92]
SPNRs (PFPV, PFBT, and PFODBT)	Semiconducting polymer nano reporters	[93]
PFPV	Poly[(9,9′-dioctyl-2,7-divinylene-fluorenylene)-alt-{2-methoxy-5-(2-ethylhexyloxy)-1,4-phenylene}]	[93]
PFBT	Poly[(9,9′-dioctylfluorenyl-2,7-diyl)-alt-(benzo[2,1,3]thiadiazol-4,7-diyl)]	[93]
PFODBT	Poly[2,7-(9,9′-dioctylfluorene)-alt-4,7-bis(thiophen-2-yl)benzo-2,1,3-thiadiazole	[93]
PBMC	Poly[3-{2-[2,5-Bis-(2-ethyl-hexyloxy)-4-propenyl-phenyl]-vinyl}-9-butyl-6-methyl-9H-carbazole]	[94]
PTD	Poly[6-(2-ethylhexyl)-4-methyl-8-(5-methylthiophen-2-yl)-6,7-dihydro-5H-[1,2,3]triazolo[4′,5′:4,5]benzo[1,2-c][1,2,5]thiadiazole]-alt-[3-methyl-6-(5-methylthiophen-2-yl)-2,5-bis(2-octyldodecyl)-2,5-dihydropyrrolo[3,4-c]pyrrole-1,4-dione]poly-thiadiazolobenzotriazole-alt-thiophene-diketopyrrolopyrrole	[95]
PCPDTBT	Poly[2,6-(4,4-bis-(2-ethylhexyl)-4H-cyclopenta[2,1-b:3,4-b′]-dithiophene)-alt-4,7-(2,1,3-benzothiadiazole]]	[96]
PDFT	Furan-containing diketopyrrolopyrrole-based semiconducting polymers	[97]
P3HT	Poly(3-hexylthiophene)	[71]

Table 1. *Cont.*

Abbreviation	Chemical Name	Ref.
NP stabilising agents (surfactants/polymers/lipids)		
PSMA	Poly(styrene-co-maleic anhydride)	[88]
(PS-PEG-COOH)	Polystyrene graft ethylene oxide functionalized with carboxylic end group	[90]
PEG-b-PPG-b-PEG	Triblock surfactant of PEG and poly-(propylene glycol) (PPG)	[93]
DSPE–PEG	1,2-Distearoyl-sn-glycero-3-phosphoethanolamine-N-[methoxy(polyethyleneglyol)]	[95]
PEG-PLGA	Poly(ethylene glycol)methyl ether-block-poly(lactide-co-glycolide) copolymer	[96]
PEG	Polyethylene glycol	[97]
CATB	Cetyltrimethylammonium bromide	[71]

Table 2. Polymer and stabilizing agents used in CPNs, their size, PLQY (%) and applications, as reported in the literature.

Conjugated Polymer	CPN Stabilizing Agent	Size (nm)	PLQY (%)	Application	Ref.
DPA-CNPPV	PSMA	17.5–22.1	10.8	NADH sensing	[88]
APNs	Protease-reactive peptide brush (via self-immolative linkers)	100–200	N.A.	Cancer and allograft	[89]
m-PBTQ4F	(PS-PEG-COOH)	22	3.2	Fluorescent imaging	[90]
Pdots-C6 (PTZTPA-BBT)	(PS-PEG-COOH)	40.8	N.A.	Fluorescent imaging	[91]
DPP-BTzTD	Self-assembling conjugated polymer	4	N.A.	Photoacoustic imaging	[92]
SPNRs	PEG-b-PPG-b-PEG	30–40	2.7 ± 0.014	Chemiluminescent imaging	[93]
PBMC	PSMA	~44	N.A.	Fe^{3+} sensing in vitro in HeLa cells	[94]
PTD	DSPE–PEG	42	N.A.	In vivo photoacoustic 3D vasculature imaging	[95]
PCPDTBT	PEG-PLGA	32 ± 0.7	2.8	Fluorescent imaging	[96]
PDFT	PEG	40	N.A.	Fluorescent imaging	[97]
PBTQ4F and CN-PPV	PSMA	~15	1.9 and 6	Fluorescent imaging	[98]
Fluorinated PBTQ	(PS-PEG-COOH)	≈22	3.2	Fluorescent imaging	[90]
Pttc-SeBTa-NIR1125/1270/1380	DSPE–PEG	35–73	0.05–0.18	Fluorescent imaging	[99]
P3HT	CATB	55.32 ± 7.3	N.A.	Fluorescent imaging, bacteria sensing, and killing	[71]

4.3. Electrochemical Sensing

Recently, intensive studies were carried out based on CPNs and their composites for electrochemical sensor development, for the highly sensitive detection, their porosity, large surface area, and the presence of binding sites originating from their functional groups for enabling biomarker capture as well as ultrasensitive and easy detection. In addition, due to their ability to convert a biological interaction into a measurable electrochemical signal, they became a material of choice for sensing applications. Consequently, CNPs can act as excellent electrode materials for immobilizing various molecular recognition elements through the most common covalent attachment method and the often-used entrapment or adsorption methods, resulting in ultrasensitive recognition with very low LOD. More importantly, functionalizing the surface of CNPs has been found to play a major role in limiting the biofouling effect, resulting in ultrasensitive detection in real samples. This anti-fouling effect is a supreme character for biomarker detection in real practical applications [100].

As a result, the fabrication of CPN-based biosensors with ultrasensitive, selective and antifouling effect is expected to be a promising diagnostic tool that can replace the conventional methods for analysis. However, a lot of future work is highly needed for the development of new CPN-based biosensors since they still have many unexplored possibilities. Another promising method for the future is the development of CPN-coated nanowires or nanotubes which recently highly improved the sensing properties of CPNs [101]. In addition, designing hybrid conducting polymer nanocomposites based on polymers and conducting inorganic nanomaterials such as carbon nanomaterials and metallic nanoparticles to produce CPNs nano-systems will result in a high-potential and synergistic effect with promising different materials, yet to be explored for biosensing.

While using conducting polymers based electrochemical sensors, the capture of a target analyte generates a measurable signal which is directly converted into an electrical signal, as shown in Figure 21. The efficiency of the biosensor depends on the strength of the bioreceptor immobilization to the conducting polymer and of the conducting polymer to the biosensor surface. This accordingly affects the selectivity of the biosensor that depends on the specific interactions between the analyte and the bioreceptor. The sensitivity of the biosensor is determined by the electrochemical signal intensity resulted from the reaction between the analyte, the bioreceptor and the conducting polymer. This electrochemical signal may result from a change in the voltage, current, conductivity/resistance, impedance, or number of electrons exchanged through an oxidation or reduction reaction, leading to the development of potentiometric, amperometric, conductimetric, impedimetric and voltametric biosensors, respectively [102].

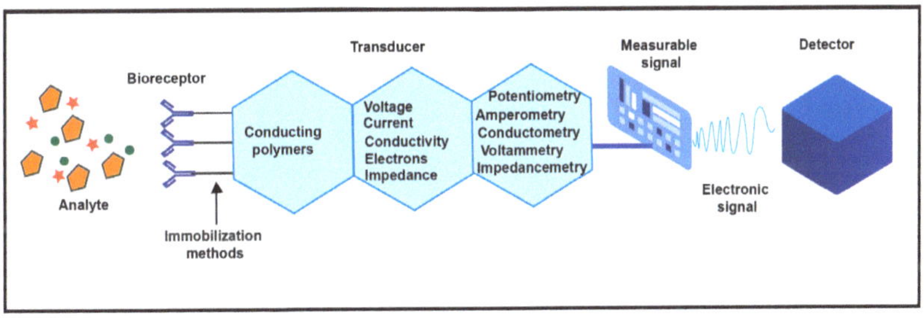

Figure 21. Detection principles of conducting polymer-based electrochemical biosensors.

Shekher et al. [103] recently demonstrated the fabrication of a biosensor for a sensitive detection of serotonin (5-HT) based on electrodeposition of 2-amino-5-mercapto-1,3,4-thiadiazole (AMT) and gold nanoparticles (nAu) on the functionalized carbon nanotube (f-CNT) on a glassy carbon electrode. An LOD of 7.8 nM was achieved by using cyclic voltammetry, square wave voltammetry, and i-t amperometric analysis. For practical utility, it was investigated using human serum with good recovery limits of 96.3 to 103.6%. More recentlym, Tseng et al. [104] investigated the electrochemical detection of Leukocyte esterase (LE) and nitrite (NIT) which are useful biomarkers for the diagnosis of urinary tract infections (UITs). A carboxylic acid-functionalized poly(3,4-ethylenedioxythiophene) (PEDOT-COOH)-modified carbon electrode was used to immobilize the LE antibody and fabricate the biosensor. Different electrochemical techniques such as differential pulse voltammetry and electrochemical impedance spectroscopy were used to characterize the biosensor. Good performances with LODs of 6.24 μM and 0.2 μg L^{-1} in a linear range of 9.1–131 μM and 0.2–590 μg L^{-1} were demonstrated for NIT and LE, respectively. In real urine specimens, a diagnostic sensitivity and specificity were 100% and 87.5%, respectively. Another novel hydrogel-based electrochemical biosensor was recently developed by Du et al. [105]. The biosensor was fabricated from the antifouling zwitterionic peptide hydrogel (CFEFKFC), modified with the conducting polymer poly(3,4-ethylenedioxythiophene) (PEDOT) and gold nanoparticles (AuNPs) through electrodeposition to detect prostate specific antigen (PSA) in human serum. The hydrogel was immobilized onto the electrode surface via the Au–S bond. It is also suitable for the immobilization of PSA antibodies through the formation of covalent amide bonds. The outstanding antifouling property of the hydrogel could effectively prevent the adsorption of nonspecific proteins, cells and other biomolecules. The results showed a linear range from 0.1 ng mL^{-1} to 100 ng mL^{-1}, with a low (LOD) down to 5.6 pg mL^{-1} [105]. Kumar A et al. developed an electrochemical biosensor based on a graphene oxide/polyaniline/polypyrrole/zinc oxide (GO/PPy/PANI/ZnO) nanocomposite, NC, for cholesterol and bilirubin detection. The results showed a sensitivity of 0.92 μA μM^{-1} cm^{-2} and 0.2 μA μM^{-1} cm^{-2} for the detection of cholesterol and bilirubin, respectively. In comparison to the existing reports, the detection sensitivity for cholesterol is almost doubled, and for bilirubin, it is twenty times greater. The sensor showed remarkable detection limits, which were 0.42 and 0.69 μM for cholesterol and bilirubin, respectively [106]. There is a series of examples of CP-based electrochemical biosensors summarized in Table 3 to highlight the recent advances in this field.

Table 3. Recent examples of conducting polymer-based electrochemical biosensors.

Active Layer	Linear Range	Sensitivity	Detection Limit	Stability	Real Samples	Ref.
CP-based glucose amperometric biosensor						
Poly(Azure A)-Pt NP	0.02–2.3 mM	42.7μA mM^{-1} cm^{-2}	7.6 μM	3 months	Fruit juice	[107]
Polycarbazole	0.01–5 mM	14.0 μA mM^{-1} cm^{-2}	0.2 μM	---	---	[108]
CP-based immunosensor		Target		Detection Mode		
Polypyrrole-polythionine	0.001–100 pg/mL	Neuron-specific enolase	0.65 pg/mL	Voltammetry	---	[109]
CP-based DNA Biosensors						
Polyaniline–MoS_2	10^{-15}–10^{-6} M	DNA	10^{-15} M	Voltammetry	---	[110]
Polyaniline–graphene	10^{-9}–10^{-6} M	Mycobacterium tuberculosis DNA	8×10^{-7} M	Voltammetry	---	[111]
CP-based Biosensors for Cancer Diagnosis		Biomarker				
Poly(triphenylamine rhodanine-3-acetic acid-co-3,4-ethoxylene dioxy thiophene)/MoS_2/Peptide	1×10^{-2} ng mL^{-1}	Matrix metalloproteinase-1	1×10^{-3} ng mL^{-1}	CV [a]	Lung cancer cells	[112]
PEDOT/peptide [b]	50–10^6 cells mL^{-1}	CTC [c]	17 cells mL^{-1}	DPV [d]	Breast cancer cells	[113]

Abbreviations: [a] cyclic voltammetry: CV; [b] poly(3,4-ethylenedioxythiophene): PEDOT; [c] circulating tumour cells: CTC; [d] differential pulse voltammetry: DPV.

5. Conclusions and Perspectives

CPNs have become an important emerging material for in vivo bioimaging and sensing, which greatly support the disease diagnosis and treatment. In this review, we survey the current state of the art of methods of preparation, advantages and disadvantages, as well as bio-applications of CPNs. Several methods of CPN preparation are discussed to demonstrate the effect of synthesis conditions on the optical, morphological, and chemical properties of CPNs. CPNs are widely used in biomedical applications due to their unique photophysical and photoelectric properties that endowed them excellent performance with high quantum yield, photostability and favorable compatibility. In addition, controllable particle size can be acquired, and these particles can easily be functionalized to be adapted to specific application requirements. Considering all the unique properties of CPNs mentioned in this review, we suggest a superior utility for practical applications compared to that of the conventional materials. As biosensors, CPNs could efficiently detect glucose, cholesterol, serotonin (5-HT), phenylalanine, prostate specific antigen, urinary tract infections and tumor markers both in buffers and human serum with excellent, rapid and accurate results in a cost-effective and eco-friendly way. CPNs are also superior for biological applications such as fluorescence, luminescence, Raman, and PA imaging due to their unique photophysical properties, biocompatibility and rapid metabolism relevant for biosafety compared to other imaging agents. Such an outstanding performance with excellent results in various biomedical applications has confirmed the significance of CPNs for bioimaging, drug delivery, and sensing applications. Regarding biomedical applications, the superior brightness of CPNs and their photostability as well as their low toxicity permits the tracking of subcellular entities, or labeled molecules over extended periods of time with high resolution. Although there are many research reports available in the open literature; however, the field is still in its infancy, and much of the effort so far has been placed on the development of new conjugated polymer structures as nanoparticle materials and their photophysical characterization. To proceed further in this field, biomedical application is required, and not just proof-of-concept study. To progress in this direction, organic and material chemists need to work side by side with biologists. CPNs have been used in diagnostic techniques with great achievements in the sensing and imaging markets; however, their clinical applications for the diagnosis of diseases and treatment are still in the research stage due to several aspects that should be taken into consideration in the future development of CPNs for biomedical applications. The major challenges facing the use of CNPs in the biomedical field can be addressed as follows:

(1) The design of novel CPNs with controllable dosages and unique properties is needed to effectively achieve the goals such as improving ability of photoenergy transfer to lower the side effects.
(2) Improved targeted drug delivery is highly needed for cancer therapy through studying the attachment of targeting ligands to the CPN surface.
(3) Further investigation of CPN biodegradation time is required to confirm its biosafety in the human body through carrying out a pharmacokinetic study upon injection of CPNs into the animal model.
(4) More efforts are still needed to make CPNs scalable through cost-effective and eco-friendly preparation methods such as a cradle-to-grave process study described in this review.
(5) Functionalization of CPNs through optimized strategies to improve their performance in biological applications is suggested such as stimulus-responsive targeting, for example, pH-responsive polymers.
(6) Synthesis of ultra-small-size CPNs is needed through optimized preparation methods to enhance its biodistribution and achieve a rapid metabolism; their current state currently limits the clinical application of nanoparticles.
(7) More efforts should be made to design nanocomposites based on CPNs for multi-faceted applications.
(8) The rapid advancement in artificial intelligence and computational simulations can be exploited to design functionalized rational CPNs for biomedical applications. Overall, more efforts are highly required to bring CPNs into real clinical use.

Author Contributions: Conceptualization, N.E. (Nada Elgiddawy), N.E. (Noha Elnagar), H.K.-Y. and A.Y.; methodology, N.E. (Nada Elgiddawy) and N.E. (Noha Elnagar); validation, N.E. (Nada Elgiddawy), N.E. (Noha Elnagar), H.K.-Y. and A.Y.; investigation, N.E. (Nada Elgiddawy), N.E. (Noha Elnagar), H.K.-Y. and A.Y.; writing—original draft preparation, N.E. (Nada Elgiddawy), H.K.-Y. and A.Y.; writing—review and editing, N.E. (Nada Elgiddawy), N.E. (Noha Elnagar), H.K.-Y. and A.Y.; visualization, N.E. (Nada Elgiddawy), N.E. (Noha Elnagar), H.K.-Y. and A.Y.; supervision, H.K.-Y. and A.Y.; project administration, H.K.-Y. funding acquisition, N.E. (Nada Elgiddawy), N.E. (Noha Elnagar) and H.K.-Y. All authors have read and agreed to the published version of the manuscript.

Funding: N. E. thanks the French Embassy in Egypt, the French Institute in Egypt (IFE) and the Agence Universitaire de la Francophonie (AUF) for Doctoral Scholarship for one year at the CNRS, Institut de Chimie Moléculaire et des Matériaux d'Orsay, Université Paris-Saclay, Orsay, France.

Data Availability Statement: Not applicable.

Conflicts of Interest: The authors declare no conflict of interest.

References

1. Braeken, Y.; Cheruku, S.; Ethirajan, A.; Maes, W. Conjugated polymer nanoparticles for bioimaging. *Materials* **2017**, *10*, 1420. [CrossRef]
2. Kuehne, A.J. Conjugated polymer nanoparticles toward in vivo theranostics–focus on targeting, imaging, therapy, and the importance of clearance. *Adv. Biosyst.* **2017**, *1*, 1700100. [CrossRef] [PubMed]
3. Wang, Y.; Feng, L.; Wang, S. Conjugated polymer nanoparticles for imaging, cell activity regulation, and therapy. *Adv. Funct. Mater.* **2019**, *29*, 1806818. [CrossRef]
4. Sun, T.; Xie, Z. Conjugated polymers and polymer dots for cell imaging. In *Fluorescent Materials for Cell Imaging*; Springer: Berlin/Heidelberg, Germany, 2020; pp. 155–180.
5. Wang, J.; Lv, F.; Liu, L.; Ma, Y.; Wang, S. Strategies to design conjugated polymer based materials for biological sensing and imaging. *Coord. Chem. Rev.* **2018**, *354*, 135–154. [CrossRef]
6. Tuncel, D.; Demir, H.V. Conjugated polymer nanoparticles. *Nanoscale* **2010**, *2*, 484–494. [CrossRef] [PubMed]

7. Feng, L.; Zhu, C.; Yuan, H.; Liu, L.; Lv, F.; Wang, S. Conjugated polymer nanoparticles: Preparation, properties, functionalization and biological applications. *Chem. Soc. Rev.* **2013**, *42*, 6620–6633. [CrossRef] [PubMed]
8. Tao, Z.; Hong, G.; Shinji, C.; Chen, C.; Diao, S.; Antaris, A.L.; Zhang, B.; Zou, Y.; Dai, H. Biological imaging using nanoparticles of small organic molecules with fluorescence emission at wavelengths longer than 1000 nm. *Angew. Chem. Int. Ed.* **2013**, *52*, 13002–13006. [CrossRef] [PubMed]
9. Asgher, M.; Qamar, S.A.; Sadaf, M.; Iqbal, H.M. Multifunctional materials conjugated with near-infrared fluorescent organic molecules and their targeted cancer bioimaging potentialities. *Biomed. Phys. Eng. Express* **2020**, *6*, 012003. [CrossRef]
10. Lin, F.; Duan, Q.-Y.; Wu, F.-G. Conjugated polymer-based photothermal therapy for killing microorganisms. *ACS Appl. Polym. Mater.* **2020**, *2*, 4331–4344. [CrossRef]
11. Li, Z.; Bai, H.; Jia, S.; Yuan, H.; Gao, L.-H.; Liang, H. Design of functional polymer nanomaterials for antimicrobial therapy and combatting resistance. *Mater. Chem. Front.* **2021**, *5*, 1236–1252. [CrossRef]
12. Liu, H.-Y.; Wu, P.-J.; Kuo, S.-Y.; Chen, C.-P.; Chang, E.-H.; Wu, C.-Y.; Chan, Y.-H. Quinoxaline-based polymer dots with ultrabright red to near-infrared fluorescence for in vivo biological imaging. *J. Am. Chem. Soc.* **2015**, *137*, 10420–10429. [CrossRef]
13. Jiang, Y.; Pu, K. Multimodal biophotonics of semiconducting polymer nanoparticles. *Acc. Chem. Res.* **2018**, *51*, 1840–1849. [CrossRef]
14. Liu, P.; Mu, X.; Zhang, X.-D.; Ming, D. The near-infrared-II fluorophores and advanced microscopy technologies development and application in bioimaging. *Bioconjugate Chem.* **2019**, *31*, 260–275. [CrossRef] [PubMed]
15. He, S.; Song, J.; Qu, J.; Cheng, Z. Crucial breakthrough of second near-infrared biological window fluorophores: Design and synthesis toward multimodal imaging and theranostics. *Chem. Soc. Rev.* **2018**, *47*, 4258–4278. [CrossRef]
16. Li, J.; Pu, K. Development of organic semiconducting materials for deep-tissue optical imaging, phototherapy and photoactivation. *Chem. Soc. Rev.* **2019**, *48*, 38–71. [CrossRef]
17. Ong, S.Y.; Zhang, C.; Dong, X.; Yao, S.Q. Recent advances in polymeric nanoparticles for enhanced fluorescence and photoacoustic imaging. *Angew. Chem. Int. Ed.* **2021**, *60*, 17797–17809. [CrossRef]
18. Arshad, F.; Pal, A.; Sk, M.P. Aggregation-induced emission in carbon dots for potential applications. *ECS J. Solid State Sci. Technol.* **2021**, *10*, 021001. [CrossRef]
19. Wang, J.; Li, J.; Wang, L.; Han, T.; Wang, D.; Tang, B.Z. AIEgen-based polymer nanocomposites for imaging-guided photothermal therapy. *ACS Appl. Polym. Mater.* **2020**, *2*, 4306–4318. [CrossRef]
20. Ansari, L.; Hallaj, S.; Hallaj, T.; Amjadi, M. Doped-carbon dots: Recent advances in their biosensing, bioimaging and therapy applications. *Colloids Surf. B Biointerfaces* **2021**, *203*, 111743. [CrossRef] [PubMed]
21. Bag, P.; Maurya, R.K.; Dadwal, A.; Sarkar, M.; Chawla, P.A.; Narang, R.K.; Kumar, B. Recent Development in Synthesis of Carbon Dots from Natural Resources and Their Applications in Biomedicine and Multi-Sensing Platform. *ChemistrySelect* **2021**, *6*, 2774–2789. [CrossRef]
22. Su, Q.; Feng, W.; Yang, D.; Li, F. Resonance energy transfer in upconversion nanoplatforms for selective biodetection. *Acc. Chem. Res.* **2017**, *50*, 32–40. [CrossRef]
23. Lyu, Y.; Pu, K. Recent advances of activatable molecular probes based on semiconducting polymer nanoparticles in sensing and imaging. *Adv. Sci.* **2017**, *4*, 1600481. [CrossRef]
24. Wu, C.; Chiu, D.T. Highly fluorescent semiconducting polymer dots for biology and medicine. *Angew. Chem. Int. Ed.* **2013**, *52*, 3086–3109. [CrossRef] [PubMed]
25. Klapper, M.; Nenov, S.; Haschick, R.; Müller, K.; Müllen, K. Oil-in-oil emulsions: A unique tool for the formation of polymer nanoparticles. *Acc. Chem. Res.* **2008**, *41*, 1190–1201. [CrossRef] [PubMed]
26. Müller, K.; Klapper, M.; Müllen, K. Synthesis of conjugated polymer nanoparticles in non-aqueous emulsions. *Macromol. Rapid Commun.* **2006**, *27*, 586–593. [CrossRef]
27. Huber, J.; Mecking, S. Processing of polyacetylene from aqueous nanoparticle dispersions. *Angew. Chem.* **2006**, *118*, 6462–6465. [CrossRef]
28. Baier, M.C.; Huber, J.; Mecking, S. Fluorescent conjugated polymer nanoparticles by polymerization in miniemulsion. *J. Am. Chem. Soc.* **2009**, *131*, 14267–14273. [CrossRef]
29. Landfester, K.; Montenegro, R.; Scherf, U.; Güntner, R.; Asawapirom, U.; Patil, S.; Neher, D.; Kietzke, T. Semiconducting polymer nanospheres in aqueous dispersion prepared by a miniemulsion process. *Adv. Mater.* **2002**, *14*, 651–655. [CrossRef]
30. Zhang, S.; Zhang, H.; Yang, S.; Zhang, X.; Li, S.; Huang, L.; Jing, Y.-n.; Xiao, L.; Zhang, Y.; Han, B. Synthesis and Application of green solvent dispersed organic semiconducting nanoparticles. *Nano Res.* **2023**. [CrossRef]
31. Rahmanudin, A.; Marcial-Hernandez, R.; Zamhuri, A.; Walton, A.S.; Tate, D.J.; Khan, R.U.; Aphichatpanichakul, S.; Foster, A.B.; Broll, S.; Turner, M.L. Organic Semiconductors Processed from Synthesis-to-Device in Water. *Adv. Sci.* **2020**, *7*, 2002010. [CrossRef]
32. MacFarlane, L.R.; Shaikh, H.; Garcia-Hernandez, J.D.; Vespa, M.; Fukui, T.; Manners, I. Functional nanoparticles through π-conjugated polymer self-assembly. *Nat. Rev. Mater.* **2021**, *6*, 7–26. [CrossRef]
33. Lin, H.; Bai, H.; Yang, Z.; Shen, Q.; Li, M.; Huang, Y.; Lv, F.; Wang, S. Conjugated polymers for biomedical applications. *Chem. Commun.* **2022**, *58*, 7232–7244. [CrossRef] [PubMed]
34. Chong, H.; Nie, C.; Zhu, C.; Yang, Q.; Liu, L.; Lv, F.; Wang, S. Conjugated polymer nanoparticles for light-activated anticancer and antibacterial activity with imaging capability. *Langmuir* **2012**, *28*, 2091–2098. [CrossRef] [PubMed]

35. Peng, J.; Han, Y. Recent advances in conjugated polythiophene-based rod–rod block copolymers: From morphology control to optoelectronic applications. *Giant* **2020**, *4*, 100039. [CrossRef]
36. Elgiddawy, N.; Ren, S.; Yassar, A.; Louis-Joseph, A.; Sauriat-Dorizon, H.; El Rouby, W.M.; El-Gendy, A.O.; Farghali, A.A.; Korri-Youssoufi, H. Dispersible conjugated polymer nanoparticles as biointerface materials for label-free bacteria detection. *ACS Appl. Mater. Interfaces* **2020**, *12*, 39979–39990. [PubMed]
37. Song, I.Y.; Kim, J.; Im, M.J.; Moon, B.J.; Park, T. Synthesis and self-assembly of thiophene-based all-conjugated amphiphilic diblock copolymers with a narrow molecular weight distribution. *Macromolecules* **2012**, *45*, 5058–5068.
38. Braeken, Y.; Cheruku, S.; Seneca, S.; Smisdom, N.; Berden, L.; Kruyfhooft, L.; Penxten, H.; Lutsen, L.; Fron, E.; Vanderzande, D. Effect of branching on the optical properties of poly (p-phenylene ethynylene) conjugated polymer nanoparticles for bioimaging. *ACS Biomater. Sci. Eng.* **2019**, *5*, 1967–1977. [CrossRef]
39. Kim, C.; Gwon, Y.J.; Kim, J.; Lee, T.S. Synthesis of fluorescent conjugated polymer nanoparticles and their immobilization on a substrate for white light emission. *Polym. Chem.* **2018**, *9*, 5671–5679. [CrossRef]
40. Foroozandeh, P.; Aziz, A.A. Insight into cellular uptake and intracellular trafficking of nanoparticles. *Nanoscale Res. Lett.* **2018**, *13*, 339. [CrossRef]
41. Bogart, L.K.; Pourroy, G.; Murphy, C.J.; Puntes, V.; Pellegrino, T.; Rosenblum, D.; Peer, D.; Lévy, R. Nanoparticles for imaging, sensing, and therapeutic intervention. *ACS Nano* **2014**, *8*, 3107–3122. [CrossRef]
42. Howarth, M.; Liu, W.; Puthenveetil, S.; Zheng, Y.; Marshall, L.F.; Schmidt, M.M.; Wittrup, K.D.; Bawendi, M.G.; Ting, A.Y. Monovalent, reduced-size quantum dots for imaging receptors on living cells. *Nat. Methods* **2008**, *5*, 397–399. [CrossRef]
43. Li, K.; Liu, B. Polymer encapsulated conjugated polymer nanoparticles for fluorescence bioimaging. *J. Mater. Chem.* **2012**, *22*, 1257–1264. [CrossRef]
44. Soo Choi, H.; Liu, W.; Misra, P.; Tanaka, E.; Zimmer, J.P.; Itty Ipe, B.; Bawendi, M.G.; Frangioni, J.V. Renal clearance of quantum dots. *Nat. Biotechnol.* **2007**, *25*, 1165–1170. [CrossRef] [PubMed]
45. Mendez, E.; Moon, J.H. Side chain and backbone structure-dependent subcellular localization and toxicity of conjugated polymer nanoparticles. *Chem. Commun.* **2013**, *49*, 6048–6050. [CrossRef]
46. Abelha, T.F.; Neumann, P.R.; Holthof, J.; Dreiss, C.A.; Alexander, C.; Green, M.; Dailey, L.A. Low molecular weight PEG–PLGA polymers provide a superior matrix for conjugated polymer nanoparticles in terms of physicochemical properties, biocompatibility and optical/photoacoustic performance. *J. Mater. Chem. B* **2019**, *7*, 5115–5124. [CrossRef] [PubMed]
47. Umezawa, K.; Yoshida, M.; Kamiya, M.; Yamasoba, T.; Urano, Y. Rational design of reversible fluorescent probes for live-cell imaging and quantification of fast glutathione dynamics. *Nat. Chem.* **2017**, *9*, 279–286. [CrossRef] [PubMed]
48. Yao, H.; Fukui, C. Size and morphology effects on the fluorescence properties of π-conjugated poly (p-phenylene) polyelectrolyte nanoparticles synthesized via polyion association. *J. Mater. Chem. C* **2016**, *4*, 2945–2953. [CrossRef]
49. Piwoński, H.; Michinobu, T.; Habuchi, S. Controlling photophysical properties of ultrasmall conjugated polymer nanoparticles through polymer chain packing. *Nat. Commun.* **2017**, *8*, 15256. [CrossRef]
50. Feng, X.; Tang, Y.; Duan, X.; Liu, L.; Wang, S. Lipid-modified conjugated polymer nanoparticles for cell imaging and transfection. *J. Mater. Chem.* **2010**, *20*, 1312–1316. [CrossRef]
51. Tang, H.; Xing, C.; Liu, L.; Yang, Q.; Wang, S. Synthesis of amphiphilic polythiophene for cell imaging and monitoring the cellular distribution of a cisplatin anticancer drug. *Small* **2011**, *7*, 1464–1470. [CrossRef] [PubMed]
52. Wu, C.; Jin, Y.; Schneider, T.; Burnham, D.R.; Smith, P.B.; Chiu, D.T. Ultrabright and bioorthogonal labeling of cellular targets using semiconducting polymer dots and click chemistry. *Angew. Chem. Int. Ed.* **2010**, *49*, 9436–9440. [CrossRef]
53. Howes, P.; Green, M.; Levitt, J.; Suhling, K.; Hughes, M. Phospholipid encapsulated semiconducting polymer nanoparticles: Their use in cell imaging and protein attachment. *J. Am. Chem. Soc.* **2010**, *132*, 3989–3996. [CrossRef] [PubMed]
54. Fernando, L.P.; Kandel, P.K.; Ackroyd, P.C.; Christensen, K.A. The relative brightness of PEG lipid-conjugated polymer nanoparticles as fluid-phase markers in live cells. *Anal. Bioanal. Chem.* **2012**, *404*, 3003–3014. [CrossRef] [PubMed]
55. Feng, X.; Lv, F.; Liu, L.; Tang, H.; Xing, C.; Yang, Q.; Wang, S. Conjugated polymer nanoparticles for drug delivery and imaging. *ACS Appl. Mater. Interfaces* **2010**, *2*, 2429–2435. [CrossRef]
56. Wu, C.; Schneider, T.; Zeigler, M.; Yu, J.; Schiro, P.G.; Burnham, D.R.; McNeill, J.D.; Chiu, D.T. Bioconjugation of ultrabright semiconducting polymer dots for specific cellular targeting. *J. Am. Chem. Soc.* **2010**, *132*, 15410–15417. [CrossRef]
57. Li, Y.; Liu, J.; Liu, B.; Tomczak, N. Highly emissive PEG-encapsulated conjugated polymer nanoparticles. *Nanoscale* **2012**, *4*, 5694–5702. [CrossRef] [PubMed]
58. Meng, Z.; Hou, W.; Zhou, H.; Zhou, L.; Chen, H.; Wu, C. Therapeutic considerations and conjugated polymer-based photosensitizers for photodynamic therapy. *Macromol. Rapid Commun.* **2018**, *39*, 1700614. [CrossRef]
59. Yuan, H.; Wang, B.; Lv, F.; Liu, L.; Wang, S. Conjugated-polymer-based energy-transfer systems for antimicrobial and anticancer applications. *Adv. Mater.* **2014**, *26*, 6978–6982. [CrossRef]
60. Wang, Y.; Tang, Y.; Zhou, Z.; Ji, E.; Lopez, G.P.; Chi, E.Y.; Schanze, K.S.; Whitten, D.G. Membrane perturbation activity of cationic phenylene ethynylene oligomers and polymers: Selectivity against model bacterial and mammalian membranes. *Langmuir* **2010**, *26*, 12509–12514. [CrossRef]
61. Liu, Y.; Qin, R.; Zaat, S.A.; Breukink, E.; Heger, M. Antibacterial photodynamic therapy: Overview of a promising approach to fight antibiotic-resistant bacterial infections. *J. Clin. Transl. Res.* **2015**, *1*, 140.

62. Lu, L.; Rininsland, F.H.; Wittenburg, S.K.; Achyuthan, K.E.; McBranch, D.W.; Whitten, D.G. Biocidal activity of a light-absorbing fluorescent conjugated polyelectrolyte. *Langmuir* **2005**, *21*, 10154–10159. [CrossRef]
63. Zhao, R.; Wang, H.; Ji, T.; Anderson, G.; Nie, G.; Zhao, Y. Biodegradable cationic ε-poly-L-lysine-conjugated polymeric nanoparticles as a new effective antibacterial agent. *Sci. Bull.* **2015**, *60*, 216–226. [CrossRef]
64. Zhang, H.; Liang, Y.; Zhao, H.; Qi, R.; Chen, Z.; Yuan, H.; Liang, H.; Wang, L. Dual-Mode Antibacterial Conjugated Polymer Nanoparticles for Photothermal and Photodynamic Therapy. *Macromol. Biosci.* **2020**, *20*, 1900301. [CrossRef]
65. Wang, Y.; Li, S.; Liu, L.; Feng, L. Photothermal-responsive conjugated polymer nanoparticles for the rapid and effective killing of bacteria. *ACS Appl. Bio Mater.* **2018**, *1*, 27–32. [CrossRef]
66. Li, J.; Koh, J.-J.; Liu, S.; Lakshminarayanan, R.; Verma, C.S.; Beuerman, R.W. Membrane active antimicrobial peptides: Translating mechanistic insights to design. *Front. Neurosci.* **2017**, *11*, 73. [CrossRef]
67. Scheberl, A.; Khalil, M.L.; Maghsoodi, F.; Strach, E.W.; Yang, J.; Chi, E.Y.; Schanze, K.S.; Reimhult, E.; Whitten, D.G. Quantitative determination of dark and light-activated antimicrobial activity of poly (phenylene ethynylene), polythiophene, and oligo (phenylene ethynylene) electrolytes. *ACS Appl. Mater. Interfaces* **2020**, *12*, 21322–21329. [CrossRef]
68. Brown, D.M.; Yang, J.; Strach, E.W.; Khalil, M.I.; Whitten, D.G. Size and substitution effect on antimicrobial activity of polythiophene polyelectrolyte derivatives under photolysis and dark conditions. *Photochem. Photobiol.* **2018**, *94*, 1116–1123. [CrossRef] [PubMed]
69. Huang, Y.; Pappas, H.C.; Zhang, L.; Wang, S.; Cai, R.; Tan, W.; Wang, S.; Whitten, D.G.; Schanze, K.S. Selective imaging and inactivation of bacteria over mammalian cells by imidazolium-substituted polythiophene. *Chem. Mater.* **2017**, *29*, 6389–6395. [CrossRef]
70. Wang, L.; Zhao, Q.; Zhang, Z.; Lu, Z.; Zhao, Y.; Tang, Y. Fluorescent conjugated polymer/quarternary ammonium salt co-assembly nanoparticles: Applications in highly effective antibacteria and bioimaging. *ACS Appl. Bio Mater.* **2018**, *1*, 1478–1486. [CrossRef] [PubMed]
71. Elgiddawy, N.; Ren, S.; Ghattas, W.; Rouby, W.M.E.; El-Gendy, A.O.; Farghali, A.A.; Yassar, A.; Korri-Youssoufi, H. Antimicrobial activity of cationic poly (3-hexylthiophene) nanoparticles coupled with dual fluorescent and electrochemical sensing: Theragnostic prospect. *Sensors* **2021**, *21*, 1715. [CrossRef] [PubMed]
72. Tagit, O.; Hildebrandt, N. Fluorescence sensing of circulating diagnostic biomarkers using molecular probes and nanoparticles. *ACS Sens.* **2017**, *2*, 31–45. [CrossRef] [PubMed]
73. Nawrot, W.; Drzozga, K.; Baluta, S.; Cabaj, J.; Malecha, K. A fluorescent biosensors for detection vital body fluids' agents. *Sensors* **2018**, *18*, 2357. [CrossRef]
74. Pan, H.; Gonuguntla, S.; Li, S.; Trau, D. *3.33 Conjugated Polymers for Biosensor Devices*; Elsevier: Amsterdam, The Netherlands, 2017.
75. Chen, X.; Hussain, S.; Chen, X.; Hao, Y.; Zhang, P.; Gao, R. Fabrication of conjugated polymer encapsulated fluorescent hybrid micelles for augmented, highly selective and step-wise detection of nitroaromatic pollutants and hepatobiliary biomarker. *Sens. Actuators B Chem.* **2023**, *377*, 133081. [CrossRef]
76. Zhang, P.; Zandieh, M.; Ding, Y.; Wu, L.; Wang, X.; Liu, J.; Li, Z. A Label-Free, Mix-and-Detect ssDNA-Binding Assay Based on Cationic Conjugated Polymers. *Biosensors* **2023**, *13*, 122. [CrossRef]
77. Tanwar, A.S.; Mehtab, M.; Kim, J.-T.; Oh, K.-J.; Iyer, P.K.; Im, Y.-H. Real-time selective pesticide detection using catalytic behavior of zwitterionic conjugated polymer. *Chem. Eng. J.* **2023**, *456*, 141002. [CrossRef]
78. Wu, C.; Bull, B.; Szymanski, C.; Christensen, K.; McNeill, J. Multicolor conjugated polymer dots for biological fluorescence imaging. *ACS Nano* **2008**, *2*, 2415–2423. [CrossRef] [PubMed]
79. Liu, Y.; Wang, Y.-M.; Zhu, W.-Y.; Zhang, C.-H.; Tang, H.; Jiang, J.-H. Conjugated polymer nanoparticles-based fluorescent biosensor for ultrasensitive detection of hydroquinone. *Anal. Chim. Acta* **2018**, *1012*, 60–65. [CrossRef]
80. Lin, N.; Zhang, Q.; Xia, X.; Liang, M.; Zhang, S.; Zheng, L.; Cao, Q.; Ding, Z. A highly zinc-selective ratiometric fluorescent probe based on AIE luminogen functionalized coordination polymer nanoparticles. *RSC Adv.* **2017**, *7*, 21446–21451. [CrossRef]
81. Zhang, C.-j.; Gao, Z.-y.; Wang, Q.-b.; Zhang, X.; Yao, J.-s.; Qiao, C.-d.; Liu, Q.-z. Highly sensitive detection of melamine based on the fluorescence resonance energy transfer between conjugated polymer nanoparticles and gold nanoparticles. *Polymers* **2018**, *10*, 873. [CrossRef]
82. Wang, C.; Sun, J.; Mei, H.; Gao, F. Organic semiconductor polymer nanodots as a new kind of off-on fluorescent probe for sulfide. *Microchim. Acta* **2017**, *184*, 445–451. [CrossRef]
83. Honeybone, D.; Peace, H.; Green, M. Infrared emitting and absorbing conjugated polymer nanoparticles as biological imaging probes. *J. Mater. Chem. C* **2023**, *11*, 7860–7871. [CrossRef]
84. Podder, S. Fluorescent Quantum Dots, A Technological Marvel for Optical Bio-imaging: A Perspective on Associated In Vivo Toxicity. In *Application of Quantum Dots in Biology and Medicine: Recent Advances*; Springer: Berlin/Heidelberg, Germany, 2022; pp. 143–163.
85. Lian, H.; Li, Y.; Saravanakumar, S.; Jiang, H.; Li, Z.; Wang, J.; Xu, L.; Zhao, W.; Han, G. Metal halide perovskite quantum dots for amphiprotic bio-imaging. *Coord. Chem. Rev.* **2022**, *452*, 214313. [CrossRef]
86. Deng, S.; Li, L.; Zhang, J.; Wang, Y.; Huang, Z.; Chen, H. Semiconducting Polymer Dots for Point-of-Care Biosensing and In Vivo Bioimaging: A Concise Review. *Biosensors* **2023**, *13*, 137. [CrossRef] [PubMed]
87. Abelha, T.F.; Dreiss, C.A.; Green, M.A.; Dailey, L.A. Conjugated polymers as nanoparticle probes for fluorescence and photoacoustic imaging. *J. Mater. Chem. B* **2020**, *8*, 592–606. [CrossRef] [PubMed]

88. Chen, H.; Yu, J.; Men, X.; Zhang, J.; Ding, Z.; Jiang, Y.; Wu, C.; Chiu, D.T. Reversible ratiometric NADH sensing using semiconducting polymer dots. *Angew. Chem.* **2021**, *133*, 12114–12119. [CrossRef]
89. Huang, J.; Chen, X.; Jiang, Y.; Zhang, C.; He, S.; Wang, H.; Pu, K. Renal clearable polyfluorophore nanosensors for early diagnosis of cancer and allograft rejection. *Nat. Mater.* **2022**, *21*, 598–607. [CrossRef]
90. Liu, Y.; Liu, J.; Chen, D.; Wang, X.; Zhang, Z.; Yang, Y.; Jiang, L.; Qi, W.; Ye, Z.; He, S. Fluorination enhances NIR-II fluorescence of polymer dots for quantitative brain tumor imaging. *Angew. Chem.* **2020**, *132*, 21235–21243. [CrossRef]
91. Men, X.; Geng, X.; Zhang, Z.; Chen, H.; Du, M.; Chen, Z.; Liu, G.; Wu, C.; Yuan, Z. Biomimetic semiconducting polymer dots for highly specific NIR-II fluorescence imaging of glioma. *Mater. Today Bio* **2022**, *16*, 100383. [CrossRef]
92. Men, X.; Wang, F.; Chen, H.; Liu, Y.; Men, X.; Yuan, Y.; Zhang, Z.; Gao, D.; Wu, C.; Yuan, Z. Ultrasmall semiconducting polymer dots with rapid clearance for second near-infrared photoacoustic imaging and photothermal cancer therapy. *Adv. Funct. Mater.* **2020**, *30*, 1909673. [CrossRef]
93. Cui, D.; Li, J.; Zhao, X.; Pu, K.; Zhang, R. Semiconducting polymer nanoreporters for near-infrared chemiluminescence imaging of immunoactivation. *Adv. Mater.* **2020**, *32*, 1906314. [CrossRef]
94. Gao, Z.-y.; Zhang, X.; Xing, S.; Lu, Q.; Yao, J.-s.; Liu, Q.-z.; Xie, R.-x.; Ding, B. Conjugated polymer nanoparticles based on carbazole for detecting ferric ion (III) with a large Stokes shift and high sensitivity and the application in cell imaging. *Dye. Pigment.* **2019**, *168*, 68–76. [CrossRef]
95. Guo, B.; Chen, J.; Chen, N.; Middha, E.; Xu, S.; Pan, Y.; Wu, M.; Li, K.; Liu, C.; Liu, B. High-resolution 3D NIR-II photoacoustic imaging of cerebral and tumor vasculatures using conjugated polymer nanoparticles as contrast agent. *Adv. Mater.* **2019**, *31*, 1808355. [CrossRef]
96. Modicano, P.; Neumann, P.R.; Schüller, M.; Holthof, J.; Kyrilis, F.L.; Hamdi, F.; Kastritis, P.L.; Mäder, K.; Dailey, L.A. Enhanced optical imaging properties of lipid nanocapsules as vehicles for fluorescent conjugated polymers. *Eur. J. Pharm. Biopharm.* **2020**, *154*, 297–308. [CrossRef]
97. Lu, F.; Zhan, C.; Gong, Y.; Tang, Y.; Xie, C.; Wang, Q.; Wang, W.; Fan, Q.; Huang, W. A General Strategy to Encapsulate Semiconducting Polymers within PEGylated Mesoporous Silica Nanoparticles for Optical Imaging and Drug Delivery. *Part. Part. Syst. Charact.* **2020**, *37*, 1900483. [CrossRef]
98. Chen, D.; Liu, Y.; Zhang, Z.; Liu, Z.; Fang, X.; He, S.; Wu, C. NIR-II fluorescence imaging reveals bone marrow retention of small polymer nanoparticles. *Nano Lett.* **2020**, *21*, 798–805. [CrossRef]
99. Liu, M.H.; Zhang, Z.; Yang, Y.C.; Chan, Y.H. Polymethine-Based Semiconducting Polymer Dots with Narrow-Band Emission and Absorption/Emission Maxima at NIR-II for Bioimaging. *Angew. Chem. Int. Ed.* **2021**, *60*, 983–989. [CrossRef] [PubMed]
100. Kappen, J.; Skorupa, M.; Krukiewicz, K. Conducting Polymers as Versatile Tools for the Electrochemical Detection of Cancer Biomarkers. *Biosensors* **2023**, *13*, 31. [CrossRef] [PubMed]
101. Çetin, M.Z.; Guven, N.; Apetrei, R.-M.; Camurlu, P. Highly sensitive detection of glucose via glucose oxidase immobilization onto conducting polymer-coated composite polyacrylonitrile nanofibers. *Enzym. Microb. Technol.* **2023**, *164*, 110178. [CrossRef]
102. Lakard, B. Electrochemical biosensors based on conducting polymers: A review. *Appl. Sci.* **2020**, *10*, 6614. [CrossRef]
103. Shekher, K.; Sampath, K.; Vandini, S.; Satyanarayana, M.; Gobi, K.V. Gold Nanoparticle Assimilated Polymer Layer on Carbon Nanotube Matrices for Sensitive Detection of Ser-otonin in Presence of Dopamine in-vitro. *Inorg. Chim. Acta* **2023**, *549*, 121399. [CrossRef]
104. Tseng, W.-T.; Chou, Y.-Y.; Wu, J.-G.; Wang, Y.-C.; Tseng, T.-N.; Pan, S.-W.; Luo, S.-C.; Ho, M.-L. An electrochemical conducting polymer-based biosensor for Leukocyte esterase and nitrite detection for diagnosing urinary tract infections: A pilot study. *Microchem. J.* **2023**, *188*, 108493. [CrossRef]
105. Du, Q.; Wang, W.; Zeng, X.; Luo, X. Antifouling zwitterionic peptide hydrogel based electrochemical biosensor for reliable detection of prostate specific antigen in human serum. *Anal. Chim. Acta* **2023**, *1239*, 340674. [CrossRef]
106. Kumar, A.; Gupta, G.H.; Singh, G.; More, N.; Keerthana, M.; Sharma, A.; Jawade, D.; Balu, A.; Kapusetti, G. Ultrahigh sensitive graphene oxide/conducting polymer composite based biosensor for cholesterol and bilirubin detection. *Biosens. Bioelectron. X* **2023**, *13*, 100290. [CrossRef]
107. Jiménez-Fiérrez, F.; González-Sánchez, M.I.; Jiménez-Pérez, R.; Iniesta, J.; Valero, E. Glucose biosensor based on disposable activated carbon electrodes modified with platinum nanoparticles electrodeposited on poly (Azure A). *Sensors* **2020**, *20*, 4489. [CrossRef] [PubMed]
108. Bagdžiūnas, G.; Palinauskas, D. Poly (9 H-carbazole) as a Organic Semiconductor for Enzymatic and Non-Enzymatic Glucose Sensors. *Biosensors* **2020**, *10*, 104. [CrossRef] [PubMed]
109. Wang, H.; Ma, Z. A cascade reaction signal-amplified amperometric immunosensor platform for ultrasensitive detection of tumour marker. *Sens. Actuators B Chem.* **2018**, *254*, 642–647. [CrossRef]
110. Dutta, S.; Chowdhury, A.D.; Biswas, S.; Park, E.Y.; Agnihotri, N.; De, A.; De, S. Development of an effective electrochemical platform for highly sensitive DNA detection using MoS2-polyaniline nanocomposites. *Biochem. Eng. J.* **2018**, *140*, 130–139. [CrossRef]
111. Mohamad, F.S.; Mat Zaid, M.H.; Abdullah, J.; Zawawi, R.M.; Lim, H.N.; Sulaiman, Y.; Abdul Rahman, N. Synthesis and characterization of polyaniline/graphene composite nanofiber and its application as an electrochemical DNA biosensor for the detection of Mycobacterium tuberculosis. *Sensors* **2017**, *17*, 2789. [CrossRef]

112. Lee, M.-H.; Lin, C.-C.; Kutner, W.; Thomas, J.L.; Lin, C.-Y.; Iskierko, Z.; Ku, Y.-S.; Lin, C.-Y.; Borowicz, P.; Sharma, P.S. Peptide-imprinted conductive polymer on continuous monolayer molybdenum disulfide transferred electrodes for electrochemical sensing of Matrix Metalloproteinase-1 in lung cancer culture medium. *Biosens. Bioelectron. X* **2023**, *13*, 100258. [CrossRef]
113. Han, R.; Li, Y.; Chen, M.; Li, W.; Ding, C.; Luo, X. Antifouling electrochemical biosensor based on the designed functional peptide and the electrodeposited conducting polymer for CTC analysis in human blood. *Anal. Chem.* **2022**, *94*, 2204–2211. [CrossRef]

microorganisms

Article

A Comparative Study of the Synthesis and Characterization of Biogenic Selenium Nanoparticles by Two Contrasting Endophytic Selenobacteria

Eulàlia Sans-Serramitjana [1,*], Carla Gallardo-Benavente [2], Francisco Melo [3], José M. Pérez-Donoso [4], Cornelia Rumpel [5], Patricio Javier Barra [1,6], Paola Durán [1,6] and María de La Luz Mora [1,*]

1 Center of Plant, Soil Interaction and Natural Resources Biotechnology, Scientific and Biotechnological Bioresource Nucleus (BIOREN-UFRO), Universidad de La Frontera, Avenida Francisco Salazar 01145, P.O. Box 54-D, Temuco 4811230, Chile; patricio.barra@ufrontera.cl (P.J.B.); paola.duran@ufrontera.cl (P.D.)

2 Centro Biotecnológico de Estudios Microbianos (CEBEM), Universidad de La Frontera, Temuco 4811230, Chile; carlagallardo6@gmail.com

3 Departamento de Física, Center for Soft Matter Research, SMAT-C, Usach, Avenida Ecuador, Estación Central, Santiago 9170124, Chile; francisco.melo@usach.cl

4 BioNanotechnology and Microbiology Lab, Center for Bioinformatics and Integrative Biology, Facultad de Ciencias de la Vida, Universidad Andres Bello, Santiago 8370133, Chile; jose.perez@unab.cl

5 Institute of Ecology and Environmental Sciences, UMR 7618, CNRS-UPMC-UPEC-INRAE-IRD, Sorbonne University, 75005 Paris, France; cornelia.rumpel@inrae.fr

6 Biocontrol Research Laboratory, Universidad de La Frontera, Temuco 4811230, Chile

* Correspondence: eulalia.sans@ufrontera.cl (E.S.-S.); mariluz.mora@ufrontera.cl (M.d.L.L.M.); Tel.: +56-45-2-325000 (E.S.S. & M.d.L.L.M.)

Abstract: The present study examined the biosynthesis and characterization of selenium nanoparticles (SeNPs) using two contrasting endophytic selenobacteria, one Gram-positive (*Bacillus* sp. E5 identified as *Bacillus paranthracis*) and one Gram-negative (*Enterobacter* sp. EC5.2 identified as *Enterobacter ludwigi*), for further use as biofortifying agents and/or for other biotechnological purposes. We demonstrated that, upon regulating culture conditions and selenite exposure time, both strains were suitable "cell factories" for producing SeNPs (B-SeNPs from *B. paranthracis* and E-SeNPs from *E. ludwigii*) with different properties. Briefly, dynamic light scattering (DLS), transmission electron microscopy (TEM), and atomic force microscopy (AFM) studies revealed that intracellular E-SeNPs (56.23 ± 4.85 nm) were smaller in diameter than B-SeNPs (83.44 ± 2.90 nm) and that both formulations were located in the surrounding medium or bound to the cell wall. AFM images indicated the absence of relevant variations in bacterial volume and shape and revealed the existence of layers of peptidoglycan surrounding the bacterial cell wall under the conditions of biosynthesis, particularly in the case of *B. paranthracis*. Raman spectroscopy, Fourier transform infrared spectroscopy (FTIR), energy-dispersive X-ray (EDS), X-ray diffraction (XRD), and X-ray photoelectron spectroscopy (XPS) showed that SeNPs were surrounded by the proteins, lipids, and polysaccharides of bacterial cells and that the numbers of the functional groups present in B-SeNPs were higher than in E-SeNPs. Thus, considering that these findings support the suitability of these two endophytic stains as potential biocatalysts to produce high-quality Se-based nanoparticles, our future efforts must be focused on the evaluation of their bioactivity, as well as on the determination of how the different features of each SeNP modulate their biological action and their stability.

Keywords: biogenic nanoparticles; selenium; selenite; *Bacillus paranthracis*; *Enterobacter ludwigii*; biofortification

check for updates

Citation: Sans-Serramitjana, E.; Gallardo-Benavente, C.; Melo, F.; Pérez-Donoso, J.M.; Rumpel, C.; Barra, P.J.; Durán, P.; Mora, M.d.L.L. A Comparative Study of the Synthesis and Characterization of Biogenic Selenium Nanoparticles by Two Contrasting Endophytic Selenobacteria. *Microorganisms* **2023**, *11*, 1600. https://doi.org/10.3390/microorganisms11061600

Academic Editor: Kamel Ahmed Abd-Elsalam

Received: 29 April 2023
Revised: 24 May 2023
Accepted: 29 May 2023
Published: 16 June 2023

1. Introduction

Selenium (Se) is an essential micronutrient with beneficial effects for human health at low concentrations due to its antioxidant, anticancer, antimicrobial, and immunoregulatory properties [1,2]. The acquisition of optimal Se levels from the daily diet can promote a reduction in many pathological disorders, such as reproductive disorders, diabetes, thyroid issues, cancer, and immune responses [2,3]. The primary dietary Se sources are the edible parts of crop plants, while the Se content in crops is highly dependent on soil Se bioavailability, which is directly influenced by their physical and agrochemical properties [4,5]. Therefore, a mineral imbalance in the soil–plant continuum can lead to Se-deficient food production with consequences for both human and animal nutrition [6]. This represents a concern, especially in Andisols from southern Chile, which represent around 60% of agricultural soils, where the most important milk, meat, and cereal crops are produced [7]. However, these soils are characterized by their low Se content (between 0.02 and 0.18 mg kg^{-1} soil) and present high acidification rates, one of the main factors limiting agricultural production [8,9].

In this context, agronomic biofortification through the application of inorganic Se fertilizers could be considered a potential tool to raise Se levels in the human diet. However, in volcanic Andisols from Chile, several studies have shown that agronomic biofortification is inadequate, since Se can form stable complexes with clays and/or can be strongly adsorbed onto the oxyhydroxides of aluminum (Al), iron (Fe), or manganese (Mn), resulting in low Se bioavailability to plants [8–10].

In view of this scenario, over the last few years, our research group has widely investigated the use of rhizospheric and endophytic microorganisms isolated from cereal crops grown in Andisols from southern Chile as a biotechnological strategy for Se biofortification [11–15]. In this regard, it has been observed that the use of endophytic bacteria frequently provides greater benefits to plants than rhizobacteria in relation to plant nutrition, the hormonal regulation of root growth, disease suppression, and Se tolerance [16–19]. In fact, Durán et al. [12] determined that endophytes are better adapted to elevated Se concentrations than rhizospheric bacteria, thus being the most effective approach for enhancing Se uptake by plants. Moreover, considering that microorganisms play a pivotal role in the Se cycle in nature [20], endophytic selenobacteria have also been proposed as a potential tool for the bioremediation of Se-contaminated soils [21].

Interestingly, Durán et al. [14] showed that plant-growth-promoting (PGP) endophytic selenobacteria grown on Se-added media are able to metabolize the most important organic Se forms, such as SeMet and SeMetSeCys, as well as Se particles (>100 nm). Nonetheless, the authors indicated that due to the large size of Se spheres, they were not able to penetrate inside the plant roots [22], thus limiting their use as biofortifying agents. In fact, many bacteria have been described to have the ability to reduce selenite to elemental Se (Se^0) [23]. However, the Se^0 form is more commonly found as Se particles (100–500 nm) than as SeNPs (<100 nm) [23].

SeNPs are generating interest in many areas of science because of their low toxicity and high biocompatibility compared to other forms of the element [24–27]. In agriculture, SeNPs are mainly used as antimicrobial, pest control, biostimulant [28], biofortifying [29], stress alleviation [30,31], and nutraceutical agents [32]. Biofortification using SeNPs has emerged in recent years as a promising alternative to conventional Se fertilizers for enriching crops [27,33]. The interest in SeNPs for the biofortification of plant food stems from the potential for their slow Se release, avoiding possible losses in agroecosystems when commercial fertilizers are used [29].

SeNPs can be synthesized by biological, physical, and chemical methods [34]. In general, the synthesis of NPs using organisms such as plants, bacteria, fungi, yeasts, and algae is less expensive and safer since it uses eco-friendly non-toxic materials [34]. Moreover, biogenic SeNPs are much more stable than those produced by non-green methods due to the natural coating of organic materials on the surface, which prevents agglomeration and uncontrolled growth [35]. The biosynthesis of SeNPs using bacteria (both anaerobic and aerobic) has been extensively explored due to their ability to colonize, adapt to, and persist in detrimental environmental niches. However, it is of great interest to search for new bacterial strains capable of producing SeNPs in order to establish a more comprehensive data bank and to identify the factors affecting the properties of SeNPs. Indeed, several studies have established that the biological functions of SeNPs depend upon various factors, such as the synthesis method, surface composition, morphology, and size. These characteristics may cause the intake of plant foods biofortified with SeNPs to have a different effect within the human body compared with other sources of Se that are conventionally used.

Amongst the bacterial populations, endophytic bacteria are considered potential candidates for NP bio-production due to their high secretion of active metabolites [36]. Their potential metabolic activity enables them to fabricate NPs with varied sizes, shapes, and high stability [37]. However, at present, there are only a few data on the biogenic synthesis of SeNPs by endophytic selenobacteria [38].

Therefore, based on the above, and given that SeNP biofortification today is proposed as an alternative to conventional Se fertilizers, the aim of this work was to explore the capability of two contrasting endophytic selenobacteria (*B. paranthracis* and *E. ludwigi*) by producing SeNPs and optimizing the synthesis conditions and comparing their physico-chemical and structural features. Here, we report for the first time the use of both endophytic isolates as a sustainable and eco-friendly system to obtain SeNPs with potential impact on human health, agriculture, and biotechnological applications through the improvement of the nutritional quality of the crops with Se, the optimization of the use of natural resources, and the increase in food security by reducing chemical fertilization.

2. Materials and Methods

2.1. Bacterial Strains and Culturing

The endophytic selenobacteria isolated from roots of wheat plants (*Bacillus* sp. E5 identified as *Bacillus paranthracis* Accession no. KF561868 and *Enterobacter* sp. EC5.2 identified as *Enterobacter ludwigi* Accession no. KF561858) were selected for Se tolerance [13]. Both isolates were routinely pre-cultured aerobically in Luria Bertani (LB) broth (Becton Dickinson and Company, Sparks, MD, USA) at 30 °C with orbital shaking at 120 rpm. When needed, LB medium was solidified by adding 15 g/L of agar (OxoidTM, Thermo Fisher Scientific, Waltham, MA, USA).

2.2. SeNP Biosynthesis

The biosynthesis of SeNPs was evaluated in selected isolates following the protocol described by Gallardo et al. [39,40] with a few modifications. Briefly, a bacterial pre-inoculum was diluted on LB medium (1:100) and grown at 30 °C with shaking at 120 rpm for 24 h. Then, cultures were harvested by centrifugation for 10 min at 7000 rpm (Centurion Scientific Pro Analytical CR4000R, Chichester, UK) and washed once with 100 mM Tris-HCl pH 8 (Santa Cruz Biotechnology, Chem Cruz, Dallas, TX, USA). The resulting bacterial pellets were resuspended in 400 mL of 100 mM Tris-HCl pH 8, adjusted to 10^8 CFU mL^{-1}, and challenged with filter-sterilized sodium selenite ($Na_2SeO_3 \times 5H_2O$; Sigma-Aldrich, St. Louis, MO, USA) to a final concentration of 5 mM as per Durán et al. [12]. The mixture was incubated at 30 °C in an orbital shaker (120 rpm) for different time periods (4, 6, 24, and 48 h). Bacterial inocula without the addition of selenite were maintained as control. The reduction of $SeO_3{}^{-2}$ to Se^0 was monitored primarily by changes in the color of bacterial suspensions. The biosynthesis of SeNPs by both *Bacillus* sp. E.5 and *Enterobacter* sp. EC5.2 was also assessed by imaging both bacterial cells grown in the presence of Na_2SeO_3 using transmission electron microscopy (TEM) and atomic force microscopy (AFM).

2.3. Determination of Total Se in Bacterial Biomass

The quantities of Se accumulated by *Bacillus* sp. E.5 and *Enterobacter* sp. EC5.2 in the presence of selenite were determined according to the methodology described by Kumpulainen et al. [41]. Briefly, the cell pellets of both Se-enriched cultures were collected by centrifugation for 15 min at 6000 rpm and rinsed twice with sterile saline at 0.85%. Then, bacterial pellets were resuspended in a protective medium (10% skim milk—Difco Laboratories, Detroit, Mich.) at a ratio of 1:2 and lyophilized overnight. All samples were weighted and, thereafter, digested in 10 mL of acid mixture (65% HNO_3, 70% $HClO_4$, and 95% H_2SO_4) and incubated overnight at room temperature. After incubation, the mixture was heated at 120 °C for 3 h and 220 °C for 5 h, and then HCl (12%) was added in order to reach 15 mL. Finally, the mixture was boiled at 120 °C for 20 min. The Se content was measured via atomic absorption spectrophotometry (AAS) with a Hydride generator (HG) 3000 (GBC Scientific Equipment Ltd, Victoria, Australia) using $NaBH_4$ solution as a reducing agent. Two Se-enriched flour samples supplied by the Department of Applied Chemistry and Microbiology of Helsinki University (Finland) were used as references.

2.4. TEM and AFM Imaging

Bacterial cultures grown with and without 5 mM Na_2SeO_3 were harvested by gentle centrifugation at 7000 rpm for 5 min. The pellets were then washed three times with deionized water and fixed in a 2.5% solution of glutaraldehyde in 0.1 M cacodylate buffer pH 7.2 for at least 1 h at room temperature. After fixation, the pellets were rinsed in 0.1 M cacodylate buffer (pH 7.2) three times for 5 min each, and then 5 µL of cells was deposited onto carbon-coated copper grids (CF300-CU, Electron Microscopy Sciences, Hatfield, PA, USA), which were then air-dried prior to visualization using a Zeiss Libra 120 plus TEM at an acceleration voltage of 120 kV.

AFM was also used to visualize both endophytic strains after 6 h exposure to Na_2SeO_3 (5 mM). Briefly, 100 µL of bacterial suspensions was deposited and incubated in the dark for 90 min on the surface of positively charged glass coverslips previously coated with a 0.2% (*w*/*v*) Poly(ethyleneimine) (Sigma-Aldrich, Dorset, UK) solution to enhance bacterial cell adhesion and to prevent the removal of the bacteria during experiments. Then, the glass coverslips were rinsed three times in Tris buffer to eliminate the excess of cells and left to air dry for 15 min. Samples were imaged in air using AFM (Witec Alpha 300, Ulm, Germany). All images were collected in AC mode at room temperature using Hi'Res-C14/Cr-Au probe of thickness 2.1 µm, length 125 µm, width 25 µm, resonance frequency 160 kHz, and a force constant of 5 N/m. The data acquired during surface scanning were converted into images of topography and amplitude. In the topography images, the shape, structure, and surface of selenite-tolerant bacteria could be observed. Furthermore, amplitude images allow the visualization of fine surface details of bacterial cells [42].

2.5. Purification of Biosynthesized SeNPs

SeNPs were separated from the culture following a published protocol with certain modifications [40]. After the incubation period, bacterial cultures (50 mL) were centrifuged for 15 min at 6000 rpm. Supernatants containing extracellular SeNPs were filtered through 0.2 µm filters to remove cellular debris. After the removal of the supernatants, the pellets containing intracellular SeNPs were washed once with deionized water and resuspended for sonication. The pellets were disrupted through ultrasonication on ice (ten cycles of 30 s of sonication with 30 s of rest). After this, the suspensions were centrifuged, followed by 30 min centrifugation at 6000 rpm to separate unbroken cells. Then, both supernatants containing intra- or extracellular SeNPs were collected and filtered through 0.22 µm filters (BioLab) to remove cellular debris. Filtered supernatants were then concentrated using a 10 kDa membrane (Amicon Ultra-15 tube, Merck Millipore Ltd., Dublin, Ireland). Concentrated SeNP suspensions were lyophilized overnight and stored at 4 °C until use.

2.6. Characterization of SeNPs

Dynamic Light Scattering Measurements (DLS)

The hydrodynamic size, zeta potential, and polydispersity index (PDI, which is an indicator of size heterogeneity) of SeNPs suspended in ultra-pure water were measured in a Zetasizer Nano ZS (Malvern Instruments Co., Malvern, UK). Polystyrene cuvettes with a path length of 10 mm at 25 °C and a refractive index of 1.6 were used. All measurements were performed in triplicate.

2.7. TEM

TEM analysis was performed to characterize the size and shape of synthesized SeNPs. TEM observations were carried out at the Center for the Development of Nanoscience and Nanotechnology (University of Santiago de Chile) using a HITACHI HT-7700 (Tokyo, Japan) instrument operated at an accelerating voltage of 120 kV with a Tungsten filament, allowing a resolution of 0.2 nm. Samples (powder) were dispersed in distilled water using an ultrasonic bath and deposited on a copper TEM grid (300 mesh) coated with Formvar/carbon. Water was subsequently evaporated at room temperature. The size of the SeNPs was determined via ImageJ software (version 1.53t) [43].

2.8. AFM

The morphology and surface topography of SeNPs were confirmed through AFM imaging. Samples were dispersed in water, and then 50 μL was added on a positively charged microscope slide (Porlab®, Nanjing, P.R China). The drop was left for 1 min for the nanoparticles (NPs) to adsorb. The glass was rinsed with plenty of deionized water and left to air dry for 15 min. The dried SeNPs were scanned by AFM (Witec Alpha 300) in AC mode at room temperature using an Hi'Res-C14/Cr-Au probe with a thickness of 2.1 μm, length of 125 μm, width of 25 μm, resonance frequency of 160 kHz, and a force constant of 5 N/m. To create the histogram, SeNPs were selected at random, and their size profile was determined. From this profile, the maximum was taken as the size of the nanoparticle.

2.9. Energy-Dispersive X-ray Spectroscopy (EDS)

The elemental composition of the purified NPs was determined with SEM coupled to EDS. SeNPs (powder) were carefully mounted on SEM stubs using adhesive tape uniformly coated with carbon (LEICA EM ACE200, Wetzlar, Germany) and then placed in a sample chamber of SEM-EDS (Zeiss EVOMA10, Oberkochen, Germany—Oxford Instruments X-act EDX system, Wycombe, UK). They were scanned at 1500× magnification and a voltage of 20 kV.

2.10. Raman Spectroscopy

Raman measurements were performed to explore the crystalline traits and lattice dynamics of the synthesized SeNPs. Solid SeNPs were placed on a clean cover slip. Normal Raman spectra were recorded using Witec Alpha 300 equipment with an excitation wavelength of 633 nm (spectral range 150–2100 cm^{-1}). The acquisition interval was 1 s, and all spectra were averaged over 32 independent runs. The acquired digital experimental spectroscopic data were processed and plotted using Origin 8.5.

2.11. X-ray Diffraction

XRD patterns were performed using a RIGAKU Multipurpose diffractometer model Smartlab, with Theta-Theta Bragg-Brentano geometry goniometer and coupled to a solid-state detector D/teX Ultra 250 (Rigaku Corporation, Akishima-shi, Japan) for data acquisition. The X-ray beam was generated using a copper tube target (Cu Kalpha λ = 1.5418Å) at 30 kV/10 mA and filtered with Ni. The measure was recorded at room temperature between 5 and 55° 2-theta (degrees) with a step of 0.02° and a scan speed of 1°/min. Optical configurations were adjusted by divergent and receiving slits on both sides, with parallel Soller slits of 5° and 5 mm, respectively. The instrumental resolution was aligned with NIST SRM660c standard (LaB6). While the XRD measurement was carried out, the sample films were positioned on a silicon wafer with an orientation of <100> (about 69°). Subsequently, the patterns obtained were slightly smoothed using the Savitzky–Golay model, and any strange peaks coming from the silicon wafer were mathematically removed. The data obtained were optimized by TOPAS Academic V6 Program by modeling the peak shapes [44].

2.12. FTIR Spectroscopy

FTIR measurements were performed to identify the functional groups involved in the synthesis of biogenic SeNPs. The FTIR spectra of the dried samples were recorded on an ALPHA FTIR-ATR Bruke Spectrometer on a wavenumber range of 400–4000 cm^{-1} at a resolution of 1 cm^{-1} and 100 scans. The baseline was corrected for each spectrum using the "automatic baseline correct" function, and the spectrum was smoothed using the software's standard "automatic smooth" function, which employs the Savitsky–Golay algorithm (95-point moving second-degree polynomial).

2.13. X-ray Photoelectron Spectroscopy

XPS measurements were performed using an X-ray photoelectron spectrometer (SPECS, Berlin, Germany). A FlexPS system from SPECS was equipped with a hemispherical analyzer PHOBIOS 150 and detector 1D-DLD, with a monochromatic X-ray source FOCUS 500 providing Ag K-alpha radiation with a characteristic energy of 2984 eV, using spot size of 500 μm and depth of 10 nm. Analyses were performed using two different points of powdered samples, and the elemental composition was analyzed using CasaXPS software (version 2.3.19, Casa Software Ltd, Las Vegas, NV, USA).

3. Results

3.1. Bacterial SeNP Synthesis

The biological synthesis of SeNPs was carried out using two endophytic selenobacteria (*Bacillus* sp. E5 and *Enterobacter* sp. EC5.2) previously selected for their ability to survive aerobically at high concentrations of selenite, as well as for their PGP traits. According to Durán et al. [13], *Bacillus* sp. E5 and *Enterobacter* sp. EC5.2 presented MIC values for selenite of 120 mM and 100 mM, respectively. Firstly, it was observed that both bacterial cultures turned red at all incubation times tested (Supplementary Figure S1). The characteristic red color has been reported to be direct proof of the selenite reduction to zero-valent elemental selenium [45]. On the contrary, no color change was observed in the control experiment carried out with the selenite-containing medium without bacterial inoculation, thus indicating the active participation of these bacteria in selenite reduction. Moreover, once cell suspensions were centrifuged, it was possible to visualize that although both the pellet and the cell-free medium were reddish yellow, the red color was found to be more intense in the pellet than in the supernatant (Supplementary Figure S1). This may suggest that both strains have the capacity for intracellular and extracellular SeNP production, maybe being more noticeable intracellularly.

Then, considering these observations, the total Se content accumulated by both endophytic strains grown in the presence of 5 mM sodium selenite (acting as a precursor of the bioreaction) was determined through HG-AAS. The results showed that both selenobacteria strains were able to accumulate high Se concentrations in comparison to the control without Se and that the red pellet of *Bacillus* sp. E5 presented a higher Se concentration than that of *Enterobacter* sp. EC5.2, with values of 12016 ± 4205 mg kg^{-1} and 3543 ± 1240 mg kg^{-1}, respectively (Supplementary Table S1).

3.2. Localization of SeNPs in Bacillus sp. E5 and Enterobacter sp. EC5.2 Cultures

The ability of both bacterial strains to convert selenite to SeNPs was also visualized by microscopy. To this end, culture samples were collected after 6 and 24 h of growth in the absence and presence of selenite 5 mM. Then, the samples were processed and analyzed by TEM. Se nanospheres were observed after 6 h of incubation in both bacterial cultures supplemented with selenite (Figure 1). These NPs were not observed in the bacterial cells grown on Tris-HCl pH 8 without the oxyanion. Interestingly, unlike *Enterobacter* sp. EC5.2, the Gram-positive bacteria *Bacillus* sp. E5 clearly presented SeNPs and rod-like structures on the edge of the bacterial cell simultaneously (Figure 1(a2)).

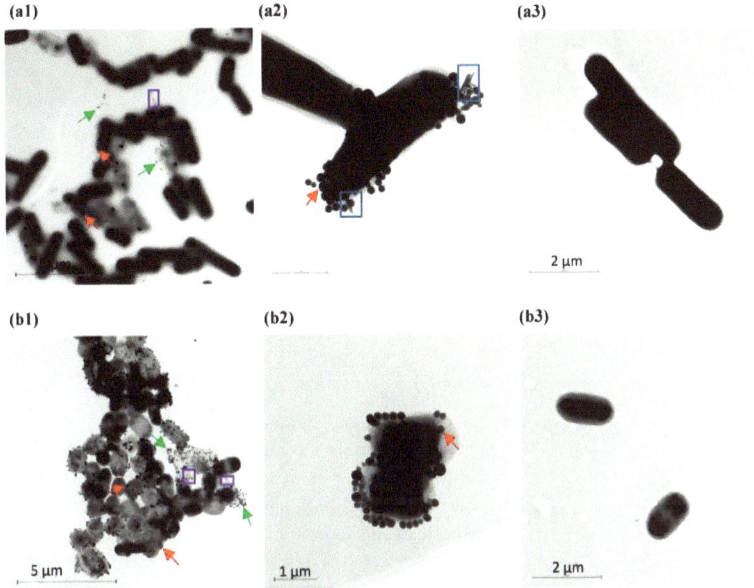

Figure 1. TEM images of cells of *Bacillus* sp. E5 (**a1**,**a2**) and *Enterobacter* sp. EC5.2 (**b1**,**b2**) after 6 h of growth in the presence of 5 mM Na_2SeO_3. Red arrows indicate SeNPs located on the cell surface, whereas green arrows indicate SeNPs located in the surrounding medium. Purple frames show components actively secreted by bacterial cells. Blue frames on image (**a2**) show nanorods (SeNRs). Images (**a3**,**b3**) show bacterial cells grown without selenite.

The micrographs showed that the SeNPs were located predominantly in the extracellular environment in the surrounding medium or attached to the cell surface (Figure 1a,b). Moreover, these NPs seemed to be entrapped on components actively secreted by bacterial cells, such as extracellular polymeric substance (EPS) (Figure 1a,b).

AFM experiments were also performed to explore the nanoscale surface morphology of both endophytic strains producing various amounts of SeNPs. AFM topography and the amplitude images of *Bacillus* sp. E5 and *Enterobacter* sp. EC5.2 after treatment with 5 mM Na_2SeO_3 for 6 h are shown in Figure 2, together with AFM images of untreated bacterial cultures. Topography images provided information about differences in surface and structure, whereas amplitude images allowed better observation of surface features and fine details [46]. Based on AFM images, bacterial cultures without selenite presented the typical rod shape of *Bacillus* spp. and *Enterobacter* spp. (Figure 2(a1,a2,b1,b2)). After 6 h of incubation with 5 mM of Na_2SeO_3, SeNPs were observed in both bacterial cultures. These images corroborated the observations obtained by TEM in the sense that SeNPs were either attached to the cell wall or located in the surrounding medium, without relevant variations in their volume and shape (Figure 2(a3,a4,b3,b4)). AFM revealed thick and robust filaments in both cases distributed randomly around the cell surface, as well as the flattened structures surrounding the cells, in which the SeNPs were trapped. They represent collapsed cell envelopes generated by the drying process [47]. The cell wall of *Bacillus* sp. E5 presented more features than that of *Enterobacter* sp. EC5.2, which may be related to the fact that Gram-positive bacteria are surrounded by layers of peptidoglycan many times thicker than are found in Gram-negative bacteria.

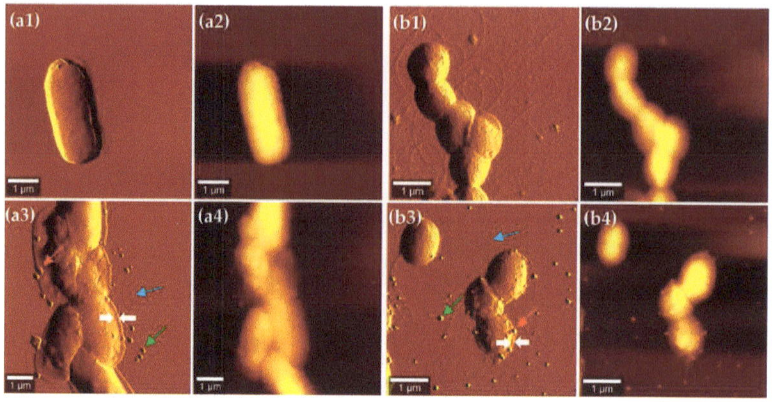

Figure 2. AFM images of *Bacillus* sp. E5 (**a**) and *Enterobacter* sp. EC5.2 (**b**) after 6 h of growth in the absence (**a1**,**a2**,**b1**,**b2**) or in the presence (**a3**,**a4**,**b3**,**b4**) of 5 mM Na_2SeO_3. The first column of each bacterial strain shows amplitude images (**a1**,**a3**,**b1**,**b3**), whereas topography images are shown in the second column (**a2**,**a4**,**b2**,**b4**). Red arrows indicate SeNPs attached to the cell wall, whereas green arrows indicate SeNPs located in the surrounding medium. The white arrows correspond to the collapsed cell envelope (see details in main text). The blue arrows correspond to bacterial filaments.

3.3. *Chemical and Physical Characteristics of Biogenic SeNPs*

Stability, Morphology, and Size of SeNPs

Intracellular and extracellular SeNPs produced by *Bacillus* sp. E5 (named B-SeNPs) and *Enterobacter* sp. EC5.2 (named E-SeNPs) after 4, 6, 24, and 48 h exposure to 5 mM selenite were purified and characterized to observe the effect of incubation time on their stability and size. The hydrodynamic size, PDI, and zeta potential of biosynthesized SeNPs were measured with DLS. Most NPs produced intracellularly by both bacteria at different incubation times presented PDI values below 0.3 (Table 1a), indicating that samples were monodispersed [48]. Interestingly, B-SeNPs obtained after 24 and 48 h incubation presented PDI values (mean $\pm$ standard deviation) of 0.33 ± 0.05 and 0.42 ± 0.06, respectively. In addition, all B-SeNP formulations had an average zeta potential (mean $\pm$ standard deviation) between -31.2 ± 1.20 and -22.9 ± 1.32 mV, whereas E-SeNPs presented zeta potential values slightly higher than those of B-SeNPs (ranging from -35.5 ± 0.9 to -27.8 ± 1.79 mV), giving them better electrostatic stability [35]. With respect to hydrodynamic size, there was a general trend showing that the average diameter of SeNPs in both formulations increased with increasing incubation time. Remarkably, both formulations reached their maximum size with incubation times of 24–48 h. Similarly, the size of extracellular SeNPs directly depended on the incubation time (Table 1b). However, DLS analysis revealed that SeNPs found in the extracellular space were larger than intracellular NPs. Furthermore, extracellularly synthesized SeNPs had high PDI values (>0.3) in every case, corresponding to a larger size distribution and signs of aggregation.

Thus, based on the results presented in Table 1, the smallest SeNPs were intracellularly synthesized at 6 h of incubation; the average diameter of B-SeNPs was 112 ± 3 nm, while that of E-SeNPs was 119 ± 2 nm. Since the size of NPs is a key factor for their chemical and biological properties, these two SeNPs were the formulations selected for further analysis. The quantitative size analysis of biosynthesized SeNPs was implemented with TEM and AFM. Firstly, TEM images revealed that B-SeNPs and E-SeNPs had a spherical shape with homogeneous size distribution (Figure 3). Moreover, TEM images determined that the average size of B-SeNPs was 83.44 ± 2.90 nm, whereas E-SeNPs presented an average size of 56.23 ± 4.85 nm (Figure 3c,d). Therefore, TEM analysis confirmed the formation of SeNPs with a size below 100 nm—the smallest particles of those produced by the Gram-negative strain.

Table 1. Particle size, PDI, and the zeta potential of intracellular (a) and extracellular (b) SeNPs biosynthesized by *Bacillus* sp. E5 (B-SeNPs) and *Enterobacter* sp. EC5.2 (E-SeNPs) at different incubation times with 5 mM of sodium selenite (Na_2SeO_3) as the sole Se source.

Nanoparticles	Particle Size (nm)	Polydispersity Index (PDI)	Zeta Potential (mV)
(a)			
B-SeNPs—4 h	114 ± 1	0.28 ± 0.02	−31.2 ± 1.20
E-SeNPs—4 h	133 ± 12	0.25 ± 0.03	−35.16 ± 0.54
B-SeNPs—6 h	112 ± 3	0.20 ± 0.02	−24.3 ± 0.62
E-SeNPs—6 h	119 ± 2	0.23 ± 0.01	−27.8 ± 1.79
B-SeNPs—24 h	198 ± 11	0.33 ± 0.05	−25.7 ± 0.09
E-SeNPs—24 h	143 ± 21	0.30 ± 0.03	−28.42 ± 0.83
B-SeNPs—48 h	195 ± 7	0.42 ± 0.06	−22.9 ± 1.32
E-SeNPs—48 h	142 ± 2	0.28 ± 0.02	−35.5 ± 0.9
(b)			
Nanoparticles	**Particle Size (nm)**	**Polydispersity Index (PDI)**	**Zeta Potential (mV)**
B-SeNPs—4 h	216 ± 16	0.34 ± 0.03	−33.13 ± 4.90
E-SeNPs—4 h	212 ± 13	0.31 ± 0.04	−28.9± 0.54
B-SeNPs—6 h	267 ± 34	0.35 ± 0.01	−27.8 ± 5.45
E-SeNPs—6 h	216 ± 30	0.35 ± 0.03	−23.5 ± 1.79
B-SeNPs—24 h	280 ± 17	0.54 ± 0.14	−35 ± 2.52
E-SeNPs—24 h	255 ± 32	0.38 ± 0.04	−34.4 ± 0.83
B-SeNPs—48 h	288 ± 18	0.51 ± 0.04	−33.4 ± 7.77
E-SeNPs—48 h	276 ± 26	0.41 ± 0.05	−31.8 ± 0.9

Data are presented as mean ± standard deviation.

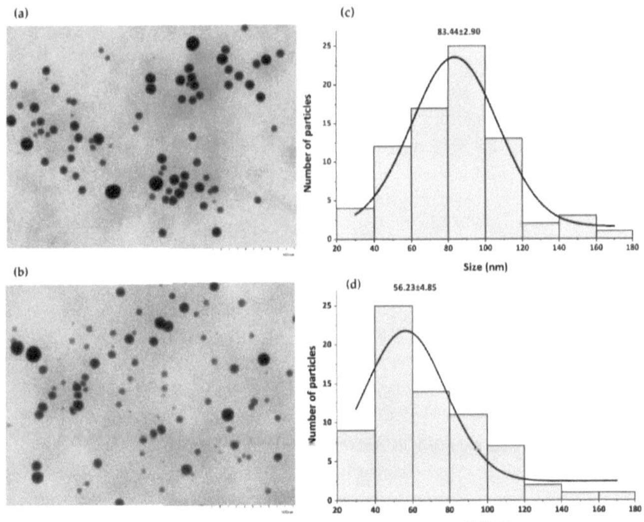

Figure 3. TEM observation of the purified SeNPs produced by *Bacillus* sp. E5 (**a**) and *Enterobacter* sp. EC5.2 (**b**) showing their spherical shape. The size distribution of SeNPs produced by *Bacillus* sp. E5 (**c**) and *Enterobacter* sp. EC5.2 (**d**).

The morphology and surface topography of SeNPs were confirmed via AFM imaging. Figure 4a,b show the 2D topographical images of B-SeNPs and E-SeNPs deposited on the glass surface, respectively, while Figure 4c,d show the diameter histograms of the SeNPs. AFM results showed that the SeNPs were spherical in shape, with average sizes of 140.66 ± 2.71 nm for B-SeNPs and 97.31 ± 4.13 nm for E-SeNPs.

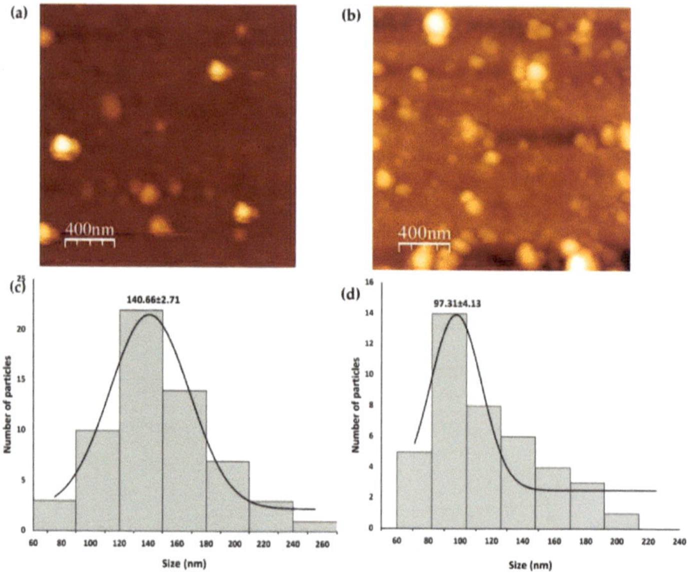

Figure 4. Characterization of SeNPs synthesized by *Bacillus* sp. E5 (B-SeNPs) and *Enterobacter* sp. EC5.2 (E-SeNPs). Two-dimensional AFM topographical image of B-SeNPs (**a**) and E-SeNPs (**b**). The graphics below (**c**,**d**) show the AFM histogram of B-SeNP and E-SeNP diameters.

When comparing the diameters of E-SeNPs measured by DLS, TEM, and AFM, it becomes apparent that the size measured by DLS was slightly larger than the size recorded by the two microscopy techniques. However, it should be noted that the diameter value of B-SeNPs obtained by AFM is the highest compared to the ones obtained by DLS and TEM.

The elemental composition of SeNPs was evaluated by EDS. Since the biogenic NPs were dried on carbon wafers, the C signal ($K\alpha$ = 0.277 KeV) was obtained for all samples analyzed (Supplementary Figure S2). However, it should be noted that the C signal could also be derived from the coating of the NPs. The EDS spectrum corroborated the presence of Se in both NPs showing the specific Se peak ($L\alpha$ = 1.379 KeV) and accounting for ≈9.50% of the total component elements (Supplementary Figure S2a,b). Moreover, other marked peaks were also detected, such as nitrogen (N; $K\alpha$ = 0.392 keV), oxygen (O; $K\alpha$ = 0.525 KeV), phosphorus (P; $K\alpha$ = 2.013 KeV), and sulfur (S; $K\alpha$ = 2.307 KeV). This could indicate that the material covering the SeNP surface was organic in nature.

3.4. *Crystallographic Structure of SeNPs*

The crystallographic structure of the SeNPs was analyzed by XRD. Based on the results shown in Figure 5, the XRD pattern of both SeNPs presented wide, overlapping curves with no sharp Bragg reflections, indicating that the red SeNPs generated by bacterial reduction are amorphous in structure [49]. The Raman spectroscopy analysis of the biosynthesized SeNPs is shown in Figure 6. The spectrum of both NPs exhibited a very strong band with a maximum at 254 cm^{-1}, attributed to the A1 stretching mode of amorphous Se. Moreover, a weak but noticeable band at 500 cm^{-1} was also observed in all the SeNPs, corresponding to homonuclear stretching S–S bonds.

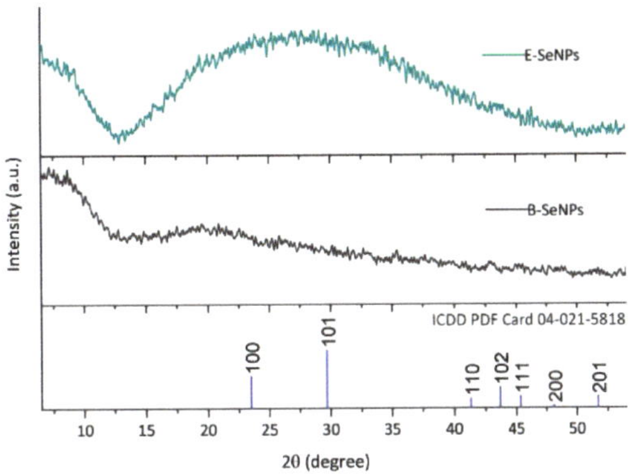

Figure 5. XRD pattern of SeNPs synthesized using *Bacillus* sp. E5 (B-SeNPs) and *Enterobacter* sp. EC5.2 (E-SeNPs).

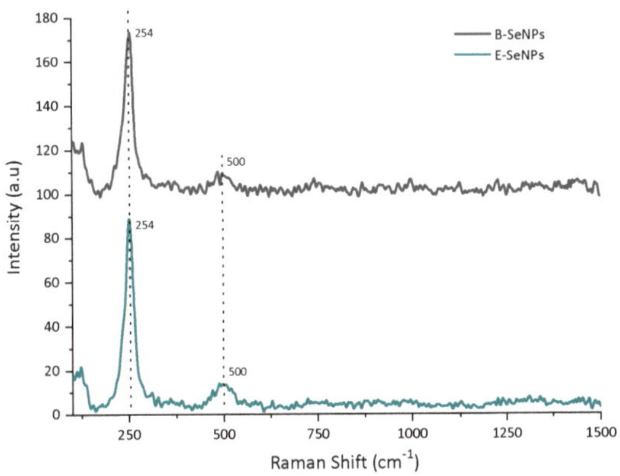

Figure 6. Raman spectrum at 532 nm excitation wavelength of SeNPs synthesized by *Bacillus* sp. E5 (B-SeNPs) and *Enterobacter* sp. EC5.2 (E-SeNPs) and incubated with 5 mM of sodium selenite within 6 h. The signal of Se–Se vibrations was observed at 254 cm^{-1}.

3.5. Biomolecules Bound to Biosynthesized SeNPs

The functional groups and biomolecules surrounding SeNPs were identified by FTIR spectroscopy analysis on bacterial cells and purified NPs. The spectrum exhibited in Figure 7 indicates the existence of biological macromolecules, such as lipids, proteins, and carbohydrates, likely acting as capping agents and most probably contributing to SeNP stability. The major absorbance bands observed in the nanoparticle and bacterial cell spectra (between 4000 and 400 cm^{-1}) were at 3280, 2927, 1625, 1592, 1450, 1390, 1233, and 1054 cm^{-1} (Supplementary Table S2). As shown in Figure 7a–c, the main characteristic peaks found in the FTIR spectrum of SeNPs were the same as those of bacterial cells in both cases, thus indicating that selenium nanoparticles are surrounded by active substances of bacterial cells. It was also observed that B-SeNPs possessed absorption peaks with higher intensity than those obtained for E-SeNPs (Figure 7c), which means an increase in the amount of the functional groups on the SeNPs produced by the Gram-positive strain *Bacillus* sp. E5. X-ray photoelectron spectroscopy (XPS) analysis was also performed to expose the bonding mechanism and element valence of SeNPs. Signals corresponding to C 1s, N 1s, and O 1s were detected (Figure 8). In addition, the peak found at 55.38 represents the 3d peak of typical Se, this signal being much weaker than that of C 1s, N 1s, and O 1s.

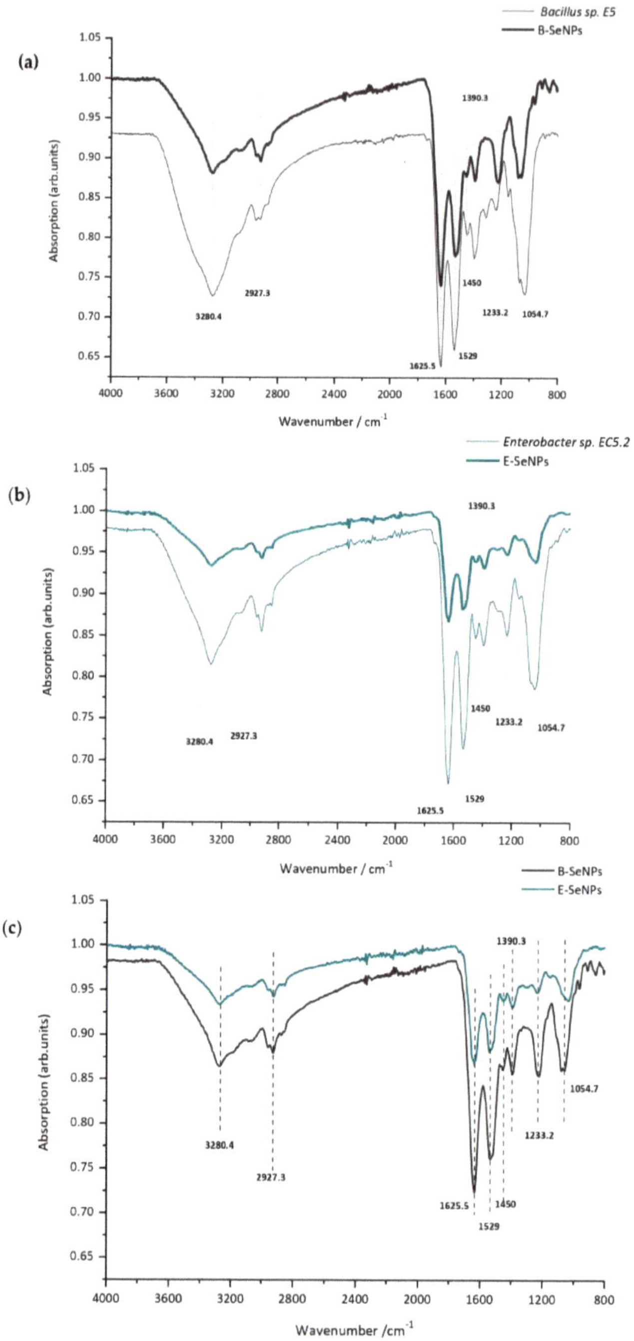

Figure 7. FTIR spectrum of bacterial cells compared to the one obtained from purified SeNPs (**a**,**b**) and the FTIR spectrum of both SeNPs synthesized by *Bacillus* sp. E5 (B-SeNPs) and *Enterobacter* sp. EC5.2 (E-SeNPs) (**c**).

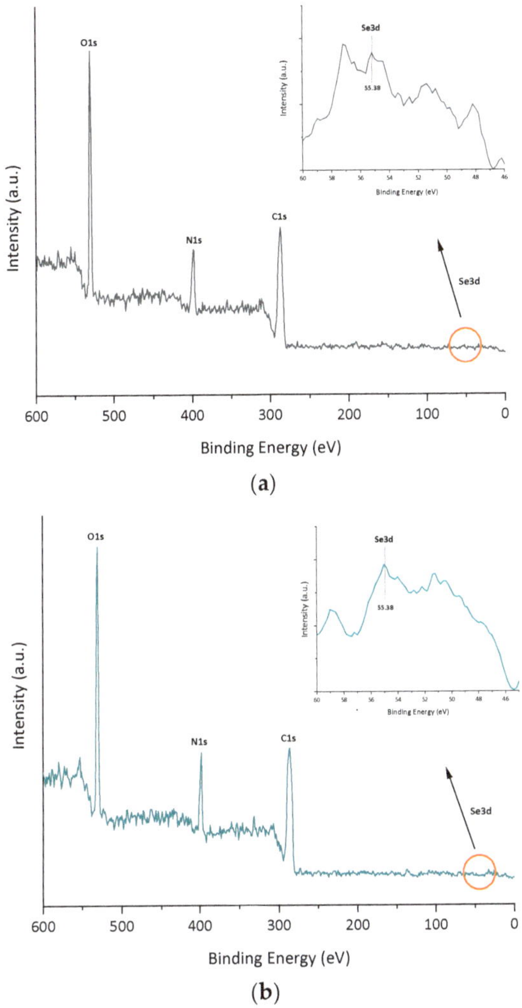

Figure 8. XPS spectra: (**a**) B-SeNPs and (**b**) E-SeNPs.

4. Discussion

In recent years, the use of SeNPs in agriculture has been widely described as a potential strategy to stimulate plant growth, biofortify crops, and control plant disease [31,32,50]. In this context, the present study strengthens the possibility of using two contrasting endophytic isolates as eco-friendly "cell factories" for producing SeNPs with different structural and physico-chemical properties for further use as biofortifying agents and/or other biotechnological purposes. Gram-positive and Gram-negative bacteria with high Se tolerance and plant-growth-promoting (PGP) attributes were used to produce SeNPs [13]. Considering that *Bacillus* sp. E5 was isolated from the tissue of wheat plants treated with Se, its Se tolerance was slightly greater (120 mM) than that observed in *Enterobacter* sp. EC5.2 (100 mM), which was isolated from the tissue of wheat plants not treated with Se [13]. However, the two bacterial strains showed high Se tolerance, similar to that reported in *Pseudomonas moraviensis* (120 mM) [51] and *Vibrio natriegens* (100 mM) [52].

Interestingly, Zhang et al. [53] reported that bacterial strains isolated from Se-rich habitats were usually quicker and more efficient than those from Se-free environments in reducing the Se oxyanion. These results are in accordance with our results, indicating that *Bacillus* sp. E5 accumulated the highest Se content in comparison to *Enterobacter* sp. EC5.2. This may suggest that the ability to incorporate the inorganic Se source into the bacterial metabolism and reduce it into other Se forms differs between the bacterial strains. It has been observed that different organic Se species, as well as SeNPs, are produced by bacteria grown on a Se-enriched medium [23]. Particularly, in cases of selenite exposure, selenomethionine (SeMet) is the main organic Se form found in bacterial inocula, whereas Se particles are the predominant Se species [14]. Interestingly, different types of bacteria produce different morphologies of Se nanostructures, and furthermore, the synthesized NPs possess different biomolecular compositions on their surface, which act as a cover or capping layer [54,55]. It has been pointed out by many authors that the biological composition of the outer organic layer plays a major role in the low toxicity and high biocompatibility of the NPs produced by bacteria [23,56,57].

Taking the above into consideration, we evaluated the bioproduction of SeNPs on the selected bacteria under optimal synthesis conditions.

The fast reduction of selenite (the appearance of the red color typical of colloidal Se) by both endophytic strains (after 6 h of incubation) could be attributed to the huge reducing power featuring metabolically active resting bacterial cells per given time [56,58], which is the kind of biotic transformation of selenite that was carried out in the present study. In fact, Presentato et al. [57,59] discovered that *Rhodococcus aetherivorans* BCP1 resting cells had a greater performance to produce TeNRs as compared to actively growing cultures.

The appearance of red color on bacterial cultures grown in the presence of selenite was accompanied by a characteristic garlic-like odor, suggesting that a volatilization step to methylated Se products occurs during the reduction process [53,58,60].

The presence of SeNPs in both intra- and extracellular compartments was observed by TEM and AFM (Figures 1 and 2), indicating that the selected strains, *Bacillus* sp. E5 and *Enterobacter* sp. EC5.2, were able to transform selenite to SeNPs. Curiously, it was visualized that *Bacillus* sp. E5 simultaneously synthesized SeNPs and a few Se nanorods (SeNRs) on the cellular surface (Figure 1). In this regard, Piacenza et al. [48] extensively reported that the different bacterial physiological states determine morphological changes in Se nanomaterials, resulting in the synthesis of both NPs and NRs.

AFM images (Figure 2) revealed no evidence of cell lysis nor a distribution of SeNPs within the surface layers. At this point, it could be hypothesized that the cell wall would play an important role in the intracellular and extracellular mechanisms of the synthesis and transport of SeNPs [61–63]. Additionally, the adhesion of SeNPs observed to the cell surface could be explained with the extended Derjaguin–Landau–Verwey–Overbeek (DLVO) theory, based on the role of electrostatic double-layer interactions in adhesion, as observed in *Pseudomonas putida* strains by Hwang et al. [64].

However, the key factors involved in the reduction of selenite in the two endophytic strains studied here have not been elucidated, and the mechanism responsible for NP transport from the cell to the exterior deserves more investigation.

The purification and subsequent analysis by DLS (Table 1a,b) of intracellular and extracellular SeNPs demonstrated that the intracellular ones have smaller sizes than those found in the extracellular space, since in the external compartment, the NPs tend to agglomerate and lose their shape, increasing in size. Similar results were reported by Torres et al. [65] for *Pantoea agglomerans*, obtaining larger SeNPs in the extracellular space. Thus, the purification involving bacterial lysis allows for obtaining the smaller nanoparticles found in the intracellular space. Then, focusing only on the intracellular SeNPs, the results obtained by DLS (Table 1a,b) showed that PDI values increased slightly in both types of SeNPs (B-SeNPs and E-SeNPs), as the incubation period increased, showing a greater hydrodynamic diameter after 24–48 h than 4–6 h of incubation. These findings are in line with previous studies that show a similar trend with increased PDI and hydrodynamic size values during prolonged reaction time [66,67]. The negative zeta potential values differed moderately between the two types of NPs, with E-SeNP values being the highest in all cases;

they would therefore be the more physically stable type. Given that the natural stability of biogenic NPs is conferred by the organic layer [63], the overall higher hydrodynamic diameter and lower zeta potential of B-SeNPs could indicate a lower stabilizing effect of Gram-positive compared to Gram-negative capping. Bulgarini et al. [55] also observed that SeNPs produced by Gram-positive bacteria had lower zeta potential values than SeNPs produced by Gram-negative bacteria. This is probably linked to the total amount of material covering the SeNP structure [40] or the composition and nature of molecules associated with these NPs [55]. The average size of SeNPs produced by *Bacillus* sp. E5 (~83 nm) and *Enterobacter* sp. EC5.2 (~56 nm), determined from TEM images (Figure 3), was consistent with the size of SeNPs synthesized by other microorganisms ranging from ~50 nm [66,68] to ~100 nm [54]. The SeNPs obtained from both strains are considered to be within the NP range (1–100 nm) [23] with potential applications in physical, chemical, and biological sciences, since they are more biocompatible with living systems and more beneficial than bigger structures (100–500 nm) [69]. The results recovered from TEM showed mainly the diameter of particle cores [70]; this explains the considerable discrepancies observed in particle sizes between TEM (size ~83 nm; ~56 nm) (Figure 3), AFM (size ~140 nm; ~97 nm) (Figure 4), and DLS (size ~112 nm; ~119 nm) (Table 1). This could be explained by that the latter two techniques might consider the organic layer surrounding the SeNPs as well as presenting a greater aggregation of particles in the samples. Moreover, a higher diameter value of B-SeNPs obtained by AFM could be justified mainly because B-SeNPs may tend to form aggregates or be covered with abundant organic material, which provokes an effect of the cantilever tip–particle interaction during imaging, resulting in the imaged particles having higher diameter values than the intact particles [71].

The EDS spectra of the spherical nanoparticles confirmed the presence of Se by the observation of the Se-specific absorption peak at 1.379 KeV in both cases (Supplementary Figure S2) [72]. Furthermore, signals for N, O, P, and S were also detected, being related to protein molecules or even associated with the presence of other biomolecules (e.g., phospholipids, nucleic acids) [73]. Interestingly, a higher percentage of S was found surrounding the structure of E-SeNPs than that found in B-SeNPs, suggesting that sulfur plays an important role in the formation of E-SeNPs [74]. This result was also corroborated by the Raman spectrum of the purified SeNPs (Figure 6), in which, aside from noting the characteristic amorphous Se band at ~254 cm^{-1}, the typical S–S stretching band (~500 cm^{-1}) was observed [75] to be more prominent in E-SeNPs. This, again, could be correlated with a substantial amount of sulfur in the SeNPs, especially in E-SeNPs. In fact, several studies [76,77] have shown that the constant presence of an S signal may be due to the involvement of RSHs in SeO_3^{2-} bioprocessing for bacterial cultures. Moreover, the XRD data also confirmed that both SeNPs are amorphous in structure (Figure 5). This result agrees with the fact that most SeNPs produced by biological reduction are typically amorphous in nature [78–80]. The amorphous nature of biogenic SeNPs is an intrinsic feature of the redox reaction for their production. Indeed, SeNPs obtained from reduction or redox processes are always amorphous in nature [35].

According to the FTIR spectroscopic data (Figure 7), the most prominent biomacromolecular composition of the two SeNPs obtained includes proteins (for which the amide I and II bands are accompanied by the weaker absorption around 1230 cm^{-1} featuring another typical protein-related band, amide III) [75,81], polysaccharides (see the characteristic carbohydrate structure at 1233 cm^{-1}), and lipids with less intense bands (featured via a combination of stretching and binding C–H vibrations with the CH3 of acyl chains) [80]. Lastly, traces of water (which can form strong H-bonds with polar biomolecular groups and thus might be not fully removed by freeze-drying; see the Materials and Methods section) could contribute to the increased stretching of the O–H region (3600–3200 cm^{-1}). As Wu et al. [81] have already reported, FTIR also corroborated that BioSeNPs inherited the active substances from bacterial cells, suggesting the important role of these biomolecules surrounding SeNPs in the synthesis and stabilization [56,57]. The higher absorption peaks observed in B-SeNPs compared to those obtained for E-SeNPs could be related to the fact that Gram-positive bacteria possessed many components actively secreted by bacterial cells, as well as a thick peptidoglycan layer (Figures 1 and 2). Finally, XPS data were obtained on the bonding mechanism and element valence of SeNPs (Figure 8), which are consistent with the EDS analysis. However, in the case of E-SeNPs, N was detected in the XPS spectrum but not in the EDS spectrum. This could be due to the overlap of C and O on the N peak in the EDS spectrum [82]. Additionally, within the XPS data, it is noteworthy that the Se 3d signal is much weaker than that of C 1s, O 1s, and N 1s. This is because XPS is a surface analysis method. Even though X-rays can penetrate SeNPs, only the photoelectrons emitted by C, O, and N in the shell layer on the surface of the sample can escape freely, while the photoelectrons emitted by Se inside the nanoparticle can do so only with difficulty, generating a weak Se 3d signal [82]. Similar results have already been published by previous studies [40,81].

5. Conclusions

Here, we report for the first time that two contrasting endophytic selenobacteria, *Bacillus* sp. E5 (*Bacillus paranthracis*) and *Enterobacter* sp. EC5.2 (*Enterobacter ludwigi*), under optimal synthesis conditions (incubation time, the concentration of oxyanion, and physiological state) can produce intra- and extracellular SeNPs. According to the size, intracellular SeNPs were chosen to analyze their structural and physico-chemical properties more deeply. Briefly, B-SeNPs have a larger size, have a lower Z-potential, and possess a more robust capping layer compared to E-SeNPs. The present findings indicate that E-SeNPs are the smallest formulations, whereas B-SeNPs are enveloped with the hugest organic layer. Thus, considering that the size and the biomolecular composition of the capping layer of biogenic NPs are key factors for their stability and bioactivity, our further efforts will be channeled into the evaluation of how the different traits of individual SeNPs modulate their biological actions.

Supplementary Materials: The following supporting information can be downloaded at https://www.mdpi.com/article/10.3390/microorganisms11061600/s1, Figure S1: Growth of *Bacillus* sp. E5 and *Enterobacter* sp. EC5.2 in Tris-HCl pH 8 in the presence of selenite (a = without selenite; b = *Bacillus* sp. E5; c = *Enterobacter* sp. EC5.2). Liquid (a–c) and solid medium (d = *Enterobacter* sp. EC5.2; e= *Bacillus* sp. E5) turned red only in the presence of 5 mM selenite. Images were obtained after culturing for 6 h (a–c) and 24 h (d,e), Figure S2: EDS graph of selenium nanoparticles (SeNPs) biosynthesized by *Bacillus* sp. E5 (a) and *Enterobacter* sp. EC5.2 (b) with 5 mM of Na_2SeO_3 for 6 h, Table S1: Total Se content in bacterial biomass of two endophytic bacterial strains (*Bacillus* sp. E5 and *Enterobacter* sp. EC5.2) isolated from wheat plants, grown with and without Se supplementation, Table S2: FTIR absorption bands and identification of *Bacillus* sp. E5 and *Enterobacter* sp. EC5.2 unchallenged cells and biogenic samples (B-SeNPs, E-SeNPs).

Author Contributions: Conceptualization, E.S.-S. and M.d.L.L.M.; methodology, E.S.-S. and C.G.-B.; validation, E.S.-S., C.G.-B. and P.D.; formal analysis, E.S.-S., P.J.B. and P.D.; investigation, E.S.-S. and M.d.L.L.M.; resources, E.S.-S., F.M., P.J.B. and P.D.; data curation, C.G.-B., J.M.P.-D. and C.R.; writing—original draft preparation, E.S.-S. and M.d.L.L.M.; writing—review and editing, J.M.P.-D., F.M. and C.R.; supervision, E.S.-S. and M.d.L.L.M.; project administration, E.S.-S.; funding acquisition, E.S.-S. and M.d.L.L.M. All authors have read and agreed to the published version of the manuscript.

Funding: This study was supported by the Agencia Nacional de Investigación y Desarrollo (ANID) from the Chilean Government (FONDECYT Postdoctoral Project No. 3210302; FONDECYT Initiation Project No. 11200377; FONDECYT Regular Project No. 1201196; MEC80180025) and Dicyt-USACH (project USA2155).

Data Availability Statement: Not applicable.

Acknowledgments: The authors gratefully acknowledge the support of the Laboratory of Crystallographic Studies (Catholic University of Temuco-UCT), FONDEQUIP EQM160152, and particularly the assistance of Héctor Pesenti in the XRD analysis.

Conflicts of Interest: The authors declare that they have no known competing financial interests or personal relationships that could have appeared to influence the work reported in this paper.

References

1. Burk, R.F.; Hill, K.E. Regulation of selenoproteins. *Annu. Rev. Nutr.* **1993**, *13*, 65–81. [CrossRef]
2. Rayman, M.P. Selenium and human health. *Lancet* **2012**, *379*, 1256–1268. [CrossRef]
3. Kipp, A.P.; Strohm, D.; Brigelius-Flohé, R.; Schomburg, L.; Bechthold, A.; Leschik-Bonnet, E.; Heseker, H. Revised reference values for selenium intake. *J. Trace Elem. Med. Biol.* **2015**, *32*, 195–199. [CrossRef]
4. El-Ramady, H.R.; Domokos-Szabolcsy, É.; Abdalla, N.A.; Alshaal, T.A.; Shalaby, T.A.; Sztrik, A.; Prokisch, J.; Fári, M. Selenium and nano-selenium in agroecosystems. *Environ. Chem. Lett.* **2014**, *12*, 495–510. [CrossRef]
5. Kabata-Pendias, A. *Trace Elements in Soils and Plants*, 4th ed.; CRC Press: Boca Raton, FL, USA, 2010.
6. Govasmark, E.; Salbu, B. Re-translocation of selenium during generative growth in wheat. *J. Plant Nutr.* **2011**, *34*, 1919–1929. [CrossRef]
7. Office of Agricultural Studies and Policies (ODEPA) of the Chilean Ministry of Agriculture. *Chilean Agriculture Overview*; ODEPA: Santiago, Chile, 2019.
8. Cartes, P.; Gianfreda, L.; Mora, M.L. Uptake of selenium and its antioxidant activity in ryegrass when applied as selenate and selenite forms. *Plant Soil* **2005**, *276*, 359–367. [CrossRef]
9. Mora, M.L.; Pinilla, L.; Rosas, A.; Cartes, P. Selenium uptake and its influence on the antioxidative system of white clover as affected by lime and phosphorus fertilization. *Plant Soil* **2008**, *303*, 139–149. [CrossRef]
10. Nakamaru, Y.M.; Altansuvd, J. Speciation and bioavailability of selenium and antimony in non-flooded and wetland soils: A review. *Chemosphere* **2014**, *111*, 366–371. [CrossRef]
11. Acuña, J.J.; Jorquera, M.A.; Barra, P.J.; Crowley, D.E.; Mora, M.L. Selenobacteria selected from the rhizosphere as a potential tool for Se biofortication of wheat crops. *Biol. Fertil. Soils* **2013**, *49*, 175–185. [CrossRef]
12. Durán, P.; Acuña, J.J.; Jorquera, M.A.; Azcón, R.; Borie, F.; Cornejo, P.; Mora, M.L. Enhanced selenium content in wheat grain by co-inoculation of selenobacteria and arbuscular mycorrhizal fungi: A preliminary study as a potential Se biofortication strategy. *J. Cereal Sci.* **2013**, *57*, 275–280. [CrossRef]
13. Durán, P.; Acuña, J.J.; Jorquera, M.A.; Azcón, R.; Paredes, C.; Rengel, Z.; Mora, M.L. Endophytic bacteria from selenium-supplemented wheat plants could be useful for plant-growth promotion, biofortification and *Gaeumannomyces graminis* biocontrol in wheat production. *Biol. Fertil. Soils* **2014**, *50*, 983–990. [CrossRef]
14. Durán, P.; Acuña, J.J.; Gianfreda, L.; Azcón, R.; Funes-Collado, V.; Mora, M.L. Endophytic selenobacteria as new inocula for selenium biofortification. *Appl. Soil Ecol.* **2015**, *96*, 319–326. [CrossRef]
15. Durán, P.; Viscardi, S.; Acuña, J.J.; Cornejo, P.; Azcón, R.; Mora, M.L. Endophytic selenobacteria and arbuscular mycorrhizal fungus for Selenium biofortification and *Gaeumannomyces graminis* biocontrol. *J. Soil Sci. Plant Nutr.* **2018**, *18*, 1021–1035. [CrossRef]
16. Lodewyckx, C.; Vangronsveld, J.; Porteous, F.; Moore, E.R.B.; Taghavi, S.; Mezgeay, M.; Van Der Lelie, D. Endophytic bacteria and their potential applications endophytic bacteria and their potential applications. *Crit. Rev. Plant Sci.* **2002**, *21*, 583–606. [CrossRef]
17. Mastretta, C.; Barac, T.; Vangronsveld, J.; Newman, L.; Taghavi, S.; van der Lelie, D. Endophytic bacteria and their potential application to improve the phytoremediation of contaminated environments. *J. Biotechnol. Genet. Eng. Rev.* **2006**, *23*, 175–207. [CrossRef] [PubMed]
18. Taghavi, S.; Garafola, C.; Monchy, S.; Newman, L.; Hoffman, A.; Weyens, N.; Barac, T.; Vangronsveld, J.; van der Lelie, D. Genome survey and characterization of endophytic bacteria exhibiting a beneficial effect on growth and development of poplar trees. *Appl. Environ. Microbiol.* **2007**, *75*, 748–757. [CrossRef] [PubMed]
19. Ryan, R.P.; Germaine, K.; Franks, A.; Ryan, D.J.; Dowling, D.N. Bacterial endophytes: Recent developments and applications. *FEMS Microbiol. Lett.* **2008**, *278*, 1–9. [CrossRef]
20. Eswayah, A.S.; Smith, T.J.; Gardiner, P.H.E. Microbial transformations of selenium species of relevance to bioremediation. *Appl. Environ. Microbiol.* **2016**, *82*, 4848–4859. [CrossRef]
21. Wang, Y.; Wang, Z.J.; Huang, J.C.; Chachar, A.; Zhou, C.; He, S. Bioremediation of selenium-contaminated soil using earthworm *Eisenia fetida*: Effects of gut bacteria in feces on the soil microbiome. *Chemosphere* **2022**, *300*, 134544. [CrossRef]
22. Kurepa, J.; Paunesku, T.; Vogt, S.; Arora, H.; Rabatic, B.M.; Lu, J.; Wanzer, M.B.; Woloschak, G.E.; Smalle, J.A. Uptake and distribution of ultrasmall anatase TiO_2 alizarin red as nanoconjugates in *Arabidopsis thaliana*. *Nano Lett.* **2010**, *10*, 2296–2302. [CrossRef]
23. Escobar-Ramirez, M.C.; Castañeda Ovando, A.; Pérez-Escalante, E.; Rodríguez-Serrano, G.M.; Ramérez-Moreno, E.; Quintero-Lira, A.; Contreras-López, E.; Añorve-Morga, J.; Jaimez-Ordaz, J.; González-Olivares, L.G. Antimicrobial Activity of Se-Nanoparticles from Bacterial Biotransformation. *Fermentation* **2021**, *7*, 130. [CrossRef]
24. Hariharan, H.; Al-Dhabi, N.A.; Karuppiah, P.; Rajaram, S.K. Microbial synthesis of selenium nanocomposite using *Saccharomyces cerevisiae* and its antimicrobial activity against pathogens causing nosocomial infection. *Chalcogenide Lett.* **2012**, *9*, 509–515.
25. Yang, F.; Tang, Q.; Zhong, X.; Bai, Y.; Chen, T.; Zhang, Y.; Li, Y.; Zheng, W. Surface decoration by Spirulina polysaccharide enhances the cellular uptake and anticancer efficacy of selenium nanoparticles. *Int. J. Nanomed.* **2012**, *7*, 835–844.
26. Forootanfar, H.; Adeli-Sardos, M.; Nikkhoo, M.; Mehrabani, M.; Amir-Heidari, B.; Shahverdi, A.R.; Shakibaie, M. Antioxidant and cytotoxic effect of biologically synthesized selenium nanoparticles in comparison to selenium dioxide. *J. Trace Elem. Med. Biol.* **2014**, *28*, 75–79. [CrossRef] [PubMed]

27. Juárez-Maldonado, A.; Ortega-Ortíz, H.; Morales-Díaz, A.B.; González-Morales, S.; Morelos-Moreno, A.; Cabrera-De la Fuente, M.; Sandoval-Rangel, A.; Cadenas-Pliego, G.; Benavides-Mendoza, A. Nanoparticles and Nanomaterials as Plant Biostimulants. *Int. J. Mol. Sci.* **2019**, *20*, 162. [CrossRef]
28. Shang, Y.; Hasan, M.K.; Ahammed, G.J.; Li, M.; Yin, H.; Zhou, J. Applications of nanotechnology in plant growth and crop protection: A review. *Molecules* **2019**, *24*, 2558. [CrossRef]
29. Schiavon, M.; Nardi, S.; Vecchia, F.; Ertani, A. Selenium biofortification in the 21st century: Status and challenges for healthy human nutrition. *Plant Soil* **2020**, *453*, 245–270. [CrossRef]
30. Zohra, E.; Ikram, M.; Omar, A.A.; Hussain, M.; Satti, S.H.; Raja Naveed, I.; Mashwani, Z.U.R.; Ehsan, M. Potential applications of biogenic selenium nanoparticles in alleviating biotic and abiotic stresses in plants: A comprehensive insight on the mechanistic approach and future perspectives. *Green Process. Synth.* **2021**, *10*, 456–475. [CrossRef]
31. Hussein, H.A.A.; Darwesh, O.M.; Mekki, B.B. Environmentally friendly nano-selenium to improve antioxidant system and growth of groundnut cultivars under sandy soil conditions. *Biocatal. Agric. Biotechnol.* **2019**, *18*, 101080. [CrossRef]
32. Li, D.; Zhou, C.; Zhang, J.; An, Q.; Wu, Y.; Li, J.; Pan, C. Nanoselenium foliar applications enhance the nutrient quality of pepper by activating the capsaicinoid synthetic pathway. *J. Agric. Food Chem.* **2020**, *68*, 9888–9895. [CrossRef]
33. Garza-García, J.J.O.; Hernández-Díaz, J.A.; Zamudio-Ojeda, A.; León-Morales, J.M.; Guerrero-Guzmán, A.; Sánchez-Chiprés, D.R.; López-Velázquez, J.C.; García-Morales, S. The Role of Selenium Nanoparticles in Agriculture and Food Technology. *Biol. Trace Elem. Res.* **2022**, *200*, 2528–2548. [CrossRef] [PubMed]
34. Bisht, N.; Phalswal, P.; Khanna, P.K. Selenium nanoparticles: A review on synthesis and biomedical applications. *Mater. Adv.* **2022**, *3*, 1415–1431. [CrossRef]
35. Piacenza, E.; Presentato, A.; Turner, R.J. Stability of biogenic metal(loid) nanomaterials related to the colloidal stabilization theory of chemical nanostructures. *Crit. Rev. Biotechnol.* **2018**, *38*, 1137–1156. [CrossRef] [PubMed]
36. Woźniak, M.; Gałązka, A.; Tyśkiewicz, R.; Jaroszuk-Ściseł, J. Endophytic Bacteria Potentially Promote Plant Growth by Synthesizing Different Metabolites and their Phenotypic/Physiological Profiles in the Biolog GEN III MicroPlateTM Test. *Int. J. Mol. Sci.* **2019**, *20*, 5283. [CrossRef]
37. Eid, A.M.; Fouda, A.; Abdel-Rahman, M.A.; Salem, S.S.; Elsaied, A.; Oelmüller, R.; Hijri, M.; Bhowmik, A.; Elkelish, A.; Hassan, S.E.-D. Harnessing bacterial endophytes for promotion of plant growth and biotechnological applications: An overview. *Plants* **2021**, *10*, 935. [CrossRef]
38. Fernández-Llamosas, H.; Castro, L.; Blázquez, M.L.; Díaz, E.; Carmona, M. Biosynthesis of selenium nanoparticles by *Azoarcus* sp. CIB. *Microb. Cell Factories* **2016**, *15*, 109. [CrossRef]
39. Gallardo, C.; Monrás, J.P.; Plaza, D.O.; Collao, B.; Saona, L.A.; Durán-Toro, V.; Venegas, F.A.; Soto, C.; Ulloa, G.; Vásquez, C.C.; et al. Low-temperature biosynthesis of fluorescent semiconductor nanoparticles (CdS) by oxidative stress resistant Antarctic bacteria. *J. Biotechnol.* **2014**, *187*, 108–115. [CrossRef]
40. Gallardo-Benavente, C.; Carrión, O.; Todd, J.D.; Pieretti, J.C.; Seabra, A.B.; Durán, N.; Rubilar, O.; Pérez-Donoso, J.M.; Quiroz, A. Biosynthesis of CdS Quantum Dots Mediated by Volatile Sulfur Compounds Released by Antarctic *Pseudomonas fragi*. *Front. Microbiol.* **2019**, *10*, 1866. [CrossRef]
41. Kumpulainen, J.; Raittila, A.M.; Lehto, J.; Koivistoinen, P. Electrothermal atomic absorption spectrometric determination of selenium in foods and diets. *J. Assoc. Off. Anal. Chem. Int.* **1983**, *66*, 1129–1135. [CrossRef]
42. Dorobantu, L.S.; Bhattacharjee, S.; Foght, J.M.; Gray, M.R. Atomic force microscopy measurement of heterogeneity in bacterial surface hydrophobicity. *Langmuir* **2008**, *24*, 4944–4951. [CrossRef]
43. Collins, T.J. ImageJ for microscopy. *Biotechniques* **2007**, *43*, 25–30. [CrossRef] [PubMed]
44. Coelho, A.A. TOPAS and TOPAS-Academic: An optimization program integrating computer algebra and crystallographic objects written in C++. *J. Appl. Cryst.* **2018**, *51*, 210–218. [CrossRef]
45. Kora, A.J.; Rastogi, L. Bacteriogenic synthesis of selenium nanoparticles by *Escherichia coli* ATCC 35218 and its structural characterisation. *IET Nanobiotechnol.* **2017**, *11*, 179–184. [CrossRef]
46. Dubrovin, E.V.; Voloshin, A.G.; Kraevsky, S.V.; Ignatyuk, T.E.; Abramchuk, S.S.; Yaminsky, I.V. Atomic force microscopy investigation of phage infection of bacteria. *Langmuir* **2008**, *24*, 13068–13074. [CrossRef]
47. Gillis, A.; Dupres, V.; Delestrait, G.; Mahillon, J.; Dufrene, Y.F. Nanoscale imaging of *Bacillus thuringiensis* flagella using atomic force microscopy. *Nanoscale* **2012**, *4*, 1585. [CrossRef]
48. Piacenza, E.; Presentato, A.; Bardelli, M.; Lampis, S.; Vallini, G.; Turner, R.J. Influence of Bacterial Physiology on Processing of Selenite, Biogenesis of Nanomaterials and Their Thermodynamic Stability. *Molecules* **2019**, *24*, 2532. [CrossRef]
49. Munawar, T.; Yasmeen, S.; Mukhtar, F.; Nadeem, M.S.; Mahmood, K.; Saqib Saif, M.; Hasan, M.; Ali, A.; Hussain, F.; Iqbal, F. $Zn_{0.9}Ce_{0.05}M_{0.05}O$ (M = Er, Y, V) Nanocrystals: Structural and Energy Bandgap Engineering of ZnO for Enhancing Photocatalytic and Antibacterial Activity. *Ceram. Int.* **2020**, *46*, 14369–14383. [CrossRef]
50. Bano, I.; Skalickova, S.; Sajjad, H.; Skladanka, J.; Horky, P. Uses of Selenium Nanoparticles in the Plant Production. *Agronomy* **2021**, *11*, 2229. [CrossRef]
51. Staicu, L.C.; Ackerson, C.J.; Cornelis, P.; Ye, L.; Berendsen, R.L.; Hunter, W.J.; Noblitt, S.D.; Henry, C.S.; Cappa, J.J.; Montenieri, R.L. *Pseudomonas moraviensis* subsp. Stanleyae, a bacterial endophyte of hyperaccumulator *Stanleya pinnata*, is capable of efficient selenite reduction to elemental selenium under aerobic conditions. *J. Appl. Microbiol.* **2015**, *119*, 400–410. [CrossRef]

52. Fernández-Llamosas, H.; Castro, L.; Blázquez, M.L.; Díaz, E.; Carmona, M. Speeding up bioproduction of selenium nanoparticles by using *Vibrio natriegens* as microbial factory. *Sci. Rep.* **2017**, *7*, 16046. [CrossRef]
53. Zhang, J.; Wang, Y.; Shao, Z.; Li, J.; Zan, S.; Zhou, S.; Yang, R. Two selenium tolerant *Lysinibacillus* sp. strains are capable of reducing selenite to elemental Se efficiently under aerobic condititions. *J. Environ. Sci.* **2018**, *77*, 238–249. [CrossRef]
54. Piacenza, E.; Presentato, A.; Ambrosi, E.; Speghini, A.; Turner, R.J.; Vallini, G.; Lampis, S. Physical–Chemical Properties of Biogenic Selenium Nanostructures Produced by *Stenotrophomonas maltophilia* SeITE02 and *Ochrobactrum* sp. MPV1. *Front. Microbiol.* **2018**, *9*, 3178. [CrossRef]
55. Bulgarini, A.; Lampis, S.; Turner, R.J.; Vallini, G. Biomolecular composition of capping layer and stability of biogenic selenium nanoparticles synthesized by five bacterial species. *Microb. Biotechnol.* **2021**, *14*, 198–212. [CrossRef]
56. Ruiz Fresneda, M.A.; Delgado Martin, J.; Gomez Bolivar, J.; Fernandez Cantos, M.V.; Bosch-Estevez, G.; Martinez Moreno, M.F.; Merroun, M.L. Green synthesis and biotransformation of amorphous Se nanospheres to trigonal 1D Se nanostructures: Impact on Se mobility within the concept of radioactive waste disposal. *Environ. Sci. Nano* **2018**, *5*, 2103–2116. [CrossRef]
57. Kora, A.J. *Bacillus cereus*, selenite-reducing bacterium from contaminated lake of an industrial area: A renewable nanofactory for the synthesis of selenium nanoparticles. *Bioresour. Bioprocess.* **2018**, *5*, 30. [CrossRef]
58. Julsing, M.K.; Kuhn, D.; Schmid, A.; Buhler, B. Resting cells of recombinant *E. coli* show high epoxidation yields on energy source and high sensitivity to product inhibition. *Biotechnol. Bioeng.* **2012**, *109*, 1109–1119. [CrossRef]
59. Presentato, A.; Piacenza, E.; Anikovskiy, M.; Cappelletti, M.; Zannoni, D.; Turner, R.J. Biosynthesis of selenium-nanoparticles and -nanorods as a product of selenite bioconversion by the aerobic bacterium *Rhodococcus aetherivorans* BCP1. *New Biotechnol.* **2018**, *41*, 1–8. [CrossRef] [PubMed]
60. Ruiz-Fresneda, M.A.; Eswayah, A.S.; Romero-González, M.; Gardiner, P.H.E.; Solari, P.L.; Merroun, M.L. Chemical and structural characterization of SeIV biotransformations by Stenotrophomonas bentonitica into Se^0 nanostructures and volatiles Se species. *Environ. Sci. Nano* **2020**, *7*, 2140–2155. [CrossRef]
61. Basaglia, M.; Toffanin, A.; Baldan, E.; Bottegal, M.; Shapleigh, J.P.; Casella, S. Selenite-reducing capacity of the copper-containing nitrite reductase of *Rhizobium sullae. FEMS Microbiol. Lett.* **2007**, *269*, 124–130. [CrossRef] [PubMed]
62. Debieux, C.M.; Dridge, E.J.; Mueller, C.M.; Splatt, P.; Paszkiewicz, K.; Knight IFlorance, H.; Love, J.; Titball, R.W.; Lewis, R.J.; Richardson, D.J.; et al. A bacterial process for selenium nanosphere assembly. *Proc. Natl. Acad. Sci. USA* **2011**, *108*, 13480–13485. [CrossRef]
63. Nancharaiah, Y.V.; Lens, P.N.L. Selenium biomineralization for biotechnological applications. *Trends Biotechnol.* **2015**, *33*, 323–330. [CrossRef]
64. Hwang, G.; Ahn, I.S.; Mhin, B.J.; Kim, J.Y. Adhesion of nano-sized particles to the surface of bacteria: Mechanistic study with the extended DLVO theory. *Colloids Surf. B Biointerfaces* **2012**, *97*, 138–144. [CrossRef] [PubMed]
65. Torres, S.K.; Campos, V.L.; León, C.G.; Rodréguez-Llamazares, S.M.; Rojas, S.M.; González, M.; Smith, C.; Mondaca, M.A. Biosynthesis of selenium nanoparticles by *Pantoea agglomerans* and their antioxidant activity. *J. Nanopart. Res.* **2012**, *14*, 1236. [CrossRef]
66. Lampis, S.; Zonaro, E.; Bertolini, C.; Bernardi, P.; Butler, C.S.; Vallini, G. Delayed formation of zero-valent selenium nanoparticles by *Bacillus mycoides* SeITE01 as a consequence of selenite reduction under aerobic conditions. *Microb. Cell Factories* **2014**, *13*, 35. [CrossRef] [PubMed]
67. Rohiwal, S.S.; Tiwari, A.P.; Verma, G.; Pawar, S.H. Preparation and evaluation of bovine serum albumin nanoparticles for ex vivo colloidal stability in biological media. *Colloids Surf. A Physicochem. Eng. Asp.* **2015**, *480*, 28–37. [CrossRef]
68. Tam, K.; Ho, C.T.; Lee, J.H.; Lai, M.; Chang, C.H.; Rheem, Y.; Chen, W.; Hur, H.G.; Myung, N.V. Growth mechanism of amorphous selenium nanoparticles synthesized by *Shewanella* sp. HN-41. *Biosci. Biotechnol. Biochem.* **2010**, *74*, 696–700. [CrossRef]
69. Jeevanandam, J.; Barhoum, A.; Chan, Y.S.; Dufresne, A.; Danquah, M.K. Review on nanoparticles and nanostructured materials: History, sources, toxicity and regulations. *Beilstein J. Nanotechnol.* **2018**, *9*, 1050–1074. [CrossRef]
70. Kale, A.; Bao, Y.; Zhou, Z.; Prevelige, P.E. Directed self-assembly of CdS quantum dots on bacteriophage P22 coat protein templates. *Nanotechnology* **2013**, *24*, 045603. [CrossRef]
71. Takechi-Haraya, Y.; Goda, Y.; Sakai-Kato, K. Atomic Force Microscopy Study on the Stiffness of Nanosized Liposomes Containing Charged Lipids. *Langmuir* **2018**, *34*, 7805–7812. [CrossRef]
72. Dhanjal, S.; Cameotra, S.S. Aerobic biogenesis of selenium nanospheres by *Bacillus cereus* isolated from coalmine soil. *Microb. Cell Factories* **2010**, *9*, 5. [CrossRef]
73. Piacenza, E.; Presentato, A.; Ferrante, F.; Cavallaro, G.; Alduina, R.; Chillura Martino, D.F. Biogenic Selenium Nanoparticles: A Fine Characterization to Unveil Their Thermodynamic Stability. *Nanomaterials* **2021**, *11*, 1195. [CrossRef] [PubMed]
74. Asghari-Paskiabi, F.; Imani, M.; Rafii-Tabar, H.; Razzaghi-Abyaneh, M. Physicochemical properties, antifungal activity and cytotoxicity of selenium sulfide nanoparticles green synthesized by *Saccharomyces cerevisiae. Biochem. Biophys. Res. Commun.* **2019**, *516*, 1078–1084. [CrossRef] [PubMed]
75. Tugarova, A.V.; Mamchenkova, P.V.; Dyatlova, Y.A.; Kamnev, A.A. FTIR and Raman spectroscopic studies of selenium nanoparticles synthesised by the bacterium *Azospirillum Thiophilum. Spectrochim. Acta A Mol. Biomol. Spectrosc.* **2018**, *192*, 458–463. [CrossRef] [PubMed]
76. Che, L.; Dong, Y.; Wu, M.; Zhao, Y.; Liu, L.; Zhou, H. Characterization of Selenite Reduction by *Lysinibacillus* sp. ZYM-1 and Photocatalytic Performance of Biogenic Selenium Nanospheres. *ACS Sustain. Chem. Eng.* **2017**, *5*, 2535–2543. [CrossRef]

77. Kamnev, A.A.; Dyatlova, Y.A.; Kenzhegulov, O.A.; Vladimirova, A.A.; Mamchenkova, P.V.; Tugarova, A.V. Fourier Transform Infrared (FTIR) Spectroscopic Analyses of Microbiological Samples and Biogenic Selenium Nanoparticles of Microbial Origin: Sample Preparation Effects. *Molecules* **2021**, *26*, 1146. [CrossRef]
78. El-Shanshoury, A.E.R.; Darwesh, O.M.; Sabae, S.Z.; Awadallah, O.A.; Hassan, S.H. Bio-manufacturing of selenium nanoparticles by *Bacillus subtilis* isolated from Qarun Lake and evaluation their activity for water remediation. *Biointerface Res. Appl. Chem.* **2020**, *10*, 5834–5842.
79. Shapaval, V.; Brandenburg, J.; Blomqvist, J.; Tafintseva, V.; Passoth, V.; Sandgren, M.; Kohler, A. Biochemical profiling, prediction of total lipid content and fatty acid profile in oleaginous yeasts by FTIR spectroscopy. *Biotechnol. Biofuels* **2019**, *12*, 140. [CrossRef]
80. Wu, Z.; Ren, Y.; Liang, Y.; Huang, L.; Yang, Y.; Zafar, A.; Hasan, M.; Yang, F.; Shu, X. Synthesis, Characterization, Immune Regulation, and Antioxidative Assessment of Yeast-Derived Selenium Nanoparticles in Cyclophosphamide-Induced Rats. *ACS Omega* **2021**, *6*, 24585–24594. [CrossRef]
81. Song, D.; Li, X.; Cheng, Y.; Xiao, X.; Lu, Z.; Wang, Y.; Wang, F. Aerobic biogenesis of selenium nanoparticles by *Enterobacter cloacae* Z0206 as a consequence of fumarate reductase mediated selenite reduction. *Sci. Rep.* **2017**, *7*, 3239. [CrossRef]
82. Korin, E.; Froumin, N.; Cohen, S. Surface Analysis of Nanocomplexes by X-ray Photoelectron Spectroscopy (XPS). *ACS Biomater. Sci. Eng.* **2017**, *3*, 882–889. [CrossRef]

microorganisms

MDPI

Article

An In Vitro Small Intestine Model Incorporating a Food Matrix and Bacterial Mock Community for Intestinal Function Testing

Mridu Malik [1,2], Jacob V. Tanzman [2,3], Sanat Kumar Dash [1,2], Cláudia N. H. Marques [2,3] and Gretchen J. Mahler [1,2,*]

1 Department of Biomedical Engineering, Binghamton University, Binghamton, NY 13902, USA
2 Binghamton Biofilm Research Center, Binghamton University, Binghamton, NY 13902, USA
3 Department of Biological Sciences, Binghamton University, Binghamton, NY 13902, USA
* Correspondence: gmahler@binghamton.edu

Abstract: Consumed food travels through the gastrointestinal tract to reach the small intestine, where it interacts with the microbiota, forming a complex relationship with the dietary components. Here we present a complex in vitro cell culture model of the small intestine that includes human cells, digestion, a simulated meal, and a microbiota represented by a bacterial community consisting of *E. coli*, *L. rhamnosus*, *S. salivarius*, *B. bifidum*, and *E. faecalis*. This model was used to determine the effects of food-grade titanium dioxide nanoparticles (TiO_2 NPs), a common food additive, on epithelial permeability, intestinal alkaline phosphatase activity, and nutrient transport across the epithelium. Physiologically relevant concentrations of TiO_2 had no effect on intestinal permeability but caused an increase in triglyceride transport as part of the food model, which was reversed in the presence of bacteria. Individual bacterial species had no effect on glucose transport, but the bacterial community increased glucose transport, suggesting a change in bacterial behavior when in a community. Bacterial entrapment within the mucus layer was reduced with TiO_2 exposure, which may be due to decreased mucus layer thickness. The combination of human cells, a synthetic meal, and a bacterial mock community provides an opportunity to understand the implications of nutritional changes on small intestinal function, including the microbiota.

Keywords: intestinal epithelium; gut microbiome; diet; titanium dioxide nanoparticles; nutrient transport; microbial community

Citation: Malik, M.; Tanzman, J.V.; Dash, S.K.; Marques, C.N.H.; Mahler, G.J. An In Vitro Small Intestine Model Incorporating a Food Matrix and Bacterial Mock Community for Intestinal Function Testing. *Microorganisms* **2023**, *11*, 1419. https://doi.org/10.3390/microorganisms11061419

Academic Editor: Kamel Ahmed Abd-Elsalam

Received: 31 March 2023
Revised: 10 May 2023
Accepted: 25 May 2023
Published: 27 May 2023

1. Introduction

The human small intestine plays a significant role in nutrient absorption and homeostasis maintenance via host–microbe interactions. Due to the difficulty of accessing the human small intestine for monitoring dietary effects on gut homeostasis and nutrient absorption, in vitro models can be used to gain this knowledge. The intestinal environment is rich in bacterial populations ranging from 10^1–10^3 bacteria per gram of contents in the duodenum to 10^4–10^7 bacteria per gram of contents in the jejunum and ileum and 10^{11}–10^{12} bacteria per gram of contents in the colon [1]. The environment of the small intestine is primarily composed of Gram-positive bacterial species such as *Lactocaseibacillus* and *Enterococcus* in the jejunum, *Streptococcus* in the duodenum and jejunum, and *Bifidum* bacilli [2] and Gram-negative *Escherichia coli* in the distal ileum [3]. The microbiota resides in the body as a community, competing for nutrients and metabolites, and are present in the mucus layer of the intestinal barrier facing the lumen. Under stressful conditions, disruption in the intestinal barrier and microbiota (dysbiosis) causes an altered permeability and disturbed intestinal function [4,5]. Intestinal alkaline phosphatase (IAP), a brush border membrane enzyme, which interacts with the microbiome to regulate the intestinal epithelial barrier, and its activity, together with intestinal permeability are often used as indicators of the integrity of the intestinal barrier.

The human small intestine, responsible for nutrient absorption and transport, is equipped with multiple transport proteins for nutrients such as glucose, proteins, and fatty acids [6]. In a healthy individual, 85% of carbohydrates, 66–95% of proteins, and all fats are absorbed in the small intestine before moving to the large intestine [7]. The gut microbiota draws its nutrition from the dietary components, and a change in diet can lead to changes in the microbial composition. For example, a change from a low-fat to a high-fat diet in mice led to a shift in their microbiota structure [8].

Nanoparticles (NPs) are used in the food industry to enhance the taste, aesthetics, or shelf life of food [9,10]. Food-grade titanium dioxide (TiO_2), referred to as E171 in the European Union (EU), contains about 36% of particles in the nanoscale [11] and is widely used in the food industry to improve the color of foods such as candies, chewing gums, pastries, and sauces [10–12]. Regular consumption of TiO_2-coated dietary products can be significant. Recently, the European Union issued a ban on the use of E171 food-grade TiO_2 in Europe due to genotoxicity concerns [13], which is expected to impact the trade of the commodity, but in the absence of potential alternatives to the additive, its use is currently inevitable. As part of the dietary intake, these nanoparticles pass through the human gastrointestinal (GI) tract to enter the intestinal environment. TiO_2 NPs can exhibit toxic effects through alteration of the gut microbiota and interference with nutrient transporters [11]. Some studies using TiO_2 NPs show its involvement in increased inflammatory cytokine production, disruption of the structure of the intestinal epithelium and mucus layer, and gut homeostasis both in vitro [14] and in vivo [15,16]. Whereas other studies using rats show little or no effect of TiO_2 ingestion but the elimination of it through the excretory system [17]. Such contradictory results call for the need to establish a standard study design to get more comparable and easily reproducible results.

This study uses a semi-synthetic food matrix supplemented with E171 food-grade TiO_2 NPs to form a physiologically relevant food model for assessing the role of NPs on GI function. An in vitro model of the intestinal barrier was created using Caco-2 and HT29-MTX-E12 cell lines. The small intestinal microbiome is represented by adding planktonic cultures of individuals or a mix of bacterial species (forming a bacterial mock community), including Gram-negative *Escherichia coli*, and Gram-positive *Lacticaseibacillus rhamnosus*, *Streptococcus salivarius*, *Bifidobacterium bifidum*, and *Enterococcus faecalis*. The intestinal monolayer with bacteria was exposed to the food matrix, with or without NP, following an in vitro digestion to assess the effects of NPs and bacterial species on intestinal permeability, IAP activity, and nutrient transport. This study presents a novel in vitro intestinal model with a multispecies community of bacteria representing the microbiome, and a standard food matrix to mimic the small intestinal environment that can be used to assess how food additives affect the GI environment.

2. Materials and Methods

2.1. Intestinal Co-Culture

The human colon carcinoma Caco-2 cell line was purchased from the American Type Culture Collection (Manassas, VA, USA) at passage 17 and used for experimentation at passage 45–55. The HT29-MTX-E12 cell line was purchased from Sigma-Aldrich, St. Louis, MO, USA, and used in experiments at passage 50–60. These cells were grown in Dulbecco's Modified Eagle Medium (DMEM, Gibco® Thermo Fisher Scientific, Waltham, MA, USA), containing 4.5 g/L glucose and 10% (v/v) heat-inactivated fetal bovine serum (HI-FBS, Gibco® Thermo Fisher Scientific, Waltham, MA, USA). The cells were maintained in 5% CO_2 at 37 °C, and the media was changed every 48 h. Once the cells reached 80% confluency, they were passaged and seeded onto polycarbonate, 0.4 μm pore size, 0.33 cm^2 membrane, 24-well Transwell® inserts for permeability, nutrient transport, and bacterial viability studies, or onto polycarbonate 96-well culture plates for brush border membrane enzyme activity assessment. Prior to seeding, the wells/inserts were coated with rat tail Type I collagen (BD Biosciences, San Jose, CA, USA) at a concentration of 8 $\mu g/cm^2$ for 1 h at room temperature. Both types of cells were stained with trypan blue, counted using a hemocytometer, and seeded at a density of 10^5 cells/cm^2 and a ratio of 75:25 (Caco-2:HT29). The co-culture of cells was grown for 14 days before experiments were conducted.

2.2. Bacterial Cell Culture

Escherichia coli (ATCC 53103), *Lacticaseibacillus rhamnosus* (ATCC 11775), *Streptococcus salivarius* SS2, *Bifidobacterium bifidum* VPI 1124, and *Enterococcus faecalis* NCTC 775 were purchased from the American Type Culture Collection. The bacterial species were cultured in brain heart infusion (BHI) broth medium supplemented with 0.5% glucose, 0.05% cysteine, and 0.1% agar. *E. coli* and *L. rhamnosus* cultures were grown at 37 °C, 5% CO_2 for 24 and 48 h, respectively. *B. bifidum* was cultured in an anaerobic GasPak at 37 °C for 48 h. *S. salivarius* cultures were incubated under normal atmospheric conditions at 37 °C for 24 h. *E. faecalis* was cultured for 12 h at 37 °C, 5% CO_2 condition. The bacterial stocks were diluted in 0.9% saline solution, plated on BHI agar medium supplemented with 0.5% glucose, 0.05% cysteine, and 1.5% agar, and grown in their respective culture conditions. The optical density (OD) of each dilution was read at a wavelength of 600 nm. The colonies formed on agar plates were quantified and associated with the OD_{600} measured for each dilution to prepare a standard curve. Prior to each experiment, bacterial concentrations were estimated using OD_{600}. Planktonic culture of the five individual bacterial species was added to the apical chamber of the Transwell plates or in the 96-well plate at a concentration of 10^3 CFU/mL in combination with the control or test conditions. Bacteria mix (BM) of the five species was prepared at a ratio of 30% *L. rhamnosus*, 30% *B. bifidum*, 20% *S. salivarius*, 15% *E. faecalis*, and 5% *E. coli*. Each bacterial species was diluted to 10^3 CFU/mL before combining in the said ratio to form the BM. The BM was combined with the control or test condition before experiments.

2.3. Food Model Preparation

The food model (FM) was based on the previously developed food matrix described by Zhang et al. [18]. The FM consists of the nutrients from a standard US diet. Sodium caseinate (Sigma, 3.44% *w*/*w*), a protein source, was dissolved in 10 mL of sterile 10 mM phosphate buffer solution, pH 7. The solution was homogenized using a homogenizer at medium speed for 45 min. The solution was substituted with 3.42% (*w*/*w*) corn oil (fat source) and another 10 mL of sterile 10 mM phosphate buffer solution, followed by homogenization at high speed for 15 min to obtain a fine emulsion. Finally, 0.7% (*w*/*w*) pectin (dietary fiber source), 5.15% (*w*/*w*) starch, 4.57% (*w*/*w*) sucrose (sugar source), and 0.534% sodium chloride (mineral source) were added to the emulsion with continuous stirring in between. The prepared nutrient emulsion was frozen at −80 °C overnight and freeze-dried using a freeze dryer for a week. The freeze-dried product was referred to as the food model and stored at −20 °C until needed. The FM was diluted and streaked on BHI agar plates to check for sterility. For experiments, 1 g of food model was subjected to in vitro digestion and added to the seeded assay plates.

2.4. Nanoparticle Preparation

E171 food-grade titanium dioxide nanoparticles (TiO_2 NPs) were purchased from Fiori Colori (Aromata Group). The TiO_2 NPs have been previously characterized [19]. TiO_2 NPs were sonicated at a concentration of 1 mg/mL (prepared using sterile 18 MΩ water) for 2 min at 10% amplitude on ice. The sonicated solution was diluted 100 times in sterile 18 MΩ water, and 5.92 μL of the diluted NP solution was used for experiments to achieve a final dose of 1.25×10^{-6} mg/cm^2 or 1.25 μg/cm^2. The calculations are based on the surface area of the small intestine and average human intake of TiO_2 [5,20]. TiO_2 was combined with 1 g of food model to represent the FM + TiO_2 condition.

2.5. In Vitro Digestion

The control and test conditions were digested in vitro in a sterile environment prior to experimental exposure. A serum-free, 5 mM glucose DMEM solution served as a control condition, while test conditions included FM, TiO_2, and FM + TiO_2. The in vitro digestion protocol has been previously described by Glahn et al. [21]. The weighed samples were dissolved in 10 mL of sterile 140 mM NaCl + 5 mL KCl with a pH of 2. The pH was then re-adjusted to 2 with sterile 0.1 M HCl. Next, 0.5 mL of a porcine pepsin solution (25 mg/mL) prepared in 0.1 M HCl was added to each sample. The samples were then incubated for 1 h at 37 °C, 5% CO_2 on a rocker (55 oscillations/min). This constituted the "gastric" phase of digestion. This was followed by "intestinal" digestion, where the pH of the samples was increased to 5.5–6 using sterilized 0.1 M $NaHCO_3$ or 1 M $NaHCO_3$. A pancreatin bile solution was prepared using 2 mg/mL pancreatin (a mixture of trypsin, amylase, lipase, ribonuclease, and protease activities) and 11 mg/mL bile extract (a mixture of glycine and taurine conjugates of hyodeoxycholic and other bile acids) and filter sterilized with a 0.45 μm filter. Next, 2.5 mL of the prepared pancreatin-bile solution was added to each sample, and the pH was brought up to 6.9–7 using sterile 0.1 M $NaHCO_3$ or 1 M $NaHCO_3$. The volume of each sample tube was made up to 15 mL by weight using a sterile 140 mM NaCl + 5 mM KCl solution, pH 6.7. The samples were then referred to as digests and used for exposure studies.

2.6. Permeability Assay

The permeability assay was used to quantify the changes in barrier integrity using Lucifer yellow (LY, Life Technologies L453), a fluorescent paracellular permeability marker. LY stock of 1 mg/mL was prepared aseptically in 18 MΩ water and stored at 4 °C until needed. A standard curve of LY was plotted using 0, 10, 20, 30, 40, and 50 μM LY concentrations and the fluorescence values for each concentration. A 100 μL of digested sample was added to the apical chamber of Transwell plates with 100 μL of 50 μM LY. Samples (100 μL) were then collected from the basolateral chamber every 15 min for the first hour and every 30 min for the next three hours and transferred into a 96-well opaque black bottom plate. The removed 100 μL volume, when sampling, was substituted by adding 100 μL DMEM to the basolateral chamber of the Transwell plate. At the end of the exposure experiment, the 96-well plate was read at 485/560 nm using a Synergy 2 plate reader, controlled by Biotek's Gen5™ Reader Control and Data Analysis Software (V1.11.5). The resulting fluorescence values of the samples and the standards were used to calculate the permeability (cm/s) across the intestinal monolayer.

2.7. Intestinal Alkaline Phosphatase (IAP) and Bradford Assays

IAP activity was determined using IAP and Bradford assay. The sample digests were added to the seeded 96-well plates and allowed a 4 h exposure at 37 °C, 5% CO_2. At the end of the 4 h exposure to digested control or test conditions, the wells were washed with 1× PBS, followed by the addition of 200 μL of 18 MΩ water. The plate was then subjected to sonication at 4 °C for 15 min. The samples were scraped off the bottom of the wells and subsequently used for the assay. A standard curve was prepared with dilutions of a standard solution of 0.25 mg/mL p-nitrophenol, prepared in p-nitrophenyl phosphate (pNPP). The assay was carried out by adding 25 μL of sample or standard solution in 85 μL of pNPP solution to a 96-well clear bottom plate, which was then incubated for 1 h at room temperature. For the Bradford assay, a standard curve was prepared using a 1 mg/mL bovine serum albumin (BSA, Rockland, NY, USA) solution prepared in 18 MΩ water. The experimental setup included a 5 μL sample or standard solution and 250 μL of Bradford reagent (Sigma) in a 96-well clear bottom plate. The plate was incubated at room temperature for 15–30 min in the dark. The IAP assay and Bradford assay plates were read at 405 nm and 595 nm, respectively, using the Synergy 2 plate reader, controlled by Biotek's Gen5™ Reader Control and Data Analysis Software.

2.8. Nutrient Transport Assay

The Caco-2 and HT29-MTX E12 cell monolayer was grown on Transwell membranes for 14 days before performing nutrient transport assays. During the 14-day period, tight junction integrity is established, the Caco-2 cells mature into enterocytes, and a mucus layer is formed by HT29-MTX E12 cells [22]. On day 15, the monolayers were exposed to digested FM and/or TiO_2 samples or control solutions, in combination with 10^3 CFU/well bacteria for 4 h. The DMEM medium used for the assays was serum free with 5 mM glucose. This ensured that the resulting protein and glucose content detected after the assays was a result of the nutrients in the food model. After the exposure period, samples were collected from the basolateral chamber for nutrient transport assays.

2.8.1. Glucose Assay

The basolateral samples were subjected to charcoal treatment due to low glucose levels [23]. Activated charcoal (2% *w*/*w*) was added to the samples and mixed on a vortex for 30 s. The samples were then centrifuged for 30 s, and the supernatant was used for the assay. Prepared samples were added to glucose hexokinase liquid stable reagent (Infinity™) at a ratio of 1:150 in a 24-well clear bottom plate. Samples were incubated for 3 min at 37 °C, and the absorbance was read at 340 nm using a Synergy 2 plate reader, controlled by Biotek's Gen5™ Reader Control and Data Analysis Software. The absorbance values along with the glucose standard (1 mg/mL in 0.1% benzoic acid) were used to calculate the amount of glucose (μg) per mL of each sample.

2.8.2. Protein Assay

The basolateral samples were subjected to charcoal treatment, as mentioned in Section 2.8.1. The protein content was assessed using a Bradford reagent. A 5 μL supernatant sample was added to a 96-well clear bottom plate with 250 μL of Bradford reagent. The protein assay was carried out as described in Section 2.6.

2.8.3. Triglyceride Assay

The fatty acid content in the basolateral samples was detected using a triglyceride assay kit (Sigma, MAK266). A triglyceride standard curve was prepared using dilutions of a 0.2 mM standard solution. A 50 μL sample or standard solution was added to a 96-well clear bottom plate, along with a 2 μL lipase solution. Samples were incubated for 20 min at room temperature. A master mix was prepared using 46 μL TG assay buffer, 2 μL TG probe, and 2 μL TG enzyme mix per well. A 50 μL of the prepared master mix was added to the plate at the end of incubation. Samples were then incubated for 30–60 min at room temperature in the dark. The absorbance values were read at 570 nm using a Synergy 2 plate reader, controlled by Biotek's Gen5™ Reader Control and Data Analysis Software. Using the standard curve, the triglyceride content (μg) per μL sample was calculated.

2.9. Bacterial Viability Quantification

The 4 h exposure to independent bacterial species and digested samples was followed by bacterial quantification. Bacterial viability was quantified using viable counts and quantitative PCR.

2.9.1. Viable Counts

A 100 μL sample was collected from the apical and basolateral chambers of the Transwell plates. The samples were serially diluted with sterile 0.9% saline solution and plated on BHI agar plates using the drop-plate technique. The bacteria that adhered to the Caco-2/HT29-MTX E12 monolayer were harvested from the Transwell inserts. The wells were washed with 1× PBS, followed by cutting of the Transwell membrane. The membrane was then transferred to a 400 μL sterile saline solution and homogenized for 30 s per sample. The homogenized samples were diluted and plated on BHI agar plates. To allow selective growth of each bacterial species in the bacteria mix, the samples were plated on MRS agar for *L. rhamnosus*, MacConkey agar for *E. coli*, ESM agar for *E. faecalis*, using the drop plate technique, blood agar for *S. salivarius*, and BSM agar for *B. bifidum* using the spread plate technique. Plates were incubated at their respective optimum growth conditions for

24–48 h, based on the bacterial species, and bacterial colonies were quantified to determine the bacterial viability (total log CFU).

2.9.2. qPCR

Samples for qPCR bacterial quantification were harvested by scraping into sterile saline, and cells were collected by centrifugation (1 min at 18,000× *g*,), the supernatant discarded, and the cell pellet stored at −80 °C. Genomic DNA (gDNA) was extracted according to manufacturer instructions (ZymoBIOMICS DNA Miniprep Kit), and samples were stored at −80 °C. qPCR was performed with the PerfeCTa® SYBR® Green FastMix® (Quantabio), the primers for each bacterial species, and an Eppendorf Mastercycler Ep Realplex instrument. Primers were used to quantify *B. bifidum* (16s rRNA, FW ACGCGAAGAACCTTACCTGG, RV 5′-ATCTCACGACACGAGCTGAC-3′), *E. faecalis* (16s rRNA, FW 5′-TACGGCGAACATACAAAGCG-3′, RV 5′-ATTTTGAACACGCGACTGG-3′), *L. rhamnosus* (16s rRNA, FW 5′-AGACACGGCCCAAACTCTAC-3′, RV 5′-CGTTGCTCCATCAGACTTGC-3′), *S. salivarius* (Hypothetical Protein, FW 5′-TGCCTGGAACTAACGGTAGC-3′, RV 5′-CAACGACTTCAGAGCCTCCC-3′), and *E. coli* (*tnaA*, FW 5′-GACTGGACCATCGAGCAGAT-3′, RV 5′-CACGCAAAGGGTTCTGCACTC-3′). Standard curves were previously generated by relating CT values to viable cell counts and allowing experimental CT values to be interpolated into equivalent cell counts. In brief, cultures of target species were grown to stationary phase and quantified by drop plating. Genomic DNA was extracted from 1 mL of stationary-phase cultures and purified. DNA was serially diluted to produce standards of decreasing DNA concentration. Quantitative PCR was then performed on these standards, and the resulting CT values were plotted against viable cell counts. Standard curves were calculated to interpolate cell counts from experimental CT values.

2.10. Statistical Analysis

The data sets were tested for normality using the Shapiro–Wilk normality test and transformed using the Box–Cox transformation with appropriate conditions to achieve a normal distribution. The LY permeability results were analyzed using a one-way ANOVA. The IAP assay, bacterial viability, and nutrient transport assay results were compared using a two-way ANOVA followed by false discovery rate (FDR) correction using GraphPad Prism 8.3.1. FDR correction was performed using the Benjamini and Hochberg procedure [24] implemented in the R [25] package multi-test [26].

3. Results

3.1. Bacteria, Food Model, and TiO_2 Affect Cell Permeability

Caco-2 and HT29-MTX E12 cells were seeded onto Transwell membranes to form an epithelial monolayer. The intestinal barrier was subjected to four different conditions: digested control (serum-free, 5 mM glucose DMEM, DM), freeze-dried food model (FM), physiologically relevant dose of TiO_2 nanoparticles, and FM + TiO_2 conditions. Permeability across the intestinal barrier was determined under these four conditions prior to the introduction of bacteria into the system to ensure that the synthetic food model was not negatively affecting the integrity of the barrier. The results show no significant changes in permeability compared to the control condition. In comparison to the food model, addition of TiO_2 to FM showed a significant decrease in permeability (Figure 1A), while *L. rhamnosus* (Figure 1B), *E. coli* (Figure 1C), *S. salivarius* (Figure 1D), and *E. faecalis* (Figure 1F) did not cause any significant changes in permeability across the four conditions compared to the control. *B. bifidum* led to a significant decrease in LY permeability across all four conditions, indicating an improved barrier function (Figure 1E).

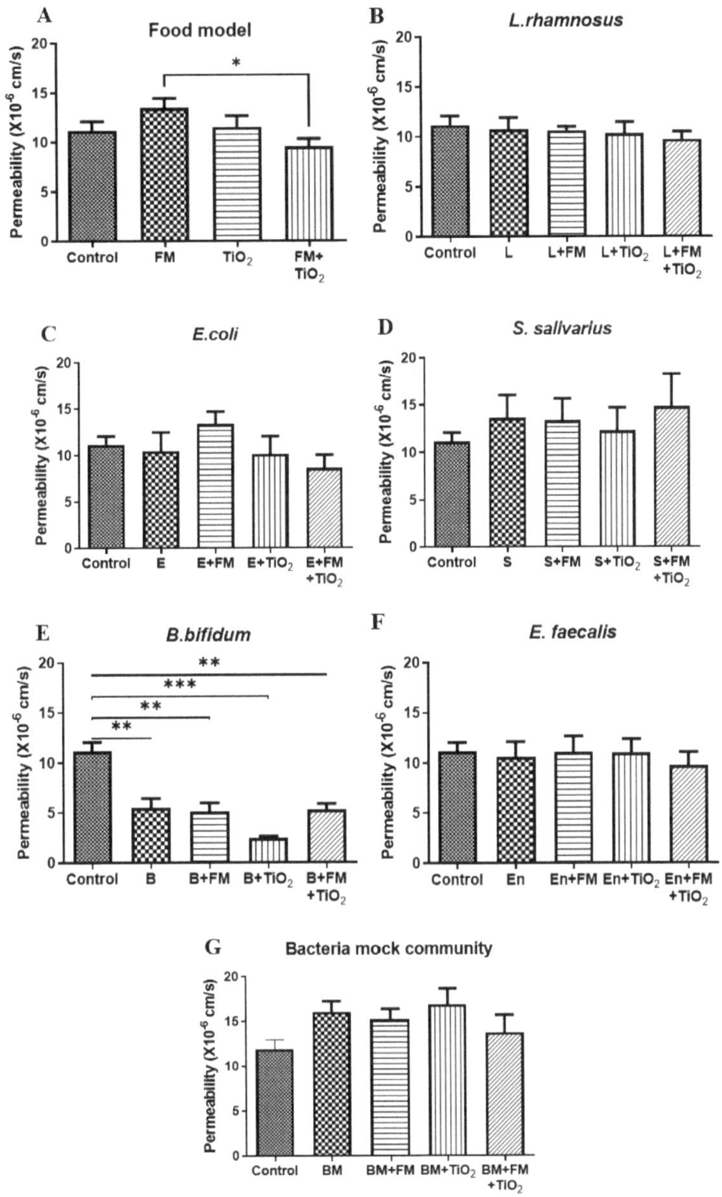

Figure 1. Permeability across the Caco-2/HT29-MTX E12 monolayer during 4 h exposure to (**A**) control (serum-free, 5 mM glucose DMEM), food model, TiO_2, and food model + TiO_2 in combination with (**B**) 10^3 CFU/well *L. rhamnosus* (L), (**C**) 10^3 CFU/well *E. coli* (E) (**D**) 10^3 CFU/well *S. salivarius* (S) (**E**) 10^3 CFU/well *B. bifidum* (B) (**F**) 10^3 CFU/well *E. faecalis* (En) (**G**) 10^3 CFU/well bacteria mix. Error bars represent SEM. FDR-adjusted *p*-values: * $p < 0.05$; ** $p < 0.01$; *** $p < 0.001$; using one-way ANOVA. n = 7. FM: food model, BM: bacteria mock community.

The five different bacteria were cultured independently and then combined in a physiologically relevant ratio to create a bacterial mock community (BM). The bacterial mix did not result in any significant changes in permeability (Figure 1G).

3.2. Bacteria Cause a Reduction in IAP Activity

The intestinal alkaline phosphatase activity was quantified after 4 h exposure. Compared to the control (serum-free, 5 mM glucose DMEM), the IAP activity following exposure to TiO_2 and/or FM was not affected. The introduction of *L. rhamnosus* (Figure 2A) and *E. coli* (Figure 2B) reduced the IAP levels compared to their respective no-bacteria conditions. *S. salivarius* was the only bacteria of the five species studied that led to an increase in IAP activity compared to the no-bacteria conditions (Figure 2C). *B. bifidum* (Figure 2D) and *E. faecalis* (Figure 2E) both caused a decrease in IAP activity with the control, TiO_2, and FM + TiO_2 conditions compared to their respective no-bacteria conditions. In addition, *B. bifidum* with TiO_2 NPs led to a reduction in IAP activity compared to the *B. bifidum* + FM condition. The bacterial mock community maintained the IAP activity compared to the no-bacteria condition, except with TiO_2, which showed a decrease with the BM but was restored to the control levels by the addition of the food model (Figure 2F).

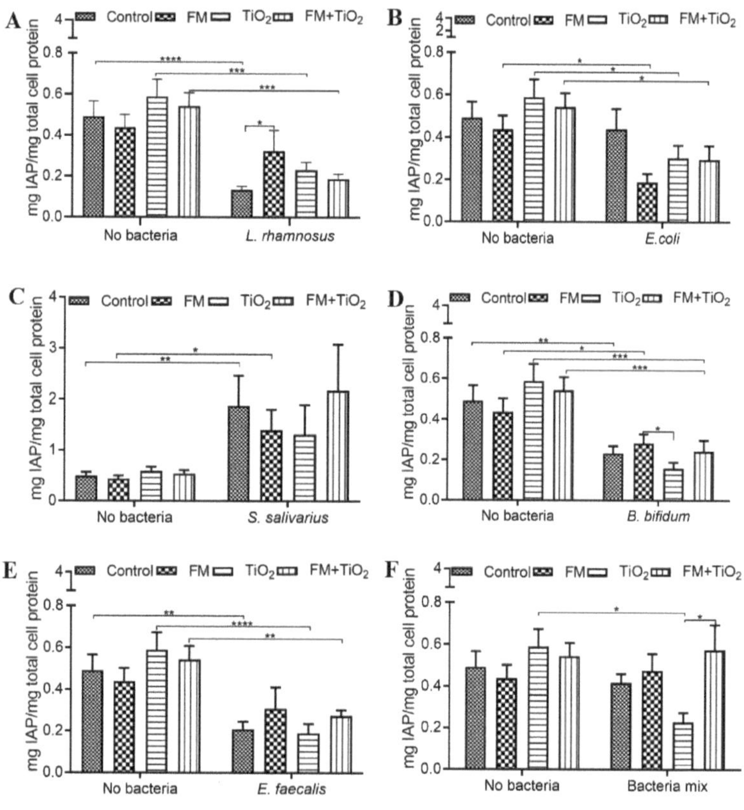

Figure 2. Intestinal alkaline phosphatase (IAP) activity following food matrix, bacteria, and/or TiO_2 nanoparticle (NP) exposure. IAP activity after a 4 h exposure to control (serum-free, 5 mM glucose DMEM), food model, TiO_2 nanoparticles, or food model + TiO_2 with planktonic bacterial cultures (**A**) 10^3 CFU/well *L. rhamnosus* (**B**) 10^3 CFU/well *E. coli* (**C**) 10^3 CFU/well *S. salivarius* (**D**) 10^3 CFU/well *B. bifidum* (**E**) 10^3 CFU/well *E. faecalis* (**F**) 10^3 CFU/well bacteria mix. Error bars represent SEM. FDR-adjusted *p*-values: * $p < 0.05$; ** $p < 0.01$; *** $p < 0.001$; **** $p < 0.0001$ using two-way ANOVA with FDR correction IAP assay. n = 12. FM: food model.

3.3. Bacterial Community Increases the Glucose Transport

The amount of glucose transported across the in vitro epithelial barrier was quantified after a 4 h exposure using a colorimetric glucose assessment assay. The control condition (serum-free, 5 mM glucose DMEM), FM, TiO_2, and FM + TiO_2 conditions displayed the same amount of glucose transported across the barrier. As shown in Figure 3A–C, the addition of *L. rhamnosus*, *E. coli*, or *S. salivarius* did not impact glucose transport. However, *B. bifidum* significantly reduced glucose transport in the FM + TiO_2 condition (Figure 3D). In the presence of *E. faecalis*, on the other hand, the amount of glucose transported reduced with TiO_2 but increased with the FM (Figure 3E). This is expected due to the additional amount of glucose added to the model as part of the FM available to be transported. In a mock community, the bacterial species increased the glucose transported compared to the respective no-bacteria conditions, with a slight reduction upon the addition of FM and TiO_2 (Figure 3F), suggesting a difference in impact due to interspecies bacterial interaction.

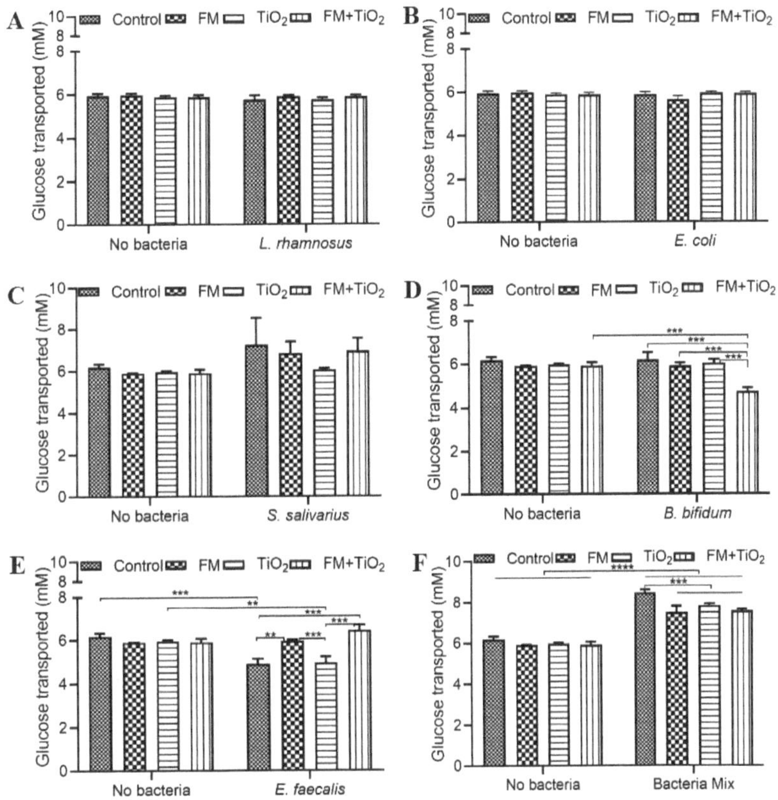

Figure 3. Glucose transport following food matrix, bacteria, and/or TiO_2 nanoparticle (NP) exposure. The amount of glucose transported across the Caco-2/HT29-MTX E12 monolayer following a 4 h exposure to control (serum-free, 5 mM glucose DMEM), food model, TiO_2 nanoparticles, or food model + TiO_2 with planktonic bacterial culture of (**A**) 10^3 CFU/well *L. rhamnosus* (**B**) 10^3 CFU/well *E. coli* (**C**) 10^3 CFU/well *S. salivarius* (**D**) 10^3 CFU/well *B. bifidum* (**E**) 10^3 CFU/well *E. faecalis* (**F**) 10^3 CFU/well bacteria mix. Error bars represent SEM. FDR-adjusted *p*-values: ** $p < 0.01$; *** $p < 0.001$; **** $p < 0.0001$ using two-way ANOVA with FDR correction. n = 3. FM: food model.

3.4. Protein Transport Remains Unchanged with NP and Bacteria

Protein transport estimation conducted using the Bradford assay revealed no significant changes in protein transport across the Caco-2/HT29-MTX E12 monolayer in response to the FM, TiO_2, or a combination of both. The inclusion of *L. rhamnosus* or *E. coli* did not impact protein transport (Figure 4A,B). *S. salivarius*, on the other hand, caused a slight increase in the amount of protein transported in FM + TiO_2, but without statistical significance (Figure 4C). *B. bifidum* showed the opposite effect by decreasing protein transport without statistical significance, as depicted in Figure 4D. The inclusion of *E. faecalis* did not change the transport of protein compared to their respective no-bacteria conditions (Figure 4E). Upon combining the bacteria into a multi-species community, their exposure resulted in a decrease in protein transport, but without statistical significance (Figure 4F). The overall lack of change in protein transport indicates the disinvolvement of NPs or bacteria in protein transport in this model.

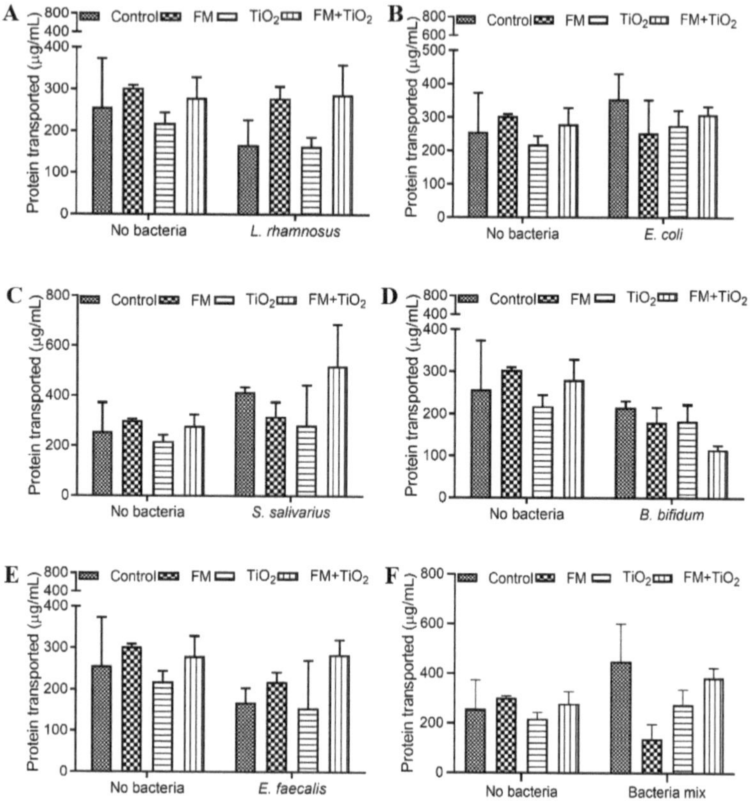

Figure 4. Protein transport following food matrix, bacteria, and/or TiO_2 nanoparticle (NP) exposure. The amount of protein transported across the Caco-2/HT29-MTX E12 monolayer following a 4 h exposure to control (serum-free, 5 mM glucose DMEM), food model, TiO_2 nanoparticles, and food model + TiO_2 in combination with planktonic bacterial culture of (**A**) 10^3 CFU/well *L. rhamnosus* (**B**) 10^3 CFU/well *E. coli* (**C**) 10^3 CFU/well *S. salivarius* (**D**) 10^3 CFU/well *B. bifidum* (**E**) 10^3 CFU/well *E. faecalis* (**F**) 10^3 CFU/well bacteria mix. Error bars represent SEM. FDR-adjusted *p*-values using two-way ANOVA with FDR correction. n = 3. FM: food model.

3.5. *Bacteria Reverse the Effect of TiO_2 on Triglyceride Transport*

The last functional aspect of the epithelial barrier studied was the transport of triglycerides (TG), as fatty acids were a part of the synthetic food model prepared in this study. The conditions with FM showed a higher TG transport compared to the fatty acid content added to the study model as part of the FM, which contributes to the negligent level of TG transport seen in conditions without the FM. The significance is depicted to showcase differences between the control, FM, TiO_2, and FM + TiO_2 conditions for no-bacteria or bacteria conditions. Figure 5A shows a significant increase in the amount of TG with FM + TiO_2 compared to the FM condition, which was reduced in the presence of *L. rhamno-*

sus. *E. coli* displayed the same result: a decrease in TG transport in the FM + TiO_2 condition when compared to the no-bacteria condition. In addition, the addition of TiO_2 to FM reduced the transport in combination with *E. coli* (Figure 5B). Compared to the no-bacteria condition, *S. salivarius* (Figure 5C), *B. bifidum* (Figure 5D), and *E. faecalis* (Figure 5E) reduced the TG transport in combination with the FM + TiO_2 condition. Within the *B. bifidum* condition, TiO_2 caused a decrease in TG transport with FM. Similar to the effect of independent bacterial species, the bacteria mix also led to a decrease in TG transport compared to the no-bacteria conditions with FM and FM + TiO_2 (Figure 5F).

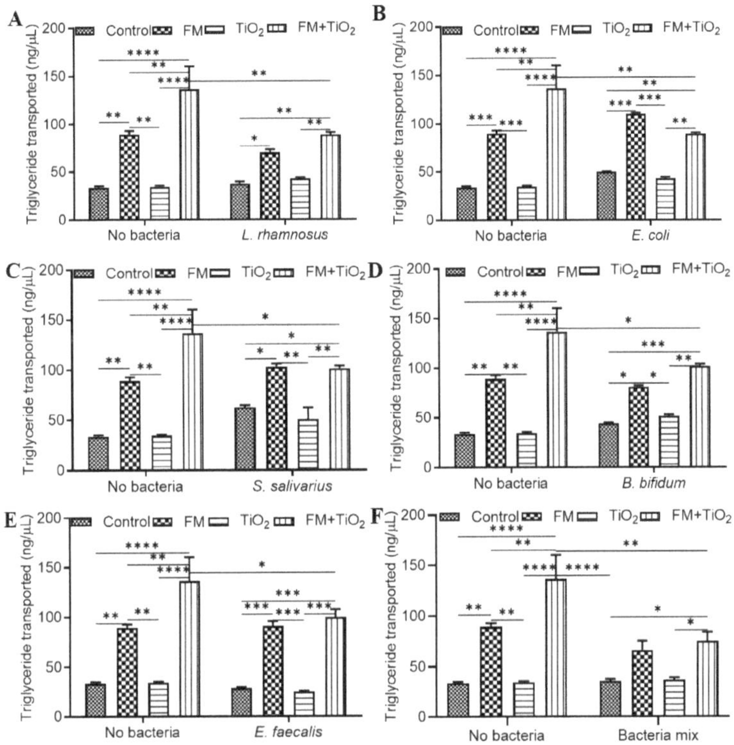

Figure 5. Triglyceride transport following food matrix, bacteria, and/or TiO_2 nanoparticle (NP) exposure. The amount of triglyceride transported across the Caco-2/HT29-MTX E12 monolayer following a 4 h exposure to control (serum-free, 5 mM glucose DMEM), food model, TiO_2 nanoparticles, or food model + TiO_2 in combination with planktonic bacterial culture of (**A**) 10^3 CFU/well *L. rhamnosus* (**B**) 10^3 CFU/well *E. coli* (**C**) 10^3 CFU/well *S. salivarius* (**D**) 10^3 CFU/well *B. bifidum* (**E**) 10^3 CFU/well *E. faecalis* (**F**) 10^3 CFU/well bacteria mix. Error bars represent SEM. FDR-adjusted *p*-values: * $p < 0.05$; ** $p < 0.01$; *** $p < 0.001$; **** $p < 0.0001$ using two-way ANOVA with FDR correction. n = 3. FM: food model.

3.6. Bacteria Adhere to the In Vitro Intestinal Monolayer

With the assessment of the integrity and function of the epithelial barrier, it became pertinent to analyze the viability of the bacterial species added to the model. The viability tests were conducted on the apical, insert, and basolateral compartments of the Transwell plates. The results revealed the number of bacterial colonies in suspension that adhered to the monolayer and those that passed across the barrier, respectively. The initial concentration introduced into the model was 10^3 CFU/well (or 3 log CFU). The viability assessment for *L. rhamnosus*, shown in Figure 6A, reveals no significant effect of the FM or TiO_2 on the apical chamber population; however, the *L. rhamnosus* colony growth on the insert seemed to decrease in the presence of the FM. *E. coli* showed a relatively higher growth rate compared to *L. rhamnosus*. While the FM or TiO_2 conditions did not impact *E. coli* growth in the apical chamber, its attachment to the insert was significantly reduced with TiO_2 but not FM (Figure 6B). *E. coli* colonies were also seen to pass into the basolateral chamber, with the highest being in combination with FM + TiO_2. As opposed to *L. rhamnosus* and *E. coli*, *S. salivarius* growth was scarce, pertaining to its strict conditions needed for optimum growth. *S. salivarius* colony growth in the apical and on the insert was favored in the absence of the FM, while only the control condition allowed the movement of the bacteria to the basolateral chamber (Figure 6C). *B. bifidum* growth, as shown in Figure 6D, was observed only on the inserts, with no significant difference between the four conditions studied. Lastly, *E. faecalis* survived and multiplied in the apical chamber and on the insert. But without any statistical difference between different conditions (Figure 6E).

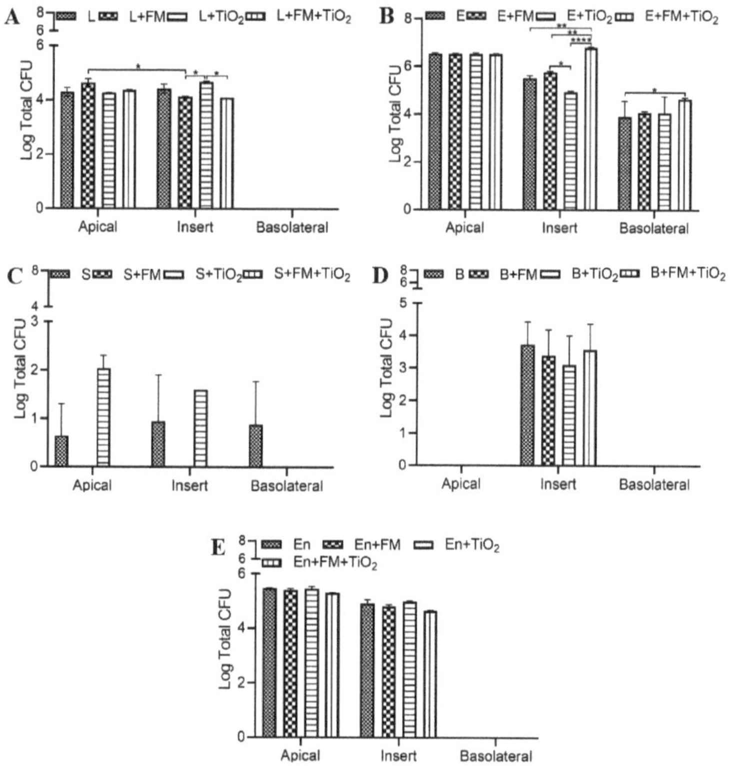

Figure 6. Bacterial viability quantification. The Caco-2/HT29-MTX E12 monolayer was exposed to control (serum-free, 5 mM glucose DMEM), food model, TiO_2 nanoparticles, and food model + TiO_2 for 4 h in combination with bacterial cultures. The bacterial viability was assessed in the apical, basolateral chamber, and insert of Transwell plates for planktonic bacterial culture of (**A**) 10^3 CFU/well *L. rhamnosus* (L), (**B**) 10^3 CFU/well *E. coli* (E), (**C**) 10^3 CFU/well *S. salivarius* (S), (**D**) 10^3 CFU/well *B. bifidum* (B), (**E**) 10^3 CFU/well *E. faecalis* (En). Error bars represent SEM. FDR-adjusted *p*-values: * $p < 0.05$; ** $p < 0.01$; **** $p < 0.0001$ using two-way ANOVA with FDR correction. n = 3. FM: food model.

3.7. Bacterial Survival as a Mock Community

The survival of each of the five bacterial species was assessed using PCR based population quantification (Figure 7). The results revealed a low viability of *S. salivarius* in the three chambers (apical, basolateral, and insert) of the Transwell, which can be attributed to its requirement for low oxygen growth conditions. *B. bifidum*, which did not survive well as a single species, was present in all three chambers. There was no significant effect of the food model or TiO_2 on the viability of bacteria in the apical chamber. The attachment of *E. faecalis* was seen to slightly increase with TiO_2, but without statistical significance. However, the presence of *E. faecalis* in the basolateral chamber increased significantly with FM + TiO_2 compared to the control condition (serum-free, 5 mM glucose DMEM). The attachment of all bacterial species on the insert shows their interaction with the mucus layer and intestinal cells and suggests their involvement in the changes seen in barrier integrity and function. The overall CFU count for bacterial survival decreased when the bacterial species was present as part of the mock community compared to an independent bacterial culture, which could be due to the limited availability of nutrients for the bacteria community.

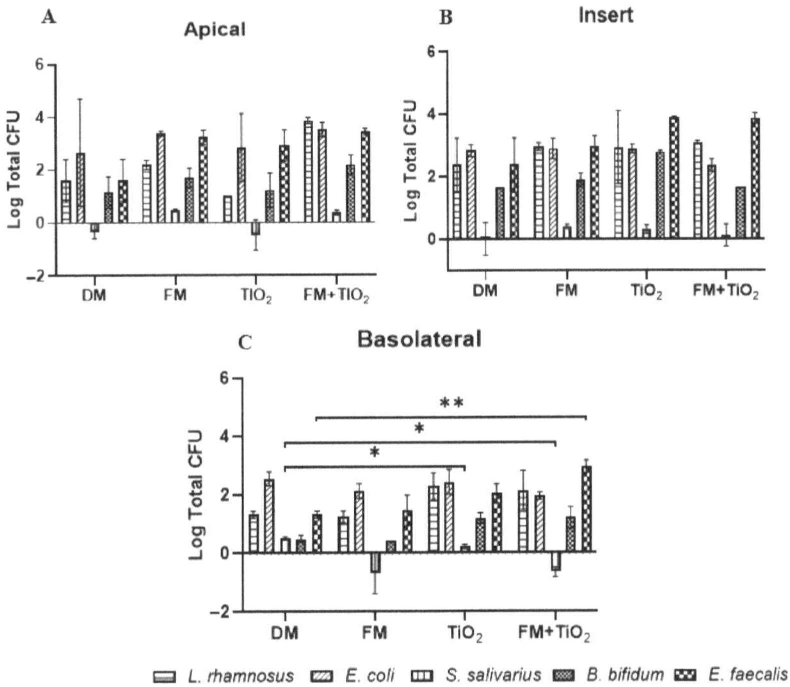

Figure 7. PCR-based bacterial quantification in a mock community. The Caco-2/HT29-MTX E12 monolayer was exposed to control (serum-free, 5 mM glucose DMEM), food model (FM), TiO_2 nanoparticles, and food model + TiO_2 for 4 h in combination with five different bacterial species as a bacteria mix with an initial concentration of 10^3 CFU/well. The bacterial viability was assessed through PCR-based CFU count in the (**A**) apical (**B**) insert and (**C**) basolateral chamber. Error bars represent SEM. * $p < 0.05$; ** $p < 0.01$; using two-way ANOVA with FDR correction. n = 2.

4. Discussion

Food-grade TiO_2 was approved by the Food and Drug Administration (FDA) in 1966 as INS171 and by the European Union in 2006 as E171. However, there have been growing health concerns with the consumption of TiO_2 NPs due to their potential impact on the intestinal tract, the gut microbiota, and their potential to cross the gut and cause an immune reaction in the body. The host GI tract provides a constant nutrient supply and stable growth environment to the microorganisms while gaining protection from the pathogens due to their competitive exclusion by the commensal bacteria, thus forming a symbiotic relationship [27].

The negative effects of TiO_2 NPs on the GI are widely published [11,14,28]; however, the ingestion of such nanoparticles as part of a diet and their interaction with the gut microbiota have yet to be fully established. The properties of the NPs can change when added to food [29] and after being absorbed by the GI, resulting in varied effects [30,31]. In this study, a normal US diet was created in the lab using all the essential nutrients, as previously described by Zhang et al. [18]. This semi-synthetic meal, referred to as the food model (FM), represented a US diet with an additional intake of products rich in TiO_2 NPs (E171 grade) at a realistic dose of 1.25 $\mu g/cm^2$, calculated based on the average human intake of TiO_2 NPs and the surface area of the small intestine [20,32]. The food model and TiO_2 were subjected to a two-step in vitro digestion to create a physiologically relevant diet, which was then introduced into a Caco-2/HT29-MTX E12 intestinal model. Prior to exposure to the intestinal monolayer, the sterility of the food model was confirmed using the drop-plate culture technique. The food model or TiO_2 did not have any significant impact on the permeability of the barrier compared to the control condition, which showed a permeability of 10^{-5} cm/s. However, when TiO_2 was incorporated into the food model, the permeability decreased by 29% compared to the food model alone (Figure 1A).

The upper small intestine is predominantly composed of Gram-positive aerobes, and the terminal ileum has an equal number of aerobes and anaerobes such as *Bifidum* bacilli [2]. The Gram-positive lactobacilli comprise the majority of the normal flora in the jejunum. Lactobacilli, along with staphylococci and streptococci, are also present in the intestine distally [33]. *L. rhamnosus* [34,35] and *Bifidobacterium* spp. [36] are well-known probiotic species with positive effects on intestinal health. *E. faecalis* is a common commensal bacterium and is one of the first colonizers in an infant [37], while *E. coli* is a Gram-negative bacterium more prevalent in the colon and can also be found in high quantities in the small intestine [3]. The bacteria mix in this study was designed to include a variety of bacterial species ranging from aerobes (*Enterococcus* and *Escherichia*) to facultative anaerobes (*Staphylococcus, Lactobacillus*) and anaerobes (*Bifidobacterium*). The broad selection of bacterial species enabled the investigation of the host–diet–microbiota interaction as close to the in vivo scenario as feasible. In this study, the presence of *B. bifidum* resulted in a significant decrease in permeability across the in vitro intestinal barrier (Figure 1E), while the other bacteria did not impact the permeability, indicating a quick response of this probiotic species towards regulating intestinal health. IAP is expressed by the enterocytes primarily in the duodenum and is heavily involved in the regulation of gut homeostasis [38]. IAP can potentially restore the commensal gut bacteria through its ability to control the pH. IAP is a crucial contributor to the maintenance of gut homeostasis and health through interactions with the intestinal microbiota and diet. While the benefits of probiotic bacterial species such as *L. rhamnosus* and *B. bifidum* are widely established, the positive influence of other commensal species such as *E. coli* is not as well known. The positive impact of *E. coli* on IAP is suggestive of such benefits. The interaction between IAP levels and *E. coli* has also been published in an in vivo model wherein wild-type mice showed higher levels of *Lactobacillaceae* and commensal *E. coli* populations compared to IAP knock-out mice. Upon oral administration of IAP, the colonization of *E. coli* increased in the IAP knock-out population [39]. The food model and TiO_2 did not cause any changes in endogenous IAP activity compared to the control condition. However, as previously reported [32], IAP was significantly reduced with the addition of *L. rhamnosus, E. coli, B. bifidum,* and *E. faecalis* (Figure 2), indicative of the changing gut environment with the addition of bacteria. The five bacterial species used in this study were combined to form a bacterial mock community (BM) at a physiologically relevant ratio of the bacterial population [40,41]. The co-culture of mammalian and bacterial cells was allowed for a short duration (4 h), during which the experiments were carried out. The BM did not significantly alter the human cell monolayer permeability (Figure 1G). As opposed to the change in IAP due to individual bacterial species, the bacterial mock community maintained the IAP at the same levels as no bacteria control in most cases (Figure 2F), suggesting that as a community, the bacteria did not negatively impact the gut environment.

The human small intestine, as the primary site of nutrient absorption, can lead to nutritional deficiencies and malnutrition due to any disruption in the intestinal barrier. Therefore, factors that influence intestinal health may also impact the body's nutritional availability. The intestinal microbiome plays a significant role in the de novo biosynthesis and bioavailability of several nutrients [42]. In this study, three nutrients were selected to determine the relationship between NPs, diet, and gut microbiota. Glucose transport across the barrier was significantly increased with BM, while protein transport did not show any changes across conditions. Triglycerides do not traverse the cellular barrier due to their hydrophobicity. Instead, they are absorbed through hydrolysis into fatty acids and monoacylglycerol in the intestinal lumen. Intestinal epithelial cells, primarily in the upper villi [43], take up the lipids and resynthesize them to triglycerides, which are secreted out in the form of chylomicrons. One of the components of the food model was fatty acids, which cause higher triglyceride transport in the presence of the FM, with a further increase with TiO_2 in the food matrix. Previous studies show a decrease in fatty acid absorption due to TiO_2 NPs [19,44]; in contrast, our results show that when NPs are part of a food matrix, their effect on TG transport changes. This suggests that while TiO_2 on its own may have a more adverse effect on intestinal permeability in this model, when consumed as part of a regular diet at a controlled level, it does not pose the same risk. However, further studies are needed to study not only TiO_2 absorption but also its accumulation in the body to assess the overall risk factors. The change in transport of each of these nutrients was different in response to the different bacteria. As part of a common community, the bacteria showed positive or no significant changes in protein and TG transport, except for the FM + TiO_2 condition. Glucose transport was enhanced with the bacterial mock community compared to the single bacteria condition (Figure 3). Microbe–microbe interactions can have combined effects on intestinal health. Literature suggests improved blood glucose levels in the presence of beneficial bacteria upon supplementation with pre- and probiotic cultures [45], which is seen in these results as well. Further supporting investigation with respect to insulin sensitivity and immune response would be needed, however, to determine this effect conclusively. Overall, the results confirm the successful incorporation of the bacterial mock community for acute exposure.

The duplication rate and the optimum growth conditions of bacteria differ from one another, but they still co-exist in the human gut. In this model, the individual bacterial populations do not cross the intestinal barrier, except for *E. coli* and *S. salivarius* (Figure 6); however, when in the bacterial mock community, all bacterial species were present in the basolateral chamber (Figure 7). Previously, a study published by Pinget et al. showed that TiO_2 causes a decrease in colonic mucin 2 expression, which causes a reduced mucus layer leading to inflammatory reactions and gut dysbiosis [5,46]. This is also observed in the current study, where *E. coli* attachment on the insert, which can be interpreted as bacterial entrapment in the mucus layer in vivo, was reduced in the presence of TiO_2 (Figure 6B). The acute exposure of the intestinal monolayer to both TiO_2 and a bacterial community could be a contributing factor to the increased presence of bacteria across the barrier. *S. salivarius* is a facultative anaerobe whose presence was undetectable when cultured planktonically in the model due to its unique growth conditions. Since the Caco-2/HT29 experiments were conducted in a 5% CO_2 incubator at 37 °C, the growth conditions were not optimal for *S. salivarius*. When cultured as part of the bacterial mock community, the *S. salivarius* species was able to survive, potentially due to microbe–microbe or microbe–mucosal interactions within the model [47]. *S. salivarius* migrated across the barrier when the monolayer was exposed to TiO_2, but this occurrence was not observed when TiO_2 was used as part of the food model, confirming the varying nature of NPs as a part of the food matrix. As a single species, *B. bifidum*, a strict anaerobe with a slow duplication rate, was detected only on the insert and not in the apical or basolateral chamber. Its presence on the inserts confirms that the effects on permeability and transport we see in this study are in fact due to its ability to attach to the monolayer. As part of the mock community, *B. bifidum* survival was enhanced, confirming the importance of using a bacterial mock community to get a more physiological representation. The attachment of all the bacterial species to the insert and their presence in the apical chamber confirm the successful incorporation of the bacterial community into the model.

5. Conclusions

The complex relationship between diet, intestinal health, and the microbiome is not fully understood or established. The impact of food additives is widely studied; however, each study consists of different experimental parameters, which prevents clear comparisons. In this work, we present a small intestinal model together with a standardized food model composed of all the components in a regular diet, which can be used as a base to study the independent contribution of dietary and environmental factors to gut health. As part of this food model, food-grade TiO_2 did not cause a negative effect on the gut permeability but led to changes in bacterial entrapment and passage across the monolayer, indicative of its involvement in gut dysbiosis. We also showed the successful incorporation of five different bacterial species as a bacterial mock community, which maintained intestinal permeability and IAP levels and improved glucose transport across the barrier. An in vitro model with the capability to incorporate dietary and microbial interactions can be used not just for nutritional studies but also to build disease or malnourishment models with the possibility of studying how changes in microbial diversity counter those conditions.

Author Contributions: Conceptualization, C.N.H.M. and G.J.M.; methodology, M.M., J.V.T., C.N.H.M. and G.J.M.; data curation, M.M. and J.V.T.; data analysis, M.M., J.V.T. and S.K.D.; writing—original draft preparation, M.M. and J.V.T.; writing—review and editing, M.M., C.N.H.M. and G.J.M.; project administration, C.N.H.M. and G.J.M.; funding acquisition, C.N.H.M. and G.J.M. All authors have read and agreed to the published version of the manuscript.

Funding: This work was supported by the National Institutes of Health (1R01ES028788).

Data Availability Statement: The data presented in this study are available on request from the corresponding author.

Conflicts of Interest: The authors declare no conflict of interest.

References

1. Patsos, G.; Corfield, A. Management of the Human Mucosal Defensive Barrier: Evidence for Glycan Legislation. *Biol. Chem.* **2009**, *390*, 581–590. [CrossRef] [PubMed]
2. Finegold, S.M. Intestinal Bacteria—The Role They Play in Normal Physiology, Pathologic Physiology, and Infection. *Calif. Med.* **1969**, *110*, 455–459.
3. Chalova, V.I.; Sirsat, S.A.; O'Bryan, C.A.; Crandall, P.G.; Ricke, S.C. Escherichia Coli, an Intestinal Microorganism, as a Biosensor for Quantification of Amino Acid Bioavailability. *Sensors* **2009**, *9*, 7038–7057. [CrossRef] [PubMed]
4. Maes, M.; Kubera, M.; Leunis, J.-C.; Berk, M. Increased IgA and IgM Responses against Gut Commensals in Chronic Depression: Further Evidence for Increased Bacterial Translocation or Leaky Gut. *J. Affect. Disord.* **2012**, *141*, 55–62. [CrossRef] [PubMed]
5. Pinget, G.; Tan, J.; Janac, B.; Kaakoush, N.O.; Angelatos, A.S.; O'Sullivan, J.; Koay, Y.C.; Sierro, F.; Davis, J.; Divakarla, S.K.; et al. Impact of the Food Additive Titanium Dioxide (E171) on Gut Microbiota-Host Interaction. *Front. Nutr.* **2019**, *6*, 57. [CrossRef]
6. Kiela, P.R.; Ghishan, F.K. Physiology of Intestinal Absorption and Secretion. *Best Pract. Res. Clin. Gastroenterol.* **2016**, *30*, 145–159. [CrossRef]
7. Chacko, A.; Cummings, J.H. Nitrogen Losses from the Human Small Bowel: Obligatory Losses and the Effect of Physical Form of Food. *Gut* **1988**, *29*, 809–815. [CrossRef]
8. Turnbaugh, P.J.; Ridaura, V.K.; Faith, J.J.; Rey, F.E.; Knight, R.; Gordon, J.I. The Effect of Diet on the Human Gut Microbiome: A Metagenomic Analysis in Humanized Gnotobiotic Mice. *Sci. Transl. Med.* **2009**, *1*, 6ra14. [CrossRef]
9. Wang, H.; Du, L.-J.; Song, Z.-M.; Chen, X.-X. Progress in the Characterization and Safety Evaluation of Engineered Inorganic Nanomaterials in Food. *Nanomedicine* **2013**, *8*, 2007–2025. [CrossRef]
10. DeLoid, G.M.; Wang, Y.; Kapronezai, K.; Lorente, L.R.; Zhang, R.; Pyrgiotakis, G.; Konduru, N.V.; Ericsson, M.; White, J.C.; De La Torre-Roche, R.; et al. An Integrated Methodology for Assessing the Impact of Food Matrix and Gastrointestinal Effects on the Biokinetics and Cellular Toxicity of Ingested Engineered Nanomaterials. *Part. Fibre Toxicol.* **2017**, *14*, 40. [CrossRef]
11. Bettini, S.; Boutet-Robinet, E.; Cartier, C.; Coméra, C.; Gaultier, E.; Dupuy, J.; Naud, N.; Taché, S.; Grysan, P.; Reguer, S.; et al. Food-Grade TiO_2 Impairs Intestinal and Systemic Immune Homeostasis, Initiates Preneoplastic Lesions and Promotes Aberrant Crypt Development in the Rat Colon. *Sci. Rep.* **2017**, *7*, 40373. [CrossRef] [PubMed]

12. Winkler, H.C.; Notter, T.; Meyer, U.; Naegeli, H. Critical Review of the Safety Assessment of Titanium Dioxide Additives in Food. *J. Nanobiotechnology* **2018**, *16*, 51. [CrossRef] [PubMed]
13. De Belder, T. *Titanium Dioxide Banned as a Food Additive in the EU*; United States Department of Agriculture, Foreign Agricultural Service USEU: Brussels, Belgium, 2022.
14. Talbot, P.; Radziwill-Bienkowska, J.M.; Kamphuis, J.B.J.; Steenkeste, K.; Bettini, S.; Robert, V.; Noordine, M.-L.; Mayeur, C.; Gaultier, E.; Langella, P.; et al. Food-Grade TiO_2 Is Trapped by Intestinal Mucus in Vitro but Does Not Impair Mucin O-Glycosylation and Short-Chain Fatty Acid Synthesis in Vivo: Implications for Gut Barrier Protection. *J. Nanobiotechnol.* **2018**, *16*, 53. [CrossRef] [PubMed]
15. Cao, X.; Han, Y.; Gu, M.; Du, H.; Song, M.; Zhu, X.; Ma, G.; Pan, C.; Wang, W.; Zhao, E.; et al. Foodborne Titanium Dioxide Nanoparticles Induce Stronger Adverse Effects in Obese Mice than Non-Obese Mice: Gut Microbiota Dysbiosis, Colonic Inflammation, and Proteome Alterations. *Small Weinh. Bergstr. Ger.* **2020**, *16*, e2001858. [CrossRef] [PubMed]
16. Kolba, N.; Guo, Z.; Olivas, F.M.; Mahler, G.J.; Tako, E. Intra-Amniotic Administration (Gallus Gallus) of TiO_2, SiO_2, and ZnO Nanoparticles Affect Brush Border Membrane Functionality and Alters Gut Microflora Populations. *Food Chem. Toxicol. Int. J. Publ. Br. Ind. Biol. Res. Assoc.* **2020**, *135*, 110896. [CrossRef]
17. Cho, W.-S.; Kang, B.-C.; Lee, J.K.; Jeong, J.; Che, J.-H.; Seok, S.H. Comparative Absorption, Distribution, and Excretion of Titanium Dioxide and Zinc Oxide Nanoparticles after Repeated Oral Administration. *Part. Fibre Toxicol.* **2013**, *10*, 9. [CrossRef]
18. Zhang, Z.; Zhang, R.; Xiao, H.; Bhattacharya, K.; Bitounis, D.; Demokritou, P.; McClements, D.J. Development of a Standardized Food Model for Studying the Impact of Food Matrix Effects on the Gastrointestinal Fate and Toxicity of Ingested Nanomaterials. *NanoImpact* **2019**, *13*, 13–25. [CrossRef]
19. Guo, Z.; Martucci, N.J.; Olivas, F.M.; Tako, E.; Mahler, G.J. Titanium Dioxide Nanoparticle Ingestion Alters Nutrient Absorption in an In Vitro Model of the Small Intestine. *NanoImpact* **2017**, *5*, 70–82. [CrossRef]
20. Malik, M.; Subedi, S.; Marques, C.N.H.; Mahler, G.J. Bacteria Remediate the Effects of Food Additives on Intestinal Function in an In Vitro Model of the Gastrointestinal Tract. *Front. Nutr.* **2020**, *7*, 131. [CrossRef] [PubMed]
21. Glahn, R.P.; Lee, O.A.; Yeung, A.; Goldman, M.I.; Miller, D.D. Caco-2 Cell Ferritin Formation Predicts Nonradiolabeled Food Iron Availability in an In Vitro Digestion/Caco-2 Cell Culture Model. *J. Nutr.* **1998**, *128*, 1555–1561. [CrossRef]
22. Cruz, N.; Qi, L.; Alvarez, X.; Berg, R.D.; Deitch, E.A. The Caco-2 Cell Monolayer System as an In Vitro Model for Studying Bacterial-Enterocyte Interactions and Bacterial Translocation. *J. Burn Care Rehabil.* **1994**, *15*, 207–212. [CrossRef] [PubMed]
23. Dayrell, R.L.C.; Cawthray, G.R.; Lambers, H.; Ranathunge, K. Using Activated Charcoal to Remove Substances Interfering with the Colorimetric Assay of Inorganic Phosphate in Plant Extracts. *Plant Soil* **2022**, *476*, 755–764. [CrossRef]
24. Benjamini, Y.; Hochberg, Y. Controlling the False Discovery Rate: A Practical and Powerful Approach to Multiple Testing. *J. R. Stat. Soc. Ser. B Methodol.* **1995**, *57*, 289–300. [CrossRef]
25. R Core Team. *R: A Language and Environment for Statistical Computing*; Global Core Biodata Resource: Copenhagen, Denmark, 2019.
26. Pollard, K.S.; Dudoit, S.; van der Laan, M.J. Multiple Testing Procedures: The Multtest Package and Applications to Genomics. In *Bioinformatics and Computational Biology Solutions Using R and Bioconductor*; Gentleman, R., Carey, V.J., Huber, W., Irizarry, R.A., Dudoit, S., Eds.; Statistics for Biology and Health; Springer: New York, NY, USA, 2005; pp. 249–271, ISBN 978-0-387-29362-2.
27. Zeuthen, L.H.; Fink, L.N.; Frokiaer, H. Epithelial Cells Prime the Immune Response to an Array of Gut-Derived Commensals towards a Tolerogenic Phenotype through Distinct Actions of Thymic Stromal Lymphopoietin and Transforming Growth Factor-Beta. *Immunology* **2008**, *123*, 197–208. [CrossRef]
28. Brun, E.; Barreau, F.; Veronesi, G.; Fayard, B.; Sorieul, S.; Chanéac, C.; Carapito, C.; Rabilloud, T.; Mabondzo, A.; Herlin-Boime, N.; et al. Titanium Dioxide Nanoparticle Impact and Translocation through Ex Vivo, in Vivo and in Vitro Gut Epithelia. *Part. Fibre Toxicol.* **2014**, *11*, 13. [CrossRef] [PubMed]
29. Li, Y.; Jiang, K.; Cao, H.; Yuan, M.; Xu, F. Influences of a Standardized Food Matrix and Gastrointestinal Fluids on the Physicochemical Properties of Titanium Dioxide Nanoparticles. *RSC Adv.* **2021**, *11*, 11568–11582. [CrossRef] [PubMed]
30. Zhang, C.; Li, Y.; Liu, L.; Gong, Y.; Xie, Y.; Cao, Y. Chemical Structures of Polyphenols That Critically Influence the Toxicity of ZnO Nanoparticles. *J. Agric. Food Chem.* **2018**, *66*, 1714–1722. [CrossRef]
31. Ersöz, N.; Çanga, E.M.; Yildirim-Elikoglu, S.; Öz, H.; Dudak, F.C. Effect of Real Food Matrix on the Behavior and Toxicity of TiO2 Nanoparticles. *J. Nanoparticle Res.* **2022**, *24*, 211. [CrossRef]
32. García-Rodríguez, A.; Moreno-Olivas, F.; Marcos, R.; Tako, E.; Marques, C.N.H.; Mahler, G.J. The Role of Metal Oxide Nanoparticles, Escherichia Coli, and Lactobacillus Rhamnosus on Small Intestinal Enzyme Activity. *Environ. Sci. Nano* **2020**, *7*, 3940–3964. [CrossRef]
33. Glynn, J. Tropical Sprue—Its Aetiology and Pathogenesis. *J. R. Soc. Med.* **1986**, *79*, 599–606. [CrossRef]
34. Ahrne, S.; Hagslatt, M.-L.J. Effect of Lactobacilli on Paracellular Permeability in the Gut. *Nutrients* **2011**, *3*, 104–117. [CrossRef] [PubMed]
35. Chen, L.; Li, H.; Li, J.; Chen, Y.; Yang, Y. Lactobacillus Rhamnosus GG Treatment Improves Intestinal Permeability and Modulates Microbiota Dysbiosis in an Experimental Model of Sepsis. *Int. J. Mol. Med.* **2019**, *43*, 1139–1148. [CrossRef]
36. Judkins, T.C.; Archer, D.L.; Kramer, D.C.; Solch, R.J. Probiotics, Nutrition, and the Small Intestine. *Curr. Gastroenterol. Rep.* **2020**, *22*, 2. [CrossRef] [PubMed]
37. Sava, I.G.; Heikens, E.; Huebner, J. Pathogenesis and Immunity in Enterococcal Infections. *Clin. Microbiol. Infect. Off. Publ. Eur. Soc. Clin. Microbiol. Infect. Dis.* **2010**, *16*, 533–540. [CrossRef] [PubMed]

38. Estaki, M.; DeCoffe, D.; Gibson, D.L. Interplay between Intestinal Alkaline Phosphatase, Diet, Gut Microbes and Immunity. *World J. Gastroenterol.* **2014**, *20*, 15650–15656. [CrossRef]
39. Malo, M.S.; Moaven, O.; Muhammad, N.; Biswas, B.; Alam, S.N.; Economopoulos, K.P.; Gul, S.S.; Hamarneh, S.R.; Malo, N.S.; Teshager, A.; et al. Intestinal Alkaline Phosphatase Promotes Gut Bacterial Growth by Reducing the Concentration of Luminal Nucleotide Triphosphates. *Am. J. Physiol. Gastrointest. Liver Physiol.* **2014**, *306*, G826–G838. [CrossRef] [PubMed]
40. Sekirov, I.; Russell, S.L.; Antunes, L.C.M.; Finlay, B.B. Gut Microbiota in Health and Disease. *Physiol. Rev.* **2010**, *90*, 859–904. [CrossRef]
41. O'Hara, A.M.; Shanahan, F. The Gut Flora as a Forgotten Organ. *EMBO Rep.* **2006**, *7*, 688–693. [CrossRef]
42. Hadadi, N.; Berweiler, V.; Wang, H.; Trajkovski, M. Intestinal Microbiota as a Route for Micronutrient Bioavailability. *Curr. Opin. Endocr. Metab. Res.* **2021**, *20*, 100285. [CrossRef]
43. Yen, C.-L.E.; Nelson, D.W.; Yen, M.-I. Intestinal Triacylglycerol Synthesis in Fat Absorption and Systemic Energy Metabolism. *J. Lipid Res.* **2015**, *56*, 489–501. [CrossRef]
44. Chen, Z.; Han, S.; Zheng, P.; Zhou, D.; Zhou, S.; Jia, G. Effect of Oral Exposure to Titanium Dioxide Nanoparticles on Lipid Metabolism in Sprague-Dawley Rats. *Nanoscale* **2020**, *12*, 5973–5986. [CrossRef] [PubMed]
45. Utzschneider, K.M.; Kratz, M.; Damman, C.J.; Hullarg, M. Mechanisms Linking the Gut Microbiome and Glucose Metabolism. J. Clin. Endocrinol. Metab. **2016**, *101*, 1445–1454. [CrossRef] [PubMed]
46. Limage, R.; Tako, E.; Kolba, N.; Guo, Z.; García-Rodríguez, A.; Marques, C.N.H.; Mahler, G.J. TiO_2 Nanoparticles and Commensal Bacteria Alter Mucus Layer Thickness and Composition in a Gastrointestinal Tract Model. *Small* **2020**, *16*, 2000601. [CrossRef] [PubMed]
47. Coyte, K.Z.; Rakoff-Nahoum, S. Understanding Competition and Cooperation within the Mammalian Gut Microbiome. *Curr. Biol.* **2019**, *29*, R538–R544. [CrossRef]

microorganisms MDPI

Review

Advances in Antimicrobial Peptide Discovery via Machine Learning and Delivery via Nanotechnology

Alexa Sowers [1,2], Guangshun Wang [3], Malcolm Xing [4] and Bingyun Li [1,*]

1 Department of Orthopaedics, School of Medicine, West Virginia University, Morgantown, WV 26506, USA
2 School of Pharmacy, West Virginia University, Morgantown, WV 26506, USA
3 Department of Pathology and Microbiology, College of Medicine, University of Nebraska Medical Center, 985900 Nebraska Medical Center, Omaha, NE 68198, USA
4 Department of Mechanical Engineering, University of Manitoba, Winnipeg, MB R3T 2N2, Canada
* Correspondence: bili@hsc.wvu.edu; Tel.: +1-304-293-1075

Abstract: Antimicrobial peptides (AMPs) have been investigated for their potential use as an alternative to antibiotics due to the increased demand for new antimicrobial agents. AMPs, widely found in nature and obtained from microorganisms, have a broad range of antimicrobial protection, allowing them to be applied in the treatment of infections caused by various pathogenic microorganisms. Since these peptides are primarily cationic, they prefer anionic bacterial membranes due to electrostatic interactions. However, the applications of AMPs are currently limited owing to their hemolytic activity, poor bioavailability, degradation from proteolytic enzymes, and high-cost production. To overcome these limitations, nanotechnology has been used to improve AMP bioavailability, permeation across barriers, and/or protection against degradation. In addition, machine learning has been investigated due to its time-saving and cost-effective algorithms to predict AMPs. There are numerous databases available to train machine learning models. In this review, we focus on nanotechnology approaches for AMP delivery and advances in AMP design via machine learning. The AMP sources, classification, structures, antimicrobial mechanisms, their role in diseases, peptide engineering technologies, currently available databases, and machine learning techniques used to predict AMPs with minimal toxicity are discussed in detail.

Keywords: antibiotic resistance; antimicrobial peptide; drug delivery; LL-37; machine learning; nanotechnology; peptide engineering; peptide database; peptide design

Citation: Sowers, A.; Wang, G.; Xing, M.; Li, B. Advances in Antimicrobial Peptide Discovery via Machine Learning and Delivery via Nanotechnology. *Microorganisms* **2023**, *11*, 1129. https://doi.org/10.3390/microorganisms11051129

Academic Editor: Kamel Ahmed Abd-Elsalam

Received: 1 March 2023
Revised: 13 April 2023
Accepted: 19 April 2023
Published: 26 April 2023

1. Introduction

The discovery of antibiotics is one of the greatest achievements in modern medicine. However, over time, antibiotics have become less effective due to the emergence of drug-resistant bacteria from the overuse and abuse of antibiotics [1]. In clinical isolates, antibiotic resistance mutations have been found in promoter regions of efflux pumps, the target of antibiotics, and binding regions of antibiotics. Antibiotic resistance, where a mutation or post-translational modification alters the target or inactivates the antibiotic [2,3], has resulted in treatment failure [3]. The Centers for Disease Control and Prevention (CDC) reports more than 2.8 million infections and over 35,000 deaths due to antibiotic resistance in the United States [4]. Therefore, discovering new antimicrobial agents to overcome this challenge is essential [5].

Antimicrobial peptides (AMPs) constitute an important source for developing a new generation of antibiotics. AMPs have a broad range of activity against microorganisms, including bacteria [6], fungi [7], parasites [8], and viruses [9,10]. These peptides play a critical role in innate immunity by responding to a variety of pathogen-associated molecular patterns through increased leukocyte recruitment to the infection site and signaling damaged tissues [11,12]. AMPs also are involved in the adaptive immune response as they have demonstrated the ability to recruit immature phagocytic and dendritic cells [13].

AMPs may exhibit other properties, including the ability to target drug-resistant microbial biofilms, kill cancer cells, and promote wound healing [14]. Biofilm inhibition is an advantageous property of AMPs, as biofilms have led to increased tolerance to various antibiotics and have accounted for about two-thirds of all human infections [13]. In the case of chronic wounds, the formation of biofilms could delay wound healing. Hence, the effective disruption of preformed biofilms by AMPs constitutes an important step toward wound healing [15–18].

AMPs have been discovered in six life kingdoms: bacteria, archaea, protists, fungi, plants, and animals [19]. Most natural AMPs (74%) originated from animals, especially amphibians and insects [20,21]. They share some common characteristics, such as cationic charge (+1 to +7), short sequences (<50 amino acids), and they are usually amphiphilic [13,22]. The net positive charge results from the abundance of arginine, lysine, and histidine residues due to the positive charges associated with these residues [23]. The presence of ~50% hydrophobic and hydrophilic residues contributes to these peptides' amphiphilicity, enabling them to interact with bacterial membranes and to enter cells [13]. These peptides can fold into different structures, including alpha helices, beta-sheets, combined alpha helix and beta-sheets, and non-alpha-beta structures [24]. Amphipathic alpha-helical structures, observed for magainin and human cathelicidin LL-37 for instance, provide a basis for understanding the mechanism of action of AMPs [25–27].

AMPs present selectivity toward bacteria over eukaryotic cells. AMPs' selectivity toward bacterial membranes over eukaryotic membranes is due to differences in the composition of the membranes. Eukaryotic membranes have primarily zwitterionic lipids, including phosphatidylcholine and sphingomyelin, which are usually neutral. In contrast, bacterial membranes are composed of anionic lipids, such as phosphatidylglycerol and cardiolipin [28]. The favored interaction between cationic AMPs and anionic bacterial membranes is responsible for peptide selectivity.

It should be pointed out that such a preferred interaction with bacteria does not mean AMPs will never interact with host cells. AMPs, such as human cathelicidin and defensins, do associate with host cell receptors (e.g., GPCRs, MrgX2, FPR2, and P2X7) at a peptide concentration much below the minimal inhibitory concentration (MIC) [29]. Such interactions are not cell damaging; rather, they induce signal transduction for immune regulation. Such a role of AMPs links them with a variety of physiological processes, including human diseases.

2. AMPs and their Relationships with Human Diseases

2.1. Role in Respiratory Diseases

AMPs provide defense mechanisms for multiple diseases, including respiratory diseases. Rai et al. reviewed the preparation of various AMP-based materials and their antimicrobial properties against infections in the brain, eye, mouth, skin, lung, etc. [30]. Due to their role in adaptive and innate immunity, AMPs can protect against pathogens in various organs [31]. The respiratory tract is lined with epithelial cells to protect against potential infections [13]. Upon infection, epithelial cells, neutrophils, and macrophages populate and can secrete AMPs, which act intracellularly to kill invading pathogens [32]. Cathelicidin LL-37, a natural human AMP, and cathelicidin-related AMP (CRAMP) demonstrated antiviral activity against influenza A virus [33,34]. LL-37 showed effectiveness against influenza A virus both in vitro and in vivo, with a 90% reduction in virus titer in vitro and a significant 70–80% reduction in lung virus three days post-infection in mice [34]. AMPs have demonstrated their role in protecting against respiratory infections, but they are also involved in non-infectious lung diseases [32].

In various respiratory diseases, such as chronic obstructive pulmonary disease (COPD) and cystic fibrosis, studies have presented different AMP expression (overexpression or down-regulation) compared to non-diseased states [13]. For instance, some AMPs, such as LL-37, are overexpressed in respiratory diseases, which may influence disease progression. Increased levels of LL-37 correlated with bronchial inflammation, which affected disease severity [35]. The overexpression of LL-37 in the airways of COPD patients stimulated the overproduction of mucus, where this mucus contributed to disease progression [36]. LL-37 is a neutrophil-derived AMP, which can further disease progression in the lungs due to its inflammatory activity and cytotoxicity at relatively high concentrations [32]. Smokers with COPD, a disease with limited airflow due to toxic particles or gases, had reduced beta defensin-2 (hBD-2) compared to smokers without COPD and ex-smokers with COPD [37,38]. Decreased concentrations of hBD-2 were also associated with disease severity in cystic fibrosis patients [35]. Diseases involving decreased hBD-2 production could correspond to an increased susceptibility to infections by different pathogens [13].

2.2. Role in Autoimmune Diseases

Studies have shown that the expression of AMPs plays a role in autoimmune disease pathogenesis. Psoriasis, rheumatoid arthritis, and Crohn's disease are autoimmune diseases that have shown altered AMP expression [39–41]. Psoriasis is an autoimmune skin condition caused by innate and adaptive immunity deviations induced by infection, injury, and stimulation [42]. Patients diagnosed with psoriasis have increased expression levels of AMPs, such as LL-37 and human β-defensins, compared to healthy individuals [43]. Studies demonstrated that LL-37 could act as an autoantigen by circulating T cells and can complex with self-RNA, which activates Toll-like receptors (TLR), leading to the exacerbation of psoriasis [44,45]. hBDs also play a role in the development of psoriasis as Rohrl et al. demonstrated the ability of hBD-2 to act as a ligand for chemokine receptor (CCR6), where this CCR6 signal induced Th17 in patients with psoriasis [46,47].

Rheumatoid arthritis is a disease that leads to chronic inflammation and joint destruction by producing disease-specific-anticitrullinated protein antibodies and complexing with citrullinated fibrinogen [48]. Francisco et al. discovered elevated liver-expressed AMP 2 (LEAP2) levels in rheumatoid arthritis patients compared to healthy individuals. This evidence indicates that LEAP2 is associated with inflammatory mediators in rheumatoid arthritis [49]. hBD-3 also played a role in the pathogenesis of rheumatoid arthritis, indicated by the destruction of the cartilage due to the activation of matrix metalloproteinases, resulting in the degradation of the extracellular matrix cartilage [50,51].

Crohn's disease is a chronic condition described by relapsing and remitting patchy inflammation along any section of the gastrointestinal tract [52]. Gutiérrez et al. discovered the upregulated expression of LL-37 in peripheral blood neutrophils in Crohn's disease patients [53]. Another study demonstrated increased LL-37 expression in the inflamed mucosa of patients with Crohn's disease. In Crohn's disease, LL-37 may enhance antibacterial properties, specifically in the inflamed mucosa, which may protect against infection [54]. Wehkamp et al. determined that Crohn's disease patients have reduced antibacterial activity in the intestinal mucosa and have shown a decreased expression in human α-defensin 5 (HD5) in the ileum. This decreased α-defensin expression may be a factor in the pathogenesis of Crohn's disease through inflammation in the ileum, leading to compromised innate immunity [55].

Altered expression levels of AMP in the gastrointestinal tract have been found in diabetes. Liang et al. determined that cathelicidin-related antimicrobial peptide (CRAMP) expression was decreased in the colon of non-obese mice, which led to the production of type I interferons. This production led to the promotion of the pancreatic autoimmune response, accelerating the progression of diabetes. However, when CRAMP expression was increased, it restored homeostasis, which prevented the diabetogenic response [56]. Although type 2 diabetes is not an autoimmune disease, this disease has been associated with a reduced expression of β-defensin [57]. Lin et al. determined that hBD-1, human beta defensin 1, was reduced in the gastric corpus in the diabetic metabolic state. Given this information, the induction of AMP expression may help resolve issues involved in intestinal barrier deficiency [58].

2.3. Role in Cancer

Vitamin D receptor (VDR) plays a role in regulating the transcriptional profile of many genes involved in physiological functions, such as regulating immune activity [13]. VDR has been shown to be one of the resistance mechanisms to pathogens. Due to the dysregulation of the VDR, AMP expression levels alter, and cancer biology is influenced through altered DNA methylation patterns [59]. Various AMPs, such as α-defensins, β-defensins, and LL-37, are involved in developing a variety of tumors and cancers, where the VDR regulates these peptides [13]. Defensins are AMPs produced by eukaryotes, where some defensin-like peptides have antiproliferative activity and apoptosis in cancer cells, demonstrated by the increased phosphorylation of MAPK p38 [60,61]. Due to this ability, defensins can be used in combination therapy to overcome chemotherapeutic resistance. For instance, Johnstone et al. determined that defensins could increase the in vitro anticancer activity of doxorubicin, an anticancer drug, against multidrug-resistant tumor cells [62]. Human neutrophil peptide-1 (HNP-1) is another alpha defensin that has shown anticancer activity in addition to antimicrobial activity [63]. Gaspar et al. established that HNP-1 induced apoptosis with defects in the cellular membrane at a low concentration of peptide [64].

Likely due to the exposure of anionic phosphatidylserine (PS), cancer cells have an increased sensitivity to cationic AMPs compared to normal cells. This increased sensitivity results from an undeveloped cytoskeleton seen in cancer cells [1]. Cancer cells have a high metabolism, which allows this change in the cytoskeleton. As a result, AMPs can interact with the cancer cell membrane [1,13]. Interestingly, LL-37 has demonstrated the ability to produce tumorigenic or anticancer effects, depending on the type of cancer involved [65]. For instance, the expression of LL-37 is down-regulated in gastric and pancreatic cancer, indicating that LL-37 can have anticancer effects. Zhang et al. showed a high concentration of LL-37 at 20 mg/kg with a 42% reduction in pancreatic tumor growth compared to the control. Results have suggested that the LL-37-induced inhibition of autophagy results in the accumulation of reactive oxygen species (ROS) and inhibits pancreatic cancer cell growth [66]. Wu et al. determined that LL-37 inhibited the proliferation of gastric cancer cells by activating bone morphogenetic protein signaling through a proteasome-dependent mechanism [67]. Of great interest is that the major antimicrobial peptide (FK-16) of LL-37, originally discovered by NMR-trim, is demonstrated to have anticancer properties as well [68,69].

LL-37 has also been shown to present tumorigenic effects in some cases. Haussen et al. showed LL-37 expression in human lung cancer cells; the overexpression of LL-37 in cancer cells resulted in an increased proliferation. This study also showed that murine models had similar results, where cells overexpressing LL-37 developed larger tumors compared to mice without the overexpression of LL-37 [70]. The overexpression of cathelicidins has also played a role in the pathogenesis of breast cancer. Weber et al. determined that the treatment of LL-37 stimulated breast cancer cell migration and their colonies had a dispersed morphology, which indicated increased metastatic potential. The overexpression of human cathelicidin protein (hCAP18) in low malignant breast cancer cell lines induced metastases in severe combined immunodeficiency mice. Therefore, the overexpression of hCAP18 and LL-37 in certain cancers may serve as a useful marker for cancer diagnosis [71].

2.4. Role in Cardiovascular Diseases

Cardiovascular disease contributes to worldwide mortality, including conditions such as atherosclerosis and heart failure [72]. Atherosclerosis is the accumulation of plaque in blood vessels and is one of the leading causes of cardiovascular diseases [73]. Edfeldt et al. discovered the expression of LL-37 in atherosclerotic lesions, which modulates inflammation expression. LL-37 activated adhesion molecules and chemokines, resulting in leukocyte recruitment and atherogenesis [74,75]. Salamah et al. determined the presence of LL-37 in platelets, which secreted upon activation and promoted thrombus formation.

The formation of a thrombus led to further complications, such as blood clots, resulting in heart failure [76].

Heart failure is an end-stage condition that results from complications associated with cardiovascular disease, which is identified as the leading cause [77]. Zhou et al. demonstrated the decreased expression of LL-37 in heart failure patients and decreased CRAMP in heart and serum samples of heart failure mice models. In their study, CRAMP supplementation suppressed cardiac hypertrophy, indicated by reduced expression levels of atrial natriuretic peptide and B-type natriuretic peptide. The presence of LL-37 in serum suggested that this AMP might be used as a biomarker for acute heart failure [78]. For instance, Bei et al. demonstrated that a lower LL-37/neutrophil ratio could predict a worse prognosis for myocardial infarction patients [79].

2.5. Role in Neurodegenerative Diseases

Neuroinflammation results from brain injuries or diseases released from activated microglia, astrocytes, and cytokines. Neurodegeneration is associated with responses due to inflammation resulting from over-activated glial cells [80]. Lee et al. showed a relative upregulation of LL-37 in the brain compared to other organs, indicating that the LL-37 expression level might be associated with chronically diseased areas in the brain involved in Alzheimer's and Parkinson's [80]. The amyloid β-protein (Aβ), another human AMP, influences Alzheimer's disease pathogenesis [81]. Soscia et al. determined that Alzheimer's disease homogenates increased higher antimicrobial activity against microbes compared to non-diseased homogenates. Although Aβ demonstrated its ability to protect Alzheimer's patients from infection, this peptide also contributed to the pathogenesis of the disease [81]. Wang et al. determined that dysregulated AMP expression might be involved in the development of Alzheimer's disease by inducing Aβ deposition. These findings indicated the potential of using AMPs as biomarkers and therapeutic agents for Alzheimer's disease [82].

3. AMPs' Antimicrobial Mechanisms

AMPs kill microorganisms primarily via selectively disrupting the membrane through membrane permeation mechanisms. These mechanisms include the barrel-stave model, carpet model, and toroidal-pore model (Figure 1) [28]. In the barrel-stave model, AMPs are embedded perpendicularly into the microbial membrane, where they self-assemble to form a transmembrane channel [83,84]. Due to their amphipathic nature, the hydrophobic faces orient themselves toward the interior of the microbial membrane, while the hydrophilic faces orient themselves to form the hydrophilic pore (Figure 1b) [85]. In the carpet model, AMPs are oriented parallel to the microbial membrane, where they are absorbed to induce membrane disruption. The membrane disrupts through micellization, degrading the membrane and resulting in microbial cell death, as shown in Figure 1c [86]. The toroidal-pore model is similar to the barrel-stave model, where a pore is formed by inserting AMPs perpendicularly into the membrane. However, this model differs from the barrel-stave model where the bacterial membrane incorporates into the pore. This incorporation induces lipid monolayers to bend, forming a continuous interface with the AMPs (Figure 1d) [28,84]. Generally, AMPs target microbial membranes through hydrophobic and electrostatic interactions, resulting in cell death due to losing cellular components and electrochemical gradients [87,88]. However, evidence suggests that membrane permeation is not the only mechanism for AMPs to eliminate bacteria. Some AMPs, such as lantibiotics, target the bacterial cell wall [89]. AMPs may also translocate across microbial membranes and bind to intracellular targets that may inhibit nucleic acid and protein synthesis, which may result in cell death (Figure 1I–III) [22]. As to Gram-negative bacteria, AMPs may act as an antagonist on the lipopolysaccharide (LPS) of the outer membrane. LPS is an endotoxin that stimulates the secretion of proinflammatory cytokines to modulate immune responses [13]. This endotoxin is essential in the integrity and stability of bacterial cell structures, providing a protective barrier from chemical attack while managing cell permeability [13,90]. AMPs have demonstrated the ability to inhibit LPS-induced cellular responses via TLR4 and may eliminate extracellular LPS through interactions with this endotoxin in various molecules using membrane permeation methods [13,91]. Such an LPS neutralization usually led to an anti-inflammatory effect.

Amino acids are essential elements that contribute to the antimicrobial activity seen in AMPs. Aspartate and glutamate are negatively charged amino acids responsible for binding divalent cations, which is essential for the antibacterial activity of some AMPs such as daptomycin. When these negatively charged amino acids bind to divalent cations, the conformation of the AMPs is altered to enhance the interaction between the AMPs and cell membranes [92,93]. Tryptophan (Trp) is a non-polar amino acid located on the protein that intersects the membrane interface region [9,94]. The preference of tryptophan residing in the membrane interface is likely the result of tryptophan's sidechain indole, rigid structure, and aromaticity, which limits access to the hydrocarbon core [95]. Because of its bulky side chain, Trp is frequently used to enhance peptide activity, especially short ones. Arginine and lysine are basic amino acids which contribute to cationic charge, a characteristic seen in amphipathic AMPs [96]. Histidine residues can function as a proton shuttle, which allows these residues to modify the net positive charge by adjusting the pH, affecting antimicrobial activity [97]. When histidine incorporates into the AMP sequence, it can provide a cationic charge at an acidic pH, but it loses a positive charge at a basic pH and reduces peptide activity [96,98]. Cysteine residues can form disulfide bonds, providing stability against chemical, thermal, and enzymatic degradation [93]. Proline residues are frequently found as a structural breaker in an alpha helix due to its properties including steric hinderance and a lack of hydrogen to participate in helix stabilization [99]. Interestingly, there are proline-rich AMPs that inhibit ribosome assembly [100].

The primary antimicrobial mechanisms of AMPs via membrane disruption have demonstrated advantages over conventional antibiotics. AMPs can kill pathogens faster than traditional antibiotics and have a narrower active concentration window to kill pathogens [101]. Typical antibiotics only act on specific intracellular targets, while AMPs utilize various intracellular and extracellular targets. Since AMPs can act rapidly on pathogens and have various killing mechanisms, it reduces the chance of bacteria developing resistance to AMPs [102].

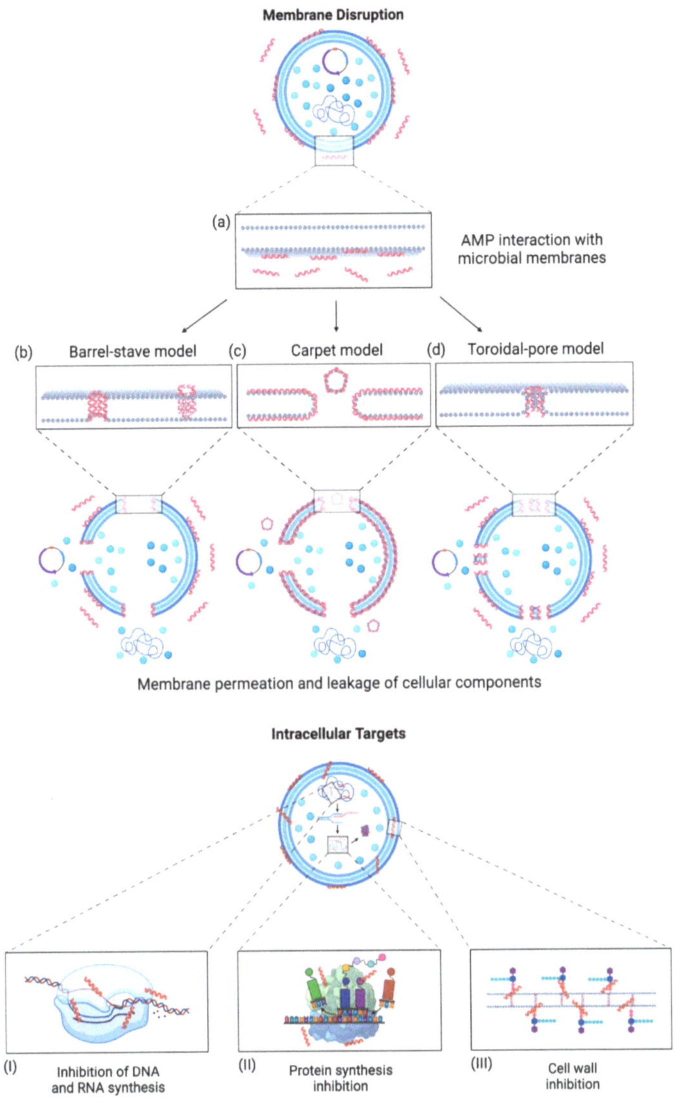

Figure 1. AMPs interact with microbial membranes and their associated mechanisms of action. The mechanisms for membrane permeation begin with (**a**) AMP interaction with microbial membranes, allowing AMPs to insert themselves. (**b**) The barrel-stave model results when AMPs insert perpendicularly into microbial membranes and self-assemble to form pores [83,84]. (**c**) The carpet model works by absorbing AMPs, resulting in disrupting the microbial membranes through micellization [86]. (**d**) The toroidal-pore model results from incorporating AMPs to form a continuous interface with the microbial membrane [28,84]. AMPs can also translocate across microbial membranes and bind to intracellular targets, resulting in cell death. These intracellular targets include (**I**) inhibiting DNA and RNA synthesis, (**II**) inhibiting protein synthesis, and (**III**) inhibition of cell wall synthesis [22]. Adapted from (Bahar et al., 2013) CC BY 3.0 [103]. Created with BioRender.com (accessed on 15 December 2022).

4. Limitations and Strategies for Clinical Applications

There is a growing interest in developing biomedical applications for AMPs due to their unique properties and different antimicrobial mechanisms. It appears that the success of AMPs to a large extent is determined by their structure and biological source. Based on a universal peptide classification scheme [24], AMPs are classified into four classes: linear (UCLL), sidechain-connected (UCSS), sidechain-backbone linked (UCSB), and backbone-connected (UCBB). The latter three belong to cyclic peptides. Because of the stability with proteases, some cyclic AMPs have been approved for clinical applications (Table 1). Linear AMPs, however, face challenges due to issues with protease degradation [104].

4.1. Clinical Applications of AMPs from Prokaryotes

Currently, there are a few Food and Drug Administration (FDA)-approved peptide antibiotics being used clinically to treat patients. Broadly, they belong to natural AMPs. Table 1 summarizes peptide molecular weight, activity, targeted pathogens, mechanism, administration routes, year of approval, and reference. The structures of these peptide antibiotics are depicted in Figure 2. They can be classified into two types: natural and engineered. Natural peptides include bacitracin [105], daptomycin [106], teicoplanin [107], vancomycin [108], colistin [109], and gramicidin [110]. Bacitracin, daptomycin, teicoplanin, and vancomycin primarily inhibit Gram-positive pathogens, while colistin is the last resort antibiotic to treat Gram-negative pathogens. Gramicidin can kill both Gram-positive and -negative pathogens. Bacitracin is a cyclic heptapeptide with its lysine sidechain attached with a short peptide segment. It is used topically to treat skin infections caused by Gram-positive bacteria, such as *Staphylococcus aureus* (*S. aureus*). Daptomycin and colistin also have cyclic structures attached with a fatty acid tail (e.g., lipopeptides). Vancomycin is a sidechain-linked peptide antibiotic for systemic use. Teicoplanin has a structure similar to vancomycin. It is a mixture of multiple molecules that differ in the fatty acid chain structure. Due to a longer half-life that allows less frequent dosing, teicoplanin has an advantage over vancomycin. Additionally, dalbavancin [111], oritavancin [112], and telavancin [113] are semi-synthetic lipoglycopeptides developed to treat Gram-positive bacterial infections where vancomycin does not work well. They all share a heptapeptide core that inhibits transglycosylation and transpeptidation. Remarkably, dalbavancin and oritavancin have a long half-life in the range of 147 to 393 h, making it possible to administer a treatment of one dose per week [114]. Gramicidin D consists of a mixture of gramicidins (80% A, 6% B, and 14% C). These linear peptides are particularly effective against Gram-positive bacteria by forming an ion channel on membranes. Gramicidin can only be used topically to treat eye, nose, throat, and wound infections since it is highly hemolytic. Note that gramicidin S (AP02243 in the APD database) is very different as it possesses a head-to-tail cyclic backbone structure. Overall, it shows higher activity against *S. aureus* and *Enterococcus faecium (E. faecium)* than Gram-negative bacteria [115].

Finally, there are also antiviral peptides (Table 1) engineered based on viral proteins to inhibit viral proteases (e.g., Boceprevir [116] and Telaprevir [117]) or fusion (e.g., Enfuvirtide [118]). Such designed peptides differ from classic AMPs.

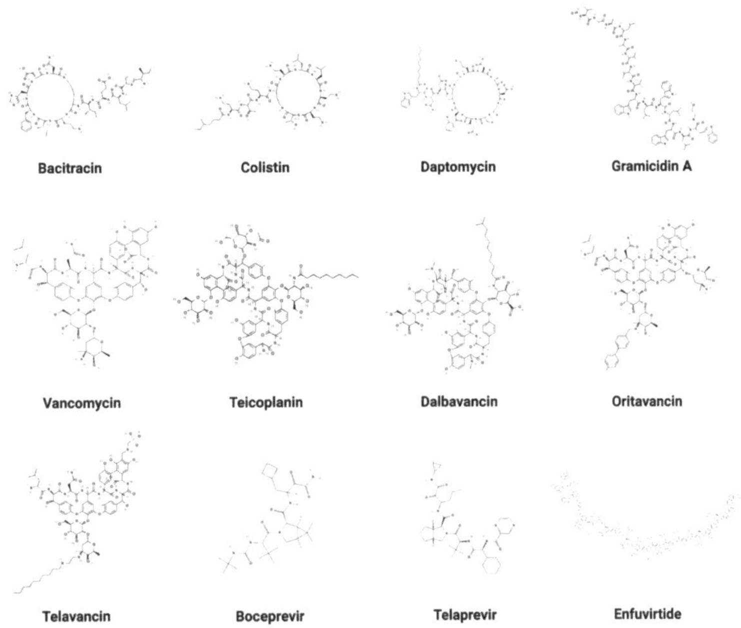

Figure 2. Two-dimensional structures of multiple FDA-approved AMPs. These AMPs include bacitracin [105], colistin [109], daptomycin [106], gramicidin A [110], vancomycin [108], teicoplanin [107], dalbavancin [111], oritavancin [112], telavancin [113], boceprevir [116], telaprevir [117], and enfuvirtide [118]. Created with BioRender.com (accessed on 15 December 2022).

4.2. Eukaryotic AMPs under Investigation for Clinical Applications

Multiple AMPs from eukaryotes, such as insects, amphibians, and mammals, have entered clinical trials but have not received FDA approval. For example, omiganan, a derivative of indolicidin, is a recently transitioned Phase III drug that reduces catheter colonization and acts as an anti-inflammatory agent against rosacea, an auto-inflammatory skin disease [13]. Pexiganan, derived from frog magainin, has also been tested to treat diabetic foot ulcers, but has not yet been approved [119]. Its antimicrobial effect is even better when it is used in combination with nisin due to different mechanisms to induce bacterial killing [120]. Iseganan, a derivative of pig cathelicidin protegrin-1, was tested clinically to treat oral mucositis [121].

More AMPs are under development. For example, DP7 (VQWRIRVAVIRK) is an AMP effective against Gram-negative, Gram-positive, and multidrug-resistant bacteria [91,122]. Studies have investigated the effectiveness of DP7 against biofilm production [91] and infections such as the severe acute respiratory syndrome (SARS) coronavirus [123]. A mutation in *Pseudomonas aeruginosa* (*P. aeruginosa*) blocked the generation of a signal molecule, resulting in hindered differentiation and biofilm formation [124]. Yin et al. investigated the antibiofilm activity of DP7 against a mutated strand of *P. aeruginosa*. Increasing the concentration of DP7 showed a reduction in biofilm biomass, which indicated that DP7 might bind to targets involved in biofilm production. This study showed a biofilm reduction in *P. aeruginosa* of 43% to 68% [91]. Zhang et al. investigated the ability of DP7 in resisting the coronavirus infection on the ACE2 receptor by the coronavirus receptor-binding domain (RBD) and spike protein (S protein). The results from their study showed that a 50% inhibitory concentration (IC50) of DP7 inhibiting SARS-CoV and SARS-CoV-2 S protein pseudovirus-infected ACE2-293T was 104 μg/mL and 74 μg/mL, respectively. An enzyme-linked immunosorbent assay (ELISA) was used to determine if DP7 could inhibit SARS-CoV-2 S-RBD from binding to the ACE2, which showed that DP7 inhibited the combination of the two [123]. There are numerous studies aimed at developing human cathelicidin LL-37 into novel antimicrobials (for a review, see ref. [125]).

4.3. Complications and Potential Solutions

The use of AMPs for clinical applications, unfortunately, faces limitations. One major issue with many AMPs is their hemolytic activity, which is the ability to burst red blood cells in an organism [126]. This AMP property is a significant concern for therapeutics, as hemolysis may cause anemia or death following treatment. AMPs may also be degraded via proteolytic enzymes found in the GI tract, blood serum proteases, and kidney drug clearance [87,126]. Such degradation results in poor oral bioavailability. Moreover, AMPs are relatively large (on average, there are 33 amino acids in the APD), which can increase their manufacturing costs [13]. Although AMPs may reduce the chance of bacteria developing resistance, there are a few resistance mechanisms reported [102,127]. These resistance mechanisms include modifying cell surfaces to reduce the negative charge on their membranes, resulting in the reduced ability of AMPs to bind to these membranes and kill microbes [128]. Other resistance mechanisms include antimicrobial efflux pumps, external trapping, proteases used to degrade peptides, AMP sequestration to block access to the cell membrane, and biofilms [129–131]. There is an increased need for techniques to minimize or remove these limitations associated with AMPs.

New technology and techniques may overcome the challenges facing the applications of AMPs. Nanotechnology, a growing field, has the potential to improve bioavailability, permeability across barriers, protect against harsh environments such as pH and enzymes, and control the release of AMPs [132]. Meanwhile, computational methods can avoid poor peptide candidates initially, as seen with conventional methods of screening peptides using growth media and determining cell toxicity [127]. Rational design, a computer-aided method, of AMPs can reduce the amino acid length of the peptides to perform their desired function against pathogens [133]. This design can reduce the length of AMPs while maintaining their antimicrobial properties, which could help reduce production costs. For example, LL-37 has been reduced to 12 amino acids (KR-12) for peptide engineering, and the recently optimized lipopeptide is composed of only eight amino acids [26,134]. Machine learning models have gained attention due to their cost-effectiveness in peptide discovery [135]. These models can mine through relationships between antimicrobial activity and biochemical features, which help predict AMPs in large-scale environments [136]. Machine learning methods can potentially lead to reduction of toxicity seen in AMPs by modifying the physicochemical features and chemical modifications responsible for toxicity [137].

Although computational methods have helped with obstacles facing AMPs, there are a few computational challenges associated with them. The availability of new sequences such as metagenomes shows their potential to find novel AMP sequences. However, predicting small genes in DNA sequences and AMP activity using homology-based methods is challenging. Due to the limited use of homology-based methods to be directly applied to AMPs, different techniques are required for the application of longer peptides [138]. Santos-Júnior developed Macrel to overcome these obstacles with the use of metagenomes to predict AMP sequences and activity. This method can process metagenomic contigs, metagenomes, or peptides. Macrel can process reads and assemble them into larger contigs, where the small open reading frames (ORFs) are extracted and classified into AMPs or rejected. Since predicting small ORFs with the current methods has issues with false positives, Macrel filters allowed the prediction of high-quality AMP sequences [138]. Huertas-Cepas et al. developed eggNOG-mapper, a tool to annotate large sets of proteins based on orthology assignments from the eggNOG database. This method showed improvements compared to homology-based techniques, revealing that orthologs are effective predictors for AMPs [139].

Table 1. FDA-approved AMP antibiotics.

Name	MW (g/mol)	Activity	Pathogen	Antimicrobial Mechanism	Administration	Year Approved	Ref.
Bacitracin	1460	Gram-positive	*Streptococcus* spp., *Staphylococcus* spp., *Clostridium* spp., *Corynebacterium* spp., and *Actinomyces* spp.	Bacitracin inhibits the synthesis of bacterial cell walls, which prevents the dephosphorylation of the P-P-phospholipid carrier. This carrier binds peptidoglycan precursors to the bacterial cell membrane. Bacitracin absorbs into injured skin to prevent mucopeptides from entering microbial cell walls.	Topical, ophthalmic, intramuscular	1984	[140,141]
Boceprevir	519.7	Antiviral	Hepatitis C	Boceprevir inhibits the hepatitis C virus (HCV) non-structural protein 3 (NS3/4A) protease, important for cleaving HCV polyprotein into its mature protein forms. HCV replication is inhibited through boceprevir binding covalently to the active site on NS3 protease.	Oral	2011	[142]
Colistin	1750	Gram-negative	*A. baumannii*, *P. aeruginosa*, *E. coli*, *Enterobacter*, *Salmonella*, *Klebsiella*, and *Shigella*	Colistin binds through electrostatic interactions to the negatively charged Gram-negative membrane component, lipid A of LPS. Colistin displacing divalent cations of magnesium and calcium and inserting its hydrophobic acyl fatty chain results in membrane lysis.	Intravenous and intramuscular	1962	[5,143,144]
Dalbavancin	1816.7	Gram-positive	*S. aureus*, *Streptococcus pyogenes*, *Streptococcus dysgalactiae*, *Enterococcus faecalis*, *Streptococcus intermedius*, *Streptococcus agalactiae*, *Streptococcus anginosus*, and *Streptococcus constellatus*	Dalbavancin inhibits cell wall synthesis through interactions with the D-alanyl-D-alanine terminus of the pentapeptide in peptidoglycan of cell wall. This binding prevents crosslinking, which is essential for building the bacterial cell wall.	Intravenous	2014	[5,145]
Daptomycin	1620.7	Gram-positive	*S. aureus*, *Enterococcus faecalis*, *Streptococcus dysgalactiae*, *Streptococcus agalactiae*, *Streptococcus pyogenes*, *Corynebacterium jeikeium*, *Enterococcus faecium*, *S. epidermis*, and *S. haemolyticus*	Daptomycin has a lipophilic acyl tail which inserts into bacterial cytoplasmic membrane, resulting in the efflux of potassium and leading to membrane depolarization. This depolarization results in the inhibition of DNA, RNA, and protein synthesis, resulting in membrane lysis.	Intravenous	2003	[5,146,147]

Table 1. *Cont.*

Name	MW (g/mol)	Activity	Pathogen	Antimicrobial Mechanism	Administration	Year Approved	Ref.
Enfuvirtide	4492	Antiviral	HIV-1	Enfuvirtide inhibits the fusion of viral and cellular membranes, which prevents HIV-1 from entering cells. Enfuvirtide binds to the first heptad-repeat (HR1) in the gp41 subunit of the viral envelope glycoprotein, preventing confirmational changes necessary for fusing viral membranes.	Subcutaneous	2003	[148]
Gramicidin D	1811.2	Gram-positive	*Streptococcus pneumoniae*, *S. aureus*, *Haemophilus influenzae*, *P. aeruginosa*, *Streptococcus agalactiae*, *E. coli*, *Neisseria meningitidis*, *Klebsiella*, *Neisseria gonorrhoeae*, and *Enterobacter*	In microbial membranes, two gramicidin molecules form a head-to-head dimeric ion channel in the center, resulting in loss of intracellular solutes, reduction in ATP, and inhibition of DNA and RNA synthesis.	Topical	1955	[5,149]
Oritavancin	1989.1	Gram-positive	*S. aureus*, *Streptococcus intermedius*, *Streptococcus pyogenes*, *Streptococcus dysgalactiae*, *Streptococcus agalactiae*, *Enterococcus faecalis* *Streptococcus anginosus*, and *Streptococcus constellatus*	Oritavancin inhibits biosynthesis of cell walls by binding to the stem peptide of peptidoglycan precursors. Oritavancin may inhibit this biosynthesis by binding to peptide bridging segments. Oritavancin also disrupts the integrity of microbial membranes, resulting in cell death.	Intravenous	2014	[5,150]
Teicoplanin	1900	Gram-positive	*S. aureus*, *S. epidermidis*, *Streptococcus pyogenes*, *Staphylococcus haemolyticus*, *Staphylococcus hominis*, *Staphylococcus saprophyticus*, *Streptococcus bovis*, *Streptococcus pneumoniae*, *Streptococcus agalactiae*, *Streptococcus mitis*, *Streptococcus milled*, *Streptococcus sanguis*, *Clostridium difficile*, *Listeria monocytogenes*, *Clostridium perfringens*, *Propionibacterium acnes*, and *Corynebacterium jeikeium*	Teicoplanin inhibits the peptidoglycan polymerization of the cell wall by binding to Ac_2-L-Lys-D-ala-D-ala, an analog of the peptidoglycan precursor in cell wall biosynthesis. As a result, teicoplanin inhibits cell wall synthesis.	Intravenous or intramuscular	1990	[151,152]
Telaprevir	679.85	Antiviral	Hepatitis C	Telaprevir inhibits the HCV NS3/4A protease, which is important for cleavage of the HCV polyprotein into mature forms necessary for viral replication.	Oral	2011	[153]

Table 1. *Cont.*

Name	MW (g/mol)	Activity	Pathogen	Antimicrobial Mechanism	Administration	Year Approved	Ref.
Telavancin	1755.6	Gram-positive	*S. aureus*, *Streptococcus pyogenes*, *Streptococcus constellatus*, *Streptococcus agalactiae*, *Enterococcus faecalis*, *Streptococcus anginosus*, and *Streptococcus intermedius*	Telavancin binds to the D-alanyl-D-alanine terminus of cell wall precursors and late-stage peptidoglycan precursors such as lipid II. This interaction inhibits the polymerization of precursors into peptidoglycan and crosslinking. The mechanism of telavancin and membrane lysis is unknown.	Intravenous	2009	[5,154,155]
Vancomycin	1485.7	Gram-positive	*S. aureus*, *Streptococcus bovis*, *S. epidermidis*, *Streptococcus pyogenes*, *Listeria monocytogenes*, *Streptococcus agalactiae*, *Clostridium* species, *Actinomyces* species, and *Lactobacillus* species.	Vancomycin inhibits the polymerization of peptidoglycans by binding to D-alanyl-D-alanine. This interaction inhibits glucosyltransferase and P-phospholipid carrier, which prevents synthesis and polymerization of the peptidoglycan layer. This inhibition results in a weakened layer and leaks intracellular components.	Intravenous and oral	1954	[5,156–158]

Spp = species plural; *S. aureus* = *Staphylococcus aureus*; *P. aeruginosa* = *Pseudomonas aeruginosa*; *E. coli* = *Escherichia coli*.

5. Machine Learning for AMP Prediction and Design

Machine learning algorithms are computational techniques that use input data to build an adaptive model that can perform a task without having to be programmed by humans. Due to the adaptiveness of these algorithms, they can modify their architecture through repetition to enhance and adapt their ability to perform a task. This adaptation is known as training, including input data and desired outcomes. Through this training, the algorithm configures itself to produce desired outcomes from not only trained inputs but also unseen data. This training learns outcomes, which is known as machine learning [159].

Machine learning can be divided into categories based on how the data are labeled, including supervised and unsupervised learning. Supervised learning can focus on pattern recognition to distinguish between different data sets based on the data. For example, AMPs and other non-AMPs are unique but still have features that can distinguish one another. Instead of coding every AMP and non-AMP, the program learns to distinguish these through repeated encounters with AMPs and non-AMPs. The input data (cationic, amphiphilic, etc.) pair with the classification label (antimicrobial or non-antimicrobial) to recognize an AMP or a non-AMP (Figure 3a) [159]. Unsupervised learning is different, where this algorithm only uses input data without output data to train the algorithm. The algorithm takes unlabeled data, finds trends or patterns, and learns features from the data. When models encounter new data, the algorithm uses the learned features to recognize and classify the data [160]. For example, peptide databases contain a variety of peptides where the algorithm finds trends in the data to distinguish AMPs from non-AMPs (Figure 3b).

When there is no experimental data, AMP prediction models can be used to find new sequences with AMP characteristics, predict the mechanism or target of AMPs, or estimate three-dimensional (3D) structures. These models can predict AMP features by mining through databases with known AMPs and scanning the literature for sequences with similar features to establish a relationship between the sequences [161]. Peptide databases contain information such as physicochemical features, modifications, mechanisms, and structures to sort the data. Databases can be helpful to researchers when developing peptides through de novo methods and optimizing known AMP sequences. Optimizing AMPs depends on utilizing the physicochemical parameters involved in disrupting microbial membranes [102]. Alpha-helical peptides, for instance, are essential for membrane permeation, where the hydrophobic and hydrophilic faces and net charge influence pore formation [162].

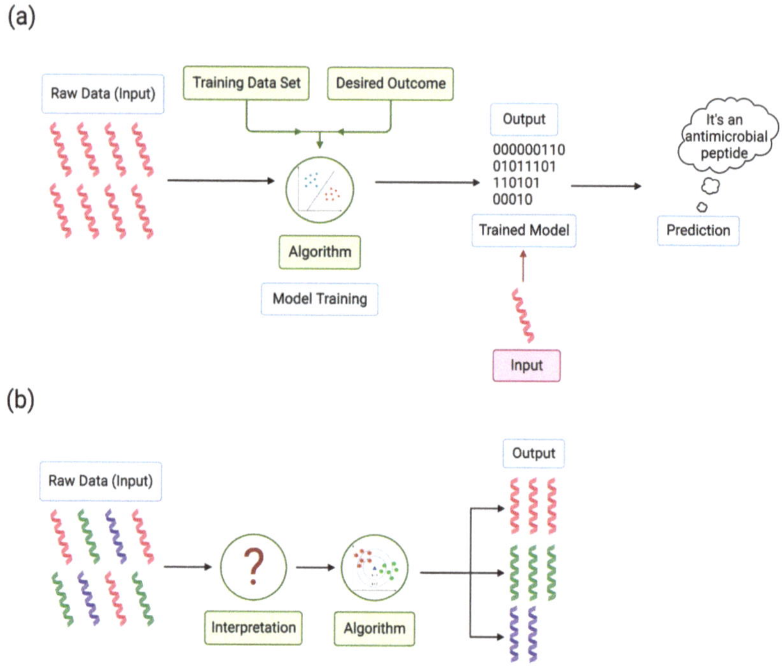

Figure 3. Schematic representation of machine learning algorithms used to develop prediction models. Machine learning algorithms are divided into two main categories. (**a**) Supervised machine learning uses a training set of known AMPs and the desired outcome, such as antimicrobial or non-antimicrobial properties. This model uses raw data of known AMPs to train the algorithm, resulting in a trained model for analyzing new data. Then, when inputting new data into the model, it can predict if that new sequence has antimicrobial or non-antimicrobial properties. (**b**) Unsupervised machine learning does not have a training data set. Instead, unsupervised learning predicts if new sequences are antimicrobial or non-antimicrobial based on any patterns or trends associated with the data. Adapted from (Ma, Y., 2018) CC BY 4.0 [163]. Created with BioRender.com (accessed on 15 December 2022).

5.1. Databases for AMP Design

Databases for the machine learning design of AMPs are classified into general and specific databases based on the categories of data collection. General databases (Table 2) provide a broad scope of AMPs, regardless of the source [164]. The AMP database (i.e., APD3), the most studied database, focuses on features such as natural AMPs with less than 200 amino acid residues, including human antimicrobial proteins, known amino acid sequences, and biological activity. This original database, which launched online in 2003, consisted of peptide search, prediction, design, and statistics interfaces [165]. There are currently 3569 AMPs in this database, where users can sort information based on charge, length, and hydrophobicity [19,166]. It contains natural, predicted, and synthetic peptides. This database has taken the lead in annotating the peptide activity (25 types), chemical modifications (27 types), peptide binding targets, and systematic classification of AMPs in numerous ways. The core data set in this database provides a foundation for us to decode the design principles of natural AMPs. The Collection of Antimicrobial Peptides (CAMP) is another database that contains 10,247 antimicrobial sequences [167]. This database includes patents, validated peptides, and predicted antimicrobial sequences based on similarity. CAMP includes sequences, structures, family signatures, AMP activity, source, target organisms, and hemolytic activity. It has also programmed the machine learning prediction interface [167,168]. Linking Antimicrobial Peptides (LAMP2) is another database currently consisting of 23,253 AMP sequences that contain both natural and synthetic AMPs with less than 100 amino acid residues. This database includes AMPs with antibacterial, antiviral, antifungal, antiparasitic, and antitumor activity. LAMP2 also contains primary structure, collection, composition, source, and function details. This database was built by merging data from other databases, such as the APD3 and CAMP [169,170]. Data Repository of Antimicrobial Peptides (DRAMP), another database, provides information about structural, sequence, clinical, physicochemical, and antimicrobial activity [171]. Peptipedia, another database built via data merging, analyzes peptide sequences with the highest number of sequences with biological activity. This database comprises 30 existing databases, including features such as estimating physicochemical and statistical properties. This database contains 92,055 peptides with machine learning techniques to classify the data for sequences [172].

Specific databases (Table 3) focus on the specific types, sources, or characteristics of AMPs instead of a broad overview. These databases can be categorized based on the biological source (plant or animal AMPs), biological activity (antibacterial), 3D structures, or molecular properties [164]. Hemolytik is a specific database with experimental information demonstrating hemolytic activities and their potencies. Since hemolysis is a significant challenge to AMPs being used for therapeutics, using this database may help overcome this clinical obstacle [173]. Thiobase is composed of thiopeptides, which are peptides produced by Gram-positive bacteria. The identification of these peptides depends on the presence of an azole-substituted nitrogen-containing six-membered ring. These peptides have shown potencies for Gram-positive bacteria, including resistant strands, which could help overcome antibacterial resistance [174]. CyBase is a database composed of plant cyclopeptides (i.e., cyclotides), where cyclic proteins derive from ribosome gene products, and cyclization is a post-translational medication [175]. A primary therapeutic obstacle to AMPs is their degradation by proteolytic enzymes; cyclotides are further stabilized by three sets of disulfide bonds have shown resistance to proteolysis [176].

Table 2. General databases including a broad scope AMPs with no limitations to the source of the data.

Database	Entries	Description	Website	References
APD	3569	This model database currently focuses on natural AMPs (ribosomal or non-ribosomal). First established criteria for data registration. First and detailed classification of over 25 AMP activity (antibacterial, antiviral, hemolytic, anticancer, spermicidal, anti-inflammatory, etc.), peptide source, properties, 3D structure, peptide binding targets, and detailed chemical modifications (>26 types).	https://aps.unmc.edu/database; accessed on 15 December 2022	[19,166]
CAMP	24,243	Experimentally validated, predicted AMP, and patents.	http://www.camp.bicnirrh.res.in/index.php; accessed on 15 December 2022	[167,168]

Table 2. *Cont.*

Database	Entries	Description	Website	References
LAMP	23,253	A database built via database merging.	http://biotechlab.fudan.edu.cn/database/lamp; accessed on 15 December 2022	[169,170]
DRAMP	22,316	Less than 100 residues, mature sequences, AMP activity demonstrated.	http://dramp.cpu-bioinfor.org; accessed on 15 December 2022	[171,177]
dbAMP	26,447	Natural, synthetic, classification based on functional activities.	https://awi.cuhk.edu.cn/dbAMP; accessed on 15 December 2022	[178]
DBAASP	18,878	Synthetic, ribosomal, non-ribosomal, monomers, multimers, multi-peptides.	https://dbaasp.org/home; accessed on 15 December 2022	[179]
MEGARes	~8000	Antimicrobial drugs, biocide, metal, and multi-compound.	http://megares.meglab.org; accessed on 15 December 2022	[180]
ADAM	7007	Natural sources, size, and sequence structure analysis.	http://bioinformatics.cs.ntou.edu.tw/adam/tool.html; accessed on 15 December 2022	[181]

Table 3. Specific databases provide information on a specific type or source of AMP instead of a broad scope.

Database	Entries	Description	Website	References
Hemolytik	~3000	Hemolytic activity	http://crdd.osdd.net/raghava/hemolytik; accessed on 15 December 2022	[173]
Thiobase	~100	Bacterial thiopeptides	https://bioinfo-mml.sjtu.edu.cn/THIOBASE/index.php; accessed on 15 December 2022	[182]
CyBase	~1771	Cyclic peptides	http://www.cybase.org.au; accessed on 15 December 2022	[175,183]
Defensins Knowledgebase	~360	Defensins	http://defensins.bii.a-star.edu.sg; accessed on 15 December 2022	[184]
Inverpep	774	Experimentally validated, AMPs from invertebrates	https://ciencias.medellin.unal.edu.co/gruposdeinvestigacion/prospeccionydisenobiomoleculas/InverPep/public/home_en; accessed on 15 December 2022	[185]
BACTIBASE	~177	Bacteriocins	http://bactibase.hammamilab.org/main.php; accessed on 15 December 2022	[186]
YADAMP	~2525	Antibacterial peptides	http://yadamp.unisa.it/searchDatabase.aspx; accessed on 15 December 2022	[187]
Peptaibol	~317	Peptaibol	http://peptaibol.cryst.bbk.ac.uk/home.shtml; accessed on 15 December 2022	[188]
DADP	~2571	Anuran defense peptide	http://split4.pmfst.hr/dadp; accessed on 15 December 2022	[189]
BaAMPs	~221	Antibiofilm peptides	http://baamps.it; accessed on 15 December 2022	[190]
CancerPPD	~3492	Anticancer peptides	http://crdd.osdd.net/raghava/cancerppd; accessed on 15 December 2022	[191]
ParaPep	~863	Antiparasitic peptides	https://webs.iiitd.edu.in/raghava/parapep/peptide.php; accessed on 15 December 2022	[192]

5.2. Machine Learning Classifications

Machine learning adopts computation approaches to predict the structure and activity of AMPs through the classification of features into peptide-level and amino-acid-level features. The peptide-level features include sequence-based and structure-based features [193]. These computational approaches use statistics to sort and learn patterns in large amounts of data. Machine learning tools use the data provided by a database, known as a training set, to allow the models to learn and select the best features for the desired outcome. Validating this hypothesis on the new data uses a test set to ensure this model is effective [102]. Machine learning or deep learning models are used in the prediction and generation of AMPs. This scheme with input features, the construction of models, and prediction is shown in Figure 4.

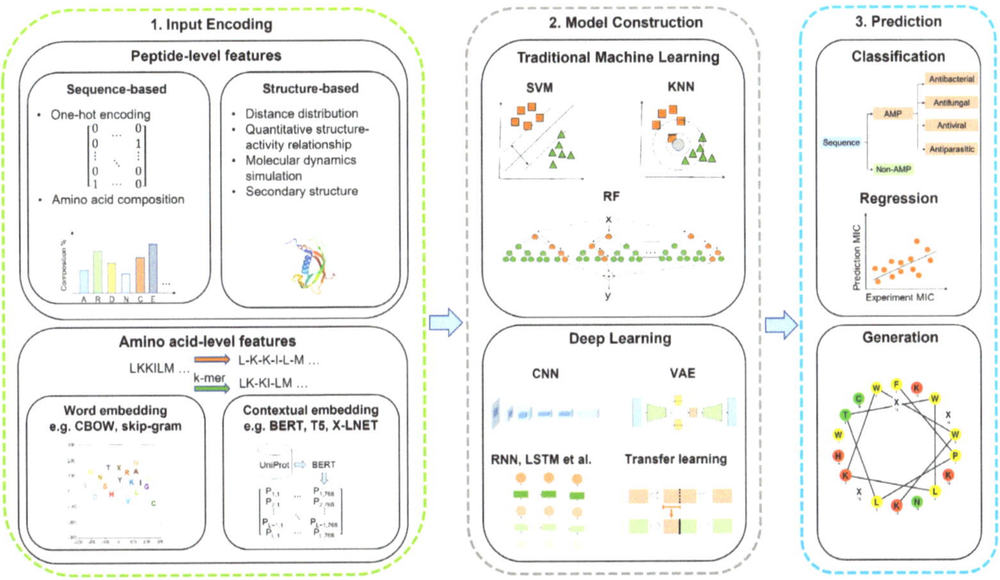

Figure 4. A general machine learning workflow of AMP discovery and design, including a summary of the major techniques in each stage of the workflow. Redistributed from (Yan et al., 2022) CC BY 4.0 [193].

5.2.1. Sequence-Based Features

Sequence-based features are dependent on the composition of amino acids or groups of amino acids to compute vectors [193]. To compute these vectors, various techniques are used, including one-hot encoding, amino acid composition, pseudo-amino acid composition, reduced amino acids, physicochemical properties, etc. [194]. One-hot encoding is a technique that represents a value as a vector of bits, where the vector length is proportional to the number of categories. In the case of AMPs, there are 20 natural amino acids which are each assigned to a bit position. The assigned bit is identified with a value of 1, while the other positions are set to 0. According to this positioning system, alanine (A) is the first amino acid alphabetically and would be positioned at the rightmost bit (00000000000000000001), while tyrosine (Y) would be assigned to the leftmost bit (10000000000000000000) [195]. Amino acid composition is a technique where the vector has a proportion of an amino acid relative to the sequence length. For AMPs, these peptides can be divided into sections with both an N and C terminus, which is used to calculate local amino acid composition. Unfortunately, the amino acid composition does not account for the sequence order effect, which can have a wide range of combinations depending on the sequence length. The pseudo-amino acid composition accounts for the sequence order by computing the correlations and occurrence number between various clusters among a pair of amino acids, resulting in a dimensional vector [193,194]. Although amino acid and pseudo-amino acid compositions have their advantages, they do not account for the specific amino acid sequence for a peptide due to limitations associated with the sequence order and sequence variations [194]. Reduced amino acid compositions group similar amino acids based on sequence correlation factors, such as physicochemical or structural properties. Physicochemical properties include hydrophobicity, charge, and a few other features that are determined through wet lab experiments. The use of physicochemical features are usually in combination with other techniques, such as pseudo-amino acid combinations, to enhance the accuracy of the techniques [194].

5.2.2. Structure-Based Features

Structure-based features look at the secondary structure of proteins or peptides, which has an impact on the antimicrobial activity of peptides. Therefore, predicting AMPs using a combination of sequence-based and structure-based features is essential. Structure-based features are computed using quantitative structure–activity relationship (QSAR), distance distribution, general structure encodings, and other techniques. QSAR is a technique used to describe amino acid sequences through chemical properties [193,194]. This technique can predict antibacterial activity by relating descriptors with peptide properties such as toxicity [196]. The structure of the peptide helps identify descriptors, which include amphiphilicity, net charge, and length. These descriptors help define classifiers and are used to classify the peptides based on their efficacy [14]. Distance distribution involves prediction based on the distribution of distances for each pair of atom types. This technique computes the distance between atom types (such as donor–donor, donor–acceptor, etc.) into a distribution function [197]. General structural encodings differ from QSAR methods in that this technique provides structure information from the whole peptide and converts this information to a numerical representation. This technique describes the peptide structure by the 3D shape and is based on the electrostatic potential. General structure encodings can be applied to predicting AMPs because the 3D composition of peptides influences the antimicrobial activity. Overall, the combination of sequence-based and structure-based features enhances the accuracy and effectiveness of the prediction [194].

5.2.3. Amino-Acid-Level Features

Amino-acid-level features look at the sequence, where each word corresponds to the one-letter code of the amino acids. These features are primarily used in deep learning algorithms, but can use embedding layers to extract representative features such as word embedding and contextualized embedding [193]. Word embedding techniques involve Word2vec and Global Vectors (GloVe). Word2vec is a shallow word embedding model that predicts words based on their context using neural models such as a continuous bag of words (CBOW) or Skip-Gram. CBOW uses three layers to predict a word, where the input layer describes the context. The hidden layer describes the projection of the words from the input and is projected into the output layer. Skip-Gram is different from CBOW in that the input layer relates to the target word, while the output layer relates to the context. Skip-Gram predicts the context of the word as opposed to the prediction of a word. GloVe is a technique based on the occurrence of words in a textual corpus. This technique involves constructing a co-occurrence matrix and factorization of the matrix to obtain vectors [198]. Contextual embedding is different from word embedding in that the embedding learned is a function of the whole input sequence, therefore allowing the same word to have various representations in different contexts. Contextual embeddings are generated through deep learning techniques and use language models, such as bidirectional encoder representation from transformers (BERT), text-to-text transfer transformer (T5), and auto-regressive model (XLNet) [193]. BERT is a technique used to pre-train deep bidirectional representations using unlabeled text, where context is conditioned in the layers. This technique is able to pre-train bidirectional representations by masking a percentage of input tokens randomly and predict those tokens through a masked language model (MLM). The pre-trained deep bidirectional representation allows the model to be fine-tuned with an additional output layer to provide a range of tasks [199]. T5 considers a task and is inputted into the model as text and trained to generate target text [200]. XLNet models extract bidirectional contextual information through the random ordering of the input order while maintaining the one-way model. This bidirectional contextual information is attained by considering the ranking order of the text, but it only predicts a partial sequence [201].

5.2.4. Machine Learning Models

Machine learning uses models or algorithms to classify data and predict an outcome based on the provided data. To predict antimicrobial activity, these models use statistics to learn the relationships between physicochemical properties and amino acid composition. Various algorithms predict the AMP activity, mechanism of action, and efficacy. Prediction models for AMPs include k-nearest neighbor (KNN), support vector machine (SVM), artificial neural network (ANN), random forest (RF), and other algorithms. These models can be trained by using features, such as amino acid composition, length, molecular weight, charge, and hydrophobicity [102].

KNN (Figure 5a) performs a classification and regression analysis of data points to find a trend. Classifying data records (t) results from retrieving its k nearest neighbors, forming a neighborhood of t [202,203]. The nearest neighbor is the data point closest to the desired data point. When the nearest items are determined, the algorithm recommends these items to the users based on the previous majority of nearest neighbors [204]. This model is dependent on selecting a k value, where the effectiveness of the classification is dependent on this value. Due to this dependence, this model is limited based on selecting the appropriate number of neighbors for the algorithm [202].

A supervised machine learning model known as SVM (Figure 5b) classifies data by finding a hyperplane to segment collected data. This model creates a hyperplane with the largest distance from the nearest training points. The maximal distance created by the hyperplane allows a strong separation between classes, resulting in a lowered generalization error [203,205]. SVM can perform a linear classification through regression analysis and non-linear classification using the kernel trick. The kernel trick works by mapping data into high-dimensional feature spaces, characterizing data in multiple dimensions [160]. Unfortunately, limitations with this classifier result from data in target classes overlapping, resulting in an uncertain classification of data [203].

ANN (Figure 5c) is a supervised machine learning model that uses various hyperparameters to approximate a relationship between input and output values. These hyperparameters include many hidden layers and units, a learning rate, and activation function [203]. ANN is composed of computational units called neurons, which are organized into layers and connected to form a network. Artificial neurons can receive signals from various sources and transform them through an activation function [206]. The general structure of an ANN consists of input, hidden, and output layers. The input layer receives incoming data, hidden layers transport incoming data into a higher-order function, and an output layer makes a prediction about the input data. Having multiple hidden layers helps obtain high non-linearity, preventing the computation of a large linear function [203,207].

By building multiple decision trees, random forests (Figure 5d) can overcome obstacles associated with decision trees, such as overfitting data due to outliers. Decision trees are algorithms used to classify an event or predict the outcomes of a variable. Decision trees primarily solve classification problems but can also perform regression analysis. Random forests function through the random sampling of the data while building several decision trees for each random sample of data [204]. Building several decision trees with controlled variation requires a combination of bootstrap aggregation and random feature selection. Bootstrap aggregation (bagging) is a method that generates and averages predictors when forming outcomes by using a majority vote to predict a class [203,208].

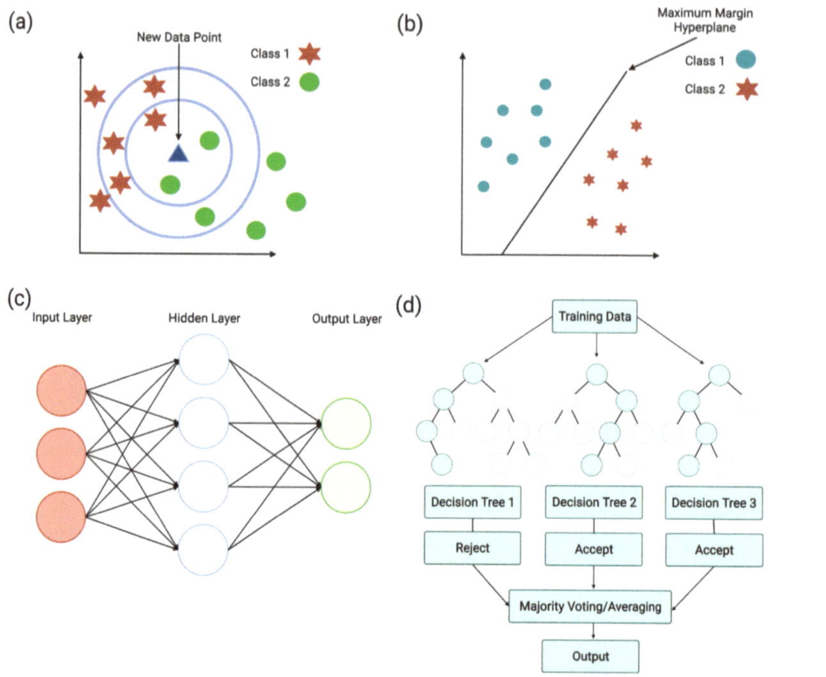

Figure 5. Machine learning uses various models to classify and predict outcomes based on data. (**a**) k-nearest neighbor (KNN) classifies data using the nearest neighbor, which is the data point closest to the desired point. When the nearest items are determined, the algorithm recommends using these items based on the previous majority vote of nearest neighbors [204]. (**b**) Support vector machine (SVM) is a machine learning algorithm that classifies data by finding a hyperplane to separate the data. A hyperplane creates a line that is the maximum distance from the nearest training points, resulting in a strong separation between classes [203,205]. (**c**) Artificial neural network (ANN) uses hyper-parameters to approximate relationships between input and output values. ANNs are composed of computational units, which receive signals and transform them to make a prediction using the network formed by layers [203,206]. (**d**) Random forests build multiple decision trees to classify an event and avoid issues such as overfitting data due to outliers [204]. Adapted from CC BY-SA 3.0 [209], CC BY 4.0 [210], CC BY-SA 3.0 [211], and CC BY-SA 4.0 [212].

5.3. *Machine Learning Prediction and Design of AMPs*

As discussed earlier, machine learning is a time-saving technique that can perform a task, such as predicting an AMP, without being programed by humans, and can modify the algorithms to adapt to the task [159]. These techniques are applied to the prediction of AMP sequences to obtain unique properties to avoid potential issues with AMPs. For instance, machine learning can be applied to develop non-hemolytic AMPs, which is a limitation in their therapeutic use. Overall, with increasing knowledge on machine learning, this technique can be used to mitigate the issues faced in wet lab experiments.

5.3.1. Predicting AMP Sequences

Machine learning methods have been used to predict AMPs soon after the establishment of the original antimicrobial peptide database in 2003, known as the Antimicrobial Peptide Database (APD) [165]. Lata et al. developed the first SVM model, named AntiBP2, to predict AMPs using the data collected from the APD. This SVM model is based on the amino acid composition of the peptide using a five-fold cross-validation technique and achieved a 92.14% accuracy [213]. Additionally, based on the APD, Xiao et al. developed a fuzzy K-nearest neighbor algorithm based on pseudo-amino acid composition. This algorithm is a multi-label classifier where the pseudo-amino acid composition components incorporate various physicochemical properties. This method was named iAMP-2L and achieved an accuracy of 86.32% when identifying AMPs and non-AMPs [214]. Torrent et al. created an ANN method that correlates physicochemical properties with antimicrobial activity. This model used the CAMP peptide database for the positive data set and the Uniprot database for the negative data set with an overall accuracy of 90% [215]. Joseph et al. developed a one-against-all classifier model using random forests and SVMs to predict sequences as antimicrobial or non-antimicrobial. In this multi-classification model named ClassAMP, random forests determined the essential features for classification. ClassAMP achieved an overall accuracy of about 95% with the test sequences [216]. Lawrence et al. developed a random forest classifier named amPEPpy to predict AMP sequences. The number of used trees in amPEPpy was optimized using the out-of-bag (OOB) error, where 128 decision trees were optimal with an OOB error of 0.036 [217].

5.3.2. Machine Learning for Peptide Design

Recently, Capecchi et al. developed a recurrent neural network (RNN) method to predict antimicrobial and hemolytic activity [218]. RNN is a branch of neural networks that incorporates feedback connections from the previous time step in the model [219]. The preference for using RNN over other models, such as SVM, random forest, and naive Bayes, is due to its better prediction of antimicrobial and hemolytic activity. The RNN model obtained an accuracy of 76% on hemolysis prediction and antimicrobial activity [218]. Sharma et al. developed a web server to predict antibiofilm peptides known as dPABBs, which used amino acid composition and selected residue features using six SVM and Weka models. dPABBs was able to generate an accuracy, sensitivity, and specificity of 95.24%, 92.50%, and 97.73%, respectively, based on the training data sets [220]. Zhang et al. designed a 12-amino-acid-long AMP called DP7 using machine learning methods based on amino acid activity. DP7 has demonstrated broad-spectrum antimicrobial activity in vitro and in vivo. The results from this study confirmed that DP7 could reduce methicillin-resistant *Staphylococcus aureus* (MRSA) bloodstream infection in mice up to day 7 post-infection at doses of 0.5, 1, and 2 mg/kg with 70, 80, and 90% reductions, respectively. DP7 also showed its effectiveness compared to current antibiotics, where the protection level of DP7 at 1 mg/kg was equivalent to 10 mg/kg of vancomycin [221]. The lower concentration indicates that DP7 is more potent than vancomycin at treating the resistant bacteria.

6. Peptide Engineering

The proteases produced by human cells and microbes may limit the bioavailability of AMPs. For instance, LL-37 can be degraded by *S. aureus* proteases, aureolysin, and the V8 protease, thereby losing its antimicrobial activity [222]. To avoid the potential inactivation of AMPs, peptide engineering techniques have been studied, including peptide backbone modifications, cyclization, terminal modifications, and substituting L-amino acids with D-amino acids or unnatural amino acids [223]. The most common terminal modifications include N-terminal acetylation and C-terminal amidation, which provide the peptides with different functions. N-terminal acetylation has shown to result in an increased stability and helical content, which can contribute to a deeper insertion into the hydrophobic region in microbial membranes [224]. C-terminal amidation has shown to enhance antimicrobial activity and reduce hemolytic properties, which are advantageous features for antimicrobial peptides [225]. Li et al. computationally designed antimicrobial peptide L163, which is effective against multidrug-resistant bacteria but is degraded by proteases. The results showed that the N-terminal acetylation of L163 had increased stability and reduced host toxicity [226]. Due to human proteases solely recognizing L-amino acids to be degraded, D-amino acids have shown enhanced stability against protease degradation [227]. Lu et al. synthesized derivatives of AMP Pep05 (KRLFKKLLKYLRKF) through the substitution of

D-amino acids and unnatural amino acids. The results showed that substitutions with D-amino acids and unnatural amino acids enhanced the ability of peptides to resist proteolytic cleavage from proteases produced by *S. aureus* and *Escherichia coli* (*E. coli*). When all the L-amino acids of Pep05 were replaced with D-amino acids, this resulted in the highest stability against proteases, but exhibited severe toxicity in vivo [228]. Wang et al. found that the partial incorporation of D-amino acids and biphenylalanines into the major antimicrobial peptide of LL-37 led to selective, stable, and potent antimicrobials against antibiotic-resistant bacteria, including MRSA [229]. An eight-amino-acid lipopeptide of LL-37 made in D-amino acids, which remained stable to five proteases, showed in vivo efficacy against MRSA [134]. More recently, White et al. developed CD4-PP, a synthetic peptide, through the dimerization and head-tail cyclization of the shortest antimicrobial region of LL-37. This cyclized form was stable against aureolysin for 6 h in comparison to LL-37, which degraded within minutes [230]. These results indicate the importance of modifying peptides to enhance the stability required for clinical applications.

7. Nanotechnologies for AMP Delivery

Nanotechnology is a growing field involved in applying structures and systems at the nanometric scale, but it has now been implemented into the delivery of drugs with a limited therapeutic efficacy [231,232]. Nanotechnology can also protect AMPs from degradation and increase their efficacy [233]. Nanoparticles range from 0.1 to 100 nm and can transport peptides across the intestinal barrier to enter the bloodstream [132,233]. Various nanomaterials used for the delivery of AMPs include metal nanoparticles (gold, silver), lipid nanoparticles (liposome), polymer nanoparticles [chitosan, hyaluronic acid, and poly(glycolide-co-lactide) or PLGA], and other nanostructures (dendrimer, carbon nanotube, and quantum dot), as shown in Figure 6 [234].

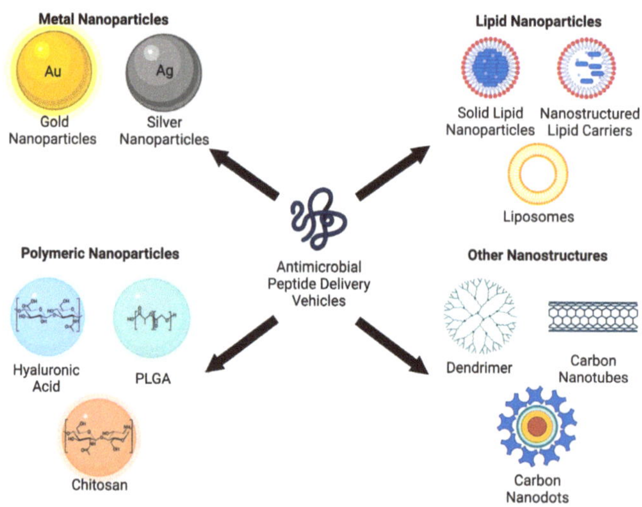

Figure 6. Nanoparticle delivery vehicles for AMPs to enhance biological stability. These delivery vehicles have different purposes based on the properties desired for delivery. Metal nanoparticles have demonstrated the ability to enhance the antimicrobial activity of AMPs but have limitations at high concentrations due to potential toxicity [235]. Lipid nanoparticles are biocompatible, which gives them the advantage of being used to avoid toxicity in tissues [236]. Polymeric nanoparticles contain a coating of molecules with specific properties such as antimicrobial activity and good bioavailability [237]. Forming other nanostructures helps combat complications, such as different structures, allowing biofilm inhibition [238]. Modified with permission from CC-BY 4.0 [235]. Created with BioRender.com (accessed on 21 December 2022).

7.1. Metal Nanoparticles

Gold nanoparticles do not have antimicrobial activity, but they function as delivery devices due to their ability to enhance the antimicrobial activity of AMPs. Gold nanoparticles may lead to increased antimicrobial activity due to the penetration of bacterial cell membranes and a high surface-to-volume ratio for drug delivery [235]. Rai et al. developed an AMP-conjugated nanoparticle with a gold core and a hydrophilic cationic peptide shell. This gold nanoparticle had a high concentration of AMPs with more significant antimicrobial activity in serum and in proteolytic environments compared to soluble AMPs [239]. Silver nanoparticles may deliver silver ions into bacterial cytoplasm and membrane and present antimicrobial properties [240]. Mei et al. synthesized silver nanoparticles conjugated with AMPs (bacitracin A and polymyxin E). The minimum inhibitory concentration (MIC) of silver nanoparticles conjugated with these AMPs was found to be significantly lower than silver nanoparticles without AMPs [241].

7.2. Lipid Nanoparticles

Liposomes are more biocompatible than metal nanoparticles because their composition is similar to cell membranes [242]. The ability of these carriers to be biodegradable and non-hemolytic gives them an advantage over other nanocarriers [236]. Li et al. developed a liposome delivery system loaded with daptomycin and clarithromycin to treat MRSA infections. This antibiotic delivery system had a reduced mutant prevention concentration and decreased mutant selection window, demonstrating its potential to reduce MRSA infections [243]. Sharaf et al. designed liposomes loaded with clarithromycin and bioflavonoid Hesperidin to provide a controlled drug release to target bacterial cell membranes. These liposome nanostructures effectively inhibited *Helicobacter pylori* growth with no cytotoxic effects ($IC_{50} < 50$ μM). These liposome carriers were biocompatible, which is essential for effectively and safely delivering drugs to their target [244]. Solid lipid nanoparticles (SLNs) were also studied due to their high drug loading and small size [245]. Severino et al. developed an SLN loaded with polymyxin B to treat bacterial infections. This polymyxin-B-loaded SLN was effective against various resistant strains of *Pseudomonas aeruginosa* (*P. aeruginosa*) [246].

7.3. Polymer Nanoparticles

PLGA nanoparticles have been well studied for drug delivery. PLGA nanoparticles have low systemic toxicity due to their ability to undergo hydrolysis, producing biodegradable monomers (lactic acid and glycolic acid) [247]. Cruz et al. developed PLGA nanoparticles loaded with an AMP, GIBIM-P5S9K, to determine the antimicrobial activity against *P. aeruginosa* and MRSA. These AMP-loaded nanoparticles showed lower hemolytic activity and higher antibacterial activity against bacterial growth at a lower concentration than free AMPs [248]. Among the polymer nanoparticles studied, chitosan is a biodegradable polymer with broad-spectrum antimicrobial activity and has been generally recognized as safe (GRAS) by the FDA. Due to these properties and being inexpensive, chitosan has been used in therapeutic applications [237,249]. Rashki et al. developed chitosan nanoparticles (CS-NP) loaded with LL-37 to improve its antibacterial and antibiofilm activity. These nanoparticles showed a 6-log reduction in colony-forming units of MRSA compared to free LL-37. The CS-NP loaded with LL-37 also inhibited MRSA biofilm formation to 68% after 2 h compared to 74% for free LL-37 [250]. Hyaluronic acid nanoparticles are another example. Hyaluronic acid is a component in the extracellular matrix that provides biocompatibility for therapeutic applications [251]. Lequeux et al. created an antimicrobial hydrogel by covalently binding nisin to hyaluronic acid. This hydrogel demonstrated an antibacterial efficacy of 99.99% for *Staphylococcus epidermidis* and 99.95% for *S. aureus* at a polymer concentration of 2 mg/mL and a nisin concentration of 0.01 mg/mL [252].

7.4. Other Nanostructures

Dendrimers are hyper-branched polymer molecules that have demonstrated effectiveness against biofilms, multidrug-resistant bacteria, and viruses [238]. Jiang et al. developed a PAMAN dendrimer conjugated with vancomycin and silver nanoparticles to kill vancomycin-resistant *Staphylococci*. This dendrimer demonstrated the ability to kill vancomycin-resistant strains of *S. aureus* with an MIC of 2 μg/mL compared to 8 μg/mL for free vancomycin [253]. Carbon nanotubes were also studied, and they can provide antimicrobial activity by oxidizing glutathione, which increases oxidative stress on microbes [254]. Qu et al. developed multi-walled carbon nanotubes, where nisin was covalently immobilized. These carbon nanotubes reduced biofilm formation by 100-fold on deposit film and 2.6-fold in the suspension. The antimicrobial activity of nisin had retained at least 90% of bacterial activity for carbon nanotubes with immobilized nisin [255]. More recently, quantum dots were studied. Quantum dots have a high transfer of electrons, allowing them to produce free radicals such as reactive oxygen species (ROS), resulting in antimicrobial activity and microbial cell death [256]. Zhao et al. developed nitrogen-doped carbon quantum dots (NCQDs) to assess their antimicrobial activity against antibiotic-resistant bacteria. These NCQDs had an MIC of 0.128 mg/mL and 0.256 mg/mL for methicillin-resistant and susceptible *S. aureus*, respectively, with no toxicity to the host's organs [257].

8. Summary and Perspective

Antibiotic resistance is a significant complication in healthcare since doctors may face situations where they have no antibiotics to use. AMPs have been shown to present a broad range of protection against pathogenic microorganisms in vitro, and more are under investigation for their use as alternatives to antibiotics. Until recently, researchers discovered new AMPs by modifying known AMPs to be individually tested against various bacteria to determine antimicrobial efficacy in wet laboratories [127]. These methods are time-consuming, costly, and require many resources [258]. This article has described the use of machine learning methods as an alternative method to identify AMPs, which is anticipated to accelerate the discovery of novel antimicrobials. AMP activity has been predicted based on 2D features, such as amino acid sequence, net charge, and hydrophobicity. Machine learning could be applied not only to predict AMP sequences but also to predict the 3D structure of AMPs. Predicting the 3D structure of AMPs through machine learning would be a helpful tool for researchers to understand the binding of molecules to AMPs, which plays a major role in drug discovery. More recently, machine learning has also been used to predict AMP sequences to target SARS-CoV-2. Using in silico models, Liscano et al. determined two AMPs (caerin 1.6 and caerin 1.10) which showed a potential to interact with the spike surface viral protein (SGP) on SARS-CoV-2 as opposed to the ACE2 protein. The SGP protein is located on the envelope protein of SARS-CoV-2 that plays a role in the binding and fusion of and entering host cells. The results from these in silico experiments show that these AMPs have the potential to block the S protein and ACE2 during viral binding and entry, but need to be experimentally validated to determine their effectiveness [259]. Nanotechnology with antimicrobial properties also has the potential to improve oral health concerns, such as endodontic infections. Currently, there are no available techniques that remove biofilm while not affecting the root dentin [260]. Nanomaterials provide unique properties, including the removal of biofilms; they also avoid demineralization and stimulate re-mineralization [261]. With the limited therapeutic techniques to treat endodontic infections, nanocarriers may be a promising approach to combat oral health concerns.

Therefore, computational technologies have great potential in designing future antibiotics. In this regard, the database filtering technology has illustrated the in silico design of novel antimicrobials with potent activity both in vitro and in vivo [262,263]. Likewise, with the accumulation of data in databases (Figure 7), machine learning algorithms may be assembled into a pipeline to predict sequences to overcome the issues faced with AMPs [19]. These obstacles include degradation by enzymes, poor oral bioavailability, and hemolytic activity. As an example, Plisson [135] et al. developed models to predict hemolytic activity, which predicted that 30% of AMPs were non-hemolytic, where 91% of the predictions were considered reliable. Based on these models, designing non-hemolytic peptides should include neutral or slightly charged sequences with an equal amount of aromatic/aliphatic residues and small amino acids. These machine learning models provided information on non-hemolytic AMP sequences, which can help researchers develop AMPs with minimal toxic effects. Together with other techniques such as peptide engineering, formulations, and nanotechnology, researchers may identify novel AMPs with improved properties and bioavailability for future applications [264].

In summary, this review focuses on nanotechnology approaches for AMP delivery, the roles of AMPs in various diseases, and recent advances in AMP design via machine learning.

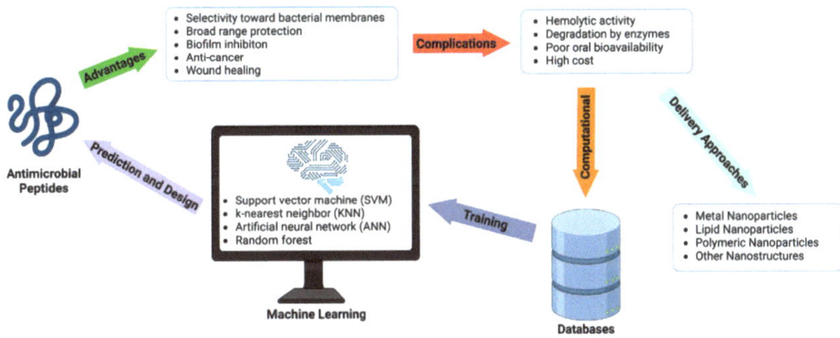

Figure 7. Overall scheme of the advantages of AMPs and different approaches to improve challenges associated with therapeutic applications. AMPs have advantages over current treatments due to their broad range protection and other properties not found in antibiotics. However, these peptides have obstacles when being applied to therapeutic use due to complications such as toxicity and instability. Delivery approaches are used to overcome some of these issues with AMPs. Machine learning has shown to be a promising approach for using models to predict and design AMPs with less therapeutic complications. Databases help train machine learning models to enhance prediction methods for AMPs. Created with BioRender.com (accessed on 21 December 2022).

Funding: This research was funded by National Institutes of Health, grant numbers GM138552 and ES032920.

Data Availability Statement: No new data were created in this study. Data sharing is not applicable to this article.

Acknowledgments: We thank the NIEHST32 Systems Toxicology (SyTOX) fellowship for the support of AS and we thank Suzanne Danley for proofreading.

Conflicts of Interest: The authors declare no conflict of interest.

References

1. Lei, J.; Sun, L.; Huang, S.; Zhu, C.; Li, P.; He, J.; Mackey, V.; Coy, D.H.; He, Q. The antimicrobial peptides and their potential clinical applications. *Am. J. Transl. Res.* **2019**, *11*, 3919–3931. [PubMed]
2. Blair, J.M.A.; Webber, M.A.; Baylay, A.J.; Ogbolu, D.O.; Piddock, L.J.V. Molecular mechanisms of antibiotic resistance. *Nat. Rev. Microbiol.* **2015**, *13*, 42–51. [CrossRef] [PubMed]
3. Lopatkin, A.J.; Yang, J.H. Digital insights into nucleotide metabolism and antibiotic treatment failure. *Front. Digit. Health* **2021**, *3*, 583468. [CrossRef] [PubMed]
4. Centers for Disease Control and Prevention. 2019 Antibiotic Resistance Threats Report | CDC. Available online: https://www.cdc.gov/drugresistance/biggest-threats.html (accessed on 15 December 2022).
5. Chen, C.H.; Lu, T.K. Development and challenges of antimicrobial peptides for therapeutic applications. *Antibiotics* **2020**, *9*, 24. [CrossRef] [PubMed]
6. Li, C.; Zhu, C.; Ren, B.; Yin, X.; Shim, S.; Gao, Y.; Zhu, J.; Zhao, P.; Liu, C.; Yu, R.; et al. Two optimized antimicrobial peptides with therapeutic potential for clinical antibiotic-resistant Staphylococcus aureus. *Eur. J. Med. Chem.* **2019**, *183*, 111686. [CrossRef]

7. Fan, L.; Wei, Y.; Chen, Y.; Jiang, S.; Xu, F.; Zhang, C.; Wang, H.; Shao, X. Epinecidin-1, a marine antifungal peptide, inhibits botrytis cinerea and delays gray mold in postharvest peaches. *Food Chem.* **2023**, *403*, 134419. [CrossRef]
8. Adade, C.; Oliveira, I.; Pais, J.; Souto-Padrón, T. Melittin peptide kills Trypanosoma cruzi parasites by inducing different cell death pathways. *Toxicon* **2013**, *69*, 227–239. [CrossRef]
9. Huan, Y.; Kong, Q.; Mou, H.; Yi, H. Antimicrobial peptides: Classification, design, application and research progress in multiple fields. *Front. Microbiol.* **2020**, *11*, 582779. [CrossRef]
10. Wachinger, M.; Kleinschmidt, A.; Winder, D.; von Pechmann, N.; Ludvigsen, A.; Neumann, M.; Holle, R.; Salmons, B.; Erfle, V.; Brack-Werner, R. Antimicrobial peptides melittin and cecropin inhibit replication of human immunodeficiency virus 1 by suppressing viral gene expression. *J. Gen. Virol.* **1998**, *79*, 731–740. [CrossRef]
11. Silva, O.; de la Fuente-Núñez, C.; Haney, E.; Fensterseifer, I.; Ribeiro, S.; Porto, W.; Brown, P.; Faria-Junior, C.; Rezende, T.; Moreno, S.; et al. An anti-infective synthetic peptide with dual antimicrobial and immunomodulatory activities. *Sci. Rep.* **2016**, *6*, 35465. [CrossRef]
12. Frohm, M.; Agerberth, B.; Ahangari, G.; Ståhle-Bäckdahl, M.; Lidén, S.; Wigzell, H.; Gudmundsson, G. The expression of the gene coding for the antibacterial peptide LL-37 is induced in human keratinocytes during inflammatory disorders. *J. Biol. Chem.* **1997**, *272*, 15258–15263. [CrossRef]
13. de la Fuente Núñez, C.; Silva, O.N.; Lu, T.K.; Franco, O.L. Antimicrobial peptides: Role in human disease and potential as immunotherapies. *Pharmacol. Ther.* **2017**, *178*, 132–140. [CrossRef]
14. Torres, M.D.T.; Cao, J.; Franco, O.L.; Lu, T.K.; Fuente-Nunez, C.d.l. Synthetic biology and computer-based frameworks for antimicrobial peptide discovery. *ACS Nano.* **2021**, *15*, 2143–2164. [CrossRef]
15. Zhao, G.; Usui, M.; Underwood, R.; Singh, P.; James, G.; Stewart, P.; Fleckman, P.; Olerud, J. Time course study of delayed wound healing in a biofilm-challenged diabetic mouse model. *Wound Repair Regen.* **2012**, *20*, 342–352. [CrossRef]
16. Fazli, M.; Bjarnsholt, T.; Kirketerp-Møller, K.; Jørgensen, A.; Andersen, C.; Givskov, M.; Tolker-Nielsen, T. Quantitative analysis of the cellular inflammatory response against biofilm bacteria in chronic wounds. *Wound Repair Regen.* **2011**, *19*, 387–391. [CrossRef]
17. Neut, D.; Tijdens-Creusen, E.; Bulstra, S.; van der Mei, H.; Busscher, H. Biofilms in chronic diabetic foot ulcers—A study of 2 cases. *Acta Orthop.* **2011**, *82*, 383–385. [CrossRef]
18. Darvishi, S.; Tavakoli, S.; Kharaziha, M.; Girault, H.; Kaminski, C.; Mela, I. Advances in the sensing and treatment of wound biofilms. *Angew. Chem. Int. Ed. Engl.* **2022**, *61*, e202112218. [CrossRef]
19. Wang, G.; Li, X.; Wang, Z. APD3: The antimicrobial peptide database as a tool for research and education. *Nucleic Acids Res.* **2015**, *44*, D1087–D1093. [CrossRef]
20. Patocka, J.; Nepovimova, E.; Klimova, B.; Wu, Q.; Kuca, K. Antimicrobial peptides: Amphibian host defense peptides. *Curr. Med. Chem.* **2019**, *26*, 5924–5946. [CrossRef]
21. Manniello, M.; Moretta, A.; Salvia, R.; Scieuzo, C.; Lucchetti, D.; Vogel, H.; Sgambato, A.; Falabella, P. Insect antimicrobial peptides: Potential weapons to counteract the antibiotic resistance. *Cell. Mol. Life Sci.* **2021**, *78*, 4259–4282. [CrossRef]
22. Lee, E.Y.; Fulan, B.M.; Wong, G.C.L.; Ferguson, A.L. Mapping membrane activity in undiscovered peptide sequence space using machine learning. *Proc. Natl. Acad. Sci. USA* **2016**, *113*, 13588–13593. [CrossRef] [PubMed]
23. Clark, S.; Jowitt, T.A.; Harris, L.K.; Knight, C.G.; Dobson, C.B. The lexicon of antimicrobial peptides: A complete set of arginine and tryptophan sequences. *Commun. Biol.* **2021**, *4*, 1–14. [CrossRef] [PubMed]
24. Wang, G. Unifying the classification of antimicrobial peptides in the antimicrobial peptide database. *Methods Enzym.* **2022**, *663*, 1–18. [CrossRef]
25. Gesell, J.; Zasloff, M.; Opella, S. Two-dimensional 1H NMR experiments show that the 23-residue magainin antibiotic peptide is an alpha-helix in dodecylphosphocholine micelles, sodium dodecylsulfate micelles, and trifluoroethanol/water solution. *J. Biomol. NMR.* **1997**, *9*, 127–135. [CrossRef]
26. Wang, G. Structures of human host defense cathelicidin LL-37 and its smallest antimicrobial peptide KR-12 in lipid micelles. *J. Biol. Chem.* **2008**, *283*, 32637–32643. [CrossRef]
27. Comune, M.; Rai, A.; Palma, P.; TondaTuro, C.; Ferreira, L. Antimicrobial and pro-angiogenic properties of soluble and nanoparticle-immobilized LL37 peptides. *Biomater. Sci.* **2021**, *9*, 8153–8159. [CrossRef]
28. Lee, E.Y.; Lee, M.W.; Fulan, B.M.; Ferguson, A.L.; Wong, G.C.L. What can machine learning do for antimicrobial peptides, and what can antimicrobial peptides do for machine learning? *Interface Focus.* **2017**, *7*, 20160153. [CrossRef]
29. Verjans, E.; Zels, S.; Luyten, W.; Landuyt, B.; Schoofs, L. Molecular mechanisms of LL-37-induced receptor activation: An overview. *Peptides* **2016**, *85*, 16–26. [CrossRef]
30. Rai, A.; Ferrão, R.; Palma, P.; Patricio, T.; Parreira, P.; Anes, E.; Tonda-Turo, C.; Martins, M.; Alves, N.; Ferreira, L. Antimicrobial peptide-based materials: Opportunities and challenges. *J. Mater. Chem. B* **2022**, *10*, 2384–2429. [CrossRef]
31. Liang, W.; Diana, J. The dual role of antimicrobial peptides in autoimmunity. *Front. Immunol.* **2020**, *11*, 2077. [CrossRef]
32. Hiemstra, P.; Amatngalim, G.; van der Does, A.; Taube, C. Antimicrobial peptides and innate lung defenses: Role in infectious and noninfectious lung diseases and therapeutic applications. *Chest* **2016**, *149*, 545–551. [CrossRef]
33. Tripathi, S.; Tecle, T.; Verma, A.; Crouch, E.; White, M.; Hartshorn, K. The human cathelicidin LL-37 inhibits influenza A viruses through a mechanism distinct from that of surfactant protein D or defensins. *J. Gen. Virol.* **2013**, *94*, 40–49. [CrossRef]
34. Barlow, P.; Svoboda, P.; Mackellar, A.; Nash, A.; York, I.; Pohl, J.; Davidson, D.; Donis, R. Antiviral activity and increased host defense against influenza infection elicited by the human cathelicidin LL-37. *PLoS ONE* **2011**, *6*, e25333. [CrossRef]

35. Chen, C.; Schaller-Bals, S.; Paul, K.; Wahn, U.; Bals, R. Beta-defensins and LL-37 in bronchoalveolar lavage fluid of patients with cystic fibrosis. *J. Cyst. Fibros.* **2004**, *3*, 45–50. [CrossRef]
36. Zhang, Y.; Jiang, Y.; Sun, C.; Wang, Q.; Yang, Z.; Pan, X.; Zhu, M.; Xiao, W. The human cathelicidin LL-37 enhances airway mucus production in chronic obstructive pulmonary disease. *Biochem. Biophys. Res. Commun.* **2014**, *443*, 103–109. [CrossRef]
37. Pace, E.; Ferraro, M.; Minervini, M.; Vitulo, P.; Pipitone, L.; Chiappara, G.; Siena, L.; Montalbano, A.; Johnson, M.; Gjomarkaj, M. Beta defensins-2 is reduced in central but not in distal airways of smoker COPD patients. *PLoS ONE* **2012**, *7*, e33601. [CrossRef]
38. Qureshi, H.; Sharafkhaneh, A.; Hanania, N. Chronic obstructive pulmonary disease exacerbations: Latest evidence and clinical implications. *Ther. Adv. Chronic Dis.* **2014**, *5*, 212–227. [CrossRef]
39. Niyonsaba, F.; Kiatsurayanon, C.; Chieosilapatham, P.; Ogawa, H. Friends or foes? Host defense (antimicrobial) peptides and proteins in human skin diseases. *Exp. Dermatol.* **2017**, *26*, 989–998. [CrossRef]
40. Bierkarre, H.; Harder, J.; Cuthbert, R.; Emery, P.; Leuschner, I.; Mrowietz, U.; Hedderich, J.; McGonagle, D.; Gläser, R. Differential expression of antimicrobial peptides in psoriasis and psoriatic arthritis as a novel contributory mechanism for skin and joint disease heterogeneity. *Scand. J. Rheumatol.* **2016**, *45*, 188–196. [CrossRef]
41. Coretti, L.; Natale, A.; Cuomo, M.; Florio, E.; Keller, S.; Lembo, F.; Chiariotti, L.; Pero, R. The interplay between defensins and microbiota in crohn's disease. *Mediators Inflamm.* **2017**, *2017*, 8392523. [CrossRef]
42. Wang, W.; Jin, H. Skin microbiome: An actor in the pathogenesis of psoriasis. *Chin. Med. J.* **2018**, *131*, 95–98. [CrossRef]
43. Ma, J.; Shao, S.; Wang, G. Antimicrobial peptides: Bridging innate and adaptive immunity in the pathogenesis of psoriasis. *Chin. Med. J.* **2020**, *133*, 2966–2975. [CrossRef] [PubMed]
44. Lande, R.; Botti, E.; Jandus, C.; Dojcinovic, D.; Fanelli, G.; Conrad, C.; Chamilos, G.; Feldmeyer, L.; Marinari, B.; Chon, S.; et al. The antimicrobial peptide LL37 is a T-cell autoantigen in psoriasis. *Nat. Commun.* **2014**, *5*, 5621. [CrossRef] [PubMed]
45. Ganguly, D.; Chamilos, G.; Lande, R.; Gregorio, J.; Meller, S.; Facchinetti, V.; Homey, B.; Barrat, F.; Zal, T.; Gilliet, M. Self-RNA-antimicrobial peptide complexes activate human dendritic cells through TLR7 and TLR8. *J. Exp. Med.* **2009**, *206*, 1983–1994. [CrossRef] [PubMed]
46. Röhrl, J.; Yang, D.; Oppenheim, J.; Hehlgans, T. Specific binding and chemotactic activity of mBD4 and its functional orthologue hBD2 to CCR6-expressing cells. *J. Biol. Chem.* **2010**, *285*, 7028–7034. [CrossRef] [PubMed]
47. Mabuchi, T.; Singh, T.; Takekoshi, T.; Jia, G.; Wu, X.; Kao, M.; Weiss, I.; Farber, J.; Hwang, S. CCR6 is required for epidermal trafficking of $\gamma\delta$-T cells in an IL-23-induced model of psoriasiform dermatitis. *J. Investig. Dermatol.* **2013**, *133*, 164–171. [CrossRef] [PubMed]
48. Ribon, M.; Seninet, S.; Mussard, J.; Sebbag, M.; Clavel, C.; Serre, G.; Boissier, M.; Semerano, L.; Decker, P. Neutrophil extracellular traps exert both pro- and anti-inflammatory actions in rheumatoid arthritis that are modulated by C1q and LL-37. *J. Autoimmun.* **2019**, *98*, 122–131. [CrossRef]
49. Francisco, V.; Tovar, S.; Conde, J.; Pino, J.; Mera, A.; Lago, F.; González-Gay, M.A.; Dieguez, C.; Gualillo, O. Levels of the novel endogenous antagonist of ghrelin receptor, liver-enriched antimicrobial peptide-2, in patients with rheumatoid arthritis. *Nutrients* **2020**, *12*, 1006. [CrossRef]
50. Vordenbäumen, S.; Schneider, M. Defensins: Potential effectors in autoimmune rheumatic disorders. *Polymers* **2011**, *3*, 1268–1281. [CrossRef]
51. Varoga, D.; Pufe, T.; Harder, J.; Schröder, J.; Mentlein, R.; Meyer-Hoffert, U.; Goldring, M.; Tillmann, B.; Hassenpflug, J.; Paulsen, F. Human beta-defensin 3 mediates tissue remodeling processes in articular cartilage by increasing levels of metalloproteinases and reducing levels of their endogenous inhibitors. *Arthritis Rheum.* **2005**, *52*, 1736–1745. [CrossRef]
52. Raftery, T.; Martineau, A.R.; Greiller, C.L.; Ghosh, S.; McNamara, D.; Bennett, K.; Meddings, J.; O'Sullivan, M. Effects of vitamin D supplementation on intestinal permeability, cathelicidin and disease markers in Crohn's disease: Results from a randomised double-blind placebo-controlled study. *UEG J.* **2015**, *3*, 294–302. [CrossRef]
53. Gutiérrez, A.; Holler, E.; Zapater, P.; Sempere, L.; Jover, R.; Pérez-Mateo, M.; Schoelmerich, J.; Such, J.; Wiest, R.; Francés, R. Antimicrobial peptide response to blood translocation of bacterial DNA in Crohn's disease is affected by NOD2/CARD15 genotype. *Inflamm. Bowel Dis.* **2011**, *17*, 1641–1650. [CrossRef]
54. Kusaka, S.; Nishida, A.; Takahashi, K.; Bamba, S.; Yasui, H.; Kawahara, M.; Inatomi, O.; Sugimoto, M.; Andoh, A. Expression of human cathelicidin peptide LL-37 in inflammatory bowel disease. *Clin. Exp. Immunol.* **2017**, *191*, 96–106. [CrossRef]
55. Wehkamp, J.; Salzman, N.; Porter, E.; Nuding, S.; Weichenthal, M.; Petras, R.; Shen, B.; Schaeffeler, E.; Schwab, M.; Linzmeier, R.; et al. Reduced Paneth cell alpha-defensins in ileal Crohn's disease. *Proc. Natl. Acad. Sci. USA* **2005**, *102*, 18129–18134. [CrossRef]
56. Liang, W.; Enée, E.; Andre-Vallee, C.; Falcone, M.; Sun, J.; Diana, J. Intestinal cathelicidin antimicrobial peptide shapes a protective neonatal gut microbiota against pancreatic autoimmunity. *Gastroenterology* **2022**, *162*, 1288–1302. [CrossRef]
57. Froy, O.; Hananel, A.; Chapnik, N.; Madar, Z. Differential effect of insulin treatment on decreased levels of beta-defensins and Toll-like receptors in diabetic rats. *Mol. Immunol.* **2007**, *44*, 796–802. [CrossRef]
58. Linn, O.; Menges, B.; Lammert, F.; Weber, S.; Krawczyk, M. Altered expression of antimicrobial peptides in the upper gastrointestinal tract of patients with diabetes mellitus. *Nutrients* **2023**, *15*, 754. [CrossRef]
59. Doig, C.L.; Singh, P.K.; Dhiman, V.K.; Thorne, J.L.; Battaglia, S.; Sobolewski, M.; Maguire, O.; O'Neill, L.P.; Turner, B.M.; McCabe, C.J.; et al. Recruitment of NCOR1 to VDR target genes is enhanced in prostate cancer cells and associates with altered DNA methylation patterns. *Carcinogenesis* **2012**, *34*, 248–256. [CrossRef]

60. Wong, J.; Ng, T. Sesquin, a potent defensin-like antimicrobial peptide from ground beans with inhibitory activities toward tumor cells and HIV-1 reverse transcriptase. *Peptides* **2005**, *26*, 1120–1126. [CrossRef]
61. Guzmán-Rodríguez, J.; López-Gómez, R.; Salgado-Garciglia, R.; Ochoa-Zarzosa, A.; López-Meza, J. The defensin from avocado (Persea americana var. drymifolia) PaDef induces apoptosis in the human breast cancer cell line MCF-7. *Biomed. Pharmacother.* **2016**, *82*, 620–627. [CrossRef]
62. Johnstone, S.; Gelmon, K.; Mayer, L.; Hancock, R.; Bally, M. In vitro characterization of the anticancer activity of membrane-active cationic peptides. I. Peptide-mediated cytotoxicity and peptide-enhanced cytotoxic activity of doxorubicin against wild-type and p-glycoprotein over-expressing tumor cell lines. *Anticancer Drug Des.* **2000**, *15*, 151–160. [PubMed]
63. McKeown, S.; Lundy, F.; Nelson, J.; Lockhart, D.; Irwin, C.; Cowan, C.; Marley, J. The cytotoxic effects of human neutrophil peptide-1 (HNP-1) and lactoferrin on oral squamous cell carcinoma (OSCC) in vitro. *Oral Oncol.* **2006**, *42*, 685–690. [CrossRef] [PubMed]
64. Gaspar, D.; Freire, J.; Pacheco, T.; Barata, J.; Castanho, M. Apoptotic human neutrophil peptide-1 anti-tumor activity revealed by cellular biomechanics. *Biochim. Biophys. Acta* **2015**, *1853*, 308–316. [CrossRef] [PubMed]
65. Wu, W.; Wang, G.; Coffelt, S.; Betancourt, A.; Lee, C.; Fan, D.; Wu, K.; Yu, J.; Sung, J.; Cho, C. Emerging roles of the host defense peptide LL-37 in human cancer and its potential therapeutic applications. *Int. J. Cancer* **2010**, *127*, 1741–1747. [CrossRef] [PubMed]
66. Zhang, Z.; Chen, W.; Zhang, S.; Bai, J.; Lau, C.; Sze, S.; Yung, K.; Ko, J. The human cathelicidin peptide LL-37 inhibits pancreatic cancer growth by suppressing autophagy and reprogramming of the tumor immune microenvironment. *Front. Pharmacol.* **2022**, *13*, 906625. [CrossRef] [PubMed]
67. Wu, W.; Sung, J.; To, K.; Yu, L.; Li, H.; Li, Z.; Chu, K.; Yu, J.; Cho, C. The host defense peptide LL-37 activates the tumor-suppressing bone morphogenetic protein signaling via inhibition of proteasome in gastric cancer cells. *J. Cell. Physiol.* **2010**, *223*, 178–186. [CrossRef] [PubMed]
68. Li, X.; Li, Y.; Han, H.; Miller, D.; Wang, G. Solution structures of human LL-37 fragments and NMR-based identification of a minimal membrane-targeting antimicrobial and anticancer region. *J. Am. Chem. Soc.* **2006**, *128*, 5776–5785. [CrossRef] [PubMed]
69. Ren, S.; Shen, J.; Cheng, A.; Lu, L.; Chan, R.; Li, Z.; Wang, X.; Wong, C.; Zhang, L.; Ng, S.; et al. FK-16 derived from the anticancer peptide LL-37 induces caspase-independent apoptosis and autophagic cell death in colon cancer cells. *PLoS ONE* **2013**, *8*, e63641. [CrossRef] [PubMed]
70. von Haussen, J.; Koczulla, R.; Shaykhiev, R.; Herr, C.; Pinkenburg, O.; Reimer, D.; Wiewrodt, R.; Biesterfeld, S.; Aigner, A.; Czubayko, F.; et al. The host defence peptide LL-37/hCAP-18 is a growth factor for lung cancer cells. *Lung Cancer* **2008**, *59*, 12–23. [CrossRef] [PubMed]
71. Weber, G.; Chamorro, C.I.; Granath, F.; Liljegren, A.; Zreika, S.; Saidak, Z.; Sandstedt, B.; Rotstein, S.; Mentaverri, R.; Sánchez, F.; et al. Human antimicrobial protein hCAP18/LL-37 promotes a metastatic phenotype in breast cancer. *Breast Cancer Res.* **2009**, *11*, 1–13. [CrossRef] [PubMed]
72. Amini, M.; Zayeri, F.; Salehi, M. Trend analysis of cardiovascular disease mortality, incidence, and mortality-to-incidence ratio: Results from global burden of disease study 2017. *BMC Public Health* **2021**, *21*, 1–12. [CrossRef] [PubMed]
73. Libby, P.; Buring, J.E.; Badimon, L.; Hansson, G.K.; Deanfield, J.; Bittencourt, M.S.; Tokgözoğlu, L.; Lewis, E.F. Atherosclerosis. *Nat. Rev. Dis. Primers* **2019**, *5*, 1–18. [CrossRef] [PubMed]
74. Li, Y. The role of antimicrobial peptides in cardiovascular physiology and disease. *Biochem. Biophys. Res. Commun.* **2009**, *390*, 363–367. [CrossRef] [PubMed]
75. Edfeldt, K.; Agerberth, B.; Rottenberg, M.; Gudmundsson, G.; Wang, X.; Mandal, K.; Xu, Q.; Yan, Z. Involvement of the antimicrobial peptide LL-37 in human atherosclerosis. *Arterioscler. Thromb. Vasc. Biol.* **2006**, *26*, 1551–1557. [CrossRef] [PubMed]
76. Salamah, M.; Ravishankar, D.; Kodji, X.; Moraes, L.; Williams, H.; Vallance, T.; Albadawi, D.; Vaiyapuri, R.; Watson, K.; Gibbins, J.; et al. The endogenous antimicrobial cathelicidin LL37 induces platelet activation and augments thrombus formation. *Blood Adv.* **2018**, *2*, 2973–2985. [CrossRef] [PubMed]
77. Hasenfuss, G. Animal models of human cardiovascular disease, heart failure and hypertrophy. *Cardiovasc. Res.* **1998**, *39*, 60–76. [CrossRef] [PubMed]
78. Zhou, Q.; Pan, L.; Xue, R.; Ni, G.; Duan, Y.; Bai, Y.; Shi, C.; Ren, Z.; Wu, C.; Li, G.; et al. The anti-microbial peptide LL-37/CRAMP levels are associated with acute heart failure and can attenuate cardiac dysfunction in multiple preclinical models of heart failure. *Theranostics* **2020**, *10*, 6167–6181. [CrossRef] [PubMed]
79. Bei, Y.; Pan, L.-L.; Zhou, Q.; Zhao, C.; Xie, Y.; Wu, C.; Meng, X.; Gu, H.; Xu, J.; Zhou, L.; et al. Cathelicidin-related antimicrobial peptide protects against myocardial ischemia/reperfusion injury. *BMC Med.* **2019**, *17*, 42. [CrossRef] [PubMed]
80. Lee, M.; Shi, X.; Barron, A.; McGeer, E.; McGeer, P. Human antimicrobial peptide LL-37 induces glial-mediated neuroinflammation. *Biochem. Pharmacol.* **2015**, *94*, 130–141. [CrossRef] [PubMed]
81. Soscia, S.; Kirby, J.; Washicosky, K.; Tucker, S.; Ingelsson, M.; Hyman, B.; Burton, M.; Goldstein, L.; Duong, S.; Tanzi, R.; et al. The alzheimer's disease-associated amyloid beta-protein is an antimicrobial peptide. *PLoS ONE* **2010**, *5*, e9505. [CrossRef] [PubMed]
82. Wang, M.; Peng, I.; Li, S.; Hu, X. Dysregulation of antimicrobial peptide expression distinguishes Alzheimer's disease from normal aging. *Aging* **2020**, *12*, 690–706. [CrossRef] [PubMed]
83. Bessin, Y.; Saint, N.; Marri, L.; Marchini, D.; Molle, G. Antibacterial activity and pore-forming properties of ceratotoxins: A mechanism of action based on the barrel stave model. *Biochim. Biophys. Acta* **2004**, *1667*, 148–156. [CrossRef] [PubMed]

84. Yang, L.; Harroun, T.A.; Weiss, T.M.; Ding, L.; Huang, H.W. Barrel-stave model or toroidal model? A case study on melittin pores. *Biophys. J.* **2001**, *81*, 1475–1485. [CrossRef] [PubMed]
85. Brogden, K.A. Antimicrobial peptides: Pore formers or metabolic inhibitors in bacteria? *Nat. Rev. Microbiol.* **2005**, *3*, 238–250. [CrossRef] [PubMed]
86. Ladokhin, A.; White, S. 'Detergent-like' permeabilization of anionic lipid vesicles by melittin. *Biochim. Biophys. Acta* **2001**, *1514*, 253–260. [CrossRef] [PubMed]
87. Wang, C.; Garlick, S.; Zloh, M. Deep learning for novel antimicrobial peptide design. *Biomolecules* **2021**, *11*, 471. [CrossRef] [PubMed]
88. Lee, M.; Lee, E.; Ferguson, A.; Wong, G. Machine learning antimicrobial peptide sequences: Some surprising variations on the theme of amphiphilic assembly. *Curr. Opin. Colloid Interface Sci.* **2018**, *38*, 204–213. [CrossRef] [PubMed]
89. Brötz, H.; Bierbaum, G.; Leopold, K.; Reynolds, P.E.; Sahl, H.-G. The lantibiotic mersacidin inhibits peptidoglycan synthesis by targeting lipid II. *Antimicrob. Agents Chemother.* **1998**, *41*, 154–160. [CrossRef] [PubMed]
90. Bertani, B.; Ruiz, N. Function and biogenesis of lipopolysaccharides. *EcoSal. Plus.* **2018**, *8*, 1–6. [CrossRef] [PubMed]
91. Yin, Q.; Wu, S.; Wu, L.; Wang, Z.; Mu, Y.; Zhang, R.; Dong, C.; Zhou, B.; Zhao, B.; Zheng, J.; et al. A novel in silico antimicrobial peptide DP7 combats MDR Pseudomonas aeruginosa and related biofilm infections. *J. Antimicrob. Chemother.* **2020**, *75*, 3248–3259. [CrossRef] [PubMed]
92. Dashper, S.G.; O'Brien-Simpson, N.M.; Cross, K.J.; Paolini, R.A.; Hoffmann, B.; Catmull, D.V.; Malkoski, M.; Reynolds, E.C. Divalent metal cations increase the activity of the antimicrobial peptide kappacin. *Antimicrob. Agents Chemother.* **2005**, *49*, 2322–2328. [CrossRef] [PubMed]
93. Li, J.; Hu, S.; Jian, W.; Xie, C.; Yang, X. Plant antimicrobial peptides: Structures, functions, and applications. *Bot. Stud.* **2021**, *62*, 5. [CrossRef] [PubMed]
94. Ippolito, J.; Alexander, R.; Christianson, D. Hydrogen bond stereochemistry in protein structure and function. *J. Mol. Biol.* **1990**, *215*, 457–471. [CrossRef] [PubMed]
95. Yau, W.; Wimley, W.; Gawrisch, K.; White, S. The preference of tryptophan for membrane interfaces. *Biochemistry* **1998**, *37*, 14713–14718. [CrossRef] [PubMed]
96. Mihailescu, M.; Sorci, M.; Seckute, J.; Silin, V.I.; Hammer, J.; Perrin, B.S., Jr.; Hernandez, J.I.; Smajic, N.; Shrestha, A.; Bogardus, K.A.; et al. Structure and function in antimicrobial piscidins: Histidine position, directionality of membrane insertion, and pH-dependent permeabilization. *J. Am. Chem. Soc.* **2019**, *141*, 9837–9853. [CrossRef] [PubMed]
97. Dong, W.; Mao, X.; Guan, Y.; Kang, Y.; Shang, D. Antimicrobial and anti-inflammatory activities of three chensinin-1 peptides containing mutation of glycine and histidine residues. *Sci. Rep.* **2017**, *7*, 40228. [CrossRef] [PubMed]
98. Lee, I.; Cho, Y.; Lehrer, R. Effects of pH and salinity on the antimicrobial properties of clavanins. *Infect. Immun.* **1997**, *65*, 2898–2903. [CrossRef] [PubMed]
99. Li, S.; Goto, N.; Williams, K.; Deber, C. Alpha-helical, but not beta-sheet, propensity of proline is determined by peptide environment. *Proc. Natl. Acad. Sci. USA* **1996**, *93*, 6676–6681. [CrossRef] [PubMed]
100. Vanhoof, G.; Goossens, F.; De Meester, I.; Hendriks, D.; Scharpé, S. Proline motifs in peptides and their biological processing. *FASEB J.* **1995**, *9*, 736–744. [CrossRef] [PubMed]
101. Li, C.; Sutherland, D.; Hammond, S.; Yang, C.; Taho, F.; Bergman, L.; Houston, S.; Warren, R.; Wong, T.; Hoang, L.; et al. AMPlify: Attentive deep learning model for discovery of novel antimicrobial peptides effective against WHO priority pathogens. *BMC Genom.* **2022**, *23*, 77. [CrossRef] [PubMed]
102. Aronica, P.G.A.; Reid, L.M.; Desai, N.; Li, J.; Fox, S.J.; Yadahalli, S.; Essex, J.W.; Verma, C.S. Computational methods and tools in antimicrobial peptide research. *J. Chem. Inf. Model.* **2021**, *61*, 3172–3196. [CrossRef] [PubMed]
103. Bahar, A.A.; Ren, D. Antimicrobial Peptides. *Pharmaceuticals* **2013**, *6*, 1543–1575. [CrossRef] [PubMed]
104. Falanga, A.; Lombardi, L.; Franci, G.; Vitiello, M.; Iovene, M.; Morelli, G.; Galdiero, M.; Galdiero, S. Marine antimicrobial peptides: Nature provides templates for the design of novel compounds against pathogenic bacteria. *Int. J. Mol. Sci.* **2016**, *17*, 785. [CrossRef] [PubMed]
105. National Center for Biotechnology Information. PubChem Compound Summary for CID 10909430, Bacitracin A. Available online: https://pubchem.ncbi.nlm.nih.gov/compound/Bacitracin-A (accessed on 15 December 2022).
106. National Center for Biotechnology Information. PubChem Compound Summary for CID 16129629, Daptomycin. Available online: https://pubchem.ncbi.nlm.nih.gov/compound/Daptomycin (accessed on 15 December 2022).
107. National Center for Biotechnology Information. PubChem Compound Summary for CID 133065662, Teicoplanin. Available online: https://pubchem.ncbi.nlm.nih.gov/compound/Teicoplanin (accessed on 15 December 2022).
108. National Center for Biotechnology Information. PubChem Compound Summary for CID 14969, Vancomycin. Available online: https://pubchem.ncbi.nlm.nih.gov/compound/Vancomycin (accessed on 15 December 2022).
109. National Center for Biotechnology Information. PubChem Compound Summary for CID 44144393, Colistin. Available online: https://pubchem.ncbi.nlm.nih.gov/compound/Colistin (accessed on 15 December 2022).
110. National Center for Biotechnology Information. PubChem Compound Summary for CID 45267103, Gramicidin A. Available online: https://pubchem.ncbi.nlm.nih.gov/compound/Gramicidin-A (accessed on 15 December 2022).
111. National Center for Biotechnology Information. PubChem Compound Summary for CID 16134627, Dalbavancin. Available online: https://pubchem.ncbi.nlm.nih.gov/compound/Dalbavancin (accessed on 15 December 2022).

112. National Center for Biotechnology Information. PubChem Compound Summary for CID 16136912, Oritavancin. Available online: https://pubchem.ncbi.nlm.nih.gov/compound/Oritavancin (accessed on 15 December 2022).
113. National Center for Biotechnology Information. PubChem Compound Summary for CID 3081362, Telavancin. Available online: https://pubchem.ncbi.nlm.nih.gov/compound/Telavancin (accessed on 15 December 2022).
114. Zhanel, G.; Calic, D.; Schweizer, F.; Zelenitsky, S.; Adam, H.; Lagacé-Wiens, P.; Rubinstein, E.; Gin, A.; Hoban, D.; Karlowsky, J. New lipoglycopeptides: A comparative review of dalbavancin, oritavancin and telavancin. *Drugs* **2010**, *70*, 859–886. [CrossRef] [PubMed]
115. Swierstra, J.; Kapoerchan, V.; Knijnenburg, A.; van Belkum, A.; Overhand, M. Structure, toxicity and antibiotic activity of gramicidin S and derivatives. *Eur. J. Clin. Microbiol. Infect. Dis.* **2016**, *35*, 763–769. [CrossRef] [PubMed]
116. National Center for Biotechnology Information. PubChem Compound Summary for CID 10324367, Boceprevir. Available online: https://pubchem.ncbi.nlm.nih.gov/compound/Boceprevir (accessed on 15 December 2022).
117. National Center for Biotechnology Information. PubChem Compound Summary for CID 3010818, Telaprevir. Available online: https://pubchem.ncbi.nlm.nih.gov/compound/Telaprevir (accessed on 15 December 2022).
118. National Center for Biotechnology Information. PubChem Compound Summary for CID 16130199, Enfuvirtide. Available online: https://pubchem.ncbi.nlm.nih.gov/compound/Enfuvirtide (accessed on 15 December 2022).
119. Drugs.Com. Locilex (Pexiganan) FDA Approval Status. Available online: https://www.drugs.com/history/locilex.html (accessed on 5 February 2023).
120. Gomes, D.; Santos, R.; Soares, R.; Reis, S.; Carvalho, S.; Rego, P.; Peleteiro, M.; Tavares, L.; Oliveira, M. Pexiganan in combination with nisin to control polymicrobial diabetic foot infections. *Antibiotics* **2020**, *9*, 128. [CrossRef] [PubMed]
121. Julliard, G. Iseganan hydrochloride in preventing oral mucositis in patients who are undergoing radiation therapy for head and neck cancer. Identifier NCT00022373 2004, February—2014, October. Available online: https://clinicaltrials.gov/ct2/show/NCT00022373 (accessed on 24 April 2022).
122. Wu, X.; Li, Z.; Li, X.; Tian, Y.; Fan, Y.; Yu, C.; Zhou, B.; Liu, Y.; Xiang, R.; Yang, L. Synergistic effects of antimicrobial peptide DP7 combined with antibiotics against multidrug-resistant bacteria. *Drug Des. Devel. Ther.* **2017**, *11*, 939–946. [CrossRef] [PubMed]
123. Zhang, R.; Jiang, X.; Qiao, J.; Wang, Z.; Tong, A.; Yang, J.; Yang, S.; Yang, L. Antimicrobial peptide DP7 with potential activity against SARS coronavirus infections. *Signal Transduct. Target. Ther.* **2021**, *6*, 140. [CrossRef] [PubMed]
124. Davies, D.G.; Parsek, M.R.; Pearson, J.P.; Iglewski, B.H.; Costerton, J.W.; Greenberg, E.P. The involvement of cell-to-cell signals in the development of a bacterial biofilm. *Science* **1998**, *280*, 295–298. [CrossRef] [PubMed]
125. Wang, G.; Narayana, J.; Mishra, B.; Zhang, Y.; Wang, F.; Wang, C.; Zarena, D.; Lushnikova, T.; Wang, X. Design of antimicrobial peptides: Progress made with human cathelicidin LL-37. *Adv. Exp. Med. Biol.* **2019**, *1117*, 215–240. [CrossRef] [PubMed]
126. Anahtar, M.; Yang, J.; Kanjilal, S. Applications of machine learning to the problem of antimicrobial resistance: An emerging model for translational research. *J. Clin. Microbiol.* **2021**, *59*, e0126020. [CrossRef] [PubMed]
127. Fjell, C.D.; Hiss, J.A.; Hancock, R.E.W.; Schneider, G. Designing antimicrobial peptides: Form follows function. *Nat. Rev. Drug Discov.* **2012**, *11*, 37–51. [CrossRef] [PubMed]
128. Peschel, A.; Sahl, H.-G. The co-evolution of host cationic antimicrobial peptides and microbial resistance. *Nat. Rev. Microbiol.* **2006**, *4*, 529–536. [CrossRef] [PubMed]
129. Nizet, V. Antimicrobial peptide resistance mechanisms of human bacterial pathogens. *Curr. Issues Mol. Biol.* **2006**, *8*, 11–26. [CrossRef] [PubMed]
130. Nawrocki, K.L.; Crispell, E.K.; McBride, S.M. Antimicrobial peptide resistance mechanisms of gram-positive bacteria. *Antibiotics* **2014**, *3*, 461–492. [CrossRef] [PubMed]
131. Band, V.I.; Weiss, D.S. Mechanisms of antimicrobial peptide resistance in gram-negative bacteria. *Antibiotics* **2014**, *4*, 18–41. [CrossRef] [PubMed]
132. Deshayes, C.; Arafath, M.; Apaire-Marchais, V.; Roger, E. Drug delivery systems for the oral administration of antimicrobial peptides: Promising tools to treat infectious diseases. *Front. Med. Technol.* **2022**, *3*, 778645. [CrossRef] [PubMed]
133. Deslouches, B.; Steckbeck, J.; Craigo, J.; Doi, Y.; Mietzner, T.; Montelaro, R. Rational design of engineered cationic antimicrobial peptides consisting exclusively of arginine and tryptophan, and their activity against multidrug-resistant pathogens. *Antimicrob. Agents Chemother.* **2013**, *57*, 2511–2521. [CrossRef]
134. Lakshmaiah Narayana, J.; Golla, R.; Mishra, B.; Wang, X.; Lushnikova, T.; Zhang, Y.; Verma, A.; Kumar, V.; Xie, J.; Wang, G. Short and robust anti-infective lipopeptides engineered based on the minimal antimicrobial peptide KR12 of human LL-37. *ACS Infect. Dis.* **2021**, *7*, 1795–1808. [CrossRef] [PubMed]
135. Plisson, F.; Ramírez-Sánchez, O.; Martínez-Hernández, C. Machine learning-guided discovery and design of non-hemolytic peptides. *Sci. Rep.* **2020**, *10*, 16581. [CrossRef] [PubMed]
136. Lin, Y.; Cai, Y.; Liu, J.; Lin, C.; Liu, X. An advanced approach to identify antimicrobial peptides and their function types for penaeus through machine learning strategies. *BMC Bioinform.* **2019**, *20*, 291. [CrossRef]
137. Vishnoi, S.; Matre, H.; Garg, P.; Pandev, S. Artificial intelligence and machine learning for protein toxicity prediction using proteomics data. *Chem. Biol. Drug Des.* **2020**, *96*, 902–920. [CrossRef]
138. Santos-Júnior, C.; Pan, S.; Zhao, X.; Coelho, L. Macrel: Antimicrobial peptide screening in genomes and metagenomes. *PeerJ* **2020**, *8*, e10555. [CrossRef]

139. Huerta-Cepas, J.; Forslund, K.; Coelho, L.P.; Szklarczyk, D.; Jensen, L.J.; von Mering, C.; Bork, P. Fast genome-wide functional annotation through orthology assignment by eggNOG-mapper. *Mol. Biol. Evol.* **2017**, *34*, 2115–2122. [CrossRef]
140. Nguyen, R.; Khanna, N.R.; Safadi, A.O.; Sun, Y. Bacitracin Topical. Available online: https://www.ncbi.nlm.nih.gov/books/NBK536993 (accessed on 18 December 2022).
141. Howard, B. *Bacitracin, xPharm: The Comprehensive Pharmacology Reference*; Elsevier: Amsterdam, The Netherlands, 2007; pp. 1–4. [CrossRef]
142. Merck & CO. Inc. Victrelis (Boceprevir). Available online: https://www.accessdata.fda.gov/drugsatfda_docs/label/2011/202258lbl.pdf (accessed on 18 November 2022).
143. Bialvaei, A.Z.; Kafil, H.S. Colistin, mechanisms and prevalence of resistance. *Curr. Med. Res. Opin.* **2015**, *31*, 707–721. [CrossRef]
144. Andrade, F.; Silva, D.; Rodrigues, A.; Pina-Vaz, C. Colistin Update on Its Mechanism of Action and Resistance, Present and Future Challenges. *Microorganisms* **2020**, *8*, 1716. [CrossRef]
145. Durata Therapeutics U.S Limited. Dalvance. Available online: https://www.accessdata.fda.gov/drugsatfda_docs/label/2016/021883s003lbl.pdf (accessed on 18 November 2022).
146. Merck Sharp & Dohme Corp. Cubicin (Daptomycin for Injection). Available online: https://www.accessdata.fda.gov/drugsatfda_docs/label/2017/021572s059lbl.pdf (accessed on 18 November 2022).
147. Cotroneo, N.; Harris, R.; Perlmutter, N.; Beveridge, T.; Silverman, J. Daptomycin exerts bactericidal activity without lysis of staphylococcus aureus. *Antimicrob. Agents Chemother.* **2008**, *52*, 2223–2225. [CrossRef]
148. Hoffmann-La Roche Inc. Fuzeon (Enfuvirtide). Available online: https://www.accessdata.fda.gov/drugsatfda_docs/label/2019/021481s033lbl.pdf (accessed on 18 November 2022).
149. Wishart, D.; Knox, C.; Guo, A.; Shrivastava, S.; Hassanali, M.; Stothard, P.; Chang, Z.; Woolsey, J. Drugbank: A comprehensive resource for in silico drug discovery and exploration. *Nucleic Acids Res.* **2006**, *34*, D668–D672. [CrossRef]
150. The Medicines Company. Orbactiv. Available online: https://www.accessdata.fda.gov/drugsatfda_docs/label/2014/206334s000lbl.pdf (accessed on 18 November 2022).
151. Somma, S.; Gastaldo, L.; Corti, A. Teicoplanin, a new antibiotic from actinoplanes teichomyceticus nov. sp. *Antimicrob. Agents Chemother.* **1984**, *26*, 917–923. [CrossRef]
152. Campoli-Richards, D.M.; Brogden, R.N.; Faulds, D. Teicoplanin. *Drugs* **1990**, *40*, 449–486. [CrossRef]
153. Vertex Pharmaceuticals. Incivek (Telaprevir). Available online: https://www.accessdata.fda.gov/drugsatfda_docs/label/2011/201917lbl.pdf (accessed on 18 November 2022).
154. Astellas Pharma Inc. Vibativ (Televancin). Available online: https://www.accessdata.fda.gov/drugsatfda_docs/label/2009/022110s000lbl.pdf (accessed on 18 November 2022).
155. Lunde, C.; Hartouni, S.; Janc, J.; Mammen, M.; Humphrey, P.; Benton, B. Telavancin disrupts the functional integrity of the bacterial membrane through targeted interaction with the cell wall precursor lipid II. *Antimicrob. Agents Chemother.* **2009**, *53*, 3375–3383. [CrossRef] [PubMed]
156. Patel, S.; Preuss, C.V.; Bernice, F. Vancomycin. Available online: https://www.ncbi.nlm.nih.gov/books/NBK459263 (accessed on 13 February 2023).
157. Watanakunakorn, C. The antibacterial action of vancomycin. *Rev. Infect. Dis.* **1981**, *3*, S210–S215. [CrossRef]
158. Xellia Pharmaceuticals USA. Vancocin. Available online: https://www.accessdata.fda.gov/drugsatfda_docs/label/2019/211962s000lbl.pdf (accessed on 18 November 2022).
159. El Naqa, I.; Murphy, M. What is machine learning? In *Machine Learning in Radiation Oncology: Theory and Applications*; Springer: Berlin/Heidelberg, Germany, 2015; pp. 3–11. [CrossRef]
160. Mahesh, B. Machine learning alorithms. *Int. J. Sci. Res.* **2020**, *9*, 381–386. [CrossRef]
161. Moretta, A.; Scieuzo, C.; Salvia, R.; Popović, Ž.; Sgambato, A.; Falabella, P. Tools in the era of multidrug resistance in bacteria: Applications for new antimicrobial peptides discovery. *Curr. Pharm. Des.* **2022**, *28*, 2856–2866. [CrossRef]
162. Drin, G.; Antonny, B. Amphipathic helices and membrane curvature. *FEBS Lett.* **2009**, *584*, 1840–1847. [CrossRef]
163. Ma, Y.; Liu, K.; Guan, Z.; Xu, X.; Qian, X.; Bao, H. Background Augmentation Generative Adversarial Networks (BAGANs): Effective Data Generation Based on GAN-Augmented 3D Synthesizing. *Symmetry* **2018**, *10*, 734. [CrossRef]
164. Liu, S.; Fan, L.; Sun, J.; Lao, X.; Zheng, H. Computational resources and tools for antimicrobial peptides. *J. Pept. Sci.* **2016**, *23*, 4–12. [CrossRef] [PubMed]
165. Wang, Z.; Wang, G. APD: The antimicrobial peptide database. *Nucleic Acids Res.* **2004**, *32*, D590–D592. [CrossRef]
166. Wang, G.; Zietz, C.; Mudgapalli, A.; Wang, S.; Wang, Z. The evolution of the antimicrobial peptide database over 18 years: Milestones and new features. *Protein Sci.* **2022**, *31*, 92–106. [CrossRef]
167. Waghu, F.; Idicula-Thomas, S. Collection of antimicrobial peptides database and its derivatives: Applications and beyond. *Protein Sci.* **2020**, *29*, 36–42. [CrossRef]
168. Thomas, S.; Karnik, S.; Baral, R.; Jayaraman, V.; Idicula-Thomas, S. CAMP: A useful resource for research on antimicrobial peptides. *Nucleic Acids Res.* **2010**, *38*, D774–D780. [CrossRef] [PubMed]
169. Zhao, X.; Wu, H.; Lu, H.; Li, G.; Huang, Q. LAMP: A database linking antimicrobial peptides. *PLoS ONE* **2013**, *8*, e66557. [CrossRef] [PubMed]
170. Ye, G.; Wu, H.; Huang, J.; Wang, W.; Ge, K.; Li, G.; Zhong, J.; Huang, Q. LAMP2: A major update of the database linking antimicrobial peptides. *Database* **2020**, *2020*, baaa061. [CrossRef] [PubMed]

171. Fan, L.; Sun, J.; Zhou, M.; Zhou, J.; Lao, X.; Zheng, H.; Xu, H. DRAMP: A comprehensive data repository of antimicrobial peptides. *Sci. Rep.* **2016**, *6*, 24482. [CrossRef]
172. Quiroz, C.; Saavedra, Y.B.; Armijo-Galdames, B.; Amado-Hinojosa, J.; Olivera-Nappa, Á.; Sanchez-Daza, A.; Medina-Ortiz, D. Peptipedia: A user-friendly web application and a comprehensive database for peptide research supported by machine learning approach. *Database* **2021**, *2021*, baab055. [CrossRef]
173. Gautam, A.; Chaudhary, K.; Singh, S.; Joshi, A.; Anand, P.; Tuknait, A.; Mathur, D.; Varshney, G.; Raghava, G. Hemolytik: A database of experimentally determined hemolytic and non-hemolytic peptides. *Nucleic Acids Res.* **2013**, *42*, D444–D449. [CrossRef]
174. Just-Baringo, X.; Albericio, F.; Alvarez, M. Thiopeptide engineering: A multidisciplinary effort towards future drugs. *Angew. Chem. Int. Ed. Engl.* **2014**, *53*, 6602–6616. [CrossRef]
175. Mulvenna, J.; Wang, C.; Craik, D. CyBase: A database of cyclic protein sequence and structure. *Nucleic Acids Res.* **2006**, *34*, D192–D194. [CrossRef]
176. Colgrave, M.L.; Craik, D.J. Thermal, chemical, and enzymatic stability of the cyclotide kalata B1: The importance of the cyclic cystine knot. *Biochemistry* **2004**, *43*, 5965–5975. [CrossRef]
177. Shi, G.; Kang, X.; Dong, F.; Liu, Y.; Zhu, N.; Hu, Y.; Xu, H.; Lao, X.; Zheng, H. DRAMP 3.0: An enhanced comprehensive data repository of antimicrobial peptides. *Nucleic Acids Res.* **2022**, *50*, D488–D496. [CrossRef]
178. Jhong, J.-H.; Yao, L.; Pang, Y.; Li, Z.; Chung, C.-R.; Wang, R.; Li, S.; Li, W.; Luo, M.; Ma, R.; et al. dbAMP 2.0: Updated resource for antimicrobial peptides with an enhanced scanning method for genomic and proteomic data. *Nucleic Acids Res.* **2022**, *50*, D460–D470. [CrossRef] [PubMed]
179. Pirtskhalava, M.; Amstrong, A.A.; Grigolava, M.; Chubinidze, M.; Alimbarashvili, E.; Vishnepolsky, B.; Gabrielian, A.; Rosenthal, A.; Hurt, D.E.; Tartakovsky, M. DBAASP v3: Database of antimicrobial/cytotoxic activity and structure of peptides as a resource for development of new therapeutics. *Nucleic Acids Res.* **2021**, *49*, D288–D297. [CrossRef]
180. Doster, E.; Lakin, S.; Dean, C.; Wolfe, C.; Young, J.; Boucher, C.; Belk, K.; Noyes, N.; Morley, P. MEGARes 2.0: A database for classification of antimicrobial drug, biocide and metal resistance determinants in metagenomic sequence data. *Nucleic Acids Res.* **2020**, *48*, D561–D569. [CrossRef] [PubMed]
181. Lee, H.; Lee, C.; Yang, J.; Lai, J.; Chang, K. A large-scale structural classification of antimicrobial peptides. *Biomed. Res. Int.* **2015**, *2015*, 475062. [CrossRef] [PubMed]
182. Li, J.; Qu, X.; He, X.; Duan, L.; Wu, G.; Bi, D.; Deng, Z.; Liu, W.; Ou, H.-Y. ThioFinder: A web-based tool for the identification of thiopeptide gene clusters in DNA sequences. *PLoS ONE* **2012**, *7*, e45878. [CrossRef]
183. Wang, C.; Kaas, Q.; Chiche, L.; Craik, D. CyBase: A database of cyclic protein sequences and structures, with applications in protein discovery and engineering. *Nucleic Acids Res.* **2007**, *36*, D206–D210. [CrossRef]
184. Seebah, S.; Suresh, A.; Zhuo, S.; Choong, Y.; Chua, H.; Chuon, D.; Beuerman, R.; Verma, C. Defensins knowledgebase: A manually curated database and information source focused on the defensins family of antimicrobial peptides. *Nucleic Acids Res.* **2007**, *35*, D265–D268. [CrossRef]
185. Gómez, E.; Giraldo, P.; Orduz, S. InverPep: A database of invertebrate antimicrobial peptides. *J. Glob. Antimicrob. Resist.* **2017**, *8*, 13–17. [CrossRef]
186. Hammami, R.; Zouhir, A.; Ben Hamida, J.; Fliss, I. BACTIBASE: A new web-accessible database for bacteriocin characterization. *BMC Microbiol.* **2007**, *7*, 89. [CrossRef]
187. Piotto, S.; Sessa, L.; Concilio, S.; Iannelli, P. YADAMP: Yet another database of antimicrobial peptides. *Int. J. Antimicrob. Agents* **2012**, *39*, 346–351. [CrossRef]
188. Whitmore, L.; Wallace, B. The peptaibol database: A database for sequences and structures of naturally occurring peptaibols. *Nucleic Acids Res.* **2004**, *32*, D593–D594. [CrossRef]
189. Novković, M.; Simunić, J.; Bojović, V.; Tossi, A.; Juretić, D. DADP: The database of anuran defense peptides. *Bioinformatics* **2012**, *28*, 1406–1407. [CrossRef]
190. Di Luca, M.; Maccari, G.; Maisetta, G.; Batoni, G. BaAMPs: The database of biofilm-active antimicrobial peptides. *Biofouling* **2015**, *31*, 193–199. [CrossRef]
191. Tyagi, A.; Tuknait, A.; Anand, P.; Gupta, S.; Sharma, M.; Mathur, D.; Joshi, A.; Singh, S.; Gautam, A.; Raghava, G. CancerPPD: A database of anticancer peptides and proteins. *Nucleic Acids Res.* **2015**, *43*, D837–D843. [CrossRef]
192. Mehta, D.; Anand, P.; Kumar, V.; Joshi, A.; Mathur, D.; Singh, S.; Tuknait, A.; Chaudhary, K.; Gautam, S.; Gautam, A.; et al. ParaPep: A web resource for experimentally validated antiparasitic peptide sequences and their structures. *Database* **2014**, *2014*, bau051. [CrossRef]
193. Yan, J.; Cai, J.; Zhang, B.; Wang, Y.; Wong, D.; Siu, S. Recent progress in the discovery and design of antimicrobial peptides using traditional machine learning and deep learning. *Antibiotics* **2022**, *11*, 1451. [CrossRef]
194. Spänig, S.; Heider, D. Encodings and models for antimicrobial peptide classification for multi-resistant pathogens. *BioData Min.* **2019**, *12*, 1–29. [CrossRef]
195. Erjavac, I.; Kalafatovic, D.; Mausa, G. Coupled encoding methods for antimicrobial peptide prediction: How sensitive is a highly accurate model? *Artif. Intell. Life Sci.* **2022**, *2*, 100034. [CrossRef]
196. Fjell, C.; Jenssen, H.; Hilpert, K.; Cheung, W.; Pante, N.; Hancock, R.; Cherkasov, A. Identification of novel antibacterial peptides by chemoinformatics and machine learning. *J. Med. Chem.* **2009**, *52*, 2006–2015. [CrossRef]

197. Sander, O.; Sing, T.; Sommer, I.; Low, A.; Cheung, P.; Harrigan, P.; Lengauer, T.; Domingues, F. Structural descriptors of gp120 V3 loop for the prediction of HIV-1 coreceptor usage. *PLoS Comput. Biol.* **2007**, *3*, e58. [CrossRef]
198. Naili, M.; Chaibi, A.; Ghezala, H. Comparative study of word embedding methods in topic segmentation. *Procedia Comput. Sci.* **2017**, *112*, 340–349. [CrossRef]
199. Devlin, J.; Chang, M.-W.; Lee, K.; Toutanova, K. BERT: Pre-training of deep bidirectional transformers for language understanding. *ACL Anthrol.* **2019**, *1*, 4171–4186. [CrossRef]
200. Raffel, C.; Shazeer, N.; Roberts, A.; Lee, K.; Narang, S.; Matena, M.; Zhou, Y.; Li, W.; Liu, P.J. Exploring the limits of transfer learning with a unified text-to-text transformer. *J. Mach. Learn. Res.* **2020**, *21*, 5485–5551.
201. Shi, F.; Kai, S.; Zheng, J.; Zhong, Y. XLNet-based prediction model for CVSS metric values. *Appl. Sci.* **2022**, *12*, 8983. [CrossRef]
202. Guo, G.; Wang, H.; Bell, D.; Bi, Y.; Greer, K. *KNN Model-Based Approach in Classification*; SpringerLink: Berlin/Heidelberg, Germany, 2003; Volume 2888.
203. Sarker, I.H. Machine learning: Algorithms, real-world applications and research directions. *SN Comput. Sci.* **2021**, *2*, 160. [CrossRef]
204. Ayyadevara, K. *Pro Machine Learning Algorithms*; Apress: Berkeley, CA, USA, 2018.
205. Gaye, B.; Zhang, D.; Wulamu, A. Improvement of support vector machine algorithm in big data background. *Math. Probl. Eng.* **2021**, *2021*, 5594899. [CrossRef]
206. Di Franco, G.; Santurro, M. Machine learning, artificial neural networks and social research. *Qual. Quant.* **2021**, *55*, 1007–1025. [CrossRef]
207. Sarker, I.H. Deep cybersecurity: A comprehensive overview from neural network and deep learning perspective. *SN Comput. Sci.* **2021**, *2*, 154. [CrossRef]
208. Breiman, L. Bagging predictors. *Mach. Learn.* **1996**, *24*, 123–140. [CrossRef]
209. Ajanki, A. Example of K-Nearest Neighbour Classification. CC BY-SA 3.0. Available online: https://commons.wikimedia.org/wiki/File:KnnClassification.svg (accessed on 21 December 2022).
210. Muzzammel, R.; Raza, A. A Support Vector Machine Learning-Based Protection Technique for MT-HVDC Systems. *Energies* **2020**, *13*, 6668. [CrossRef]
211. Burnett, C. Illustration of the Topology of a Generic Artificial Neural Network (ANN). CC BY-SA 3.0. Available online: https://commons.wikimedia.org/wiki/File:Artificial_neural_network.svg (accessed on 21 December 2022).
212. Jagannath, V. Random Forest. CC BY-SA 4.0. Available online: https://commons.wikimedia.org/wiki/File:Random_forest_diagram_complete.png (accessed on 21 December 2022).
213. Lata, S.; Mishra, N.K.; Raghava, G.P. AntiBP2: Improved version of antibacterial peptide prediction. *BMC Bioinform.* **2010**, *11*, S19. [CrossRef]
214. Xiao, X.; Wang, P.; Lin, W.; Jia, J.; Chou, K. iAMP-2L: A two-level multi-label classifier for identifying antimicrobial peptides and their functional types. *Anal. Biochem.* **2013**, *436*, 168–177. [CrossRef]
215. Torrent, M.; Andreu, D.; Nogués, V.; Boix, E. Connecting peptide physicochemical and antimicrobial properties by a rational prediction model. *PLoS ONE* **2011**, *6*, e16968. [CrossRef]
216. Joseph, S.; Karnik, S.; Nilawe, P.; Jayaraman, V.; Idicula-Thomas, S. ClassAMP: A prediction tool for classification of antimicrobial peptides. *IEEE/ACM Trans. Comput. Biol. Bioinform.* **2012**, *9*, 1535–1538. [CrossRef]
217. Lawrence, T.J.; Carper, D.L.; Spangler, M.K.; Carrell, A.A.; Rush, T.A.; Minter, S.J.; Weston, D.J.; Labbé, J.L. amPEPpy 1.0: A portable and accurate antimicrobial peptide prediction tool. *Bioinformatics* **2021**, *37*, 2058–2060. [CrossRef]
218. Capecchi, A.; Cai, X.; Personne, H.; Köhler, T.; Delden, C.V.; Reymond, J.-L. Machine learning designs non-hemolytic antimicrobial peptides. *Chem. Sci.* **2021**, *12*, 9221–9232. [CrossRef]
219. Fausett, L. *Fundamentals of Neural Networks*; Pearson: London, UK, 1944.
220. Sharma, A.; Gupta, P.; Kumar, R.; Bhardwaj, A. dPABBs: A novel in silico approach for predicting and designing anti-biofilm peptides. *Sci. Rep.* **2016**, *6*, 21839. [CrossRef]
221. Zhang, R.; Wang, Z.; Tian, Y.; Yin, Q.; Cheng, X.; Lian, M.; Zhou, B.; Zhang, X.; Yang, L. Efficacy of antimicrobial peptide DP7, designed by machine-learning method, against methicillin-resistant staphylococcus aureus. *Front. Microbiol.* **2019**, *10*, 1175. [CrossRef]
222. Sieprawska-Lupa, M.; Mydel, P.; Krawczyk, K.; Wójcik, K.; Puklo, M.; Lupa, B.; Suder, P.; Silberring, J.; Reed, M.; Pohl, J.; et al. Degradation of human antimicrobial peptide LL-37 by staphylococcus aureus-derived proteinases. *Antimicrob. Agents Chemother.* **2004**, *48*, 4673–4679. [CrossRef]
223. Shi, S.; Di, L. *Strategies to Optimize Peptide Stability and Prolong Half-Life*; Springer: Berlin/Heidelberg, Germany, 2022; Volume 47.
224. Alvares, D.; Wilke, N.; Ruggiero, N.J. Effect of N-terminal acetylation on lytic activity and lipid-packing perturbation induced in model membranes by a mastoparan-like peptide. *Biochim. Biophys. Acta Biomembr.* **2018**, *1860*, 737–748. [CrossRef]
225. Zhang, F.; Guo, Z.; Chen, Y.; Li, L.; Yu, H.; Wang, Y. Effects of C-terminal amidation and heptapeptide ring on the biological activities and advanced structure of amurin-9KY, a novel antimicrobial peptide identified from the brown frog, Rana kunyuensis. *Zool. Res.* **2019**, *40*, 198–204. [CrossRef] [PubMed]
226. Li, D.; Yang, Y.; Li, R.; Huang, L.; Wang, Z.; Deng, Q.; Dong, S. N-terminal acetylation of antimicrobial peptide L163 improves its stability against protease degradation. *J. Pept. Sci.* **2021**, *27*, e3337. [CrossRef] [PubMed]

227. Kang, S.-J.; Nam, S.H.; Lee, B.-J. Engineering approaches for the development of antimicrobial peptide-based antibiotics. *Antibiotics* **2022**, *11*, 1338. [CrossRef] [PubMed]
228. Lu, J.; Xu, H.; Xia, J.; Ma, J.; Xu, J.; Li, Y.; Feng, J. D- and unnatural amino acid substituted antimicrobial peptides with improved proteolytic resistance and their proteolytic degradation characteristics. *Front. Microbiol.* **2020**, *11*, 563030. [CrossRef] [PubMed]
229. Wang, G.; Hanke, M.; Mishra, B.; Lushnikova, T.; Heim, C.; Chittezham, T.V.; Bayles, K.; Kielian, T. Transformation of human cathelicidin LL-37 into selective, stable, and potent antimicrobial compounds. *ACS Chem. Biol.* **2014**, *9*, 1997–2002. [CrossRef]
230. White, J.; Muhammad, T.; Alsheim, E.; Mohanty, S.; Blasi-Romero, A.; Gunasekera, S.; Stromstedt, A.; Ferraz, N.; Goransson, U.; Brauner, A. A stable cyclized antimicrobial peptide derived from LL-37 with host immunomodulatory effects and activity against uropathogens. *Cell. Mol. Life Sci.* **2022**, *79*, 411. [CrossRef]
231. Fadaka, A.O.; Sibuyi, N.R.S.; Madiehe, A.M.; Meyer, M. Nanotechnology-based delivery systems for antimicrobial peptides. *Pharmaceutics* **2021**, *13*, 1795. [CrossRef]
232. Emilian Leucuta, S. Nanotechnology for delivery of drugs and biomedical applications. *Pharmacology* **2010**, *5*, 257–280. [CrossRef]
233. Martin-Serrano, Á.; Gómez, R.; Ortega, P.; de la Mata, F. Nanosystems as vehicles for the delivery of antimicrobial peptides. *Pharmaceutics* **2019**, *11*, 448. [CrossRef]
234. Thapa, R.K.; Diep, D.B.; Tønnesen, H.H. Nanomedicine-based antimicrobial peptide delivery for bacterial infections: Recent advances and future prospects. *J. Pharm. Investig.* **2021**, *51*, 377–398. [CrossRef]
235. Zhang, Y.; Shareena Dasari, T.; Deng, H.; Yu, H. Antimicrobial Activity of Gold Nanoparticles and Ionic Gold. *J. Environ. Sci. Health C. Environ. Carcinog. Ecotoxicol. Rev.* **2015**, *33*, 286–327. [CrossRef]
236. Singh, B.; Garg, T.; Goyal, A.; Rath, G. Recent advancements in the cardiovascular drug carriers. *Artif. Cells Nanomed. Biotechnol.* **2016**, *44*, 216–225. [CrossRef]
237. Ahmad, N.; Wee, C.; Wai, L.; Zin, N.; Azmi, F. Biomimetic amphiphilic chitosan nanoparticles: Synthesis, characterization, and antimicrobial activity. *Carbohydr. Polym.* **2021**, *254*, 117299. [CrossRef]
238. Falanga, A.; Genio, V.D.; Galdiero, S. Peptides and dendrimers: How to combat viral and bacterial infections. *Pharmaceutics* **2021**, *13*, 101. [CrossRef]
239. Rai, A.; Pinto, S.; Velho, T.; Ferreira, A.; Moita, C.; Trivedi, U.; Evangelista, M.; Comune, M.; Rumbaugh, K.; Simões, P.; et al. One-step synthesis of high-density peptide-conjugated gold nanoparticles with antimicrobial efficacy in a systemic infection model. *Biomaterials* **2016**, *85*, 99–110. [CrossRef]
240. Xiu, Z.-M.; Zhang, Q.-B.; Puppala, H.L.; Colvin, V.L.; Alvarez, P.J.J. Negligible particle-specific antibacterial activity of silver nanoparticles. *Nano Lett.* **2012**, *12*, 4271–4275. [CrossRef]
241. Mei, L.; Lu, Z.; Zhang, W.; Wu, Z.; Zhang, X.; Wang, Y.; Luo, Y.; Li, C.; Jia, Y. Bioconjugated nanoparticles for attachment and penetration into pathogenic bacteria. *Biomaterials* **2013**, *34*, 10328–10337. [CrossRef]
242. Li, Z.; Tan, S.; Li, S.; Shen, Q.; Wang, K. Cancer drug delivery in the nano era: An overview and perspectives. *Oncol. Rep.* **2017**, *38*, 611–624. [CrossRef]
243. Li, Y.; Su, T.; Zhang, Y.; Huang, X.; Li, J.; Li, C. Liposomal co-delivery of daptomycin and clarithromycin at an optimized ratio for treatment of methicillin-resistant staphylococcus aureus infection. *Drug Deliv.* **2015**, *22*, 627–637. [CrossRef]
244. Sharaf, M.; Arif, M.; Khan, S.; Abdalla, M.; Shabana, S.; Chi, Z.; Liu, C. Co-delivery of hesperidin and clarithromycin in a nanostructured lipid carrier for the eradication of Helicobacter pylori in vitro. *Bioorg. Chem.* **2021**, *112*, 104896. [CrossRef]
245. Mukherjee, S.; Ray, S.; Thakur, R. Solid lipid nanoparticles: A modern formulation approach in drug delivery system. *Indian. J. Pharm. Sci.* **2009**, *71*, 349–358. [CrossRef] [PubMed]
246. Severino, P.; Silveira, E.; Loureiro, K.; Chaud, M.; Antonini, D.; Lancellotti, M.; Sarmento, V.; da Silva, C.; Santana, M.; Souto, E. Antimicrobial activity of polymyxin-loaded solid lipid nanoparticles (PLX-SLN): Characterization of physicochemical properties and in vitro efficacy. *Eur. J. Pharm. Sci.* **2017**, *106*, 177–184. [CrossRef]
247. Kumari, A.; Yadav, S.; Yadav, S. Biodegradable polymeric nanoparticles based drug delivery systems. *Colloids Surf. B Biointerfaces* **2010**, *75*, 1–18. [CrossRef] [PubMed]
248. Cruz, J.; Flórez, J.; Torres, R.; Urquiza, M.; Gutiérrez, J.A.; Guzmán, F.; Ortiz, C.C. Antimicrobial activity of a new synthetic peptide loaded in polylactic acid or poly(lactic-co-glycolic) acid nanoparticles against Pseudomonas aeruginosa, Escherichia coli O157:H7 and methicillin resistant staphylococcus aureus (MRSA). *Nanotechnology* **2017**, *28*, 135102. [CrossRef]
249. Rabea, E.; Badawy, M.; Stevens, C.; Smagghe, G.; Steurbaut, W. Chitosan as antimicrobial agent: Applications and mode of action. *Biomacromolecules* **2003**, *4*, 1457–1465. [CrossRef] [PubMed]
250. Rashki, S.; Safardoust-Hojanghan, H.; Mirzaei, H.; Abdulsahib, W.; Mahdi, M.; Salavati-Niasari, M.; Khaledi, A.; Khorshidi, A.; Gholam, M.; Seyyed. Delivery LL37 by chitosan nanoparticles for enhanced antibacterial and antibiofilm efficacy. *Carbohydr. Polym.* **2022**, *291*, 119634. [CrossRef]
251. Zhu, J.; Li, F.; Wang, X.; Yu, J.; Wu, D. Hyaluronic acid and polyethylene glycol hybrid hydrogel encapsulating nanogel with hemostasis and sustainable antibacterial property for wound healing. *ACS Appl. Mater. Interfaces* **2018**, *10*, 13304–13316. [CrossRef]
252. Lequeux, I.; Ducasse, E.; Jouenne, T.; Thebault, P. Addition of antimicrobial properties to hyaluronic acid by grafting of antimicrobial peptide. *Eur. Polym. J.* **2014**, *51*, 182–190. [CrossRef]
253. Jiang, G.; Liu, S.; Yu, T.; Wu, R.; Ren, Y.; van der Mei, H.; Liu, J.; Busscher, H. PAMAM dendrimers with dual-conjugated vancomycin and Ag-nanoparticles do not induce bacterial resistance and kill vancomycin-resistant staphylococci. *Acta Biomater.* **2021**, *123*, 230–243. [CrossRef]

254. He, H.; Pham-Huy, L.; Dramou, P.; Xiao, D.; Zuo, P.; Pham-Huy, C. Carbon nanotubes: Applications in pharmacy and medicine. *Biomed. Res. Int.* **2013**, *2013*, 578290. [CrossRef]
255. Qi, X.; Poernomo, G.; Wang, K.; Chen, Y.; Chan-Park, M.; Xu, R.; Chang, M. Covalent immobilization of nisin on multi-walled carbon nanotubes: Superior antimicrobial and anti-biofilm properties. *Nanoscale* **2011**, *3*, 1874–1880. [CrossRef] [PubMed]
256. Rajendiran, K.; Zhao, Z.; Pei, D.-S.; Fu, A. Antimicrobial activity and mechanism of functionalized quantum dots. *Polymers* **2019**, *11*, 1670. [CrossRef] [PubMed]
257. Zhao, C.; Wang, X.; Wu, L.; Wu, W.; Zheng, Y.; Lin, L.; Weng, S.; Lin, X. Nitrogen-doped carbon quantum dots as an antimicrobial agent against Staphylococcus for the treatment of infected wounds. *Colloids Surf. B Biointerfaces* **2019**, *179*, 17–27. [CrossRef] [PubMed]
258. Lata, S.; Sharma, B.; Raghava, G. Analysis and prediction of antibacterial peptides. *BMC Bioinform.* **2007**, *8*, 263. [CrossRef]
259. Liscano, Y.; Oñate-Garzón, J.; Ocampo-Ibáñez, I. In silico discovery of antimicrobial peptides as an alternative to control SARS-CoV-2. *Molecules* **2020**, *25*, 5535. [CrossRef]
260. Diogo, P.; Faustino, M.A.F.; Palma, P.J.; Rai, A.; Neves, M.G.P.M.S.; Santos, J.M. May carriers at nanoscale improve the endodontic's future? *Adv. Drug Deliv. Rev.* **2023**, *195*, 114731. [CrossRef]
261. Kishen, A.; Shrestha, A. Nanoparticles for dentin tissue stabilization. In *Nanotechnology in Endodontics*; Springer: Berlin/Heidelberg, Germany, 2015; pp. 121–138. [CrossRef]
262. Mishra, B.; Wang, G. Ab initio design of potent anti-MRSA peptides based on database filtering technology. *J. Am. Chem. Soc.* **2012**, *134*, 12426–12429. [CrossRef]
263. Mishra, B.; Lakshmaiah Narayana, J.; Lushnikova, T.; Wang, X.; Wang, G. Low cationicity is important for systemic in vivo efficacy of database-derived peptides against drug-resistant gram-positive pathogens. *Proc. Natl. Acad. Sci. USA* **2019**, *116*, 13517–13522. [CrossRef]
264. Lai, Z.; Yuan, X.; Chen, H.; Zhu, Y.; Dong, N.; Shan, A. Strategies employed in the design of antimicrobial peptides with enhanced proteolytic stability. *Biotechnol. Adv.* **2022**, *59*, 107962. [CrossRef]

microorganisms

Article

Biogenic Silver Nanoparticles Produced by Soil Rare Actinomycetes and Their Significant Effect on *Aspergillus*-derived mycotoxins

Mohamed N. Abd El-Ghany [1,*], Salwa A. Hamdi [2], Shereen M. Korany [3,4], Reham M. Elbaz [3,5], Ahmed N. Emam [6,7] and Mohamed G. Farahat [1,8,*]

1 Botany and Microbiology Department, Faculty of Science, Cairo University, Giza 12613, Egypt
2 Zoology Department, Faculty of Science, Cairo University, Giza 12613, Egypt; salwa_abdelhamid@cu.edu.eg
3 Botany and Microbiology Department, Faculty of Science, Helwan University, Cairo 11795, Egypt; smkorany@pnu.edu.sa (S.M.K.); reham_elbaz@science.helwan.edu.eg (R.M.E.)
4 Department of Biology, College of Science, Princess Nourah bint Abdulrahman University, P.O. Box 84428, Riyadh 11671, Saudi Arabia
5 Department of Biology, Faculty of Science, University of Bisha, P.O. Box 551, Bisha 61922, Saudi Arabia
6 Refractories, Ceramics and Building Materials Department, Advanced Materials Technology & Mineral Resources Research Institute, National Research Centre (NRC), El Bohouth St., Dokki, Cairo 12622, Egypt; an.emam@nrc.sci.eg
7 Nanomedicine & Tissue Engineering Research Lab, Medical Research Centre of Excellence, National Research Centre, El Bohouth St., Dokki, Cairo 12622, Egypt
8 Biotechnology Department, Faculty of Nanotechnology for Postgraduate Studies, Cairo University, Sheikh Zayed Branch Campus, Giza 12588, Egypt
* Correspondence: mabdelghany@sci.cu.edu.eg (M.N.A.E.-G.); farahat@cu.edu.eg (M.G.F.)

Abstract: The current investigation addressed the green synthesis of silver nanoparticles (AgNPs) using newly isolated silver-resistant rare actinomycetes, *Glutamicibacter nicotianae* SNPRA1 and *Leucobacter aridicollis* SNPRA2, and investigated their impact on the mycotoxigenic fungi *Aspergillus flavus* ATCC 11498 and *Aspergillus ochraceus* ATCC 60532. The formation of AgNPs was evidenced by the reaction's color change to brownish and the appearance of the characteristic surface plasmon resonance. The transmission electron microscopy of biogenic AgNPs produced by *G. nicotianae* SNPRA1 and *L. aridicollis* SNPRA2 (designated Gn-AgNPs and La-AgNPs, respectively) revealed the generation of monodispersed spherical nanoparticles with average sizes of 8.48 ± 1.72 nm and 9.67 ± 2.64 nm, respectively. Furthermore, the XRD patterns reflected their crystallinity and the FTIR spectra demonstrated the presence of proteins as capping agents. Both bioinspired AgNPs exhibited a remarkable inhibitory effect on the conidial germination of the investigated mycotoxigenic fungi. The bioinspired AgNPs caused an increase in DNA and protein leakage, suggesting the disruption of membrane permeability and integrity. Interestingly, the biogenic AgNPs completely inhibited the production of total aflatoxins and ochratoxin A at concentrations less than 8 μg/mL. At the same time, cytotoxicity investigations revealed the low toxicity of the biogenic AgNPs against the human skin fibroblast (HSF) cell line. Both biogenic AgNPs exhibited feasible biocompatibility with HSF cells at concentrations up to 10 μg/mL and their IC_{50} values were 31.78 and 25.83 μg/mL for Gn-AgNPs and La-AgNPs, respectively. The present work sheds light on the antifungal prospect of the biogenic AgNPs produced by rare actinomycetes against mycotoxigenic fungi as promising candidates to combat mycotoxin formation in food chains at nontoxic doses.

Keywords: biosynthesis; silver nanoparticles; mycotoxin production; antifungal activity; mycotoxin inhibition; rare actinomycetes; spore germination

Citation: Abd El-Ghany, M.N.; Hamdi, S.A.; Korany, S.M.; Elbaz, R.M.; Emam, A.N.; Farahat, M.G. Biogenic Silver Nanoparticles Produced by Soil Rare Actinomycetes and Their Significant Effect on *Aspergillus*-derived mycotoxins. *Microorganisms* **2023**, *11*, 1006. https://doi.org/10.3390/microorganisms11041006

Academic Editor: Kamel Ahmed Abd-Elsalam

Received: 4 March 2023
Revised: 9 April 2023
Accepted: 11 April 2023
Published: 12 April 2023

1. Introduction

Nanotechnology is a rapidly growing, modern multidisciplinary science that involves the production and manipulation of matter at the nanoscale level. Compared with bulk materials, nanoparticles (less than 100 nm in at least one dimension) possess distinctive chemical and physical properties because of their vast surface area to volume proportion and high surface energy [1]. Owing to their admirable properties, metal nanoparticles have emerged as exploratory areas in physics, chemistry, materials science, drug delivery, the food sector, and biomedicine [2–4]. Among the metal nanoparticles, nanosilver has received considerable attention for its exceptional antibacterial, antifungal, anticancer, and antioxidant potential [5–10].

Generally, silver nanoparticles (AgNPs) are produced either by a "top-down" or "bottom-up" strategy. In the "top-down" approach, the bulk materials are broken down into small nanoparticles using various physical techniques, including ball milling and laser ablation [11–13], which are energy-consuming approaches that make them more capital-intensive. In the "bottom-up" strategy, AgNPs are synthesized via chemical reactions in which the self-assembly of atoms into nuclei occurs, further developing them into nanoscale particles. Fabrication of AgNPs by chemical methods involves the use of reducing agents, such as sodium borohydride, N,N-dimethyl formamide, and Tollens' reagent [14]. Though being the most accustomed method for the synthesis of AgNPs, chemical methods are thought to cause environmental hazards due to using various perilous chemicals attributed to carcinogenicity, genotoxicity, and cytotoxicity [15]. Hence, the green synthesis of nanoparticles provides an alternative approach to conventional chemical and physical methods. The biosynthesis processes include the usage of bacteria, fungi, algae, or plant extracts, offering an eco-friendly green solution that reduces the utilization of toxic chemicals [16–23].

Aspergillus species are ubiquitous fungi that produce various life-threatening mycotoxins such as aflatoxins, ochratoxins, patulin, and citrinin in the contaminated food chains [24]. Aflatoxin is one of the carcinogenic and mutagenic mycotoxins predominantly produced by *A. flavus*, *A. parasiticus*, and *A. nomius*, whereas ochratoxin A (OTA) is a teratogenic and nephrogenic mycotoxin produced mainly by *A. ochraceus*, *A. westerdijkiae*, and *A. steynii* [25]. Of all the mycotoxins, aflatoxins are one of the most toxic and carcinogenic groups comprising four main types, namely, AFB1, AFB2, AFG1, and AFG2 [26]. On the other hand, OTA is responsible for various health issues due to its carcinogenicity, mutagenicity, hepatotoxicity, genotoxicity, immunotoxicity, embryotoxicity, and testicular toxicity [27].

Although the biosynthesis of AgNPs using various microorganisms is well-documented, there is limited research in the literature addressing the biosynthesis of AgNPs using actinomycetes (actinobacteria). The vast majority of described AgNP-producing actinomycetes belong to the genus *Streptomyces* [28–35], while the production of AgNPs by non-*Streptomyces* species (rare actinomycetes) is relatively unexplored. Herein, we address for the first time the biosynthesis of AgNPs by two rare actinomycetes, *G. nicotianae* SNPRA1 and *L. aridicollis* SNPRA2, and evaluate their antifungal and anti-mycotoxin efficacy on mycotoxigenic fungi at nontoxic doses.

2. Materials and Methods

2.1. Isolation of Actinobacteria

Soil samples were collected from the Wadi El-Natrun depression, Western desert, Egypt. The collected samples were subjected to dry heating at 100 °C for 15 min, which was followed by treatment with 1.5% phenol for 30 min at 30 °C. Subsequently, the physicochemically treated samples were serially diluted in sterile saline (NaCl, 9 g/L) and aliquots of each dilution (100 µL) were plated on humic-acid–vitamin agar (Kisan Biotech, Seoul, South Korea) plates supplemented with nalidixic acid (25 µg/mL) (Merck, Darmstadt, Germany) and cycloheximide (25 µg/mL) (Merck, Darmstadt, Germany) [36]. After incubation at 30 °C for 14 days, the developed colonies of different shapes were selected and purified by repeated subculturing. Afterwards, the purified colonies were subcultured on ISP2 agar (BD BBL and Difco, Franklin Lakes, NJ, USA) supplemented with various concentrations of $AgNO_3$ (1–5 mM) and incubated at 30 °C for 14 days. Accordingly, the resistant isolates were picked and assessed for the production of biogenic AgNPs.

2.2. Identification and Phylogenetic Analysis

The promising AgNP-producing strains SNPRA1 and SNPRA2 were identified based on their 16S rRNA gene sequences. Briefly, the 16S rRNA gene was amplified by polymerase chain reaction (PCR) using the universal primers 27F and 1492R as described elsewhere [37]. The nucleotide sequences of the purified amplicons were sequenced using an ABI 3730XL sequence analyzer (Applied Biosystems, Waltham, MA, USA). A sequence similarity search was performed using The NCBI BLASTN and the EzTaxon-e server database [38]. A phylogenetic tree was constructed in the context of the 16S rRNA gene sequences using the neighbor-joining method of the MEGA-X software (version 10.0.5), and the bootstrap analysis was performed based on 1,000 replicates.

2.3. Biosynthesis of AgNPs

The biosynthesis of AgNPs was conducted by using $AgNO_3$ (Alfa Aesar, Ward Hill, MA, USA) as a metal precursor and the actinomycete biomass extract as the reducing and stabilizing agent. Typically, each actinobacterial isolate was inoculated in a conical flask containing 50 mL of ISP2 broth (g/L: glucose 4.0, yeast extract 4.0, and malt extract 10.0) and incubated at 30 °C and 150 rpm for 96 h. Subsequently, the culture was centrifuged at 12,000× *g* for 15 min, and the collected cells were washed twice with sterile distilled water. Then, the washed biomass was resuspended in 50 mL of ultrapure water (18.2 Mcm) and incubated on a shaking incubator (150 rpm) at 30 °C for 72 h. After osmotic lysis, the cell lysates were filtered through Whatman No. 1 filter papers yielding actinomycete biomass extracts. After that, the extract was mixed with an equal volume of 2 mM $AgNO_3$ solution, and the mixture was incubated at 30 °C and 150 rpm for 48 h under dark conditions. The synthesis was monitored by visual inspection for the color change into brown as the initial indicator of AgNP biosynthesis, which was confirmed by UV–Vis absorbance of the reaction mixture. For purification of the biogenic AgNPs, the brownish mixture was centrifuged at 40,000× *g* for 15 min using a Sorvall LYNX 6000 centrifuge (Thermo Scientific, Waltham, MA, USA). The AgNP pellets were resuspended in ultrapure water and centrifuged again at 40,000× *g* for 15 min. To remove any undesirable impurities, the pellets were washed three more times. Finally, the nanosilver pellets were dried in a vacuum oven at 50 °C for 24 h. Based on the potent reduction of $AgNO_3$ into AgNPs, two proficient isolates designated SNPRA1 and SNPRA2 were selected and used for further characterization.

2.4. Characterization of Biogenic AgNPs

UV–Vis absorbance spectra of the biogenic AgNPs were recorded in the 300–700 nm range with a resolution of 1 nm using an Epoch 2 UV–Vis spectrophotometer (BioTek, Santa Clara, CA, USA). The size and shape of biogenic AgNPs were investigated by high-resolution transmission electron microscopy (HRTEM) using a JEM-2100 transmission electron microscope (JEOL, Tokyo, Japan) with a 200 kV acceleration voltage. The X-ray diffraction (XRD) patterns of biogenic AgNPs were analyzed on a D8 Discover X-ray diffractometer (Bruker, Karlsruhe, Germany). The Cu Kα radiation with a current of 30 mA and applied voltage of 40 kV was used as a light source. The 2θ values were measured with a scan speed of 0.5°/min in a range from 20 to 90°. Fourier-transform infrared (FTIR) spectroscopy was performed using a Nicolet 6700 FT-IR spectrometer (Thermo Scientific, Waltham, MA, USA), and the scanning spectra were recorded within the range of 4000 to 400 cm^{-1}.

2.5. Antifungal Activity

The inhibitory effect of the biogenic AgNPs on the conidial germination of the mycotoxigenic fungi *A. flavus* (ATCC 11498) and *A. ochraceus* (ATCC 60532) was investigated in vitro. In brief, conidia of *A. flavus* and *A. ochraceus*, cultivated on potato dextrose agar (PDA) plates (Condalab, Madrid, Spain) at 28 °C for 7 days, were harvested, and suspended in sterile saline solution (NaCl, 0.9 %). After filtration through sterile muslin, the conidial count was adjusted to 2×10^6 conidia/mL using a hemocytometer. Subsequently, 50 µL of the prepared conidial suspension was transferred to each well of a 96-well microtiter plate containing 100 µL of potato dextrose broth (PDB) (Condalab, Madrid, Spain) amended with different concentrations of AgNPs. Wells containing AgNP-free PDB served as the negative control. Then, the microtiter plates were incubated at 28 °C for 16 h. Afterward, the percentage of germinated conidia was calculated by analyzing 100 conidia under a DM500 optical microscope (Leica, Heerbrugg, Switzerland). A conidium was considered germinated if the germ tube length was equal to or longer than the conidial length. The minimum inhibitory concentration (MIC) values of AgNPs were estimated statistically using the GraphPad Prism software (version 7.0.0). The MIC values indicating the lowest concentration of biogenic AgNPs inhibiting conidial germination were expressed as µg/mL. Moreover, leakage of proteins and DNA in AgNPs-treated *A. flavus* and *A. ochraceus* was investigated following the method described elsewhere [39].

2.6. Effect of Biogenic AgNPs on Mycotoxin Production

The impact of biogenic AgNPs on the production of total aflatoxins and OTA by *A. flavus* and *A. ochraceus*, respectively, was assessed. Practically, the conidia of each investigated mycotoxigenic fungi were inoculated into yeast extract sucrose (YES) broth ((20% sucrose and 2% yeast extract (Merck, Darmstadt, Germany)) containing various concentrations of actinomycete-mediated biogenic AgNPs [40]. Experiments containing sterile distilled water instead of the biogenic AgNPs were considered to be the negative control. After incubation for 10 days at 28 °C, the produced total aflatoxins (AFLAs) and OTA were determined using Celer® AFLA and Celer® OCHRA quantitative ELISA test kits (Eurofins technologies, Budapest, Hungary), respectively.

2.7. Cytotoxicity Assay

The cytotoxicity of the as-prepared biogenic AgNPs was investigated in vitro using human skin fibroblast (HSF) cells. The HSF cell line was kindly provided by Nawah Scientific Inc. (Mokatam, Cairo, Egypt). The HSF cells were cultured and maintained in the Dulbecco's modified eagle medium (DMEM) containing 10% heat-inactivated fetal bovine serum (Gibco; Thermo Fisher Scientific, Loughborough, UK), streptomycin (100 µg/mL), and penicillin (100 units/mL). The cell line was cultured under 5% CO_2 in a water-jacketed incubator (BINDER GmbH, Tuttlingen, Germany) in a high-humidity atmosphere at 37 °C. The potential cytotoxicity of the bio-fabricated AgNPs was assayed against HSF cells using the sulforhodamine B (SRB) method [41].

2.8. Statistical Analysis

The presented data are the means of triplicate assays. The analysis of variance (ANOVA) was conducted, and the significant differences were compared via Duncan's test using the IBM SPSS software (version 22), with the critical difference set at a 5% level of probability. The statistical linear regression and the half-maximal inhibitory concentration (IC_{50}) were executed using GraphPad Prism software (version 7.0.0).

3. Results

3.1. Isolation of Actinobacteria

Fifty-eight actinobacterial isolates were recovered from arid soil samples obtained from the Egyptian desert and screened for their capability to tolerate $AgNO_3$. The preliminary screening revealed that the assessed actinomycete isolates tolerated the presence of Ag^+ up to 3 mM, but none tolerated 4 mM of $AgNO_3$. Of the screened fifty-eight isolates, nineteen were tolerant to 1 mM of $AgNO_3$, eleven isolates were tolerant to 2 mM, and five isolates were tolerant to 3 mM. Consequently, the most tolerant five isolates were selected for further investigations.

3.2. Identification and Phylogenetic Analysis

The selected strains were identified based on their 16S rRNA gene sequences. The BLAST results of the 16S rRNA gene revealed that the strain SNPRA1 shared a 99.78% identity with the *G. nicotianae* strain DSM 20123 (accession number: NR_026190.1), a 99.34% identity with the *G. mysorens* strain LMG 16219 (accession number: NR_114924.1), and a 99.26% identity with the *G. halophytocola* strain KLBMP 5180 (accession number: NR_156872.1). On the other hand, the strain SNPRA2 shared a 99.43% identity with the *L. aridicollis* strain L-9 (accession number: NR_042288.1), a 99.07% identity with the *L. komagatae* strain IFO 15245 (accession number: NR_114929.1), and a 98.63% with the *L. denitrificans* strain M1T8B10 (accession number: NR_108568.1). Based on their 16S rRNA gene sequences, the strains SNPRA1 and SNPRA2 were putatively identified as *G. nicotianae* and *L. aridicollis*, respectively. The 16S rRNA gene sequences of the strains SNPRA1 and SNPRA2 were submitted into the GenBank and accession numbers OQ148401 and OQ148402 were assigned, respectively. The neighbor-joining phylogenetic tree was constructed in the context of the 16S rRNA gene sequences (Figure 1).

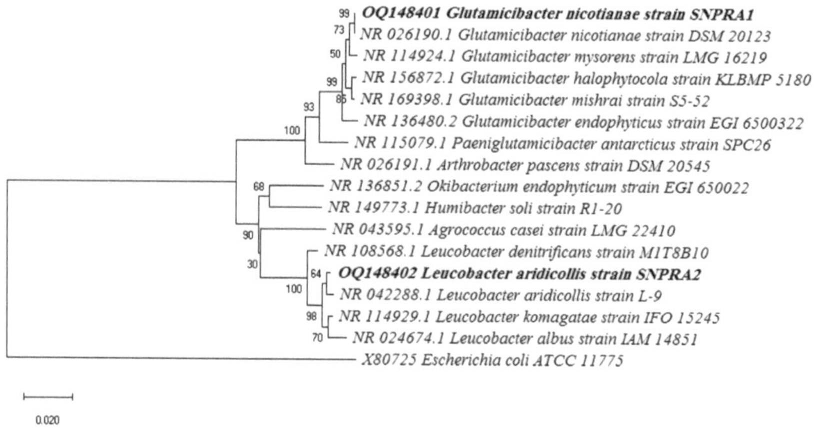

Figure 1. Phylogenetic tree showing the relationships between *G. nicotianae* SNPRA1 and *L. aridicollis* SNPRA2 and the most closely related species.

3.3. Biosynthesis of AgNPs

The biomass extracts of the selected isolates were evaluated for their ability to synthesize AgNPs. Of the five Ag-tolerant isolates, only two strains exhibited the color of the $AgNO_3$ solution changing into brown, and this formation of AgNPs was confirmed by spectrophotometric scanning of the reaction mixture. As a result, the two strains, designated SNPRA1 and SNPRA2, demonstrating the characteristic surface plasmon resonance (SPR) at 400-430 nm (indicating the potential for the production of AgNPs) were selected for further investigations. The control experiments without biomass extracts did not produce any change in color after the same incubation period.

3.4. Characterization of Biogenic AgNPs

The biosynthesis of AgNPs was determined by the reaction's color change to brownish, which is the primary indication of AgNP formation. The biogenic AgNPs produced by the biomass extracts of *G. nicotianae* SNPRA1 and *L. aridicollis* SNPRA2 were designated as Gn-AgNPs and La-AgNPs, respectively. UV–Vis spectra of the biogenic Gn-AgNPs and La-AgNPs revealed characteristic absorption peaks located at 405 and 416 nm (Figure 2), respectively, which were in agreement with the surface plasmon resonance reported for AgNPs. In addition, the morphological properties based on TEM analysis confirmed the generation of monodispersed spherical AgNPs by both actinomycete biomass extracts. The average sizes of the Gn-AgNPs and La-AgNPs, as determined by TEM, were 8.48 ± 1.72 nm and 9.67 ± 2.64 nm, respectively (Figure 3). Furthermore, the crystallographic structures of the biogenic Gn-AgNPs and La-AgNPs were confirmed by XRD analysis. The XRD patterns revealed four intense peaks in the whole spectra, with 2θ peak values around 38.26°, 44.47°, 64.71°, and 77.73° (Figure 4). The spectra correlated to the main characteristic peaks of AgNPs (COD card No./file No. 1509146) corresponding to the lattice plane clusters of (111), (200), (220), and (311). The obtained lattice plane values confirmed the face-centered cubic (FCC) nature of the biogenic AgNPs. Finally, the surface properties of the as-prepared spherical-like shaped La-AgNPs and Gn-AgNPs were investigated via diffused reflectance as a function of wavenumber by FTIR. The obtained spectra showed a wide variety of surface functional groups that decorated the silver nanoparticles' surfaces, which was based on the use of actinomycetes in the formation of the biogenic silver nanoparticles (Figure 5).

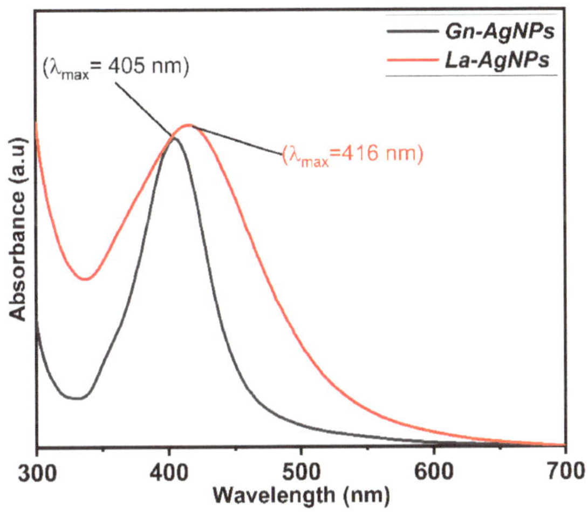

Figure 2. UV–Vis spectra of Gn-AgNPs and La-AgNPs produced by *G. nicotianae* SNPRA1 and *L. aridicollis* SNPRA2, respectively.

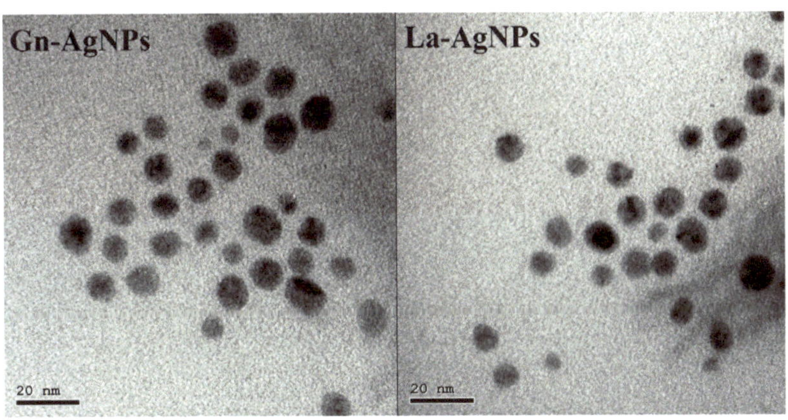

Figure 3. Transmission electron micrographs showing the size and shape of Gn-AgNPs and La-AgNPs produced by *G. nicotianae* SNPRA1 and *L. aridicollis* SNPRA2, respectively. The average sizes of Gn-AgNPs and La-AgNPs were 8.48 ± 1.72 nm and 9.67 ± 2.64 nm, respectively.

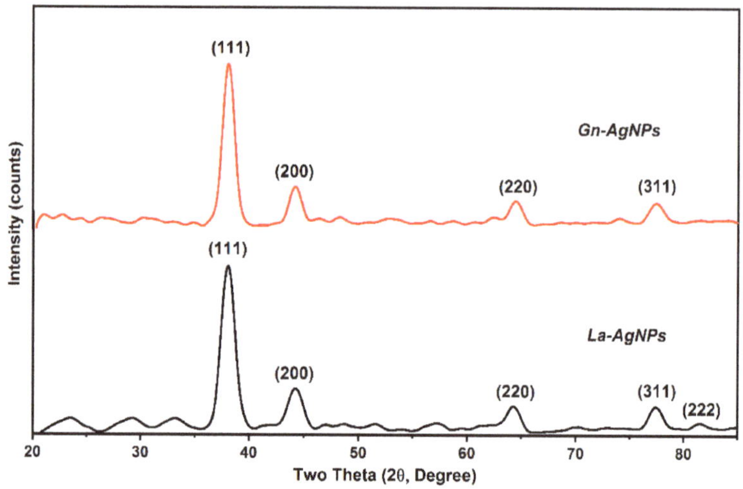

Figure 4. XRD spectra of the biogenic Gn-AgNPs and La-AgNPs produced by *G. nicotianae* SNPRA1 and *L. aridicollis* SNPRA2, respectively.

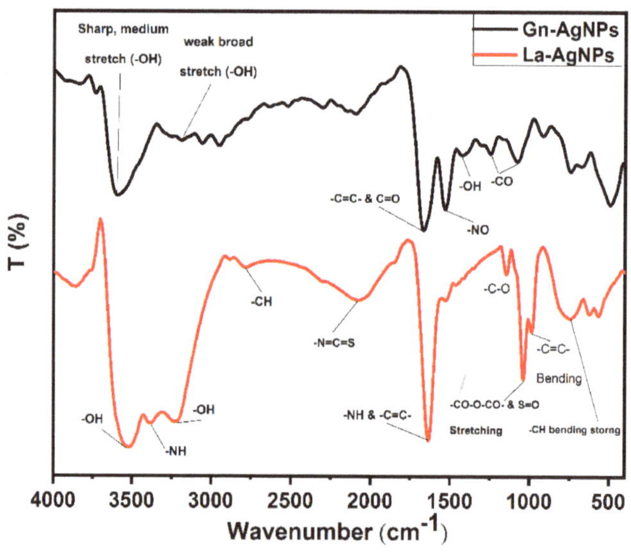

Figure 5. FTIR spectra of the biogenic Gn-AgNPs and La-AgNPs produced by *G. nicotianae* SNPRA1 and *L. aridicollis* SNPRA2, respectively.

3.5. Antifungal Activity

The antifungal potential of the actinomycete-mediated biogenic AgNPs was evaluated against two mycotoxin-producing fungi. Our findings provide evidence of the inhibitory effect of Gn-AgNPs and La-AgNPs in the conidial germination of both mycotoxigenic fungi (Figure S1). The MIC values of Gn-AgNPs and La-AgNPs against *A. flavus* and *A. ochraceus* were statistically deduced (Figure 6). The results revealed a slightly superior inhibitory effect of Gn-AgNPs compared with that of La-AgNPs. The deduced MIC values of Gn-AgNPs and La-AgNPs against *A. flavus* were 32.34 and 37.35 µg/mL, respectively. In the case of *A. ochraceus*, the deduced MIC values of Gn-AgNPs and La-AgNPs were

29.15 and 35.06 μg/mL, respectively. Furthermore, the present results indicate that the exposure of *A. flavus* and *A. ochraceus* conidia to the as-prepared biogenic AgNPs led to the significant leakage of DNA and proteins (Figure 7).

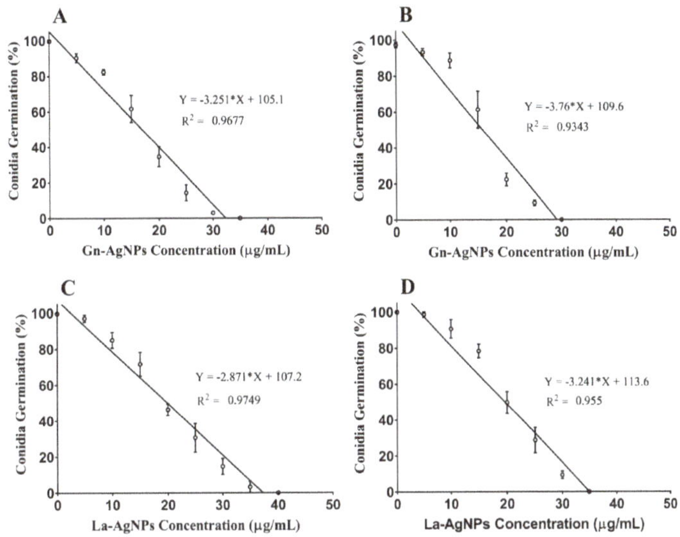

Figure 6. Effect of the biogenic AgNPs on conidial germination. (**A**) *A. flavus* treated with Gn-AgNPs, (**B**) *A. ochraceus* treated with Gn-AgNPs, (**C**) *A. flavus* treated with La-AgNPs, (**D**) *A. ochraceus* treated with La-AgNPs. Error bars represent the standard deviation of triplicate assays.

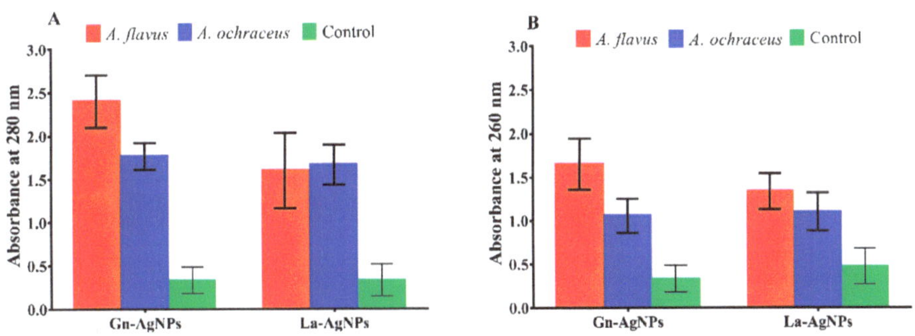

Figure 7. Effect of the biogenic AgNPs on the leakage of proteins (**A**) and DNA (**B**) from the conidia of *A. flavus* and *A. ochraceus* compared with the untreated control.

3.6. Effect of Biogenic AgNPs on Mycotoxin Production

The anti-mycotoxin activity of actinobacteria-mediated biogenic AgNPs was evaluated against aflatoxigenic *A. flavus* ATCC 11498 and ochratoxigenic *A. ochraceus* ATCC 60532. The anti-mycotoxin activity of AgNPs was determined by comparing the relative toxin production of the treated fungi with that of the untreated control. The results revealed an obvious declining trend in the total AFLA and OTA contents in the treated mycotoxigenic cultures (Figure 8). In this study, Gn-AgNPs significantly decreased the total AFLA and OTA content in treated *A. flavus* ATCC 11498 and *A. ochraceus* ATCC 60532, respectively, in a dose-dependent manner. Apparently, Gn-AgNPs completely inhibited the total AFLA and OTA production at concentrations of 5.64 and 4.59 μg/mL, respectively. Likewise, La-AgNPs completely inhibited the total AFLA and OTA production by *A. flavus* ATCC 11498 and *A. ochraceus* ATCC 60532 at concentrations of 7.74 and 5.48 μg/mL, respectively.

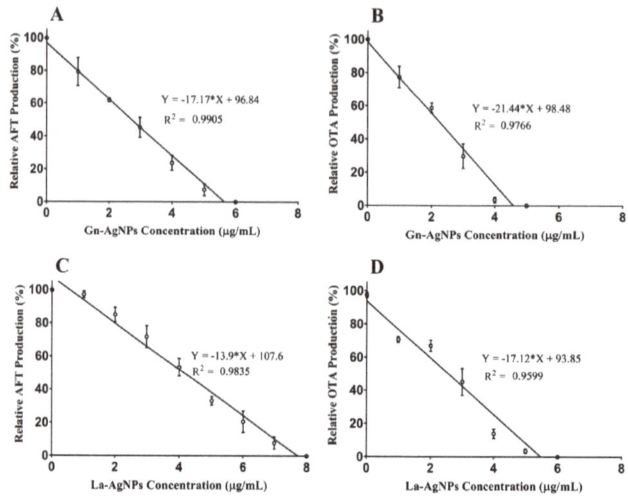

Figure 8. Linear regression model fit showing the anti-mycotoxin activity of the biogenic AgNPs. (**A**) *A. flavus* treated with Gn-AgNPs, (**B**) *A. ochraceus* treated with Gn-AgNPs, (**C**) *A. flavus* treated with La-AgNPs, (**D**) *A. ochraceus* treated with La-AgNPs. Error bars represent the standard deviation of triplicate assays.

3.7. Cytotoxicity Assay

In the present investigation, the potential cytotoxicity of the actinomycete-mediated biogenic AgNPs was tested in vitro against HSF cell lines. The results obtained from the SRB assay clarified the biocompatibility of both investigated AgNPs with HSF cells at concentrations up to 10 μg/mL (Figure 9). Regarding their cytotoxicity, Gn-AgNPs did not exhibit any significant differences in the viability of the treated HSF cells when exposed to concentrations up to 15 μg/mL, as compared with the untreated control. A further increase in the Gn-AgNP concentration significantly reduced the cell viability. Indeed, the La-AgNPs demonstrated a slightly higher cytotoxicity effect on HSF cells. However, no significant difference in the HSF cell viability was observed up to 10 μg/mL, as compared with the untreated control. The results revealed that the IC_{50} values of Gn-AgNPs and La-AgNPs against HSF cells were 31.78 and 25.83 μg/mL, respectively.

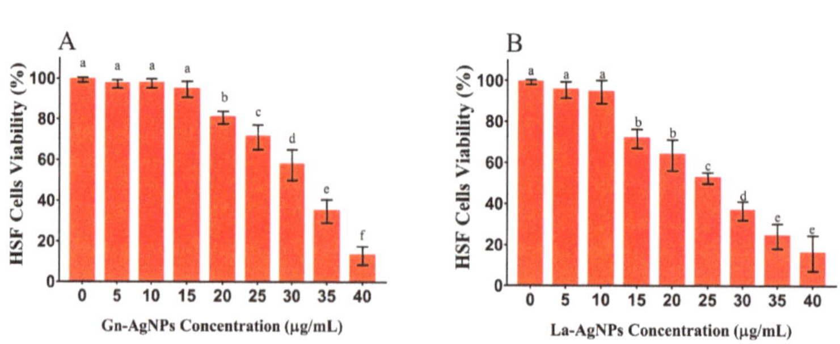

Figure 9. SRB cell viability assay for the determination of the cytotoxicity of the biogenic Gn-AgNPs (**A**) and La-AgNPs (**B**) against HSF cell lines. Columns headed by the same letter were not significantly different according to Duncan's multiple range test ($p < 0.05$). Error bars represent standard deviations.

4. Discussion

Recently, the green synthesis of nanoparticles has attracted more attention as an emerging eco-friendly strategy due to its high efficiency, nontoxicity, and environmental protection [41]. Consequently, our study sheds light on the pronounced antifungal and anti-mycotoxin potential of biogenic AgNPs produced by two rare actinomycetes. In this investigation, we used an integrated method to isolate non-*Streptomyces* actinomycetes via the pretreatment of soil samples at 100 °C for 15 min, which was followed by treatment with 1.5% phenol for 30 min at 30 °C. These physical and chemical pretreatments could eliminate most fungi, streptomycetes, and other common bacteria. It is believed that pretreatment of environmental samples by drying and heating stimulates the isolation of rare actinomycetes. In a previous study, the pretreatment of samples at 100 °C for 15 min enabled the isolation of rare actinomycetes belonging to the genera *Pseudonocardia, Blastococcus, Nocardiopsis, Actinocorallia, Micromonospora, Dactylosporangium*, and *Streptosporangium* [42]. Subramani and Aalbersberg reviewed how the dry heating of samples, followed by treatment with certain chemicals such as 1.5% phenol, 0.01% benzethonium chloride, 0.05% sodium dodecylsulfate (SDS), and 0.03% chlorhexidine gluconate, drastically eliminated most microorganisms and unwanted actinomycetes propagules belonging to the *Streptomyces* species [43]. Moreover, it has been suggested that the pretreatment of the soil samples with 1.5% phenol (30 °C for 30 min) inhibits the growth of fungi, bacteria, and other common actinomycetes by denaturing their proteins or by disrupting their cell membrane, and facilitates the recovery of phenol-resistant actinomycetes [44]. Likewise, various rare actinomycetes belonging to the genera *Gordonia, Nonomuraea, Actinoplanes, Microbispora*, and *Micromonospora* were selectively isolated from soil samples and subjected to harsher pretreatments, including moist (50 °C for 6 min) and dry (120 °C for 1 h) heating and 1.5 % phenol [45]. Herein, the biomass extracts of Ag-resistant rare actinobacteria *G. nicotianae* SNPRA1 and *L. aridicollis* SNPRA2 reduced $AgNO_3$ and yielded biogenic AgNPs under dark conditions. To the best of our knowledge, there is limited data on the biogenic AgNPs from desert-derived non-*Streptomyces* actinomycetes, and to date, no published reports describe the fabrication of AgNPs using microorganisms belonging to the *Glutamicibacter* and *Leucobacter* genera. Nonetheless, previous reports indicated the metal resistance in *Glutamicibacter* and *Leucobacter* spp. It has been suggested that heavy metals exert selective pressure on the exposed microbial communities and lead to the evolution of metal resistance determinants to sequester and transform these compounds [46–48]. In this context, it has been reported that metal-resistant *G. nicotianae* MSSRFPD35 exhibited the potential to grow and degrade phenol in the presence of several heavy metals such as Pb, Ni, Cd, Co, and Cu [49]. Likewise, various species of the genus *Leucobacter*, which belongs to the phylum *Actinobacteria*, can tolerate a wide variety of heavy metals such as Cd, Cr, Cu, As, and Pb [50–52]. In agreement with our findings, the biomass extract from the Ag-tolerant actinobacterium *Nocardiopsis dasonvillei* KY772427 was used as a reducing agent for the biosynthesis of bioactive AgNPs [53]. It has commonly been assumed that the biosynthesis of metal nanoparticles is correlated to the capability of microorganisms for tolerating heavy metals [54]. The exact mechanism of the biosynthesis of metal nanoparticles by microorganisms is not fully understood; however, it has been suggested that microbial extracts could contain biomolecules such as polyphenols, polysaccharides, proteins, vitamins, and enzymes that have the potential to reduce silver salts and convert them into AgNPs [55]. Moreover, a recent study proved the capability of nicotinamide adenine dinucleotide phosphate (NADPH) to reduce silver nitrate as the sole reducing agent forming AgNPs [56].

In this work, the emergence of the brown color denoted the reduction in $AgNO_3$ and the fabrication of AgNPs [32,35,57,58]. The biosynthesis of Gn-AgNPs and La-AgNPs was affirmed by the appearance of single absorption peaks at 405 and 416 nm, respectively, which could be attributed to the characteristic SPR of AgNPs. Generally, distinctive AgNPs demonstrate characteristic SPR at wavelengths ranging from 400 to 450 nm [33,59–63]. The current study demonstrated the formation of pure Gn-AgNPs and La-AgNPs with a crystalline nature through XRD patterns that illustrated the distinguished characteristic diffraction peaks, which were consistent with previous reports on AgNPs [30,64–68]. In addition, the FTIR spectra revealed the presence of biomolecules, especially proteins, on the surface of Gn-AgNPs and La-AgNPs. In addition to their reducing properties, the biomass extracts of *G. nicotianae* SNPRA1 and *L. aridicollis* SNPRA2 seem to play a crucial role as stabilizing and capping agents that may confer a bifold action regarding the fabrication of biogenic AgNPs. This finding was in line with the previous literature and concords with the presence of protein capping on the surface of biogenic AgNPs that stabilizes the nanoparticles in aqueous environments [69–73]. It has been assumed that proteins can bind to nanoparticles as capping agents through their cysteine residues and/or free amine groups [74]. It is worth mentioning that the biomass extracts of *G. nicotianae* SNPRA1 and *L. aridicollis* SNPRA2 reduced Ag^+ to Ag^0 under dark conditions without photocatalysis. On the contrary, the biosynthesis of AgNPs by the actinomycete *Sinomonas mesophila* MPKL 26 was achieved only after exposure to sunlight [75]. Additionally, the sunlight-assisted biosynthesis of AgNPs by various fungal and plant extracts is well-documented [76–81].

In this study, the biogenic Gn-AgNPs and La-AgNPs exhibited a remarkable inhibitory effect on the conidial germination of the investigated mycotoxigenic fungi, which was accomplished with impaired membrane integrity evidenced by the leakage of DNA and cellular proteins. In the same way, Khalil and coworkers proved that biogenic AgNPs produced by *Penicillium chrysogenum* NG85 and *Fusarium chlamydosporum* NG30 exert their antifungal activity against *A. flavus* and *A. ochraceous* via cellular membrane damage [39]. Aspergilli and many fungi produce conidia "asexual spores" for dispersion or to survive hostile conditions [82,83]. As conidial germination is the first step for the development of *Aspergillus* spp., attempts to prevent their germination could diminish the adverse impact of these widespread fungi. These findings are in harmony with previous studies suggesting that AgNPs provoke the permeability of membranes, causing the leakage of DNA and proteins [84–87]. Similarly, biogenic AgNPs produced by the *Pseudomonas poae* strain CO showed a significant inhibitory effect on the spore germination of *F. graminearum* [88]. Hence, the inhibition of spore/conidial germination by AgNPs could be attributed to reducing the fungal propagation, and thus, their risk. Additionally, biogenic AgNPs produced by *F. oxysporum* inhibited the sclerotial germination of *Stromatinia cepivora* [89,90]. It has been suggested that the antifungal activity of nanosilver may be attributed to the attachment of AgNPs to the cell wall and the anchoring of the cell membranes causing damage and leakage to the intracellular content, which eventually leads to cell death [55,91]. Furthermore, AgNPs were reported to cause surface protein damage, nucleic acid damage, and the blockage of proton pumps [92]. In addition, several possible modes of action have been elucidated regarding the antifungal activity of AgNPs. The extracellular accumulation of AgNPs is thought to cause a dynamic release of Ag^+ that penetrates the cell, leading to an accumulation of intracellular reactive oxygen species (ROS) which hurt the proteins of the membrane and affect the reactions of electron transport, triggering apoptosis [93]. Furthermore, Ag^+ and AgNPs modulate the transcriptome and metabolome, altering the essential functions of fungal cells. It has been reported that Ag^+ and AgNPs cause the downregulation of the tricarboxylic acid cycle genes and other genes involved in ergosterol synthesis and lipid metabolism, leading to structural alternations principally at the level of biological membranes [94].

Mycotoxigenic fungi such as *A. flavus* and *A. ochraceus* have a detrimental effect on economic plants such as rice plants because they cause biotic stress and significantly decrease the plant's physiological activity [95]. Additionally, aflatoxins and OTA are carcinogenic secondary metabolites that are mainly produced by *A. flavus* and *A. ochraceus*, respectively [96]. These mycotoxins may be considered more dangerous than the fungi themselves, making it necessary to study the effect of AgNPs on the mycotoxin production by *A. flavus* and *A. ochraceus* [97]. Thus, the determination of the MIC of the AgNPs that inhibit mycotoxin production completely is highly recommended, and its value should be lower than that required for fungal growth inhibition [98,99]. Our results revealed that the MIC values required for the complete inhibition of total aflatoxin production by *A. flavus* or OTA production by *A. ochraceus* were much lower than that required for the total inhibition of fungal spore germination. In this regard, it has been reported that the biogenic AgNPs produced by *A. terreus* and *P. expansum* exhibited a 58.87 and 52.18% reduction of OTA production, respectively [100]. In good agreement with these findings, the deduced MIC values for the complete inhibition of total aflatoxin production by *A. flavus* using biosynthesized AgNPs from *F. chlamydosporum* NG30 and *P. chrysogenum* NG85 were 5.8 or 5.5 μg/mL, respectively. Meanwhile, the MIC values found for the previously synthesized AgNPs from *F. chlamydosporum* NG30 and *P. chrysogenum* NG85 that inhibited the production of OTA by *A. ochraceus* were 6.3 and 6.1 μg/mL, respectively [39]. Although having superior antimicrobial activity, nanosilver is causing concerns regarding its negative impact on human health due to its potential toxicity [101–103]. Thus, we evaluated the cytotoxicity of the as-prepared biogenic AgNPs against HSF cell lines. Based on the obtained results, no significant cytotoxic effect was observed in HSF cells upon exposure to Gn-AgNPs and La-AgNPs in concentrations up to 10 and 15 μg/mL, with IC_{50} values of 31.78 and 25.83 μg/mL, respectively. At the same time, Gn-AgNPs and La-AgNPs completely inhibited mycotoxin production at concentrations lower than 8 μg/mL. By comparing the effective anti-mycotoxin concentrations with cytotoxic ones, we suggest that Gn-AgNPs and La-AgNPs could be used as safe anti-mycotoxin agents at nontoxic doses.

5. Conclusions

The silver-resistant rare actinomycetes, *G. nicotianae* SNPRA1 and *L. aridicollis* SNPRA2, could be promising candidates for the biosynthesis of AgNPs in a facilitated eco-friendly process. The actinomycete-mediated AgNPs could disrupt the life cycle of the mycotoxigenic fungi by inhibiting their conidial germination, which is the first step in fungal propagation. The data presented here provide evidence that biogenic AgNPs induce the leakage of cellular proteins and DNA, reflecting the disruption of membrane integrity. Furthermore, the bioinspired AgNPs completely inhibited the production of total aflatoxins and ochratoxin A in *A. flavus* and *A. ochraceus* at nontoxic doses. Owing to their biocompatibility and low toxicity, the as-prepared biogenic AgNPs produced by desert-derived actinomycetes could be used as potent antifungals and anti-mycotoxins at nontoxic doses.

Supplementary Materials: The following supporting information can be downloaded at: https://www.mdpi.com/article/10.3390/microorganisms11041006/s1, Figure S1: Effect of the biogenic AgNPs on the conidial germination. The percentage of germinated conidia was calculated by analyzing 100 conidia under an optical microscope. A conidium was considered germinated if the germ tube length was equal to or longer than the conidial length.

Author Contributions: Conceptualization, M.N.A.E.-G. and M.G.F.; Methodology, M.N.A.E.-G., A.N.E. and M.G.F.; Validation, M.G.F.; Formal analysis, M.G.F.; Investigation, M.N.A.E.-G., A.N.E. and M.G.F.; Resources, S.A.H., S.M.K., R.M.E. and M.G.F.; Data curation, M.G.F.; Writing—original draft, M.N.A.E.-G.; Writing—review & editing, M.G.F.; Funding acquisition, S.A.H., S.M.K. and R.M.E. All authors have read and agreed to the published version of the manuscript.

Funding: This research received no external funding.

Data Availability Statement: The essential data supporting the reported results are contained in this study. All other data are available upon request from the corresponding authors.

Conflicts of Interest: The authors declare no conflict of interest.

References

1. Khan, Y.; Sadia, H.; Ali Shah, S.Z.; Khan, M.N.; Shah, A.A.; Ullah, N.; Ullah, M.F.; Bibi, H.; Bafakeeh, O.T.; Khedher, N.B.; et al. Classification, Synthetic, and Characterization Approaches to Nanoparticles, and Their Applications in Various Fields of Nanotechnology: A Review. *Catalysts* **2022**, *12*, 1386. [CrossRef]
2. Chandrakala, V.; Aruna, V.; Angajala, G. Review on Metal Nanoparticles as Nanocarriers: Current Challenges and Perspectives in Drug Delivery Systems. *Emergent Mater.* **2022**, *5*, 1593–1615. [CrossRef]

3. Saravanan, A.; Kumar, P.S.; Karishma, S.; Vo, D.V.N.; Jeevanantham, S.; Yaashikaa, P.R.; George, C.S. A Review on Biosynthesis of Metal Nanoparticles and Its Environmental Applications. *Chemosphere* **2021**, *264*, 128580. [CrossRef] [PubMed]
4. Couto, C.; Almeida, A. Metallic Nanoparticles in the Food Sector: A Mini-Review. *Foods* **2022**, *11*, 402. [CrossRef]
5. Dua, T.K.; Giri, S.; Nandi, G.; Sahu, R.; Shaw, T.K.; Paul, P. Green Synthesis of Silver Nanoparticles Using Eupatorium Adenophorum Leaf Extract: Characterizations, Antioxidant, Antibacterial, and Photocatalytic Activities. *Food Chem. Adv.* **2023**, *2*, 100205. [CrossRef]
6. Thomas, T.; Thalla, A.K. Synthesis of Silver Nanoparticles Using Myristica Fragrans Seed Shell: Assessment of Antibacterial, Antioxidant Properties and Photocatalytic Degradation of Dyes. *J. Environ. Chem. Eng.* **2023**, *109585*, 109585. [CrossRef]
7. Shyamalagowri, S.; Charles, P.; Manjunathan, J.; Kamaraj, M.; Anitha, R.; Pugazhendhi, A. In Vitro Anticancer Activity of Silver Nanoparticles Phyto-Fabricated by Hylocereus Undatus Peel Extracts on Human Liver Carcinoma (HepG2) Cell Lines. *Process Biochem.* **2022**, *116*, 17–25. [CrossRef]
8. Saxena, J.; Ayushi, K.M. Evaluation of Sclerotinia Sclerotiorum MTCC 8785 as a Biological Agent for the Synthesis of Silver Nanoparticles and Assessment of Their Antifungal Potential against Trichoderma Harzianum MTCC 801. *Environ. Res.* **2023**, *216*, 114752. [CrossRef] [PubMed]
9. Shantkriti, S.; Pradeep, M.; Unish, K.K.; Das, V.; Nidhin, S.; Gugan, K.; Murugan, A. Bioynthesis of Silver Nanoparticles Using Dunaliella Salina and Its Antibacterial Applications. *Appl. Surf. Sci. Adv.* **2023**, *13*, 100377. [CrossRef]
10. Mourad, R.; Helaly, F.; Darwesh, O.; Sawy, S. El Antimicrobial and Physicomechanical Natures of Silver Nanoparticles Incorporated into Silicone-Hydrogel Films. *Contact Lens Anterior Eye* **2019**, *42*, 325–333. [CrossRef]
11. Mamdouh, S.; Mahmoud, A.; Samir, A.; Mobarak, M.; Mohamed, T. Using Femtosecond Laser Pulses to Investigate the Nonlinear Optical Properties of Silver Nanoparticles Colloids in Distilled Water Synthesized by Laser Ablation. *Phys. B Condens. Matter* **2022**, *631*, 413727. [CrossRef]
12. Sportelli, M.C.; Clemente, M.; Izzi, M.; Volpe, A.; Ancona, A.; Picca, R.A.; Palazzo, G.; Cioffi, N. Exceptionally Stable Silver Nanoparticles Synthesized by Laser Ablation in Alcoholic Organic Solvent. *Colloids Surf. A Physicochem. Eng. Asp.* **2018**, *559*, 148–158. [CrossRef]
13. Jeevanandam, J.; Krishnan, S.; Hii, Y.S.; Pan, S.; Chan, Y.S.; Acquah, C.; Danquah, M.K.; Rodrigues, J. Synthesis Approach-Dependent Antiviral Properties of Silver Nanoparticles and Nanocomposites. *J. Nanostructure Chem.* **2022**, *12*, 809–831. [CrossRef]
14. Vigneswari, S.; Amelia, T.S.M.; Hazwan, M.H.; Mouriya, G.K.; Bhubalan, K.; Amirul, A.A.A.; Ramakrishna, S. Transformation of Biowaste for Medical Applications: Incorporation of Biologically Derived Silver Nanoparticles as Antimicrobial Coating. *Antibiotics* **2021**, *10*, 229. [CrossRef]
15. Gahlawat, G.; Choudhury, A.R. A Review on the Biosynthesis of Metal and Metal Salt Nanoparticles by Microbes. *RSC Adv.* **2019**, *9*, 12944–12967. [CrossRef] [PubMed]
16. Mondal, A.H.; Yadav, D.; Mitra, S.; Mukhopadhyay, K. Biosynthesis of Silver Nanoparticles Using Culture Supernatant of *Shewanella* Sp. ARY1 and Their Antibacterial Activity. *Int. J. Nanomed.* **2020**, *15*, 8295–8310. [CrossRef]
17. Alfryyan, N.; Kordy, M.G.M.; Abdel-Gabbar, M.; Soliman, H.A.; Shaban, M. Characterization of the Biosynthesized Intracellular and Extracellular Plasmonic Silver Nanoparticles Using Bacillus Cereus and Their Catalytic Reduction of Methylene Blue. *Sci. Rep.* **2022**, *12*, 12495. [CrossRef] [PubMed]
18. Hanna, A.L.; Hamouda, H.M.; Goda, H.A.; Sadik, M.W.; Moghanm, F.S.; Ghoneim, A.M.; Alenezi, M.A.; Alnomasy, S.F.; Alam, P.; Elsayed, T.R. Biosynthesis and Characterization of Silver Nanoparticles Produced by Phormidium Ambiguum and Desertifilum Tharense Cyanobacteria. *Bioinorg. Chem. Appl.* **2022**, *2022*, 9072508. [CrossRef] [PubMed]
19. Cepoi, L.; Rudi, L.; Zinicovscaia, I.; Chiriac, T.; Miscu, V.; Rudic, V. Biochemical Changes in Microalga Porphyridium Cruentum Associated with Silver Nanoparticles Biosynthesis. *Arch. Microbiol.* **2021**, *203*, 1547–1554. [CrossRef]
20. Silva, L.P.C.; Oliveira, J.P.; Keijok, W.J.; da Silva, A.R.; Aguiar, A.R.; Guimarães, M.C.C.; Ferraz, C.M.; Araújo, J.V.; Tobias, F.L.; Braga, F.R. Extracellular Biosynthesis of Silver Nanoparticles Using the Cell-Free Filtrate of Nematophagous Fungus Duddingtonia Flagrans. *Int. J. Nanomed.* **2017**, *12*, 6373–6381. [CrossRef]
21. Saad, A.M.; El-Saadony, M.T.; El-Tahan, A.M.; Sayed, S.; Moustafa, M.A.M.; Taha, A.E.; Taha, T.F.; Ramadan, M.M. Polyphenolic Extracts from Pomegranate and Watermelon Wastes as Substrate to Fabricate Sustainable Silver Nanoparticles with Larvicidal Effect against Spodoptera Littoralis. *Saudi J. Biol. Sci.* **2021**, *28*, 5674–5683. [CrossRef] [PubMed]
22. Xing, Y.; Liao, X.; Liu, X.; Li, W.; Huang, R.; Tang, J.; Xu, Q.; Li, X.; Yu, J. Characterization and Antimicrobial Activity of Silver Nanoparticles Synthesized with the Peel Extract of Mango. *Materials* **2021**, *14*, 5878. [CrossRef]
23. Devi, N.; Rani, K.; Kharb, P.; Kaushik, P. Bio-Fabrication of Euryale Ferox (Makhana) Leaf Silver Nanoparticles and Their Antibacterial, Antioxidant and Cytotoxic Potential. *Plants* **2022**, *11*, 2766. [CrossRef]
24. Navale, V.; Vamkudoth, K.R.; Ajmera, S.; Dhuri, V. Aspergillus Derived Mycotoxins in Food and the Environment: Prevalence, Detection, and Toxicity. *Toxicol Rep.* **2021**, *8*, 1008–1030. [CrossRef] [PubMed]
25. Chuaysrinule, C.; Maneeboon, T.; Roopkham, C.; Mahakarnchanakul, W. Occurrence of Aflatoxin- and Ochratoxin A-Producing Aspergillus Species in Thai Dried Chilli. *J. Agric. Food Res.* **2020**, *2*, 100054. [CrossRef]
26. Massomo, S.M.S. Aspergillus Flavus and Aflatoxin Contamination in the Maize Value Chain and What Needs to Be Done in Tanzania. *Sci. Afr.* **2020**, *10*, e00606. [CrossRef]

27. Kumar, P.; Mahato, D.K.; Sharma, B.; Borah, R.; Haque, S.; Mahmud, M.M.C.; Shah, A.K.; Rawal, D.; Bora, H.; Bui, S. Ochratoxins in Food and Feed: Occurrence and Its Impact on Human Health and Management Strategies. *Toxicon* **2020**, *187*, 151–162. [CrossRef]
28. Ma, L.; Liu, J.; Su, W.; Zeng, X.; Liu, X.; Li, W.; Deng, J.; Tang, J. Extracellular Biosynthesis, Characterization and Cytotoxic Effect of Silver Nanoparticles by Streptomyces Coelicoflavus KS-3. *J. Nanosci. Nanotechnol.* **2018**, *18*, 8133–8141. [CrossRef]
29. Bakhtiari-Sardari, A.; Mashreghi, M.; Eshghi, H.; Behnam-Rasouli, F.; Lashani, E.; Shahnavaz, B. Comparative Evaluation of Silver Nanoparticles Biosynthesis by Two Cold-Tolerant Streptomyces Strains and Their Biological Activities. *Biotechnol. Lett.* **2020**, *42*, 1985–1999. [CrossRef]
30. Al-Dhabi, N.A.; Ghilan, A.K.M.; Arasu, M.V.; Duraipandiyan, V. Green Biosynthesis of Silver Nanoparticles Produced from Marine *Streptomyces* Sp. Al-Dhabi-89 and Their Potential Applications against Wound Infection and Drug Resistant Clinical Pathogens. *J. Photochem. Photobiol. B* **2018**, *189*, 176–184. [CrossRef]
31. Elsilk, S.E.; Khalil, M.A.; Aboshady, T.A.; Alsalmi, F.A.; Ali, S.S. Streptomyces Rochei MS-37 as a Novel Marine Actinobacterium for Green Biosynthesis of Silver Nanoparticles and Their Biomedical Applications. *Molecules* **2022**, *27*, 7296. [CrossRef] [PubMed]
32. Abd-Elhady, H.M.; Ashor, M.A.; Hazem, A.; Saleh, F.M.; Selim, S.; el Nahhas, N.; Abdel-Hafez, S.H.; Sayed, S.; Hassan, E.A. Biosynthesis and Characterization of Extracellular Silver Nanoparticles from Streptomyces Aizuneusis: Antimicrobial, Anti Larval, and Anticancer Activities. *Molecules* **2021**, *27*, 212. [CrossRef] [PubMed]
33. Zwar, I.P.; Trotta, C.D.V.; Ziotti, A.B.S.; Lima Neto, M.; Araújo, W.L.; de Melo, I.S.; Ottoni, C.A.; de Souza, A.O. Biosynthesis of Silver Nanoparticles Using Actinomycetes, Phytotoxicity on Rice Seeds, and Potential Application in the Biocontrol of Phytopathogens. *J. Basic Microbiol.* **2023**, *63*, 64–74. [CrossRef]
34. Elnady, A.; Sorour, N.M.; Abbas, R.N. Characterization, Cytotoxicity, and Genotoxicity Properties of Novel Biomediated Nanosized-Silver by Egyptian Streptomyces Roseolus for Safe Antimicrobial Applications. *World J. Microbiol. Biotechnol.* **2022**, *38*, 47. [CrossRef]
35. Mechouche, M.S.; Merouane, F.; Messaad, C.E.H.; Golzadeh, N.; Vasseghian, Y.; Berkani, M. Biosynthesis, Characterization, and Evaluation of Antibacterial and Photocatalytic Methylene Blue Dye Degradation Activities of Silver Nanoparticles from Streptomyces Tuirus Strain. *Environ. Res.* **2022**, *204*, 112360. [CrossRef]
36. Hayakawa, M.; Nonomura, H. Humic Acid-Vitamin Agar, a New Medium for the Selective Isolation of Soil Actinomycetes. *J. Ferment. Technol.* **1987**, *65*, 501–509. [CrossRef]
37. Kamel, Z.; Mohamed, N.M.; Farahat, M.G. Optimization of Culture Conditions for Production of B-Galactosidase by Bacillus Megaterium NM56 Isolated from Raw Milk. *Res. J. Pharm. Biol. Chem. Sci.* **2016**, *7*, 366–376.
38. Yoon, S.H.; Ha, S.M.; Kwon, S.; Lim, J.; Kim, Y.; Seo, H.; Chun, J. Introducing EzBioCloud: A Taxonomically United Database of 16S RRNA Gene Sequences and Whole-Genome Assemblies. *Int. J. Syst. Evol. Microbiol.* **2017**, *67*, 1613–1617. [CrossRef] [PubMed]
39. Khalil, N.M.; Abd El-Ghany, M.N.; Rodríguez-Couto, S. Antifungal and Anti-Mycotoxin Efficacy of Biogenic Silver Nanoparticles Produced by Fusarium Chlamydosporum and Penicillium Chrysogenum at Non-Cytotoxic Doses. *Chemosphere* **2019**, *218*, 477–486. [CrossRef]
40. Davis, N.D.; Diener, U.L.; Eldridge, D.W. Production of Aflatoxins B1 and G1 by Aspergillus Flavus in a Semisynthetic Medium. *Appl. Microbiol.* **1966**, *14*, 378–380. [CrossRef] [PubMed]
41. Abd El-Ghany, M.N.; Hamdi, S.A.; Korany, S.M.; Elbaz, R.M.; Farahat, M.G. Biosynthesis of Novel Tellurium Nanorods by *Gayadomonas* Sp. TNPM15 Isolated from Mangrove Sediments and Assessment of Their Impact on Spore Germination and Ultrastructure of Phytopathogenic Fungi. *Microorganisms* **2023**, *11*, 558. [CrossRef] [PubMed]
42. Qin, S.; Li, J.; Chen, H.H.; Zhao, G.Z.; Zhu, W.Y.; Jiang, C.L.; Xu, L.H.; Li, W.J. Isolation, Diversity, and Antimicrobial Activity of Rare Actinobacteria from Medicinal Plants of Tropical Rain Forests in Xishuangbanna China. *Appl. Environ. Microbiol.* **2009**, *75*, 6176–6186. [CrossRef]
43. Subramani, R.; Aalbersberg, W. Culturable Rare Actinomycetes: Diversity, Isolation and Marine Natural Product Discovery. *Appl. Microbiol. Biotechnol.* **2013**, *97*, 9291–9321. [CrossRef]
44. Singh, V.; Haque, S.; Singh, H.; Verma, J.; Vibha, K.; Singh, R.; Jawed, A.; Tripathi, C.K.M. Isolation, Screening, and Identification of Novel Isolates of Actinomycetes from India for Antimicrobial Applications. *Front. Microbiol.* **2016**, *7*, 1921. [CrossRef] [PubMed]
45. Niyomvong, N.; Pathom-Aree, W.; Thamchaipenet, A.; Duangmal, K. Actinomycetes from Tropical Limestone Caves. *Chiang Mai J. Sci.* **2012**, *39*, 373–388.
46. González Henao, S.; Ghneim-Herrera, T. Heavy Metals in Soils and the Remediation Potential of Bacteria Associated with the Plant Microbiome. *Front. Environ. Sci.* **2021**, *9*, 15. [CrossRef]
47. Hao, X.; Zhu, J.; Rensing, C.; Liu, Y.; Gao, S.; Chen, W.; Huang, Q.; Liu, Y.R. Recent Advances in Exploring the Heavy Metal(Loid) Resistant Microbiome. *Comput. Struct. Biotechnol. J.* **2021**, *19*, 94–109. [CrossRef]
48. Fadl, M.G.; Farahat, M.G.; Mohamed, Z.K. Optimum Biosorption and Resistance of Uranium by Metal-Resistant Bacteria Isolated from Rock Ore. *Geomicrobiol. J.* **2022**, *39*, 689–696. [CrossRef]
49. Duraisamy, P.; Sekar, J.; Arunkumar, A.D.; Ramalingam, P.V. Kinetics of Phenol Biodegradation by Heavy Metal Tolerant Rhizobacteria Glutamicibacter Nicotianae MSSRFPD35 from Distillery Effluent Contaminated Soils. *Front. Microbiol.* **2020**, *11*, 1573. [CrossRef]
50. Sturm, G.; Brunner, S.; Suvorova, E.; Dempwolff, F.; Reiner, J.; Graumann, P.; Bernier-Latmani, R.; Majzlan, J.; Gescher, J. Chromate Resistance Mechanisms in Leucobacter Chromiiresistens. *Appl. Environ. Microbiol.* **2018**, *84*, 02208–02218. [CrossRef]

51. Tahri Joutey, N.; Bahafid, W.; Sayel, H.; Nassef, S.; El Ghachtouli, N. Leucobacter Chromiireducens CRB2, a New Strain with High Cr(VI) Reduction Potential Isolated from Tannery-Contaminated Soil (Fez, Morocco). *Ann. Microbiol.* **2016**, *66*, 425–436. [CrossRef]
52. Hyun, D.W.; Sung, H.; Kim, P.S.; Yun, J.H.; Bae, J.W. Leucobacter Coleopterorum Sp. Nov., Leucobacter Insecticola Sp. Nov., and Leucobacter Viscericola Sp. Nov., Isolated from the Intestine of the Diving Beetles, Cybister Brevis and Cybister Lewisianus, and Emended Description of the Genus Leucobacter. *J. Microbiol.* **2021**, *59*, 360–368. [CrossRef]
53. Khalil, M.A.; El-Shanshoury, A.E.R.R.; Alghamdi, M.A.; Alsalmi, F.A.; Mohamed, S.F.; Sun, J.; Ali, S.S. Biosynthesis of Silver Nanoparticles by Marine Actinobacterium Nocardiopsis Dassonvillei and Exploring Their Therapeutic Potentials. *Front. Microbiol.* **2022**, *12*, 4117. [CrossRef] [PubMed]
54. Bahrulolum, H.; Nooraei, S.; Javanshir, N.; Tarrahimofrad, H.; Mirbagheri, V.S.; Easton, A.J.; Ahmadian, G. Green Synthesis of Metal Nanoparticles Using Microorganisms and Their Application in the Agrifood Sector. *J. Nanobiotechnol.* **2021**, *19*, 1–26. [CrossRef]
55. Roy, A.; Bulut, O.; Some, S.; Mandal, A.K.; Yilmaz, M.D. Green Synthesis of Silver Nanoparticles: Biomolecule-Nanoparticle Organizations Targeting Antimicrobial Activity. *RSC Adv.* **2019**, *9*, 2673–2702. [CrossRef] [PubMed]
56. Hietzschold, S.; Walter, A.; Davis, C.; Taylor, A.A.; Sepunaru, L. Does Nitrate Reductase Play a Role in Silver Nanoparticle Synthesis? Evidence for NADPH as the Sole Reducing Agent. *ACS Sustain. Chem. Eng.* **2019**, *7*, 8070–8076. [CrossRef]
57. Tag, H.M.; Saddiq, A.A.; Alkinani, M.; Hagagy, N. Biosynthesis of Silver Nanoparticles Using Haloferax Sp. NRS1: Image Analysis, Characterization, In Vitro Thrombolysis and Cytotoxicity. *AMB Express* **2021**, *11*, 75. [CrossRef]
58. Alam, A.; Tanveer, F.; Khalil, A.T.; Zohra, T.; Khamlich, S.; Alam, M.M.; Salman, M.; Ali, M.; Ikram, A.; Shinwari, Z.K.; et al. Silver Nanoparticles Biosynthesized from Secondary Metabolite Producing Marine Actinobacteria and Evaluation of Their Biomedical Potential. *Antonie Van Leeuwenhoek Int. J. Gen. Mol. Microbiol.* **2021**, *114*, 1497–1516. [CrossRef]
59. El-Bendary, M.A.; Afifi, S.S.; Moharam, M.E.; Abo El-Ola, S.M.; Salama, A.; Omara, E.A.; Shaheen, M.N.F.; Hamed, A.A.; Gawdat, N.A. Biosynthesis of Silver Nanoparticles Using Isolated Bacillus Subtilis: Characterization, Antimicrobial Activity, Cytotoxicity, and Their Performance as Antimicrobial Agent for Textile Materials. *Prep. Biochem. Biotechnol.* **2020**, *51*, 54–68. [CrossRef]
60. Huq, M.A.; Akter, S. Biosynthesis, Characterization and Antibacterial Application of Novel Silver Nanoparticles against Drug Resistant Pathogenic Klebsiella Pneumoniae and Salmonella Enteritidis. *Molecules* **2021**, *26*, 5996. [CrossRef]
61. Talabani, R.F.; Hamad, S.M.; Barzinjy, A.A.; Demir, U. Biosynthesis of Silver Nanoparticles and Their Applications in Harvesting Sunlight for Solar Thermal Generation. *Nanomaterials* **2021**, *11*, 2421. [CrossRef]
62. Wang, X.; Lee, S.Y.; Akter, S.; Huq, M.A. Probiotic-Mediated Biosynthesis of Silver Nanoparticles and Their Antibacterial Applications against Pathogenic Strains of Escherichia Coli O157:H7. *Polymers* **2022**, *14*, 1834. [CrossRef] [PubMed]
63. Nagaraja, S.K.; Niazi, S.K.; Bepari, A.; Assiri, R.A.; Nayaka, S. Leonotis Nepetifolia Flower Bud Extract Mediated Green Synthesis of Silver Nanoparticles, Their Characterization, and In Vitro Evaluation of Biological Applications. *Materials* **2022**, *15*, 8990. [CrossRef] [PubMed]
64. Malik, M.; Iqbal, M.A.; Malik, M.; Raza, M.A.; Shahid, W.; Choi, J.R.; Pham, P.V. Biosynthesis and Characterizations of Silver Nanoparticles from Annona Squamosa Leaf and Fruit Extracts for Size-Dependent Biomedical Applications. *Nanomaterials* **2022**, *12*, 616. [CrossRef] [PubMed]
65. Mahmoudi, F.; Mahmoudi, F.; Gollo, K.H.; Amini, M.M. Biosynthesis of Novel Silver Nanoparticles Using Eryngium Thyrsoideum Boiss Extract and Comparison of Their Antidiabetic Activity with Chemical Synthesized Silver Nanoparticles in Diabetic Rats. *Biol. Trace Elem. Res.* **2021**, *199*, 1967–1978. [CrossRef]
66. Bold, B.E.; Urnukhsaikhan, E.; Mishig-Ochir, T. Biosynthesis of Silver Nanoparticles with Antibacterial, Antioxidant, Anti-Inflammatory Properties and Their Burn Wound Healing Efficacy. *Front. Chem.* **2022**, *10*, 1022. [CrossRef]
67. Weng, X.; Yang, K.; Owens, G.; Chen, Z. Biosynthesis of Silver Nanoparticles Using Three Different Fruit Extracts: Characterization, Formation Mechanism and Estrogen Removal. *J. Environ. Manag.* **2022**, *316*, 115224. [CrossRef]
68. Alhomaidi, E.; Jasim, S.A.; Amin, H.I.M.; Lima Nobre, M.A.; Khatami, M.; Jalil, A.T.; Hussain Dilfy, S. Biosynthesis of Silver Nanoparticles Using Lawsonia Inermis and Their Biomedical Application. *IET Nanobiotechnol.* **2022**, *16*, 284–294. [CrossRef]
69. Ogunleye, G.E.; Adebayo-Tayo, B.C.; Oyinlola, K.A. Biological Evaluation of Extracellular Mycosynthesized Silver Nanoparticles by Trichoderma Asperellum. *BioMetals* **2022**, *36*, 97–109. [CrossRef]
70. Bhadani, R.V.; Gajera, H.P.; Hirpara, D.G.; Savaliya, D.D.; Anuj, S.A. Biosynthesis and Characterization of Extracellular Metabolites-Based Nanoparticles to Control the Whitefly. *Arch. Microbiol.* **2022**, *204*, 311. [CrossRef]
71. Li, P.-J.; Pan, J.-J.; Tao, L.-J.; Li, X.; Su, D.-L.; Shan, Y.; Li, H.-Y.; Li, P.-J.; Pan, J.-J.; Tao, L.-J.; et al. Green Synthesis of Silver Nanoparticles by Extracellular Extracts from Aspergillus Japonicus PJ01. *Molecules* **2021**, *26*, 4479. [CrossRef]
72. Abdelmoneim, H.M.; Taha, T.H.; Elnouby, M.S.; AbuShady, H.M. Extracellular Biosynthesis, OVAT/Statistical Optimization, and Characterization of Silver Nanoparticles (AgNPs) Using Leclercia Adecarboxylata THHM and Its Antimicrobial Activity. *Microb. Cell Factories* **2022**, *21*, 277. [CrossRef] [PubMed]
73. Murugaiah, H.; Teh, C.L.; Loh, K.C.; Mohamad Yahya, A.R.; Md Noh, N.A.; Abu Bakar, N.H.H.; Kernain, D.; Hashim, R.; Bustami, Y. Study of Antibacterial and Anticancer Properties of BioAgNPs Synthesized Using *Streptomyces* Sp. PBD-311B and the Application of BioAgNP-CNC/Alg as an Antibacterial Hydrogel Film against *P. Aeruginosa* USM-AR2 and MRSA. *Molecules* **2021**, *26*, 6414. [CrossRef] [PubMed]

74. Gurunathan, S.; Kalishwaralal, K.; Vaidyanathan, R.; Venkataraman, D.; Pandian, S.R.K.; Muniyandi, J.; Hariharan, N.; Eom, S.H. Biosynthesis, Purification and Characterization of Silver Nanoparticles Using *Escherichia coli*. *Colloids Surf. B Biointerfaces* **2009**, *74*, 328–335. [CrossRef] [PubMed]
75. Manikprabhu, D.; Cheng, J.; Chen, W.; Sunkara, A.K.; Mane, S.B.; Kumar, R.; Das, M.; Hozzein, W.N.; Duan, Y.Q.; Li, W.J. Sunlight Mediated Synthesis of Silver Nanoparticles by a Novel Actinobacterium (*Sinomonas Mesophila* MPKL 26) and Its Antimicrobial Activity against Multi Drug Resistant Staphylococcus Aureus. *J. Photochem. Photobiol. B* **2016**, *158*, 202–205. [CrossRef]
76. Arshad, H.; Sadaf, S.; Hassan, U. De-Novo Fabrication of Sunlight Irradiated Silver Nanoparticles and Their Efficacy against E. Coli and S. Epidermidis. *Sci. Rep.* **2022**, *12*, 676. [CrossRef]
77. Satheesh, V.; Mohamed, J.M.M.; El-Sherbiny, M.; Othman, G.; Al-Serwi, R.H.; Thilagar, S. Sunlight-Assisted Green Synthesis of Silver Nanoparticles Using Zizania Latifolia Extract: Toward Antimicrobial Applications. *Biomass Convers Biorefin.* **2022**, *12*, 1–11. [CrossRef]
78. Nallal, V.U.M.; Prabha, K.; VethaPotheher, I.; Ravindran, B.; Baazeem, A.; Chang, S.W.; Otunola, G.A.; Razia, M. Sunlight-Driven Rapid and Facile Synthesis of Silver Nanoparticles Using Allium Ampeloprasum Extract with Enhanced Antioxidant and Antifungal Activity. *Saudi J. Biol. Sci.* **2021**, *28*, 3660–3668. [CrossRef]
79. Patil, H.V.; Nithin, K.S.; Sachhidananda, S.; Chandrashekara, K.T.; Kumar, S. Mycofabrication of Bioactive Silver Nanoparticle: Photo Catalysed Synthesis and Characterization to Attest Its Augmented Bio-Efficacy. *Arab. J. Chem.* **2019**, *12*, 4596–4611. [CrossRef]
80. Sonbol, H.; Mohammed, A.E.; Korany, S.M. Soil Fungi as Biomediator in Silver Nanoparticles Formation and Antimicrobial Efficacy. *Int. J. Nanomed.* **2022**, *17*, 2843–2863. [CrossRef]
81. Bhatnagar, S.; Ogbonna, C.N.; Ogbonna, J.C.; Aoyagi, H. Effect of Physicochemical Factors on Extracellular Fungal Pigment-Mediated Biofabrication of Silver Nanoparticles. *Green Chem. Lett. Rev.* **2022**, *15*, 274–284. [CrossRef]
82. Baltussen, T.J.H.; Zoll, J.; Verweij, P.E.; Melchers, W.J.G. Molecular Mechanisms of Conidial Germination in Aspergillus Spp. *Microbiol. Mol. Biol. Rev.* **2020**, *84*, e00049-e19. [CrossRef] [PubMed]
83. Osherov, N.; May, G.S. The Molecular Mechanisms of Conidial Germination. *FEMS Microbiol. Lett.* **2001**, *199*, 153–160. [CrossRef]
84. Zhou, L.; Zhao, X.; Li, M.; Lu, Y.; Ai, C.; Jiang, C.; Liu, Y.; Pan, Z.; Shi, J. Antifungal Activity of Silver Nanoparticles Synthesized by Iturin against Candida Albicans in Vitro and in Vivo. *Appl. Microbiol. Biotechnol.* **2021**, *105*, 3759–3770. [CrossRef]
85. Li, L.; Pan, H.; Deng, L.; Qian, G.; Wang, Z.; Li, W.; Zhong, C. The Antifungal Activity and Mechanism of Silver Nanoparticles against Four Pathogens Causing Kiwifruit Post-Harvest Rot. *Front. Microbiol.* **2022**, *13*, 3418. [CrossRef] [PubMed]
86. Khan, S.; Singh, S.; Gaikwad, S.; Nawani, N.; Junnarkar, M.; Pawar, S.V. Optimization of Process Parameters for the Synthesis of Silver Nanoparticles from Piper Betle Leaf Aqueous Extract, and Evaluation of Their Antiphytofungal Activity. *Environ. Sci. Pollut. Res.* **2020**, *27*, 27221–27233. [CrossRef] [PubMed]
87. Gajera, H.P.; Hirpara, D.G.; Bhadani, R.V.; Golakiya, B.A. Green Synthesis and Characterization of Nanosilver Derived from Extracellular Metabolites of Potent Bacillus Subtilis for Antifungal and Eco-Friendly Action against Phytopathogen. *BioMetals* **2022**, *35*, 479–497. [CrossRef]
88. Ibrahim, E.; Zhang, M.; Zhang, Y.; Hossain, A.; Qiu, W.; Chen, Y.; Wang, Y.; Wu, W.; Sun, G.; Li, B. Green-Synthesization of Silver Nanoparticles Using Endophytic Bacteria Isolated from Garlic and Its Antifungal Activity against Wheat Fusarium Head Blight Pathogen Fusarium Graminearum. *Nanomaterials* **2020**, *10*, 219. [CrossRef]
89. Darwesh, O.M.; Elshahawy, I.E. Silver Nanoparticles Inactivate Sclerotial Formation in Controlling White Rot Disease in Onion and Garlic Caused by the Soil Borne Fungus Stromatinia Cepivora. *Eur. J. Plant Pathol.* **2021**, *160*, 917–934. [CrossRef]
90. Darwesh, O.M.; Barakat, K.M.; Mattar, M.Z.; Sabae, S.Z.; Hassan, S.H. Production of Antimicrobial Blue Green Pigment Pyocyanin by Marine Pseudomonas Aeruginosa. *Biointerface Res. Appl. Chem.* **2019**, *9*, 4334–4339. [CrossRef]
91. Ameen, F.; Al-Homaidan, A.A.; Al-Sabri, A.; Almansob, A.; AlNAdhari, S. Anti-Oxidant, Anti-Fungal and Cytotoxic Effects of Silver Nanoparticles Synthesized Using Marine Fungus Cladosporium Halotolerans. *Appl. Nanosci.* **2021**, *13*, 623–631. [CrossRef]
92. Mansoor, S.; Zahoor, I.; Baba, T.R.; Padder, S.A.; Bhat, Z.A.; Koul, A.M.; Jiang, L. Fabrication of Silver Nanoparticles against Fungal Pathogens. *Front. Nanotechnol.* **2021**, *3*, 679358. [CrossRef]
93. Mussin, J.; Giusiano, G. Biogenic Silver Nanoparticles as Antifungal Agents. *Front. Chem.* **2022**, *10*, 1270. [CrossRef] [PubMed]
94. Barros, D.; Pradhan, A.; Pascoal, C.; Cássio, F. Transcriptomics Reveals the Action Mechanisms and Cellular Targets of Citrate-Coated Silver Nanoparticles in a Ubiquitous Aquatic Fungus. *Environ. Pollut.* **2021**, *268*, 115913. [CrossRef]
95. Sultana, T.; Javed, B.; Raja, N.I.; Mashwani, Z.U.R. Silver Nanoparticles Elicited Physiological, Biochemical, and Antioxidant Modifications in Rice Plants to Control Aspergillus Flavus. *Green Process. Synth.* **2021**, *10*, 314–324. [CrossRef]
96. Fouad, F.A.H.; Youssef, D.G.; Shahat, F.M.; Abd El-Ghany, M.N. Role of Microorganisms in Biodegradation of Pollutants. In *Handbook of Biodegradable Materials*; Springer: Berlin/Heidelberg, Germany, 2023; pp. 221–260. [CrossRef]
97. Nasirian, V.; Chabok, A.; Barati, A.; Rafienia, M.; Arabi, M.S.; Shamsipur, M. Ultrasensitive Aflatoxin B1 Assay Based on FRET from Aptamer Labelled Fluorescent Polymer Dots to Silver Nanoparticles Labeled with Complementary DNA. *Microchim. Acta* **2017**, *184*, 4655–4662. [CrossRef]
98. Dizaj, S.M.; Lotfipour, F.; Barzegar-Jalali, M.; Zarrintan, M.H.; Adibkia, K. Antimicrobial Activity of the Metals and Metal Oxide Nanoparticles. *Mater. Sci. Eng. C* **2014**, *44*, 278–284. [CrossRef]
99. Cao, J.; Feng, Y.; He, S.; Lin, X. Silver Nanoparticles Deteriorate the Mutual Interaction between Maize (*Zea mays* L.) and Arbuscular Mycorrhizal Fungi: A Soil Microcosm Study. *Appl. Soil Ecol.* **2017**, *119*, 307–316. [CrossRef]

100. Ammar, H.A.M.; El-Desouky, T.A. Green Synthesis of Nanosilver Particles by Aspergillus Terreus HA1N and Penicillium Expansum HA2N and Its Antifungal Activity against Mycotoxigenic Fungi. *J. Appl. Microbiol.* **2016**, *121*, 89–100. [CrossRef]
101. Zhang, J.; Wang, F.; Yalamarty, S.S.K.; Filipczak, N.; Jin, Y.; Li, X. Nano Silver-Induced Toxicity and Associated Mechanisms. *Int. J. Nanomed.* **2022**, *17*, 1851–1864. [CrossRef]
102. Noga, M.; Milan, J.; Frydrych, A.; Jurowski, K. Toxicological Aspects, Safety Assessment, and Green Toxicology of Silver Nanoparticles (AgNPs)—Critical Review: State of the Art. *Int. J. Mol. Sci.* **2023**, *24*, 5133. [CrossRef] [PubMed]
103. Dos Santos, C.A.; Seckler, M.M.; Ingle, A.P.; Gupta, I.; Galdiero, S.; Galdiero, M.; Gade, A.; Rai, M. Silver Nanoparticles: Therapeutical Uses, Toxicity, and Safety Issues. *J. Pharm. Sci.* **2014**, *103*, 1931–1944. [CrossRef] [PubMed]

microorganisms

Article

Antimicrobial Activity of Green Synthesized Silver Nanoparticles Using Waste Leaves of *Hyphaene thebaica* (Doum Palm)

Nadiyah M. Alabdallah [1,2,*] and Essam Kotb [1,2]

1 Basic and Applied Scientific Research Center (BASRC), Imam Abdulrahman Bin Faisal University (IAU), P.O. Box 1982, Dammam 31441, Saudi Arabia
2 Department of Biology, College of Science, Imam Abdulrahman Bin Faisal University (IAU), P.O. Box 1982, Dammam 31441, Saudi Arabia
* Correspondence: nmalabdallah@iau.edu.sa

Abstract: Silver nanoparticles (AgNPs) were biosynthesized for the first time from waste leaves extract of local doum palms in Tabuk, Saudi Arabia. The transmission electron microscope (TEM) revealed a spherical shape with a particle size from 18 to 33 nm. The d-spacing is about 2.6 Å, which confirms a face-centered cubic crystalline building. The biosynthesized AgNPs were evaluated as an antimicrobial agent against several pathogenic bacteria, including *Escherichia coli* ATCC 25922, *Staphylococcus aureus* ATCC 29213, and *Pseudomonas aeruginosa* ATCC 27853. The highest action was exerted against *S. aureus* ATCC 29213 (MIC = 1.5 μg/mL). Interestingly, AgNPs also showed anticandidal activity against the pathogenic yeasts *Candida albicans* ATCC 14053 (MIC = 24 μg/mL) and *Candida tropicalis* ATCC 13803 (MIC = 96 μg/mL). Scanning electron microscope (SEM) revealed deep morphological changes in *Candida* spp. due to the treatment of the AgNPs. Scarce pseudohyphae, perforation, exterior roughness, irregularly shaped cells, and production of protective exopolysaccharide (EPS) were the main features. In conclusion, the process of biosynthesis of AgNPs from the aqueous leaf extract of *Hyphaene thebaica* is environmentally compatible and induces the biosynthesis of tiny AgNPs that could be a promising candidate in biomedical applications, including antimicrobials against some pathogenic bacteria and yeasts.

Keywords: silver nanoparticles; antibacterial; anticandidal; characterization; doum palm

Citation: Alabdallah, N.M.; Kotb, E. Antimicrobial Activity of Green Synthesized Silver Nanoparticles Using Waste Leaves of *Hyphaene thebaica* (Doum Palm). *Microorganisms* **2023**, *11*, 807. https://doi.org/10.3390/microorganisms11030807

Academic Editors: Juan Carlos Gutiérrez and Kamel Ahmed Abd-Elsalam

Received: 25 January 2023
Revised: 4 March 2023
Accepted: 6 March 2023
Published: 22 March 2023

1. Introduction

Multidrug resistance is a huge health concern worldwide as it results in high mortality and morbidity. This dilemma resulted from the overuse and misuse of antibiotics in various medical and non-medical applications. The resulting drug-resistant strains can cause serious infections of high morbidity [1]. For instance, the Gram-positive bacterium *S. aureus* evolved methicillin-resistant strains called MRSA strains; these also became resistant to all known β-lactam antibiotics, including cephalosporins, penicillins, and carbapenems [2]. Other strains of MRSA became unaffected by exposure to vancomycin, linezolid, and daptomycin, making *S. aureus* a leading cause of death worldwide. These resistant strains can produce severe diseases such as endocarditis, osteomyelitis, pneumonia, and sepsis if they reach the circulatory system [2]. In the case of Gram-negative bacteria, such as *P. aeruginosa*, it is an opportunistic pathogen causing a high level of morbidity in cystic fibrosis and immunocompromised people. Control of *P. aeruginosa* is very challenging because of its remarkable ability to neglect the action antibiotics due to its biofilm-mediated resistance, high level of intrinsic resistance, and acquired resistance mechanisms [3]. The pathogenic bacterium *E. coli* has two major contributors to its intrinsic resistance against antibiotics. The first contributor is its impermeable outer membrane. The second contributor is its ability to express many efflux pumps that successfully lower the accumulation of drugs intracellularly [4]. According to surveillance data from the centers for disease control and prevention (CDC), antifungal resistance of pathogenic *Candida* is also an increasing problem over the past 20 years. *Candida* infections may resist antifungal drugs such as fluconazole and echinocandins, making them difficult to treat, causing severe infections in our mouth, skin, vagina, and life-threatening circulatory diseases. The most common human pathogen belonging to *Candida* is *C. albicans* [5].

Relatively recently, nanotechnology attracted significant attention in the field of antimicrobials and other applications in medicine and the environment. The secret behind their distinctive physicochemical and biological properties relies on their nano size, which gives them a high surface-to-volume ratio [6]. Contrary to antibiotics, nanoparticles can escape the counteraction of bacterial cells due to exposure by breaking the bacterial resistance mechanisms such as β-lactamases, the expression of efflux pumps, impermeability of the outer membrane, and the modification of the target site or target process [6]. In addition, nanoparticles use different antimicrobial actions comprising interruption of cell membrane consistency and induction of reactive oxygen species (ROS) production, which inhibits RNA synthesis and protein synthesis [6]. Nanoparticles prepared from noble metals such as silver have been broadly studied for biological applications and other concerns. The significance of AgNPs as antimicrobial agents is superior to other noble metals due to their antimicrobial properties, fast healing of wounds, and additional applications [7]. Due to this, the demand for AgNPs has grown. Estimates showed that around 320 tons of AgNPs are made every year around the world by chemical, physical and biological methods [6,7].

Nanoparticles can be produced utilizing top-down and bottom-up techniques. Top-down methods comprise bulk metal reduction with advanced ablations such as sputtering, mechanical grinding, thermal decomposition and green synthesis techniques. In contrast, the bottom-up procedures involve constructing nanoparticles atom by atom [8]. The chemical and physical methods are effective in the production of nanoparticles; however, the chemical methods may include toxic chemicals as stabilizing agents, which may cause toxicity upon use [7]. In addition, their biocompatibility is very low and possesses latent biological risks. Further steps may be needed to avoid particle agglomeration. In addition, they are hazardous upon leakage into the environment. The physical methods, on the other hand, expend much energy, are expensive and give low yields [9].

Recently, the green synthesis of metal nanoparticles has arisen, which is centered on the utilization of plant extracts as bioreductants. This approach provides safer, easier, faster, more eco-friendly and economical nanoparticles when compared with the physical and chemical approaches and safe alternatives for antimicrobial applications [7,10]. Successful biosynthesis of AgNPs from many plant extracts has been carried out by many authors. An effort was made using *H. thebaica* fruit extracts. *H. thebaica* is traditionally recognized as a gingerbread tree or doum tree in the Arab region [10]. It is one of the most important medicinal desert plants. It contains high amounts of polyphenols, minerals, vitamins, saponins, tannins, essential oils, linoleic acid, coumarins, flavonoids, hydroxycinnamic acids, carbohydrates and fibers. These have very important health benefits, including the treatment of hematuria, bleeding, dyslipidemia, as an antihypertensive, diaphoretic, diuretic, or to lower blood pressure [9,10]. The extract of doum fruits exhibited antimicrobial and antioxidant properties in many studies. Although the biological activities of doum fruits have been investigated in different studies, other parts, such as leaves, have received less attention from researchers [7–9]. This research aimed at the biosynthesis of small-sized AgNPs from the aqueous waste leaf extract of *H. thebaica* for the first time and explored both their antibacterial and anticandidal activity against some pathogenic strains of bacteria and yeasts.

2. Materials and Methods

2.1. Plant Material, Microbial Strains, Media, and Reagents

Waste plant leaves were gathered from a local garden in Tabuk, Saudi Arabia, and washed with distilled water thoroughly three times to remove any dust and undesirable surface particles. The ATCC microbial strains were helpfully gifted from King Fahd University Hospital, Khobar, Saudi Arabia. Silver nitrate ACS reagent, ≥99.0% purity, Mueller-Hinton agar (MHA) and yeast extract peptone dextrose (YPD) broth were purchased from Sigma Aldrich. Other chemicals were obtained from local suppliers in the analytical grade form.

2.2. Green Synthesis of AgNPs

AgNPs were fabricated greenly using waste leaves extracted from the doum palm [*H. thebaica* (L.) Mart.] as capping, stabilizing, reducing agents and silver nitrate following the methods of Ahmed et al. [11] and Lima et al. [12]. The waste leaves were collected and thoroughly washed with distilled water. Then 10 g of sundried powder was mixed with 100 mL of deionized water in a 500 mL beaker and incubated in a shaking incubator at 35 °C for 3 h at a speed of 100 rpm. Thereafter, it was filtrated using Whatman filter paper 11.0 µm. Next, 100 mL of 10 µM $AgNO_3$ solution was blended with 50 mL of the aqueous extract from the leaves of the *H. thebaica* (L.) Mart. at 2:1 ratio (*v*/*v*) and incubated at 40 °C until the color changed to brown.

2.3. Characterization of AgNPs

Several analytical techniques were used to analyze the properties of AgNPs. The UV–visible spectroscopy method was utilized in order to carry out the beginnings of the characterization process for the AgNPs. The UV–Vis spectrometer from Shimadzu-UV 1800 was used to measure the absorbance of the colloidal sample in the range of 200–800 nm, with distilled water as a comparison. The FTIR RX1-Perkin Elmer was used to analyze the functional group responsible for the silver nanoparticles in the wavelength range of 5000–100 cm^{-1}. The TEM was applied to determine the exact morphology and particle size of the biosynthesized AgNPs.

2.4. In-Vitro Antimicrobial Activity of AgNPs

The biosynthesized nanoparticles stock suspension was set up at a dilution of 128 µg/mL sterile Dimethyl sulfoxide (DMSO). It was then homogenized by sonication for 30 min. The initial antimicrobial effect of the fabricated sample was assessed by the measurement of formed inhibition zones (mm) in MHA according to the standard Beecher and Wong [13] method. Seeded clinical microbial strains were *P. aeruginosa* ATCC 27853, *E. coli* ATCC 25922, *S. aureus* ATCC 29213, *C. albicans* ATCC 14053 and *C. tropicalis* ATCC 13803. The MHA was adjusted to pH 7.0 and then sterilized by autoclaving for 20 min at 121 °C. After settling the medium in Petri dishes, the selected bacteria and yeasts at 1.0×10^8 CFU/mL were swabbed on the surface of an agar layer. Thereafter, filter paper disks of 5 mm diameter were positioned onto the agar surface. The target AgNPs were then applied at 5 µL volume per disk. Incubation of the cultures was carried out for 24 h at 37 °C. The width of the clearing zones was evaluated to verify the antimicrobial potential of AgNPs prepared from doum palm [14].

For minimal inhibitory concentration (MIC) determination, the broth dilution method was adopted [15]. The initial dilution was adjusted to 128 µg/mL in a total volume of 4 mL of sterile Brain Heart Infusion Broth (BHIB) in test tubes. Two-fold dilution series were then made to achieve 64 µg/mL, 32 µg/mL, 16 µg/mL, 8 µg/mL, 4 µg/mL, 2 µg/mL and 1 µg/mL dilutions. Two hundred microliters of bacterial and/or yeast suspensions at 1.0×10^8 CFU/mL were introduced in separate tubes. Tubes not containing NPs served as positive controls (100% microbial growth), whereas tubes without microbial inoculations served as negative controls (0% microbial growth). Incubation was carried out for 24 h at 37 °C under static conditions. After incubation, 440 µL filter-sterilized resazurin was added, and the tubes were further incubated overnight under the previous condition. MIC was defined as the average of the lowest concentration of AgNPs inhibiting the growth and the highest concentration of AgNPs allowing the growth of tested microbes [15].

For minimum bactericidal concentration (MBC) and/or minimum fungicidal concentration MFC determination, 100 μL from the tube contents above the value of MIC were plated on the surface of MHA. At the MBC values, no colony growth was observed.

All microbial nutrient media in this study were made following the producer's standards. Microbial strains were preserved in Heart Infusion Agar (HIA) at 4 °C and subcultured on a nutrient medium for 24 h at 37 °C before proceeding with the antimicrobial assessments.

2.5. Tolerance Level

The tolerance of the tested microbes against AgNPs was assessed according to Mondal et al. [15] following the formula: Tolerance = MBC (or MFC)/MIC. When the tolerance is equal to or more than 16, it indicates a static effect of NPs, whereas it is considered a cidal effect when it is equal to or less than 4.

2.6. Anticandidal Activity under SEM

The exterior and topology of the fungal cells of both *C. albicans* ATCC 14053 and *C. tropicalis* ATCC 13803 after treatment with AgNPs were examined with the help of SEM. Initially, YPD broth was formulated from yeast extract (10 g/L, *w*/*v*), peptone (20 g/L, *w*/*v*) and dextrose sugar (20 g/L, *w*/*v*) at pH 6.5. The medium was sterilized for 15 min at 15 lbs pressure. Two loopfuls of yeast cells were used to inoculate a 25 mL medium. Treated cultures were exposed to a AgNP suspension at 5 μg/mL concentration. Blanks were prepared by replacing the AgNP suspension with distilled deionized water. Incubation was done at 37 °C at a speed of 180 rpm for 24 h to allow cell growth.

For the processing of cell examination under SEM, cells were harvested by centrifugation at 5000 rpm for 15 min. Cells were then fixed with 10% formaldehyde for 24 h. Dehydration of cells was carried out in ascending concentrations of ethanol (10%, 30%, 50%, 70%, 90% and absolute ethanol). Exposure time to alcohol was set to 10 min except for the absolute ethanol treatment, which was left for 1 h. Drying was then performed at 35 °C for 24 h. The examination was performed using VEGA3 TESCAN SEM.

2.7. Statistical Analysis

All the measurements and treatments were performed in triplicates unless otherwise stated. The statistical analysis was performed using the SPSS Statistics V24 software. The final readings were represented in the form of averages ± standard deviations.

3. Results

3.1. Characterization of AgNPs

AgNPs in the current research were fabricated by mixing the leaf extracts of *H. thebaica* (L.) Mart with $AgNO_3$ at 40 °C until a brown color was obtained. In UV–visible spectroscopy analysis, silver nanostructures were suspended in deionized water and revealed in the 200–800 nm range. The UV–visible investigation showed a strong peak at 400 nm (Figure 1).

Next, FTIR spectroscopy was employed to determine the functional groups presenting the synthesized silver nanoparticles. The FTIR pattern showed strong IR bands at 3451, 1767, 1603 and 686 cm^{-1} (Figure 2).

The TEM investigation has demonstrated that AgNPs prepared by the *H. thebaica* leaf extract were spherical with a dimension of 18–33 nm (Figure 3a). The distance between fringes (d-spacing) was about 2.6 Å, which confirms a face-centered cubic crystalline structure (Figure 3b).

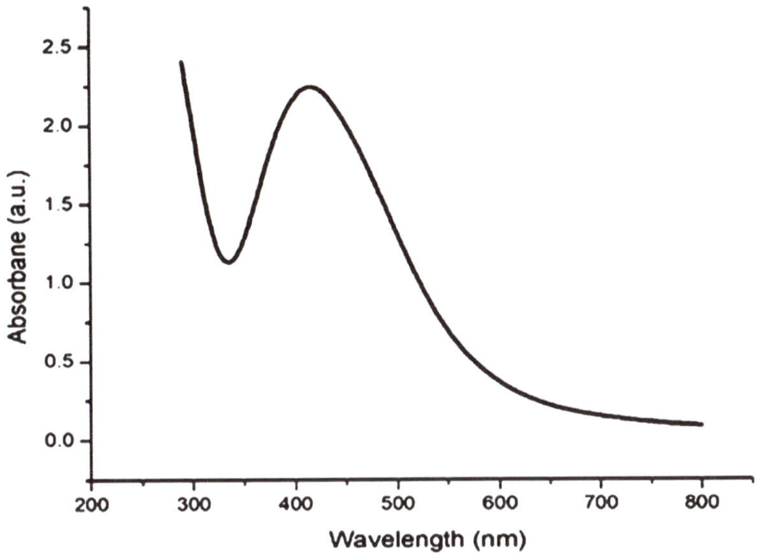

Figure 1. The UV spectrum of the greenly synthesized silver nanoparticles.

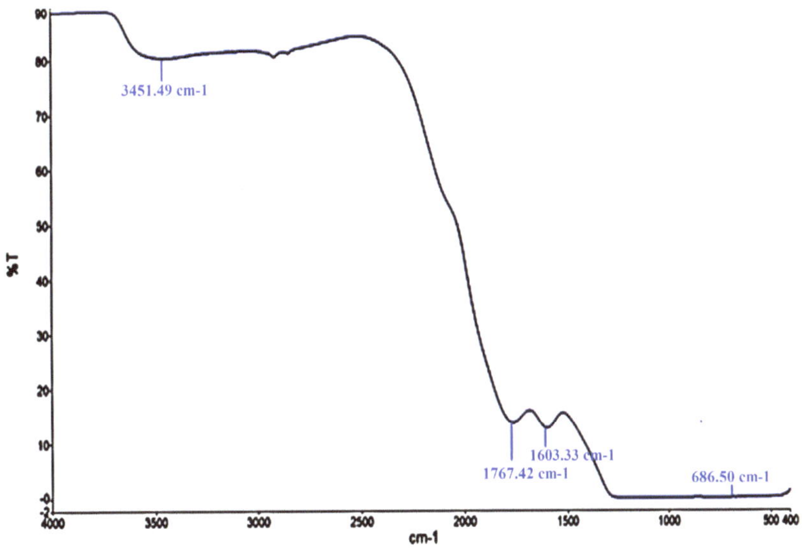

Figure 2. The FTIR pattern of biosynthesized silver nanoparticles from the aqueous leaf extract of *H. thebaica*.

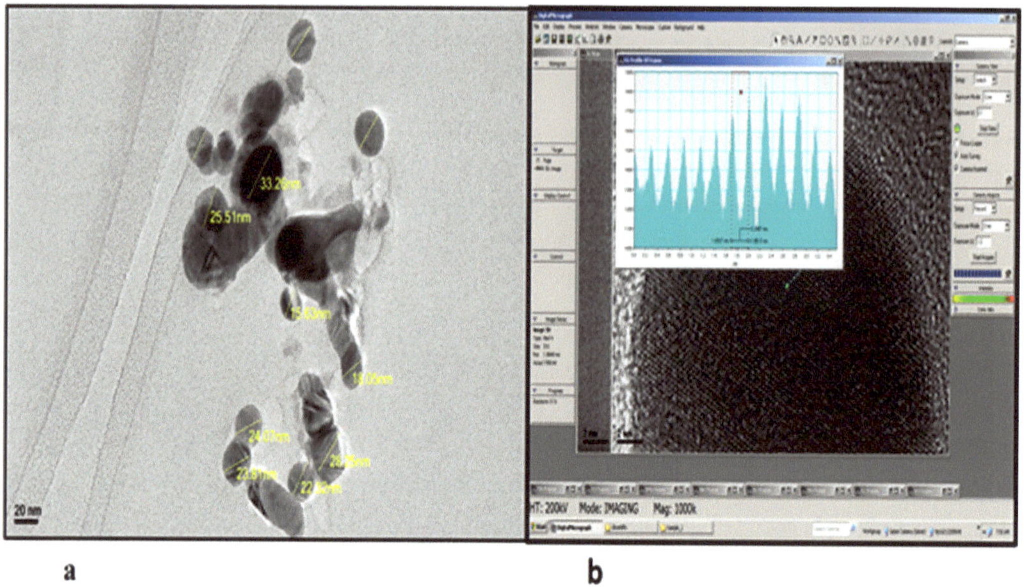

Figure 3. The TEM investigation (**a**) the particle size of silver nanoparticles (**b**) the AgNPs d-spacing was 2.6 Å, with a crystalline structure.

3.2. Antimicrobial Study

From Table 1 and Figure 4, it appears that AgNPs have activity against all the tested microorganisms. The MICs ranged from 1.5 to 96 µg/mL. The highest action was exerted against the Gram-positive bacterium *S. aureus* ATCC 29213 (MIC = 1.5 µg/mL). While AgNPs have exerted lower activity against the Gram-negative bacteria *E. coli* ATCC 25922 (MIC = 3.0 µg/mL) and *P. aeruginosa* ATCC 27853 (MIC = 6.0 µg/mL). Interestingly, the NPs have exerted antifungal activity against the unicellular pathogenic fungi *C. albicans* ATCC 14053 (MIC = 24 µg/mL) and *C. tropicalis* ATCC 13803 (MIC = 96 µg/mL). The results of the MIC experiment coincided with those observed with the inhibition diameters in Figure 4.

Table 1. The antimicrobial activity of the newly prepared AgNPs. The inhibition zones in mm were measured after 24 h of incubation at 37 °C. The minimum inhibitory concentration (MIC), minimum bactericidal concentration (MBC), and minimum fungicidal concentration (MFC) were assessed by the broth dilution method.

Microbe	Antimicrobial Activity			
	Inhibition Zone (mm)	**MIC (µg/mL)**	**MBC (or MFC) (µg/mL)**	**MBC (or MFC)/MIC**
S. aureus ATCC 29213	18.0 ± 0.4	1.5	4	2.67
E. coli ATCC 25922	14.5 ± 1.0	3	4	1.33
P. aeruginosa ATCC 27853	15.0 ± 0.5	6	16	2.67
C. albicans ATCC 14053	11.0 ± 0.8	24	32	1.33
C. tropicalis ATCC 13803	6.5 ± 0.1	96	256	2.67

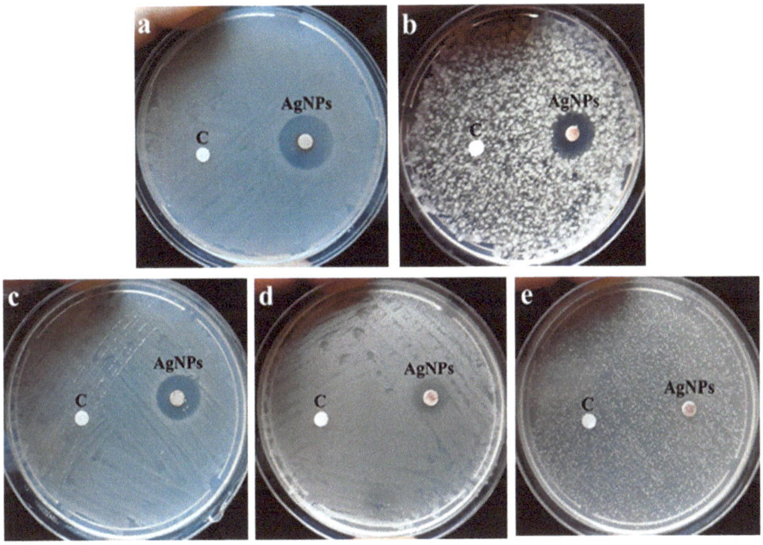

Figure 4. The antimicrobial activity of the newly prepared AgNPs against cultures of different pathogenic microbes. The inhibition zones in mm were measured after 24 h of incubation at 37 °C. Filter paper disks of 5 mm were positioned on the surface of Mueller Hinton agar then 5 μL of 128 μg/mL AgNPs prepared from doum palm waste leaves were applied onto each disk. (**a**) represents *S. aureus* ATCC 29213, (**b**) represents *P. aeruginosa* ATCC 27853, (**c**) represents *E. coli* ATCC 25922, (**d**) represents *C. albicans* ATCC 14053, whereas (**e**) represents *C. tropicalis* ATCC 13803. Disk C corresponds to the control treatment (solvent alone).

Furthermore, the values of MBCs of the greenly synthesized AgNPs were successfully evaluated against the five microorganisms. The values ranged from 4 to 256 μg/mL. The best MBC value was obtained against the Gram-positive bacterium *S. aureus* ATCC 29213 (4 μg/mL), whereas the lowest values were obtained against Gram-negative bacteria and pathogenic yeasts.

The ultrastructural differences under SEM revealed that *C. tropicalis* exposed to AgNPs at 5 μg/mL concentration suffered scarce pseudohyphae (Figure 5b,d,f), perforation (Figure 5b,d,f), exterior roughness (Figure 5b,d,f), irregularly shaped cells (Figure 5f) and EPS formation (Figure 5f). In parallel, untreated cells have shown well-established clusters with prominent pseudohyphae along with the dominant yeast cells (Figure 5a,c,e). The produced EPS from *C. tropicalis* is a means of cell protection against conventional antifungals and may also be a reason for the diminished susceptibility of cells against AgNPs (Figure 5f). In addition, this may be the reason for the high MIC when compared with *C. albicans* (96 μg/mL vs. 24 μg/mL, Table 1).

In the case of *C. albicans*, SEM revealed that AgNPs induced intensive cell damage, deformation, and irregularly shaped cells (Figure 6b,d) when compared with untreated cells showing well-established cell clusters (Figure 6a,c).

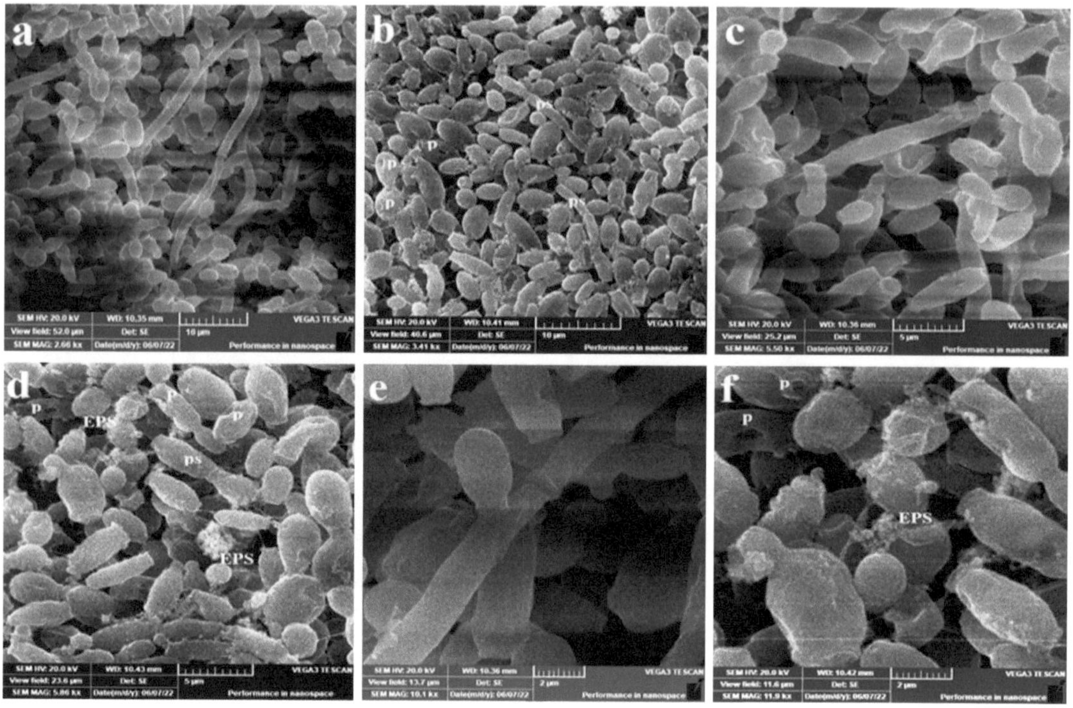

Figure 5. The SEM micrographs of *C. tropicalis* ATCC 13803. (**b**,**d**,**f**) represent treated cells with AgNPs at 5 µg/mL concentration in YPD broth for 24 h. (**a**,**c**,**e**) represent the untreated cells. Treated cells show reduced pseudohyphae (ps), perforation (p), roughened irregularly shaped cells and exopolysaccharide (EPS) production.

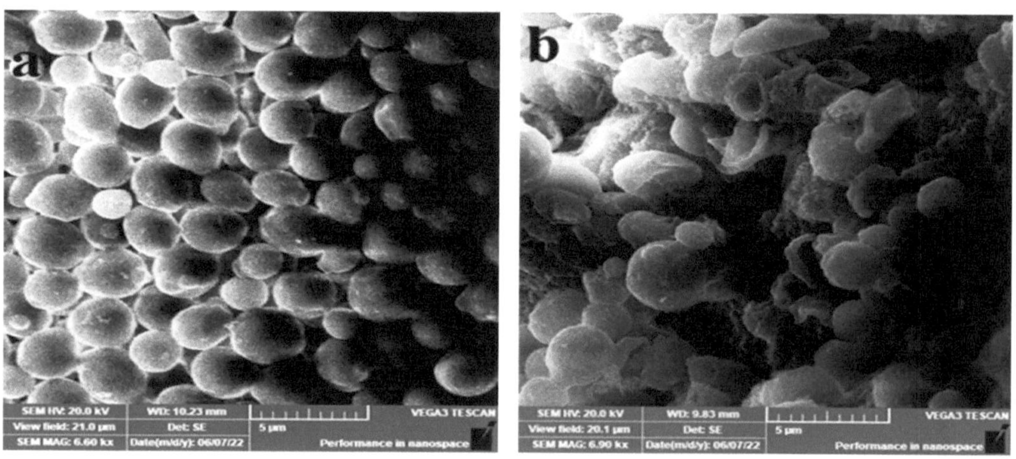

Figure 6. *Cont.*

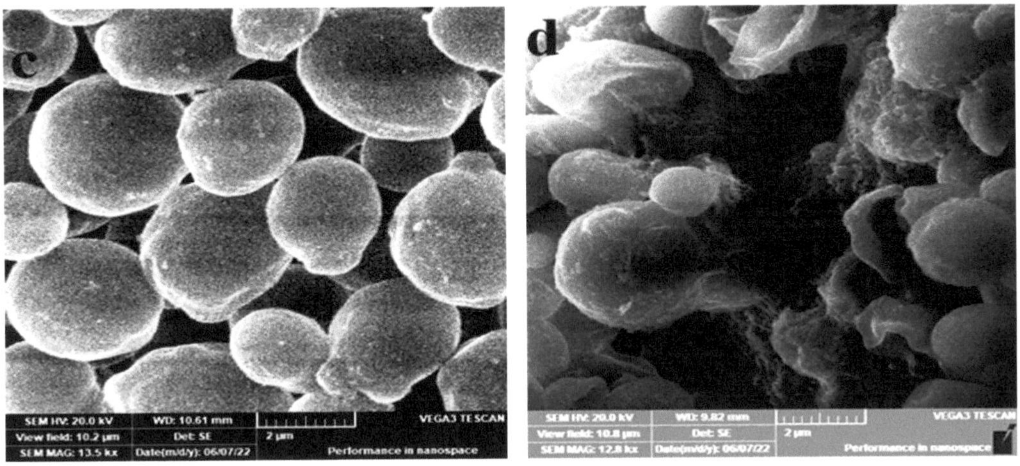

Figure 6. The SEM micrographs of *C. albicans* ATCC 14053. Panels b and d represent treated cells with AgNPs at 5 µg/mL concentration in YPD broth for 24 h. Panels a and c represent the untreated cells. Treated cells show intensive cell damage, deformation and irregularly shaped cells (**b**,**d**) when compared with untreated cells showing well-established cell clusters (**a**,**c**).

4. Discussion

The formation of AgNPs with the help of plant extracts was completed in many studies to inhibit the growth of pathogenic microorganisms, the progression of cancer cells as well as other applications. Their biological activity was found to differ with their shape, size and the plant part used. The antibacterial application of AgNPs was tested for many of the prepared AgNPs, whereas the anticandidal activity was studied in a few reports. The present NPs were found to exert a prominent antimicrobial efficiency against five of the most common pathogenic bacteria and *Candida* spp.

Doum palm trees are common in the Arab region and were selected in this study. The production of AgNPs by the waste leaves of doum palm extract was confirmed by the color change of the reaction mixture towards orange to brown colors, which was an indication of the reduction in Ag^+ ions to Ag° [16,17]. This color change is the initial, observable sign of AgNP production because the excitation of the AgNP's surface plasmon induces a color shift in solutions [18]. The production of AgNPs from $AgNO_3$ is verified by several analytical protocols such as FTIR spectroscopy, UV–visible spectrophotometry and TEM. UV–visible spectroscopy is an applicable technique that verifies the biosynthesis of AgNPs [19]. In accordance with our result, Singh et al. [20] also stated that the UV spectra of silver nanoparticles were maximal at the wavelength of 400 nm.

In the FTIR analysis, the band at 3451 is attributed to the O–H mode vibration, band 1767 corresponds to carbonyl stretching vibrations, band 1603 shows N–H band primary amines in a protein, and the band at 686 is assigned to the O–H vibration mode. These results collectively indicate that these AgNPs were surrounded by phytochemical compounds, which reduced the metal. The O-H, N-H and C=O groups were adsorbed on the AgNP's surface and engaged in the bioreduction of $AgNO_3$. In parallel, the FTIR spectrum of the formed AgNPs with the help of *H. thebaica* fruit extract. Mohamed et al. [9] revealed vibrational bands of =C-H, -C-O-C, -C-O, C=O, O=C=O and C-H, indicating that the synthesized nanoparticles were adsorbed by plant components such as proteins, phenols, flavonoids, terpenoids, organic acids and other active constituents. The findings of the FTIR study have generally recognized the existence of carbonyl groups from the amino acid residues of plants, therefore, indicating the role played by plant proteins and other phytochemicals in the biosynthesis of AgNPs. It is difficult to figure out the role of a specific phytochemical [21]. Referring to the literature, AgNP preparation from *H. thebaica* fruit extract gave larger particle sizes reaching 70 nm [9]. In addition, the preparation of AgNPs from *Origanum vulgare* aqueous leaf extract gave a particle size of 136 nm. For this, their antimicrobial potential was lower when compared with the current nanoparticles because the efficacy of AgNPs was found to be dependent on physical characteristics, such as size and shape. The smaller the size of the AgNP, the better its ability to enter the cell through the cell wall and cell membrane, hindering the structural molecules and the critical metabolic pathways at the cellular and subcellular level, therefore inhibiting the bacterial growth, cell division and eventually can cause cell death [22]. Regarding other studies, the preparation of AgNPs from *H. thebaica* was carried out in one paper using fruit extract [9]. In this research, we used the leaf extract of *H. thebaica* for the first time. This gave a smaller size and more potent nanoparticles. AgNPs from *H. thebaica* fruit extract gave particle sizes of 4.5–70 nm with lower inhibition to both bacteria and some molds. The induced nanoparticles from fruit extract did not obstruct the growth of Gram-negative bacteria such as *E. coli* and *K. pneumonia* at concentrations less than 1 mg/mL, whereas the Gram-positive bacteria were more susceptible [9].

When compared with AgNPs formed by other plants, a previous study reported the biosynthesis of larger particle sizes (136 $\pm$ 10.09 nm) from *Origanum vulgare* aqueous leaf extract. These exerted antibacterial activity against *E. coli*, *Aeromonas hydrophilla*, *Salmonella paratyphi*, and *Shigella dysenteriae* with inhibition halos greater than 10 mm in diameter [23]. Another study biosynthesized AgNPs from the aqueous leaf extract of *Origanum vulgare* with particle sizes from 2 to 25 nm but with some degree of agglomeration that may lower the internalization of individual particles [24]. The biosynthesized AgNPs from the fungus *Phoma exigua* had a particle size of 22 nm [25].

The mechanism of action of nanoparticles is not yet clearly understood; some explanations have been proposed. They attach to microbial cells easily because of the negative charge on their cells; subsequently, they penetrate the cell membrane causing structural changes. The membrane permeability and proton motive force are also adversely affected [26]. Other researchers suggest the release of silver ions from AgNPs which interact with thiol groups on the sulfur-containing amino acids leading to the inactivation of enzymes, especially the respiratory chain enzymes generating ROS such as H_2O_2 and hydroxyl radicals, which induce oxidative stress in microbes. This results in DNA damage, the depletion of GSH and lipid peroxidation [27], and eventually results in dysfunction and cell death [28].

The broad-spectrum activity of AgNPs prepared from doum palm against both Gram-positive bacteria and Gram-negative bacteria (Table 1 and Figure 4) agrees with most previous investigations [29]. The superior activity of the synthesized AgNPs against Gram-positive bacteria may be attributed to the ability of Gram-negative bacteria to express efflux pumps (EPs) in their cell membranes [30,31]. Recently, the AcrAB multidrug efflux system and RND-type efflux pump have been discovered in the Gram-negative bacteria used in this study (*E. coli* ATCC 25922 and *P. aeruginosa* ATCC 27853) [32,33]. In fact, EPs enhance the resistance of Gram-negative bacteria by acting as an export system to a wide range of antimicrobials, including antibiotics and nanoparticles. This mechanism allows Gram-negative bacteria to pump out the antimicrobial agent in less time than it requires for them to be diffused toward the inside of cells; consequently, the intrabacterial concentration becomes less than the effective value [30]. The different potentials of AgNPs against Gram-positives and Gram-negatives may also be attributed to the variation in cell wall composition and width. In Gram-negatives, the cell wall is very rich in lipids, and its thickness is only 3–4 nm compared with the 30 nm thickened cell wall in Gram-positives [34].

In order to visualize the association between the tested microorganisms, MIC, MBC, MFC and the tolerance level (ratio MBC/MIC) of AgNPs were evaluated (Table 1). AgNPs could be considered as having strong antimicrobial activity against all the microbial strains, as they scored excellent MIC values (1.5–96.0 µg/mL). In addition, AgNPs showed prominent bactericidal and fungicidal activity (4.0–256 µg/mL). The MBC (or MFC)/MIC ratio was between 1.33–267; it is established in the literature that an antimicrobial compound is considered bactericidal/fungicidal when the ratio of MBC (or MFC)/MIC $\leq$ 4 [35,36].

The SEM study has revealed ultrastructural changes in both *C. tropicalis* and *C. albicans* exposed when treated with AgNPs. In the case of *C. tropicalis*, scarce pseudohyphae (Figure 5b,d,f), perforation (Figure 5b,d,f), exterior roughness (Figure 5b,d,f), irregularly shaped cells (Figure 5f) and EPS formation (Figure 5f) were the main features. The EPS may be considered a means of cell protection against AgNPs and may also be a reason for the diminished susceptibility of cells to the antimicrobials (Figure 5f). In addition, this may be the reason for the high value of MIC when compared with *C. albicans* (96 μg/mL vs. 24 μg/mL, Table 1). In the case of *C. albicans*, SEM revealed that AgNPs induced intensive cell damage, deformation and irregularly shaped cells (Figure 6b,d). This indicates the ability of AgNPs prepared from doum palm to damage the cell wall and hamper pseudohyphae formation in dimorphic yeasts. This is especially significant as both effects are core virulence tools in the genus *Candida*.

5. Conclusions

The extract of the doum palm leaf was successfully used as a reductant in the synthesis of silver nanoparticles. AgNPs showed excellent antibacterial and anticandidal properties against some pathogenic strains of both bacteria and yeasts. The Gram-positive bacterium *S. aureus* ATCC 29213 (MIC = 1.5 μg/mL) showed the highest degree of sensitivity. For this, our study recommends the use of AgNPs as an antimicrobial agent. Much more research on the cyto and systemic toxicities of very small metal NPs (which can concentrate in fat cells, neuronal junctions, muscles, brain, organs and so on) is required before wide-scale application in medicine.

Author Contributions: Conceptualization, N.M.A. and E.K.; methodology, N.M.A.; software, N.M.A.; validation, N.M.A. and E.K.; formal analysis, N.M.A. and E.K.; investigation, N.M.A.; resources, N.M.A.; data curation, N.M.A.; writing original draft, N.M.A.; writing—review and editing, N.M.A.; visualization, N.M.A. and E.K.; supervision, E.K.; project administration, N.M.A. and E.K.; funding acquisition, N.M.A. and E.K. All authors have read and agreed to the published version of the manuscript.

Funding: The authors extend their appreciation to the Deputyship for Research & Innovation, Ministry of Education in Saudi Arabia for funding this research work through project number IF-2020-028-BASRC at Imam Abdulrahman bin Faisal University (IAU)/Basic and Applied Scientific Research Center (BASRC).

Institutional Review Board Statement: Not applicable.

Informed Consent Statement: Not applicable.

Data Availability Statement: Not applicable.

Acknowledgments: The authors would like to extend their sincere appreciation to Imam Abdulrahman bin Faisal University (IAU) and Basic and Applied Scientific Research Center (BASRC), Saudi Arabia.

Conflicts of Interest: The authors declared no conflict of interest among them.

References

1. Pulingam, T.; Parumasivam, T.; Gazzali, A.M.; Sulaiman, A.M.; Chee, J.Y.; Lakshmanan, M.; Chin, C.F.; Sudesh, K. Antimicrobial resistance: Prevalence, economic burden, mechanisms of resistance and strategies to overcome. *Eur. J. Pharm. Sci.* **2021**, *170*, 106103. [CrossRef] [PubMed]
2. Elmowalid, G.A.E.; Ahmad, A.A.M.; El-Hamid, M.I.A.; Ibrahim, D.; Wahdan, A.; El Oksh, A.S.A.; Yonis, A.E.; Elkady, M.A.; Ismail, T.A.; Alkhedaide, A.Q.; et al. *Nigella sativa* extract potentially inhibited methicillin resistant *Staphylococcus aureus* induced infection in rabbits: Potential immunomodulatory and growth promoting properties. *Animals* **2022**, *12*, 2635. [CrossRef] [PubMed]
3. Pang, Z.; Raudonis, R.; Glick, B.; Lin, T.; Cheng, Z. Antibiotic resistance in *Pseudomonas aeruginosa*: Mechanisms and alternative therapeutic strategies. *Biotechnol. Adv.* **2019**, *37*, 177–192. [CrossRef] [PubMed]

4. Umapathi, A.; Nagaraju, N.P.; Madhyastha, H.; Jain, D.; Srinivas, S.P.; Rotello, V.M.; Daima, H.K. Highly efficient and selective antimicrobial isonicotinylhydrazide-coated polyoxometalate-functionalized silver nanoparticles. *Colloids Surf. B Biointerfaces* **2019**, *184*, 110522. [CrossRef]
5. Toda, M.; Williams, S.R.; Berkow, E.L.; Farley, M.M.; Harrison, L.H.; Bonner, L. Population-based active surveillance for culture-confirmed candidemia—Four sites, United States, 2012–2016. *MMWR Surveill Summ.* **2019**, *68*, 1. [CrossRef]
6. Mathur, P.; Jha, S.; Ramteke, S.; Jain, N. Pharmaceutical aspects of silver nanoparticles. *Artif. Cells Nanomed. Biotechnol.* **2018**, *46*, 115–126. [CrossRef]
7. Siddiqi, K.S.; Husen, A.; Rao, R.A. A review on biosynthesis of silver nanoparticles and their biocidal properties. *J. Nanobiotec.* **2018**, *16*, 14. [CrossRef]
8. Abid, N.; Khan, A.; Shujait, S.; Maqbool, M. Synthesis of nanomaterials using various top-down and bottom-up approaches, influencing factors, advantages, and disadvantages: A review. *Adv. Colloid Int Sci.* **2021**, *300*, 102597. [CrossRef]
9. Mohamed, H.; Afridi, S.; Khalil, A.; Zia, D.; Iqbal, J.; Ullah, I.; Shinwari, Z.; Maaza, M. Biosynthesis of silver nanoparticles from *Hyphaene thebaica* fruits and their in vitro pharmacognostic potential. *Mater. Res. Express.* **2019**, *6*, 1050c9. [CrossRef]
10. Khalil, N.A.; Motaal, A.A.; Meselhy, K. Renin and angiotensin converting enzyme inhibition of standardized bioactive fractions of *Hyphaene thebaica* L. Mart Growing in Egypt. *Pharmacog. J.* **2018**, *10*, 622–627. [CrossRef]
11. Ahmed, S.; Ahmad, M.; Swami, B.L.; Ikram, S. Green synthesis of silver nanoparticles using *Azadirachta indica* aqueous leaf extract. *J. Radiat. Res. Appl. Sci.* **2016**, *9*, 1–7. [CrossRef]
12. Lima, A.K.O.; Vasconcelos, A.A.; Kobayashi, R.K.T.; Nakazato, G.; de Campos Braga, H.; Taube, P.S. Green synthesis: Characterization and biological activity of silver nanoparticles using aqueous extracts of plants from the Arecaceae family. *Acta Sci. Tech.* **2021**, *43*, e52011. [CrossRef]
13. Beecher, D.J.; Wong, A.C.L. Identification of hemolysin BL-producing *Bacillus cereus* isolates by a discontinuous hemolytic pattern in blood agar. *Appl. Environ. Microbiol.* **1994**, *60*, 1646–1651. [CrossRef] [PubMed]
14. Palomino, J.C.; Martin, A.; Camacho, M.; Guerra, H.; Swings, J.; Portaels, F. Resazurin microtiter assay plate: Simple and inexpensive method for detection of drug resistance in *Mycobacterium tuberculosis*. *Antimicrob Agents Chemother.* **2002**, *46*, 2720–2722. [CrossRef]
15. Mondal, A.H.; Yadav, D.; Mitra, S.; Mukhopadhyay, K. Biosynthesis of silver nanoparticles using culture supernatant of *Shewanella* sp. Ary1 and their antibacterial activity. *Int. J. Nanomed.* **2020**, *15*, 8295–8310. [CrossRef] [PubMed]
16. Chandran, S.P.; Chaudhary, M.; Pasricha, R.; Ahmad, A.; Sastry, M. Synthesis of gold nanotriangles and silver nanoparticles using *Aloe vera* plant extract. *Biotechnol. Prog.* **2006**, *22*, 577–583. [CrossRef] [PubMed]
17. Khalil, M.M.; Ismail, E.H.; El-Baghdady, K.Z.; Mohamed, D. Green synthesis of silver nanoparticles using olive leaf extract and its antibacterial activity. *Arabian J. Chem.* **2014**, *7*, 1131–1139. [CrossRef]
18. Kumar, P.; Selvi, S.S.; Govindaraju, M. Seaweed-mediated biosynthesis of silver nanoparticles using *Gracilaria corticata* for its antifungal activity against *Candida* spp. *Appl. Nanosci.* **2013**, *3*, 495–500. [CrossRef]
19. Zou, J.; Xu, T.; Hou, B.; Wu, D.; Sun, Y. Controlled growth of silver nanoparticles in a hydrothermal process. *Chin. Particuol.* **2007**, *5*, 206–212. [CrossRef]
20. Singh, S.; Bharti, A.; Meena, V.K. Green synthesis of multi-shaped silver nanoparticles: Optical, morphological and antibacterial properties. *J. Mater. Sci. Mater. Electron.* **2015**, *26*, 3638–3648. [CrossRef]
21. Kumawat, M.; Madhyastha, H.; Singh, M.; Revaprasadu, N.; Srinivas, S.P.; Daima, H.K. Double functionalized haemocompatible silver nanoparticles control cell inflammatory homeostasis. *PLoS ONE* **2022**, *17*, e0276296. [CrossRef] [PubMed]
22. Madhyastha, H.; Madhyastha, R.; Thakur, A.; Kentaro, S.; Dev, A.; Singh, S.; Kumar, H.; Acevedo, O.; Nakajima, Y.; Daima, H.K.; et al. c-Phycocyanin primed silver nano conjugates: Studies on red blood cell stress resilience mechanism. *Colloids Surf. B Biointerfaces* **2020**, *194*, 111211. [CrossRef] [PubMed]
23. Shaik, M.R.; Khan, M.; Kuniyil, M.; Al-Warthan, A.; Alkhathlan, H.Z.; Siddiqui, M.R.H.; Shaik, J.P.; Ahamed, A.; Mahmood, A.; Khan, M. Plant-extract-assisted green synthesis of silver nanoparticles using *Origanum vulgare* L. extract and their microbicidal activities. *Sustainability* **2018**, *10*, 913. [CrossRef]
24. Shende, S.; Gade, A.; Rai, M. Large-scale synthesis and antibacterial activity of fungal-derived silver nanoparticles. *Environ. Chem Lett.* **2017**, *15*, 427–434. [CrossRef]
25. Khalandi, B.; Asadi, N.; Milani, M. A review on potential role of silver nanoparticles and possible mechanisms of their actions on bacteria. *Drug Res.* **2017**, *67*, 70–76. [CrossRef]
26. Shende, S.; Ingle, A.P.; Gade, A.; Rai, M. Green synthesis of copper nanoparticles by *Citrus medica* Linn. (Idilimbu) juice and its antimicrobial activity. *World J. Microbiol. Biotechnol.* **2015**, *31*, 865–873. [CrossRef]
27. Lee, W.; Kim, K.J.; Lee, D.G. A novel mechanism for the antibacterial effect of silver nanoparticles on *Escherichia coli*. *Biometals* **2014**, *27*, 1191–1201. [CrossRef]
28. Rai, M.; Kon, K.; Ingle, A.; Duran, N.; Galdiero, S.; Galdiero, M. Broad-spectrum bioactivities of silver nanoparticles: The emerging trends and future prospects. *Appl. Microbiol. Biotechnol.* **2014**, *98*, 1951–1961. [CrossRef]
29. Khameneh, B.; Iranshahy, M.; Soheili, V.; Bazzaz, B.S.F. Review on plant antimicrobials: A mechanistic viewpoint. *Antimicrob. Resist. Infect. Control.* **2019**, *8*, 118. [CrossRef] [PubMed]
30. Mahamoud, A.; Chevalier, J.; Alibert-Franco, S.; Kern, W.V.; Pagès, J.M. Antibiotic efflux pumps in Gram-negative bacteria: The inhibitor response strategy. *J. Antimicrob. Chemother.* **2007**, *59*, 1223–1229. [CrossRef]

31. Padilla, E.; Llobet, E.; Doménech-Sánchez, A.; Martínez-Martínez, L.; Bengoechea, J.A.; Albertí, S. *Klebsiella pneumoniae* AcrAB efflux pump contributes to antimicrobial resistance and virulence. *Antimicrob. Agents Chemother.* **2010**, *54*, 177–183. [CrossRef]
32. Ni, R.T.; Onishi, M.; Mizusawa, M.; Kitagawa, R.; Kishino, T.; Matsubara, F.; Tsuchiya, T.; Kuroda, T.; Ogawa, W. The role of RND-type efflux pumps in multidrug-resistant mutants of *Klebsiella pneumoniae*. *Sci. Rep.* **2020**, *10*, 10876. [CrossRef] [PubMed]
33. Chatterjee, T.; Chatterjee, B.K.; Majumdar, D.; Chakrabarti, P. Antibacterial effect of silver nanoparticles and the modeling of bacterial growth kinetics using a modified Gompertz model. *Biochim. Biophys. Acta* **2015**, *2*, 299–306. [CrossRef] [PubMed]
34. Oussou, K.R.; Coffi, K.; Nathalie, G.S.; Gerard, K.; Mireille, D.; Yao, T.N.; Gille, F.; Jean-Claude, C.H. Activités antibactériennes des huiles essentielles de trois plantes aromatiques de Côte d'Ivoire. *Comptes Rendus. Chim.* **2008**, *7*, 1081–1086. [CrossRef]
35. Teke, G.N.; Kuiate, J.R.; Kuete, V.; Teponno, R.B.; Tapondjou, L.A.; Tane, P.; Giacinti, G.; Vilarem, G. Bio guided isolation of potential antimicrobial and antioxidant agents from the stem bark of *Trilepisium madagascariense*. *S. Afr. J. Bot.* **2011**, *77*, 319–327. [CrossRef]
36. Pierce, C.G.; Srinivasan, A.; Uppuluri, P.; Ramasubramanian, A.K.; López-Ribot, J.L. Antifungal therapy with an emphasis on biofilms. *Curr. Opin. Pharmacol.* **2013**, *13*, 726–730. [CrossRef] [PubMed]

microorganisms

MDPI

Article

Histopathological, Immunohistochemical, Biochemical, and In Silico Molecular Docking Study of Fungal-Mediated Selenium Oxide Nanoparticles on *Biomphalaria alexandrina* (Ehrenberg, 1831) Snails

Amina M. Ibrahim [1], Mahassen Ghazy [2], Heba El-Sayed [3], Rehab M. Abd El-Hameed [3], Rehab G. Khalil [4], Shereen M. Korany [5], Abeer S. Aloufi [5,*], Olfat A. Hammam [6] and Mostafa Y. Morad [7]

1 Medical Malacology Department, Theodor Bilharz Research Institute, Giza 12411, Egypt
2 Water Pollution Research Department, National Research Centre, El Tahrir Street, Dokki, Giza 12622, Egypt
3 Botany and Microbiology Department, Faculty of Science, Helwan University, Helwan 11795, Egypt
4 Immunology Division, Faculty of Science, Beni-Suef University, Beni-Suef 62521, Egypt
5 Department of Biology, College of Science, Princess Nourah bint Abdulrahman University, P.O. Box 84428, Riyadh 11671, Saudi Arabia
6 Pathology Departments, Theodor Bilharz Research Institute, Giza 12411, Egypt
7 Zoology and Entomology Department, Faculty of Science, Helwan University, Helwan 11795, Egypt
* Correspondence: asaloufi@pnu.edu.sa

Abstract: *Daphnia magna* and freshwater snails are used as delicate bioindicators of contaminated aquatic habitats. Due to their distinctive characteristics, selenium oxide nanoparticles (SeONPs) have received interest regarding their possible implications on aquatic environments. The current study attempted to investigate the probable mechanisms of fungal-mediated selenium nanoparticles' ecotoxicological effects on freshwater *Biomphalaria alexandrina* snails and *Daphnia magna*. SeONPs revealed a toxicological impact on *D. magna*, with a half-lethal concentration (LC_{50}) of 1.62 mg/L after 24 h and 1.08 mg/L after 48 h. Survival, fecundity, and reproductive rate were decreased in *B. alexandrina* snails exposed to SeONPs. Furthermore, the aspartate aminotransferase (AST) and alanine aminotransferase (ALT) levels were markedly elevated, while albumin and total protein levels decreased. Histopathological damage in the hermaphrodite and digestive glands was detected by light, electron microscopy, and immunohistochemistry studies. The molecular docking study revealed interactions of selenium oxide with the ALT and AST. In conclusion, *B. alexandrina* snails and *D. magna* could be employed as bioindicators of selenium nanomaterial pollution in aquatic ecosystems. This study emphasizes the possible ecological effects of releasing SeONPs into aquatic habitats, which could serve as motivation for regulatory organizations to monitor and control the use and disposal of SeONPs in industry.

Keywords: *Biomphalaria alexandrina*; selenium oxide nanoparticles; *Penicillium chrysogenum*; *Daphnia magna*; reproductive rate; biochemical; immunohistopathology; molecular docking

Citation: Ibrahim, A.M.; Ghazy, M.; El-Sayed, H.; Abd El-Hameed, R.M.; Khalil, R.G.; Korany, S.M.; Aloufi, A.S.; Hammam, O.A.; Morad, M.Y. Histopathological, Immunohistochemical, Biochemical, and In Silico Molecular Docking Study of Fungal-Mediated Selenium Oxide Nanoparticles on *Biomphalaria alexandrina* (Ehrenberg, 1831) Snails. *Microorganisms* **2023**, *11*, 811. https://doi.org/10.3390/microorganisms11030811

Academic Editor: Kamel Ahmed Abd-Elsalam

Received: 27 February 2023
Revised: 15 March 2023
Accepted: 17 March 2023
Published: 22 March 2023

1. Introduction

Nanotechnology development in a wide range of industries, including chemical, electronics, metal industry, biomedicine, cosmetics, and household purposes, poses a potential threat to the natural environment. Nanoparticles (NPs) are frequently utilized in a variety of industrial applications and are released into the environment through wastewater from houses and different industries. They are extremely harmful to aquatic life [1].

One of those elements that has attracted interest in the industrial and biomedical sector is elemental selenium (Se), which can be found in the form of selenium nanoparticles (SeNPs). Selenium is a vital element for life and can be found in nature in four different forms: selenate (+VI, SeO_4^{2-}), selenite (+IV, SeO_3^{2-}), elemental selenium (0, Se^0), and selenide ($-$II, Se^{2-}) [2]. Se toxicity varies with dosage and chemical types; dangerous Se concentrations have an adverse impact on the ecosystem [3,4]. Due to their antioxidant characteristics and ability to be integrated into selenoproteins, SeNPs have been used in several therapeutic applications, including the cancer treatment and diabetes. [5,6]. They have also been shown to suppress antibiotic-resistant bacterial strains at levels as low as 1 ppm [6]. Furthermore, SeNPs can be employed in glass production and in screening mammography detectors, and have been applied in a number of industries, including inverters and solar cells [7,8]. SeNPs are naturally occurring in the ecosystem, and living organisms, such as plants, fungi, or soil bacteria, can produce biogenic SeNPs by reducing selenite or selenate ions to these nanoparticles. Synthesized SeNPs are then probable to exist and have been distributed to numerous aquatic ecosystems. Although it has widely used applications, selenium nanoparticles' potential for bioaccumulation makes it an effective harmful element for natural ecosystems and public health since it releases ROS and causes toxicity [9]. While selenium is naturally found in the crust of the earth, around 40% of selenium emissions into the atmosphere and aquatic habitats are caused by various industrial processes, particularly those associated with mining [8]. However, little is known about its toxicity and how functionalization affects aquatic invertebrates [10].

The second-most abundant animal group, the Mollusca phylum, is often regarded as a great bioindicator of ecosystem health since they are sensitive, accessible, and common. They are also commonly employed in environmental studies. The OECD has designed and approved *Lymnaea stagnalis*, a freshwater pulmonate snail, as a viable candidate species for assessing the effects of chemical pollutants on reproduction in Testing Series, No. 121 [11,12]. Snails (Phylum Mollusca, Class Gastropoda) have gained great attention due to their role in ecosystems and disease transmission [13]. *Biomphalaria alexandrina* (Ehrenberg, 1831) is a common freshwater snail in Egypt [14]. It is a host for the trematode parasite *Schistosoma manosni*, which causes human schistosomiasis [15]. It is a well-known model for investigating the toxicity and toxicokinetics of inorganic materials in aquatic ecosystems [16]. These species serve as an important biomonitor for pollution caused by heavy metals [17].

Freshwater zooplankton belonging to the Cladocera order, *Daphnia magna* (Straus, 1820), is found in the aquatic environments of Egypt [18]. As a non-target organism, it is a highly useful model to evaluate the chronic and acute toxicity in freshwater ecosystems using *B. alexandrina* habitat [19,20]. In addition, it is a highly sensitive animal to many minerals and organic contaminants [21]. Its laboratory culture is simple, and it has a parthenogenetic reproduction mode and a high fertility rate [19].

The enzymes known as aspartate aminotransferase (AST) and alanine aminotransferase (ALT) are mostly located in the liver, red blood cells, heart, kidneys, and pancreatic cells [22]. The enzyme activity in the liver is frequently utilized as indicators for hazardous substances. Tissue damage is directly correlated with blood levels of the AST and ALT enzymes. The tissue damages increase or decrease depending on the concentration of the enzymes present [23]. The morphology and structure of the cell may be harmed by an increase in nanoparticles content. The inflammation and necrosis brought on by hepatocyte injury may increase a cell's permeability. Via the cell membrane, AST and ALT enzymes are released into the body, increasing their concentration in the blood [13]. As a result, the AST and ALT enzymes are signs of liver injury [23]. Xenobiotically induced variations in cellular or biochemical components could occur due to ingested molluscicides [24]. Aspartate and Alanine aminotransferases (AST and ALT) perform a critical role in many stressful conditions [25]. These enzymes AST and ALT were used as biomarker for water pollution in *Helisoma duryi*, *B. alexandrina*, and *Lymnaea natalensis* snails [26].

Therefore, the current study aimed to describe the ecotoxicity of fungal-synthesized SeONPs in aquatic systems as they were naturally synthesized there by different aquatic microorganisms on the freshwater snail *B. alexandrina* and how it affects survival, the concentrations of alanine aminotransferase and aspartate aminotransferase and the structure of the digestive and hermaphrodite glands in order to evaluate the snail's potential as a bioindicator for SeONPs pollutant. Moreover, in silico molecular docking studies were conducted to analyze the interactions between the ALT and AST and nano-selenium oxide. *D. magna*, which is regarded as a sensitive organism for the detection of aquatic environmental pollution, was used to confirm the nanotoxicity.

2. Materials and Methods

2.1. Fungal Strain and Selenium Nanoparticles Preparation and Characterization

The strain *Penicillium chrysogenum* MZ945518 was previously isolated from the southern coast of Alexandria, Egypt, and identified using molecular techniques. This fungal strain was used to synthesize SeONPs according to [27]. *P. chrysogenum* was developed in culture flasks containing potato dextrose broth and kept for 7 days at 25 °C. Centrifugation at 10,000 rpm for 10 min separated the mycelia from the culture filtrate. (Sigma, 3-16 PK, Osterode am Harz, Germany). Following that, 15 mL of culture supernatant was mixed with 3 mM sodium selenite (Na_2SeO_3), which was then incubated at 40 °C for 30 min until the red color of SeONPs as a soluble form in the liquid was formed. Finally, the myco-fabricated SeONPs were cleaned by spinning them at 10,000 rpm 3 times in double-distilled water for 10 minutes, followed by 48 h of drying at 60 °C.

The myco-synthesized SeONPs were characterized with a Zetasizer analyzer, UV–Visible spectrophotometer, X-ray diffraction, and transmission electron microscope (TEM), according to [27]. A UV-visible spectrophotometer (PerkinElmer Life and Analytical Sciences, CT, Akron, OH, USA) was used to measure the absorption spectra of SeONPs at wavelengths between 400 nm to 800 nm. The average diameter size, distribution, and the zeta potential charges were calculated using Zetasizer Dynamic Light Scattering (DLS) (Zetasizer Nano ZN, Malvern Panalytical Ltd., Malvern, UK), at an angular position of 173° at 25 °C. To assess the crystalline dimensions of the particles, XRD was conducted on a Bruker D8 Discover Diffractometer, USA, with Cu-K radiation (λ = 1.54060 Angstrom). The relative intensity data was collected over a 2θ range of 5100°. The chart yielded 2θ values and relative intensities (I/Io), and core minerals were identified using JCPDS carts. Transmission electron images were also taken using a high-resolution transmission electron microscope (HR-TEM; JEOL 2100, Tokyo, Japan) fitted with an electron diffraction pattern. The elemental structure of SeONPs and their functional groups were investigated using Fourier transform infrared spectroscopy (FTIR; PerkinElmer, OH, USA).

2.2. Snails

Adult *B. alexandrina* snails (9–11 mm) were kept and adapted at the Theodor Bilharz Research Institute (TBRI) in Giza, Egypt. For snail fecundity, growth, and shell length, snails (10 snails/L) were kept in plastic aquaria (16 × 23 × 9 cm) with 30 mg/L $CaCO_3$ (small pieces of chalk) added [28]. In the aquaria, dechlorinated tap water was used with a pH of 7 ± 0.2 and a temperature of 25 ± 2 °C with a 12 h/12 h photoperiod. Blue-green algae (*Nostoc muscorum*), oven-dried lettuce leaves, and Tetramin were fed to the snails. Egg masses were collected using small pieces of 33 foam sheets [29].

2.3. Daphnia Magna

Daphnia magna has been cultivated in a synthetic freshwater medium in the Hydrobiology Laboratory of the National Research Center's Water Pollution Research Department.

In 1 L glass beakers containing gravid females (10–20) on a regular basis, the media was changed three times per week. Every day, the offspring were gathered to be employed in the toxicity trials. *Scenedesmus obliquus*, a green microalga, was given to these animals three times per week at a concentration of 14 × 10^7 coenobia/mL. The photoperiod used to maintain daphnids was 16 h of light and 8 h of darkness at a temperature of 22 ± 2 °C.

2.4. Bioassays

2.4.1. Toxicological Impact on *D. magna*

Daphnids aged 24 h to 30 h were divided into groups and subjected to varying doses of SeONPs (0.67, 0.83, 1.00, and 1.33 mg/L, respectively) in 500 mL beakers containing 100 mL of medium in triplicates without the addition of food. The number of living organisms was counted after 24 h and 48 h. A control group was also run at the same time. In this group, 10 daphnids were 24 h old, and were kept in a 500 mL beaker with 10 mL of free SeONPs medium. Using a dissecting microscope, daphnids were examined, and those with no movement of any internal organs were considered dead [30]. The medial lethal concentrations of *D. magna* were determined using probit analysis.

2.4.2. Toxicological Impact on *B. alexandrina* Snails

Sixty adult *B. alexandrina* snails (measuring 9–11 mm in length) were used in the study. Thirty snails in triplicate (10 snails/L) were subjected to 44.15 mg/L of SeONPs for 24 h, then recovered for two weeks [27]. Thirty non-exposed control snails were tested alongside the exposed snails.

Snails' Egg-Laying Capacity Analysis

Snail survival (Lx as a ratio of the correct one at the time of exposure in weeks (x)), egg-laying capacity (Fecundity) (Mx: mean number of eggs/snail/week), and reproductive rates R_0 (the sum of LxMx during the experimental period) of *B. alexandrina* snails were all recorded weekly alongside the control group [31].

Tissue Preparation and Biochemical Analysis

The soft tissues of control and treated snails were crushed with two slides before being removed and cleaned from the hard shell parts with forceps. After the soft tissues were formed, they were weighed, and 1 g of each group's dry tissue was homogenized in a glass Dounce homogenizer with 10 mL of phosphate buffer.

The tissue homogenates were centrifuged for 10 min at 3000 rpm, and the supernatants were utilized for the biochemical assay. The concentrations of alanine aminotransferase (ALT) and aspartate aminotransferase (AST) were assessed using the [32] method. The optical density of the mixture was recorded with water serving as the reference blank at 505 nm. A standard curve was used to calculate the Units per Liter. Total protein was measured according to the method described in [33]. To calculate the total protein concentration, the absorbance of the sample (A_{sample}) and standard ($A_{standard}$) against reagent blank at 550 nm. (520–570 nm). The following formula was used, Protein Concentration (g/100 mL) = ($A_{sample}/A_{standard}$) × 5. To calculate the albumin concentration, the absorbance of the sample (A_{sample}) and standard ($A_{standard}$) against reagent blank at 630 nm. The following formula was used, albumin concentration (g/100 mL) = ($A_{sample}/A_{standard}$) × 4. Albumin was analyzed according to [34]. In brief, three tubes were set up, and each received 2.0 mL of the albumin reagent (citrate buffer, pH 4.2, bromcresol green, detergent, and preservative). An amount of 10 µL of the tissue homogenate solution (the sample) were put into the first tube. As a blank, the second tube was used, which only contained albumin reagent. The third tube, containing 2.0 mL of albumin reagent, received 10 µL of albumin as a positive control (standard). The mixture was incubated for 5 min at 37 °C. The absorbance of the sample (A_{sample}) and standard ($A_{standard}$) against the reagent blank at 630 nm are used to determine the albumin concentration. Albumin concentration (g/100 mL) = ($A_{sample}/A_{standard}$) × 4) was calculated.

Histopathological and Immunohistochemical Analysis

Adult *B. alexandrina* snails (9–11 mm) were treated with 44.15 mg/L SeONPs for 24 h before recovering for two weeks. According to Carleton et al. [35], dissection and processing of the digestive and hermaphrodite glands were completed. Briefly, the dissected parts were put in 10% formalin for 12 h, then for the dehydration process, different ethanol concentrations (80%, 90%, and 100%, for 3 h each, respectively) were used. After that, they were cleared in xylene and placed in paraffin. Five micrometer sections were cut using a microtome, fixed on slides, dewaxed in xylene, stained with hematoxylin and eosin, and then covered with glass slips using Canada balsam. A Zeiss microscope was used to examine the prepared slides (Carl Zeiss Microscopy GmbH 07,745 Jena, Germany). For immunohistochemical experiments, 4 μm-thick slices of snails were cut from paraffin blocks and placed on positively charged slides (SuperFrost Plus, Menzel-Glaser, Schoemperlenstraße, Germany). On an automated platform (DakoAutostainer Link 48), the slides were stained using anti-mouse proliferating cell antigen (PCNA) and anti-Cyclin D1 antibodies (Santa Cruz Biotechnology, Dallas, TX, USA) at appropriate working dilution of 1:100 [36]. In ten microscopic areas, the ratio of positively stained brown nuclear (PCNA, Cyclin D1) was determined (under Zeiss light microscopy at ×400).

Electron Microscope Study

After *B. alexandrina* snails were exposed to 44.15 mg/L SeONPs for 24 h, the digestive gland was removed and cut into small pieces. Then, it was fixed in 2.5% glutaraldehyde in 0.2 M sodium cacodylate solution for three hours at room temperature (pH 7.4), cleaned in the same buffer for four to eight hours at room temperature, and finally post-fixed for two hours at the same temperature in 1% OsO_4 solution in 0.2 (pH 7.4). Prior to being implanted in an Epon-Araldite solution, tissues were dried in a succession of ethanol concentrations (70%, 80%, 90%, and 95%, and two changes of absolute ethanol, respectively). Ultrathin pieces were made with a diamond knife, stained with uranyl acetate and lead citrate, and examined using an 80 kV JEOL 100CX-II TEM at the Faculty of Agriculture, Cairo University, Egypt.

The Molecular Docking Interaction Study

To study the effect of SeO_2 (SeONPs) in silico, two enzymes secreted from the hepatopancreas of treated snails, ALT and AST, were selected. The structures of these two enzymes were obtained from the Protein Data Bank (PDB) and encoded (ID: 1XI9) for *Pyrococcus furiosus* ALT and (ID: 1AAM) for *Escherichia coli* AST. The Molecular Operating Environment software was used to perform the molecular docking (MOE 2014.09). After determining the best order for the enzymes, hydrogens were added, and partial charges were calculated in order to reduce the energy of the ligand (SeO_2) molecule.

2.5. Statistical Analysis

Lethal concentration levels were established using probit analysis [37]. Using the Student's *t*-test, the averages of the experimental and control groups were compared [38].

3. Results

3.1. Selenium Nanoparticles Green Synthesis

In the current research, the biomolecules in the culture filtrate of *Penicillium chrysogenum* MZ945518 were shown to be effective at reducing Se ions and producing SeONPs. The transformation of the fungus filtrate's color from colorless to ruby red after the addition of the precursor solution (Na_2SeO_3) served as an indicator for the synthesis of SeONPs (Figure 1a).

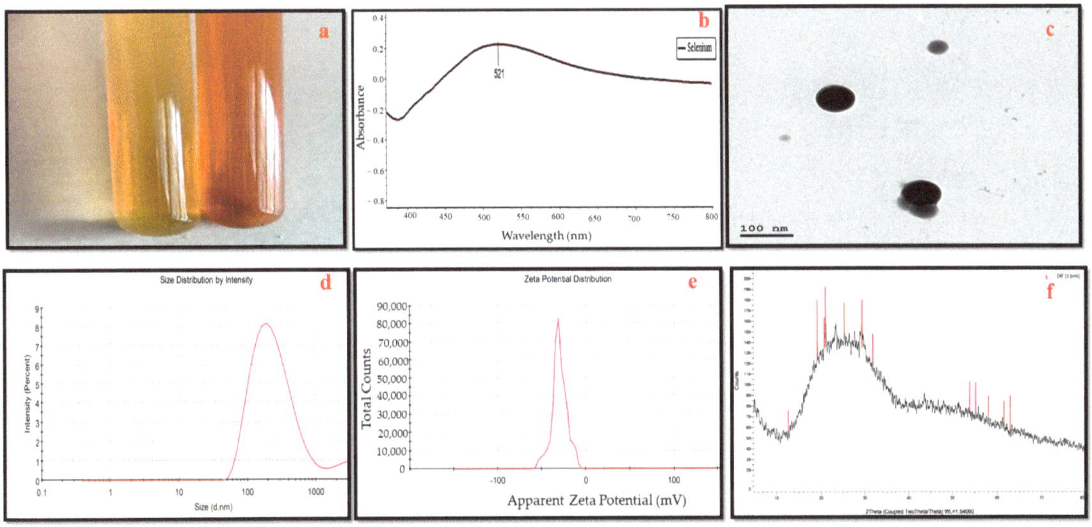

Figure 1. SeONPs Characterization. (**a**) brick-red color of the biogenic selenium oxide nanoparticles. (**b**) UV–Vis spectrum with maximum absorbance at 521 nm. (**c**) TEM photograph of the spherical SeONPs. (**d**) Size distribution. (**e**) Zeta potential. (**f**) XRD pattern of SeONPs.

3.2. Selenium Nanoparticles' Characteristics

In addition to color change, the fabrication of SeONPs was confirmed by determining the maximum SPR with UV-vis spectroscopy at a wavelength between 200 nm and 800 nm. The maximum absorbance for SeONPs produced by *P. chrysogenum* was seen as a single absorbance peak at 521 nm (Figure 1b). Furthermore, the TEM analysis also showed that the developed nanoparticles were round and had crystalline diameters between 44 and 77 nm (Figure 1c). The obtained nano-formulation of the myco-synthesized SeONPs had 204 nm average hydrodynamic diameters and a zeta potential of −30.1 mV (Figure 1d,e). Figure 1f shows X-ray diffraction patterns of the SeONPs, which showed a broad pattern with no Bragg peaks.

3.3. Toxicological Impact Assessment

In the current investigation, *D. magna* was used as a bioindicator for SeONPs as it is sensitive to aquatic pollution. According to the recent findings, the half-lethal concentration (LC_{50}) of SeONPs for *D. magna* after 24 h was 1.62 mg/L, and after 48 h, it was 0.67 mg/L (Table 1).

Table 1. Toxicity of SeONPs on *Daphnia magna* after 24 h and 48 h of exposure.

Tested Animals	LC_{10} (mg/L)	LC_{25} (mg/L)	LC_{50} (mg/L)	LC_{90} (mg/L)
***D. magna*/24 h**	0.70	1.04	1.62	3.77
***D. magna*/48 h**	0.42	0.53	0.67	1.08

Our previous work showed that a molluscicidal effect against adult *B. alexandrina* snails was observed after being exposed to SeONPs for 96 h at LC_{25}, 44.15 mg/L [27]. According to the results of the current investigation, *B. alexandrina* snails exposed to 44.15 mg/L of SeONPs for 24 h per week for two weeks had considerably lower survival rates (Lx) and fecundities (MX) than those in the control group. Furthermore, exposed snails had a significantly lower reproductive (Ro) rate than the control group (Figure 2A–C).

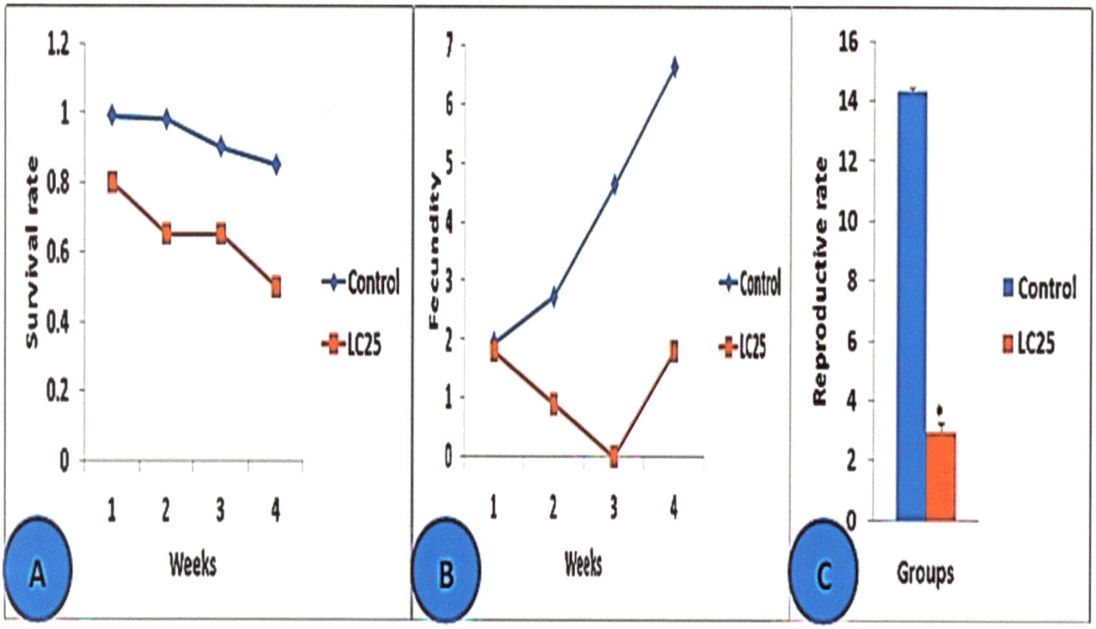

Figure 2. *B. alexandrina* snail survival, fecundity, and reproductive rate after two weeks of exposure to LC25 (44.15 mg/L) SeONPs and two weeks of recovery. (**A**) The rate of survival (Lx). (**B**) The fecundity (Mx). (**C**) *B. alexandrina* snail reproductive rate. * = significant difference from control at $p < 0.05$.

The SeONPs' effect on some biochemical studies of *B. alexandrina* was assessed. The current investigation found that the concentrations of total protein and albumin were significantly reduced ($p < 0.05$) in the snails subjected to 44.15 mg/L SeONPs, while the levels of transaminases (AST and ALT) were significantly elevated ($p < 0.01$) in comparison to the control group (Figure 3).

The light microscopy showed that the digestive gland of control *B. alexandrina* snails consisted of many follicles, and each follicle was lined by a basement membrane and has three cell types: digestive, excretory, and calcium cells (Figure 4A). The digestive gland tissues were damaged after being exposed to 44.15 mg/L of SeONPs. Such effects included follicle and basement membrane degeneration, and the destruction of digestive, excretory, and calcium cells, and an increase in follicle lumen (Figure 4B). The normal hermaphrodite gland of *B. alexandrina* is primarily made up of female oogenic cells and male reproductive cells (Figure 4C). Gonadal cells of exposed snails to 44.15 mg/L SeONPs experienced severe damage, including connective tissue deformation, mature ova destruction, and sperm degradation (Figure 4D). Immunohistochemistry of the normal hermaphrodite gland showed positive expression of both PCNA and Cyclin D1 in the ovum, oocytes, and interstitial cells as brownish positive nuclei (Figure 5A–D).

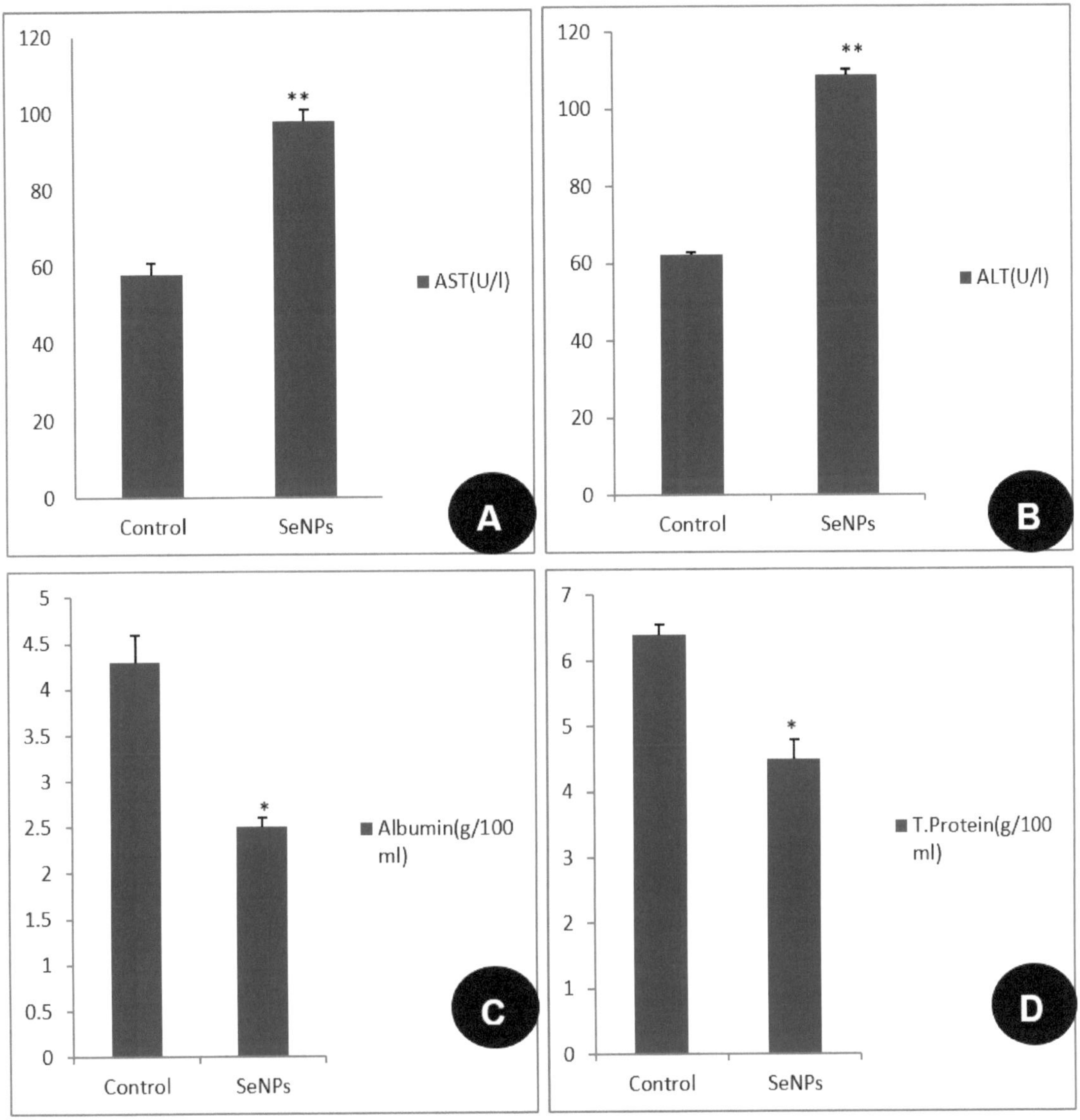

Figure 3. The effect of SeONPs on *B. alexandrina snail* transaminases (AST and ALT)) (**A**,**B**), (**C**) total protein, and (**D**)albumin concentrations. * = significant difference from control at $p < 0.05$. ** = highly significant compared to control at $p < 0.01$.

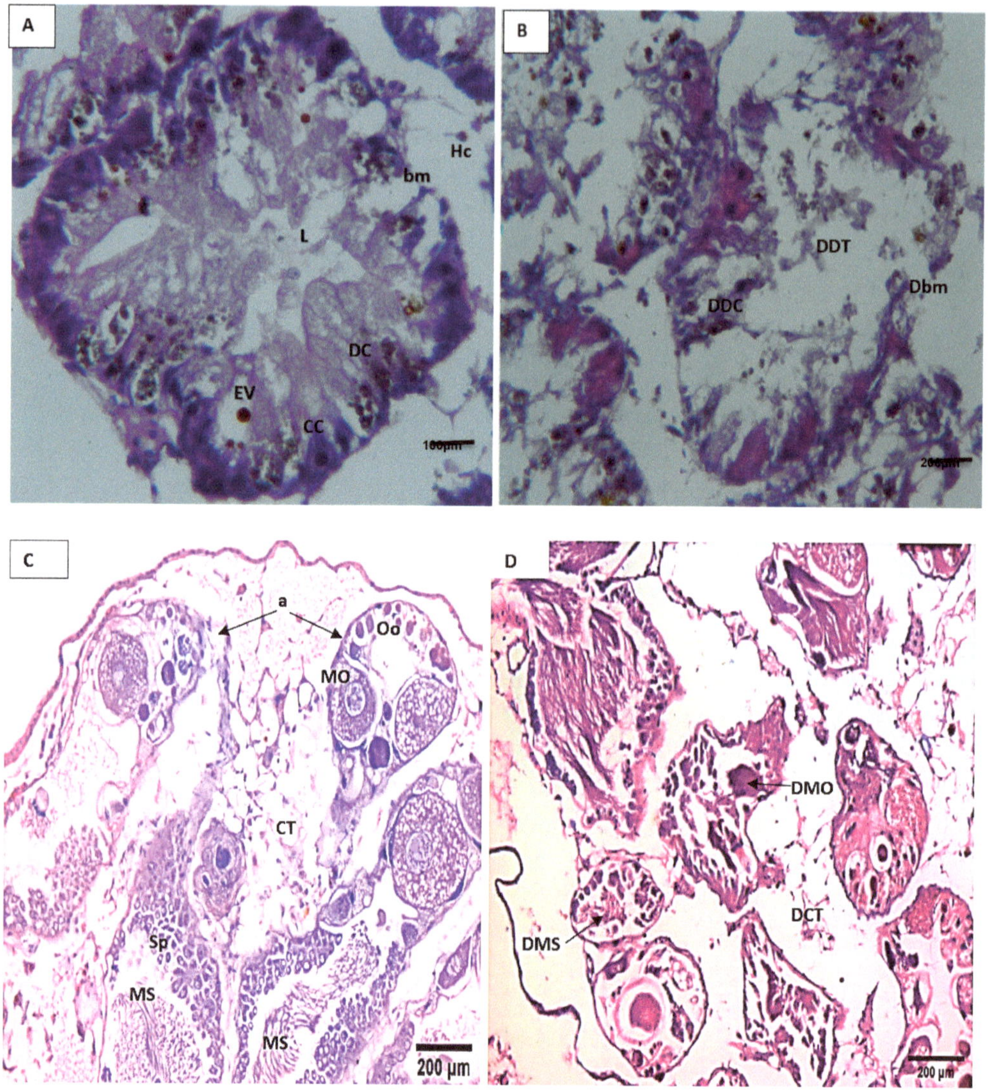

Figure 4. Histopathological sections of *B. alexanderina* snails' digestive gland and ovotestis region. (**A**) The digestive gland lumen (L) of an untreated *B. alexanderina* digestive gland was surrounded by the basement membrane (bm) and three different cells: calcium cells (CC), digestive cells (DC), and excretory vesicle (Ev). (**B**) Exposed *B. alexanderina* snails to 44.15 mg/L of SeONPs, showing degenerated digestive tubules (DDT), basement membrane (Dbm), and digestive cells (DDC). (**C**) Section from an untreated snail showing the hermaphrodite gland with many acini (a) containing mature sperm (MS) at the center of acini and spermatocytes (Sp), while female acini showing mature ovum (MO) and Oocytes (Oo) and well-formed connective tissue between acini (CT). (**D**) Section from exposed snail to 44.15 mg/L of SeONPs showing the hermaphrodite gland with deformed connective tissue (DCT), and destructed mature ovum (DMO) and sperms (DMS).

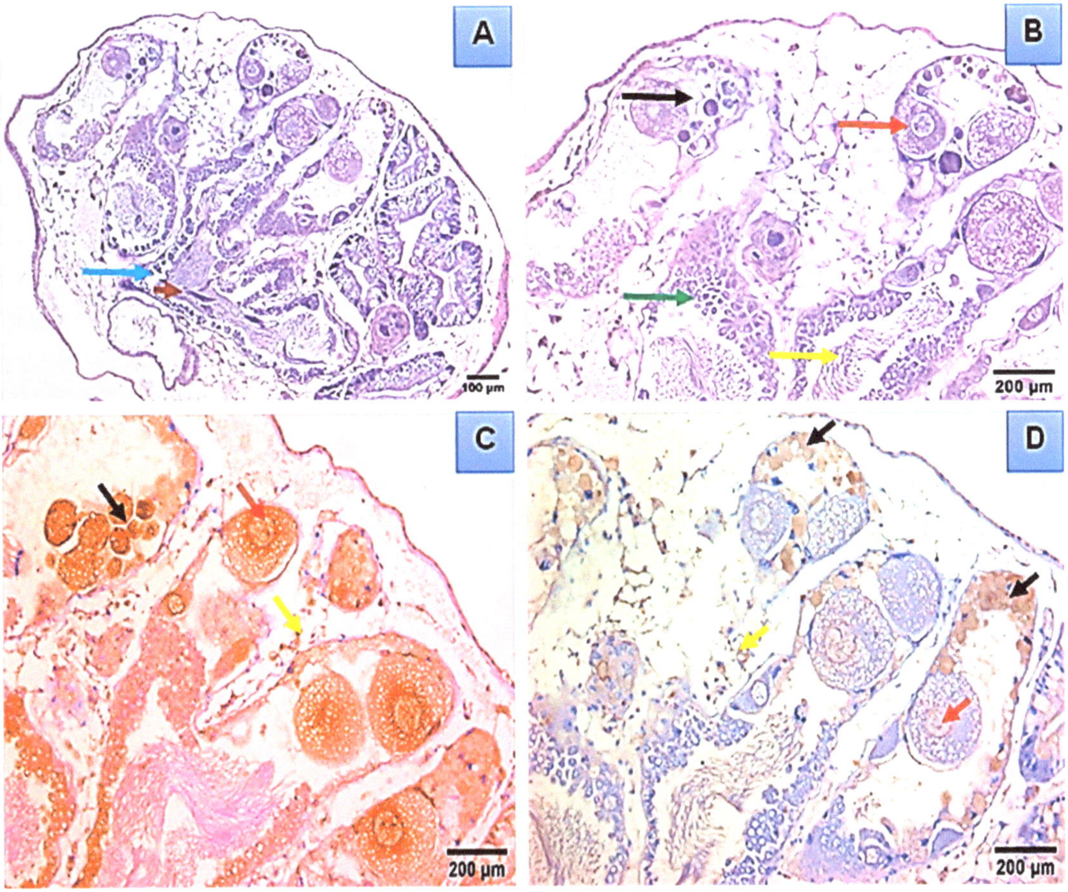

Figure 5. The hermaphrodite gland of *B. alexanderina* snails by light microscope. (**A**,**B**) mature ovum (red arrow), oocytes (black arrow), sperms (yellow arrow), spermatocytes (green arrow), primary oocytes (blue arrow), and connective tissue (dark red arrow). (**C**) Control hermaphrodite gland, demonstrating PCNA expression in the ovum (red arrow) and in oocytes (black arrow), in the interstitial cells (yellow arrow), as brownish positive nuclei. (**D**) Normal control showing cyclin D1 expression in the ovum (red arrow) and in oocytes (black arrow), in the interstitial cells (yellow arrow) as brownish positive nuclei.

There were negative changes in the snails' hermaphrodite glands after 44.15 mg/L SeONPs exposure (Figure 6A,B). PCNA immunohistochemistry revealed that at least 70% of the nuclei in sperm and ova were brownish in color (Figure 6C). Cyclin D1 was also found to be expressed in sperm, ova, and interstitial cells (Figure 6D).

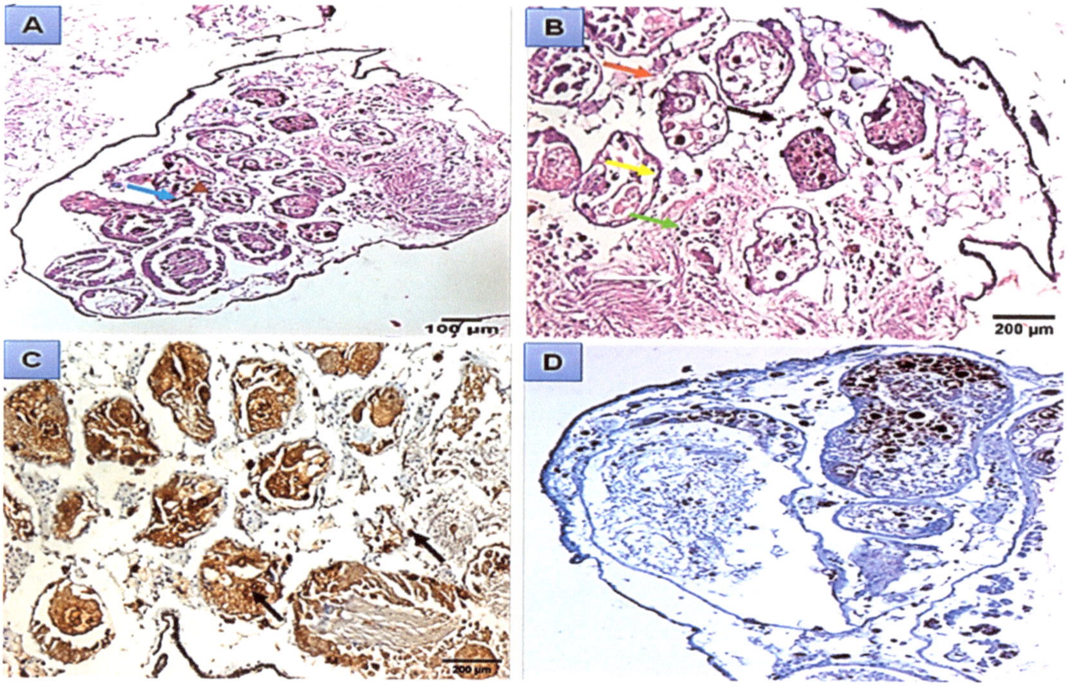

Figure 6. Light micrographs of the hermaphrodite gland of *B. alexanderina* snails exposed to 44.15 mg/L of SeONPs. (**A**,**B**) Degenerated ovum (blue arrow), degenerated oocytes (head arrow), Oocytes (yellow arrow), Sperms (white arrow), Spermatocytes (green arrow), connective tissue (black arrow), space between acini (red arrow). (**C**) Section in the hermaphrodite gland demonstrating positive expression of PCNA (70%) in the ova, interstitial cells, and oocytes (black arrow). (**D**) Section in the hermaphrodite gland revealing positive expression of cyclin D1 (30%) in interstitial cells, ova, and oocytes (black arrow).

Based on an electron microscopy study, the digestive gland of *B. alexandrina* control snails has numerous microvilli at the apex of the digestive cells, and normal nuclei located near the basement membrane (Figure 7A). Tight junction at the upper lateral zone bound adjacent digestive cells. Many mitochondria were distributed near the nucleus of the digestive cell and the rough endoplasmic reticulum (Figure 7B). The excretory cells of the digestive glands showed two excretory vacuoles and many rootlets of the microvilli (Figure 7C). The digestive cells of *B. alexandrina* snails showed partial destruction of the apical microvilli and the presence of some surface blebs after being exposed to 44.15 mg/L of SeONPs (Figure 7D). Partially destroyed microvilli rootlets appeared under the destroyed microvilli (Figure 7E).

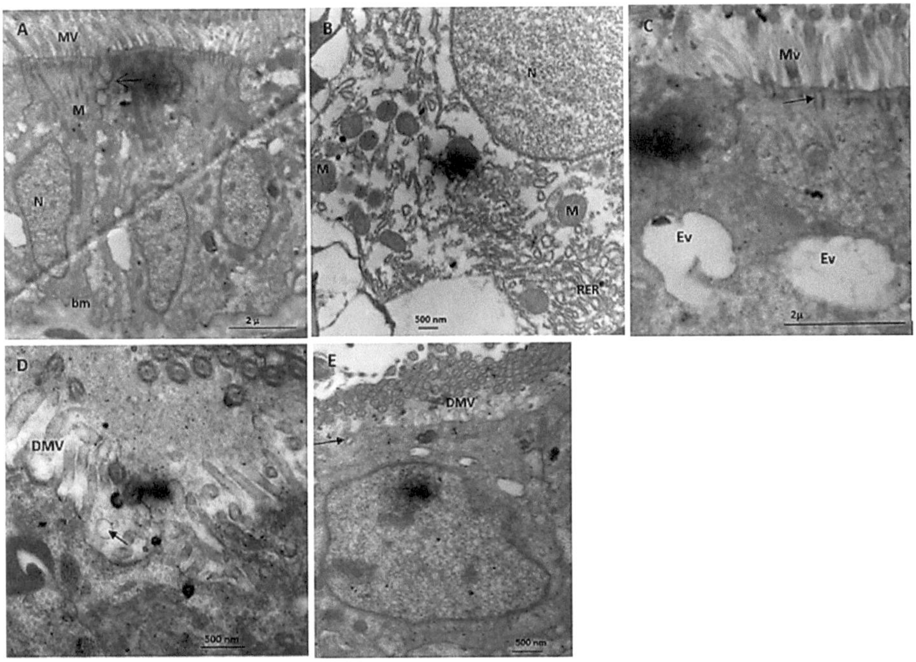

Figure 7. Electron micrographs of *B. alexandrina* snails' digestive glands. (**A**) Control adjacent digestive cells showing an apical region with microvilli (MV), mitochondria (M), and nuclei (N) near the basement membrane (bm). Note that a tight junction binds two neighboring cells (arrows). (**B**) Part of the digestive cell showing the nucleus (N), numerous mitochondria (M), and rough endoplasmic reticulum (RER). (**C**) Excretory cells of untreated *B. alexandrina* snails showing excretory vacuole (Ev) and microvilli (MV) at the apical region. Arrow refers to the rootlet of a microvillus. (**D**) The apical region of treated digestive cells with 44.15 mg/L of SeONPs showing destroyed microvilli (DMV) and presence of surface bleb (arrow). (**E**) Treated digestive cell with LC_{25} SeONPs showing destroyed microvilli (DMV) and destroyed rootlet of a microvillus (arrow).

3.4. Docking Study

ALT and AST, the two hepatopancreatic enzymes, were chosen to investigate the action of selenium oxide (SeONPs). The docking interaction demonstrated a high efficiency for SeO_2 against the ALT and AST receptor binding sites (Figure 8). The docking score, also called interaction-free energy, was used to examine the enzymes' effect. Selenium oxide was able to dock with these enzymes through their H-interaction scores (−1.0, −3.8, and −2.3 Kcal/mol) against ALT and AST, respectively (Table 2).

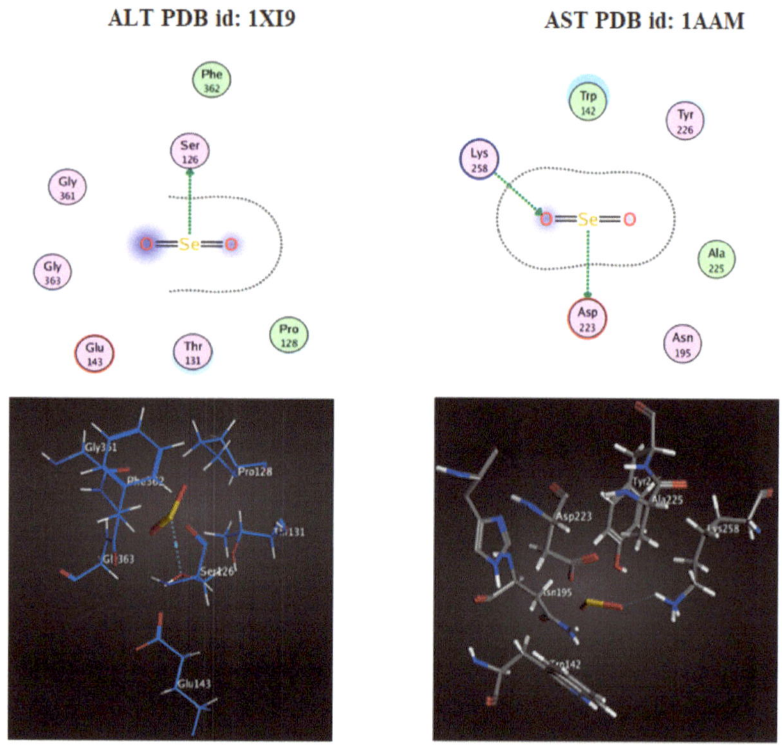

Figure 8. 3D and 2D docked interaction map for the SeO_2 (SeONPs) with the binding sites of ALT and AST.

Table 2. In silico docking study of hepatopancreas enzymes, ALT, and AST with the ligand selenium oxide.

PDB ID	Docking Score (Kcal/mol)	Interaction Type	Amino Acid Residue Involved in Docking
ALT (1XI9)	−1.0	H-donor	SER 126
AST (1AAM)	−3.8 −2.3	H-donor H-acceptor	ASP 223 LYS 258

4. Discussion

Selenium is a vital mineral for health that controls a number of cellular functions through selenium proteins. A variety of disorders, including infectious diseases, can be prevented with selenium. Despite all these advantages, selenium in high dosages might have undesirable side effects. Reports now concentrate on employing nanoparticles to avoid excessive dosages of selenium while maintaining biological benefits [39]. Nano-selenium (SeNPs) has many advantages, such as its small size and higher activity, in addition to the ability to use selenium in its zero oxidation state (Se^0) [40]. The aquatic invertebrates, such as gastropods, bivalves, and crustaceans are good model organisms for studying the nanomaterials toxicological effects [41]. Additionally, *Daphnia magna* is extremely susceptible to contaminants and serves as a model organism for bioindicators in ecotoxicology [18,42]. This model may help detect the acute toxicity of tested materials with *Daphnia* spp., because it is simple, quick, and has a distinct ending [43].

The present results indicated that SeONPs have a toxicological effect on *D. magna*. SeONPs had a half-lethal concentration (LC_{50}) of 1.62 mg/L after 24 h and 1.08 mg/L after 48 h on *D. magna*. These findings were in a good accordance with Dunbar et al. [44], who studied the toxic impacts of sodium selenite on *Daphnia magna Straus* and stated that the acceptable maximum toxicant concentration of 1.73 mg/L (<2.0 mg/L Se). Additionally, In their study of the toxicity of 50 metals on *Daphnia magna*, Okamoto et al. [45] found that 13 elements (Al, Sc, Cr, Co, Ni, Zn, Se, Rb, Y, Rh, Pt, Tl, and Pb) had half lethal values between 100 $\mu g\ L^{-1}$ and 1000 $\mu g\ L^{-1}$. Boyum [46] reported similar findings and indicated that the acute 48 h LC_{50} of sodium selenate for *Daphnia pulicaria* and *Daphnia magna* was 1.01mg Se/1 and 0.25 mg Se/1, respectively. For *D. magna* and *D. pulicaria*, the sodium selenite 48 h LC/sub-50/values were 0.45 mg Se/1and 0.006 mg Se/1, respectively. Daphnids collected Se quantities in the water that might be poisonous and negatively impact fish or seabirds along the food chain [47]. Similarly, ref. [48] found that SeNPs were toxic to the microalgae *Chaetoceros gracilis* in a time and concentration dependent manner. Moreover, ref. [49] observed that the use of nano-selenium improved the growth performance of the Nile *tilapia* in 1 mg/kg concentration group while this growth performance was low in 2 mg/kg Se-Nps concentration, thus they concluded that SeONps supplementation may lead to toxic effects in over-dose concentrations. Additionally, Ibrahim et al. [50] concluded that the nanomaterials could negatively affect the membrane permeability and structure, leading to the death of the aquatic creatures.

Morad et al. [27] stated the molluscicidal activity of the myco-synthesized SeNPs against *Biomphalaria alexandrina* snails with sublethal concentration (LC_{25}) 44.15 mg/L. The current investigation revealed that exposure of *B. alexandrina* snails to 44.15 mg/L of SeONPs decreased the survival rate, fecundity, and reproductive (R_0) rate compared to the control group. These parameters were the most significant toxicological endpoint tests for evaluating the nanomaterials' toxicity [50]. Nano-metals caused deleterious effects in the reproductive system, kidney, liver, brain, and other body systems [51] through inducing of releasing of reactive oxygen that caused damages of cells decreased fertility, delayed in organism development, and subsequently the death of this snail [52,53]. Ibrahim and Sayed [54] correlated these reductions with the severe damages in the digestive and hermaphrodite gland cells of treated snails. Oliveira-Filho et al. [55] stated that silver nanoparticles caused deleterious effects on reproduction and survival rates of *B. glabrata* snails.

The present results showed that the activities of the transaminases (AST and ALT) in snails exposed to 44.15 mg/L of SeONPs were significantly increased, while there was a reduction in total protein and albumin concentrations compared with the control group. The increase in ALT and AST after exposure was due to the great damages of SeNPs on the hepatic cells [25,56]. The decrease in protein might be due to the animals' effort to restore normal levels after the destructive effect of SeONPs on protein biosynthesis [57]. The major cause of the decline in albumin level after exposure may be due to the deleterious effect of SeNPs on liver parenchyma [52].

The digestive gland is composed of "digestive" cells that are responsible for the absorption and digestion of food and "secretory" cells that produce digestive enzymes and calcareous concretions. In between these two cell types, there are scattered undifferentiated cells [58]. The present results showed histopathological damage in the digestive and hermaphrodite glands, where, after the exposure to SeONPs, there was degeneration of the digestive cells and excretory cells. Additionally, there was an increase in the lumen of tubules. Regarding the hermaphrodite gland of *B. alexandrina*, it was comprised of the male reproductive cells and the female oogenic cells. Exposing snails to 44.15 mg/L of SeONPs caused great damage to gonadal cells; the connective tissue was deformed, the mature ova were destroyed, and the sperm were degenerated. These histopathological alterations might be related to the direct toxic effects of the heavy metals on the different organs of the target animals [59]. Similarly, Abdel-Tawab et al. [16] reported severe damages in the digestive and hermaphrodite glands of *B. alexandrina* snails after exposure of to 314.5 mg/L of Cerium oxide nano-composite; furthermore, there were degeneration and rupture of sperms and ova. Additionally, in the digestive gland, there were ruptures, degeneration, and vacuolation of some digestive cells, in addition to a marked increase in the number of the secretory cells.

Specific antibody-antigen interactions were used in immunohistochemistry to demonstrate antigens in tissue sections [60–62]. Cyclin D1 is a cell cycle progression regulator. Its upregulation may be associated with cancer [63–65]. PCNA could be used as an important marker for hepatotoxicity, with more PCNA-positive cells indicating severe and necrotic cell damage [66], with more PCNA-positive cells indicating severe and necrotic cell damage [67].

Immunohistochemistry of a normal hermaphrodite gland showed that both PCNA and Cyclin D1 were expressed as brownish nuclei with a low percentage. PCNA expression was ±70% positive in sperm and ova after 44.15 mg/L SeNPs exposure. Cyclin D1 was also found to be expressed (±30%) in sperm, ova, and interstitial cells. Similarly, Ibrahim et al. [36] revealed that after being exposed to LC_{25} of *Nerium oleander* methanolic extract, there was degeneration and destruction in some ova and sperm, and immunohistochemistry for PCNA expression was positive in ±30% of sperm and ova.

Numerous cilia were visible in electron photomicrographs of the digestive glands of *B. alexandrina* snails, as were microvilli at the apex of the digestive cells, normal nuclei based near the basement membrane, rough endoplasmic reticulum, and many mitochondria. The excretory cells showed two excretory vacuoles and many rootlets of microvilli. The digestive cells of *B. alexandrina* snails showed partial destruction of the apical microvilli and the presence of some surface blebs after being exposed to 44.15 mg/L of SeONPs. These results are in accordance with Yousef and EI-Kassas [68], who reported cytoplasmic fragmentation, vacuolation, and toxic agent accumulation in the cytoplasm of *B. alexandrina* digestive and excretory cells after exposure to *E. splendens*-plant. In addition, Hamed et al. [69] discovered ultrastructural changes in the digestive gland of *Eobania vermiculata* snails after topical application of two carbamate molluscicides. Disruption and reduction in microvilli, and the formation of surface blabs, cytoplasmic vacuolization, and increases in the number of calcium and excretory cells, were among the changes. Receptor-ligand interactions using molecular docking is an interesting method for predicting the role of ligand compounds in the inhibition or activation actions of some enzymes. The enzymatic mechanisms included ALT (PDB id: 1XI9 [70]) and AST (PDB id: 1AAM [70]. ALT and AST are secreted in the body to metabolize protein and produce energy; in addition, they are important indicators of liver disorders [71]. The current findings showed potential SeO_2 effect that reacted with ALT and AST and caused interfering with their active amino acid catalytic sites.

5. Conclusions

The present study investigated the toxicity of mycosynthesized selenium oxide nanoparticles (SeONPs) in many aspects regarding the intermediate host *Biomphlaria alexandrina* of *Schistosoma mansoni* parasite. The *B. alexandrina* snails' rates of survival, fecundity, and reproduction may be positively reduced by SeONPs. The biochemical analysis revealed a significant increase in the digestive gland enzymes, such as AST and ALT, while there was a noticeable reduction in the concentration of albumin and total protein. The use of in silico molecular docking revealed a potential SeONPs impact that interacted with the active amino acid catalytic sites of ALT and AST. The histological and immunohistochemical results showed great defects in the structure of digestive and ovotestis glands after the exposure to SeONPs. In addition, the transmission electron microscope examination revealed abnormalities in the digestive gland cells' fine structure. As an indicator model of freshwater pollution, *D. magna* showed high sensitivity to low concentrations of SeONPs. The results of the present investigation represent, to our knowledge, the first ecotoxicological evaluations of SeONPs on the freshwater snail *B. alexandrina*. In conclusion, this snail can be used as a biomonitor for SeONPs as a freshwater pollutant. To determine if it is currently necessary to regulate and/or manage the use and discharge of SeONPs in the Nile River, more research is required to assess the current contamination levels of SeONPs in Egyptian aquatic ecosystems.

Author Contributions: Conceptualization, M.Y.M., H.E.-S. and A.M.I.; methodology, A.M.I., M.Y.M., H.E.-S., M.G. and O.A.H.; software, M.Y.M., H.E.-S., A.S.A. and A.M.I.; validation, M.Y.M., H.E.-S. and A.M.I.; formal analysis, R.M.A.E.-H., R.G.K., S.M.K. and A.S.A.; investigation, M.Y.M., H.E.-S. and A.M.I.; resources, M.Y.M., H.E.-S., A.M.I., M.G. and O.A.H.; data curation, R.M.A.E.-H., R.G.K., S.M.K. and A.S.A.; funding acquisition, S.M.K. and A.S.A.; writing—original draft preparation, M.Y.M., H.E.-S., M.G., O.A.H. and A.M.I.; writing—review and editing, M.Y.M., H.E.-S., R.M.A.E.-H., S.M.K., A.S.A. and A.M.I. All authors have read and agreed to the published version of the manuscript.

Funding: Princess Nourah bint Abdulrahman University Researchers Supporting Project number (PNURSP2023R214), Princess Nourah bint Abdulrahman University, Riyadh, Saudi Arabia.

Institutional Review Board Statement: Not applicable.

Informed Consent Statement: Not applicable.

Data Availability Statement: All data generated or analyzed during this study are included in this article.

Acknowledgments: We are grateful to Princess Nourah bint Abdulrahman University, Riyadh, Saudi Arabia.

Conflicts of Interest: The authors declare no conflict of interest.

References

1. Murthy, M.K.; Mohanty, C.S.; Swain, P.; Pattanayak, R. Assessment of Toxicity in the Freshwater Tadpole Polypedates Maculatus Exposed to Silver and Zinc Oxide Nanoparticles: A Multi-Biomarker Approach. *Chemosphere* **2022**, *293*, 133511. [CrossRef] [PubMed]
2. Pyrzyńska, K. Determination of Selenium Species in Environmental Samples. *Microchim. Acta* **2002**, *140*, 55–62. [CrossRef]
3. Lemly, A.D. Aquatic Selenium Pollution Is a Global Environmental Safety Issue. *Ecotoxicol. Environ. Saf.* **2004**, *59*, 44–56. [CrossRef] [PubMed]
4. Santos, S.; Ungureanu, G.; Boaventura, R.; Botelho, C. Selenium Contaminated Waters: An Overview of Analytical Methods, Treatment Options and Recent Advances in Sorption Methods. *Sci. Total Environ.* **2015**, *521–522*, 246–260. [CrossRef]
5. Filipović, N.; Ušjak, D.; Milenković, M.T.; Zheng, K.; Liverani, L.; Boccaccini, A.R.; Stevanović, M.M. Comparative Study of the Antimicrobial Activity of Selenium Nanoparticles with Different Surface Chemistry and Structure. *Front. Bioeng. Biotechnol.* **2021**, *8*, 624621. [CrossRef]
6. Nayak, V.; Singh, K.R.B.; Singh, A.K.; Singh, R.P. Potentialities of Selenium Nanoparticles in Biomedical Science. *New J. Chem.* **2021**, *45*, 2849–2878. [CrossRef]
7. Hashem, A.H.; Khalil, A.M.A.; Reyad, A.M.; Salem, S.S. Biomedical Applications of Mycosynthesized Selenium Nanoparticles Using *Penicillium expansum* ATTC 36200. *Biol. Trace Elem. Res.* **2021**, *199*, 3998–4008. [CrossRef]
8. Tan, L.C.; Nancharaiah, Y.V.; van Hullebusch, E.D.; Lens, P.N.L. Selenium: Environmental Significance, Pollution, and Biological Treatment Technologies. *Biotechnol. Adv.* **2016**, *34*, 886–907. [CrossRef]
9. Xia, T.; Li, N.; Nel, A.E. Potential Health Impact of Nanoparticles. *Annu. Rev. Public Health* **2009**, *30*, 137–150. [CrossRef]
10. Caixeta, M.B.; Araújo, P.S.; Gonçalves, B.B.; Silva, L.D.; Grano-Maldonado, M.I.; Rocha, T.L. Toxicity of Engineered Nanomaterials to Aquatic and Land Snails: A Scientometric and Systematic Review. *Chemosphere* **2020**, *260*, 127654. [CrossRef]
11. Liu, W.; Chen, Y.; Leng, X.; Stoll, S. Embryonic Exposure to Selenium Nanoparticles Delays Growth and Hatching in the Freshwater Snail *Lymnaea stagnalis*. *Chemosphere* **2022**, *307*, 136147. [CrossRef] [PubMed]
12. OECD. *Detailed Review Paper (DRP) on Molluscs Life-Cycle Toxicity Testing*; OECD: Paris, France, 2010; 182p.
13. Loker, E.S.; Bayne, C.J. Molluscan Immunobiology: Challenges in the Anthropocene Epoch. *Adv. Comp. Immunol.* **2018**, 343–407. [CrossRef]
14. Mossalem, H.S.; Ibrahim, A.M. The Ameliorative Potential of the Ethanol Extract of the Plant Ocimum Basilicum on Biomphalaria Alexandrina Snails Exposed to the Insecticide Bestacid. *Egypt. J. Aquat. Biol. Fish.* **2019**, *23*, 161–172. [CrossRef]
15. Abu El Einin, H.M.; Ali, R.E.; Gad El-Karim, R.M.; Youssef, A.A.; Abdel-Hamid, H.; Habib, M.R. Biomphalaria Alexandrina: A Model Organism for Assessing the Endocrine Disrupting Effect of 17β-Estradiol. *Environ. Sci. Pollut. Res.* **2019**, *26*, 23328–23336. [CrossRef]
16. Abdel-Tawab, H.; Ibrahim, A.M.; Hussein, T.; Mohamed, F. Mechanism of Action and Toxicological Evaluation of Engineered Layered Double Hydroxide Nanomaterials in Biomphalaria Alexandrina Snails. *Environ. Sci. Pollut. Res.* **2022**, *29*, 11765–11779. [CrossRef]

17. Sharaf El-Din, A.T.; Mohamed, A.M.; Mohamed, A.H.; El-Hommossany, K.M.; Habib, M.R. Relationship between Some Heavy Metals and Schistosoma Mansoni Infection Rates in Biomphalaria Alexandrina Snails Collected from Different Egyptian Localities. *World Appl. Sci. J.* **2010**, *11*, 38–43.
18. Ghazy, M.M. Aquatic Fauna as a Water Quality Monitoring Device. Ph.D. Thesis, Faculty of Science, Cairo University, Cairo, Egypt, 1997; 169p.
19. Montassir, L.; Berrebaan, I.; Mellouki, F.; Zkhiri, F.; Boughribil, S.; Bessi, H. Acute Toxicity and Reprotoxicity of Aqueous Extract of a Moroccan Plant (*Tetraclinis articulata*) on Freshwater Cladoceran Daphnia Magna. *J. Mater. Environ. Sci.* **2017**, *8*, 770–776.
20. Petraretti, M.; Siciliano, A.; Carraturo, F.; Cimmino, A.; De Natale, A.; Guida, M.; Pollio, A.; Evidente, A.; Masi, M. An Ecotoxicological Evaluation of Four Fungal Metabolites with Potential Application as Biocides for the Conservation of Cultural Heritage. *Toxins* **2022**, *14*, 407. [CrossRef]
21. Seremet, O.; Olaru, O.; Gutu, C.; Nitulescu, G.; Ilie, M.; Negres, S.; Zbarcea, C.; Purdel, C.; Spandidos, D.; Tsatsakis, A.; et al. Toxicity of Plant Extracts Containing Pyrrolizidine Alkaloids Using Alternative Invertebrate Models. *Mol. Med. Rep.* **2018**, *17*, 7757–7763. [CrossRef]
22. Huang, X.-J.; Choi, Y.-K.; Im, H.-S.; Yarimaga, O.; Yoon, E.; Kim, H.-S. Aspartate Aminotransferase (AST/GOT) and Alanine Aminotransferase (ALT/GPT) Detection Techniques. *Sensors* **2006**, *6*, 756–782. [CrossRef]
23. Rasool, A.; Zulfajri, M.; Gulzar, A.; Hanafiah, M.M.; Unnisa, S.A.; Mahboob, M. In Vitro Effects of Cobalt Nanoparticles on Aspartate Aminotransferase and Alanine Aminotransferase Activities of Wistar Rats. *Biotechnol. Rep.* **2020**, *26*, e00453. [CrossRef] [PubMed]
24. Abdel-Ghaffar, F.; Ahmed, A.K.; Bakry, F.; Rabei, I.; Ibrahim, A. The Impact of Three Herbicides on Biological and Histological Aspects of Biomphalaria Alexandrina, Intermediate Host of Schistosoma Mansoni. *Malacologia* **2016**, *59*, 197–210. [CrossRef]
25. Ibrahim, A.M.; Abdalla, A.M. Impact of Moringa Oleifera Seed Aqueous Extract on Some Biological, Biochemical, and Histological Aspects of Biomphalaria Alexandrina Snails. *Environ. Sci. Pollut. Res.* **2017**, *24*, 28072–28078. [CrossRef] [PubMed]
26. Morad, M.Y.; El-Sayed, H.; El-Khadragy, M.F.; Abdelsalam, A.; Ahmed, E.Z.; Ibrahim, A.M. Metabolomic Profiling, Antibacterial, and Molluscicidal Properties of the Medicinal Plants Calotropis Procera and Atriplex Halimus: In Silico Molecular Docking Study. *Plants* **2023**, *12*, 477. [CrossRef] [PubMed]
27. Morad, M.Y.; El-Sayed, H.; Elhenawy, A.A.; Korany, S.M.; Aloufi, A.S.; Ibrahim, A.M. Myco-Synthesized Molluscicidal and Larvicidal Selenium Nanoparticles: A New Strategy to Control Biomphalaria Alexandrina Snails and Larvae of Schistosoma Mansoni with an In Silico Study on Induced Oxidative Stress. *J. Fungi* **2022**, *8*, 262. [CrossRef]
28. Eveland, L.K.; Haseeb, M.A. Laboratory Rearing of Biomphalaria Glabrata Snails and Maintenance of Larval Schistosomes In Vivo and In Vitro. In *Biomphalaria Snails and Larval Trematodes*; Springer: New York, NY, USA, 2011; pp. 33–55.
29. Pellegrino, J.; Goncalves, M. A Simple Method for Collecting Egg Clutches of Biomphalaria Glabrata (*Australorbis glabratus*) and for Rearing Newly Hatched Snails. *J. Parasitol.* **1965**, *51*, 1014. [CrossRef]
30. Behie, L.A.; Zajic, J.E.; Berk, D.; Brouzes, R.J.P.; Naish, V.A. Standard Primary Toxicity Bioassay Using Daphnia Magna as a Test Species. *Water Pollut. Res. Can.* **1976**, *12*, 27–49. [CrossRef]
31. El-Gindy, M.S.; Radhawy, I.A. Effect of Low Concentrations of Sodium Pentachlorophenate on the Fecundity and Egg Viability of Bulinus Truncatus from Central Iraq. *Bull. Endem. Dis.* **1965**, *7*, 44–54.
32. Reitman, S.; Frankel, S. A Colorimetric Method for the Determination of Serum Glutamic Oxalacetic and Glutamic Pyruvic Transaminases. *Am. J. Clin. Pathol.* **1957**, *28*, 56–63. [CrossRef]
33. Doumas, B.T. Standards for Total Serum Protein Assays: A Collaborative Study. *Clin. Chem.* **1975**, *21*, 1159–1166. [CrossRef]
34. Gustafsson, J.E. Improved Specificity of Serum Albumin Determination and Estimation of "Acute Phase Reactants" by Use of the Bromcresol Green Reaction. *Clin. Chem.* **1976**, *22*, 616–622. [CrossRef] [PubMed]
35. Carleton, M.; Drury, R.A.B.; Willington, E.A.; Cammeron, R. *Carleton's Histological Technique*; Ulster Med J. Oxford University Press: Oxford, UK, 1967; Volume 36, p. 172.
36. Ibrahim, A.M.; Ahmed, A.K.; Hammam, O.A.; Abdel-Ghaffar, F. Immunotoxical, Neurotoxical, Histopathological and Immunohistopathological Alterations of Nerium Oleander and Tecoma Stans Methanolic Extract on Biomphalaria Alexandrina Snails. *Acta Trop.* **2022**, *230*, 106405. [CrossRef] [PubMed]
37. Finney, D.J. *Probit Analysis*; Cambridge University Press: London, UK, 1971; Volume 60, ISBN 1520-6017.
38. Murray, R. Speigel: Statistical Estimation Theory. In *Schaum's Outline of Theory and Problems of Statistics in S.I. Units*, 1st ed.; McGraw Hill B. Company: Singapore, 1981; 156p.
39. Liao, W.; Yu, Z.; Lin, Z.; Lei, Z.; Ning, Z.; Regenstein, J.M.; Yang, J.; Ren, J. Biofunctionalization of Selenium Nanoparticle with Dictyophora Indusiata Polysaccharide and Its Antiproliferative Activity through Death-Receptor and Mitochondria-Mediated Apoptotic Pathways. *Sci. Rep.* **2015**, *5*, 18629. [CrossRef] [PubMed]
40. Ghazi, D.A. The Contribution of Nano-Selenium in Alleviation of Salinity Adverse Effects on Coriander Plants. *J. Soil Sci. Agric. Eng.* **2018**, *9*, 753–760. [CrossRef]
41. Wang, T.; Liu, W. Emerging Investigator Series: Metal Nanoparticles in Freshwater: Transformation, Bioavailability and Effects on Invertebrates. *Environ. Sci. Nano* **2022**, *9*, 2237–2263. [CrossRef]
42. Ibrahim, A.M.; Saleh, H.A.; Zayed, K.M. Colchicum Ritchii Flower: A New Molluscicidal Plant for Biomphalaria Alexandrina Snails and the Infective Stages of Schistosoma Mansoni Alexandrina Snails and the Infective Stages of Schistosoma Mansoni. *Molluscan Res.* **2021**, *41*, 289–297. [CrossRef]

43. Hylton, C.A.; Tsui, M.T.K. Alteration of Acute Toxicity of Inorganic and Methyl Mercury to Daphnia Magna by Dietary Addition. *Sci. Rep.* **2021**, *11*, 22865. [CrossRef]
44. Dunbar, A.M.; Lazorchak, J.M.; Waller, W.T. Acute and Chronic Toxicity of Sodium Selenate to Daphnia Magna Straus. *Environ. Toxicol. Chem.* **1983**, *2*, 239–244. [CrossRef]
45. Okamoto, A.; Yamamuro, M.; Tatarazako, N. Acute Toxicity of 50 Metals to Daphnia Magna. *J. Appl. Toxicol.* **2015**, *35*, 824–830. [CrossRef]
46. Boyum, K.W. The Toxic Effect of Selenium on the Zooplankton, Daphnia Magna and Daphnia Pulicaria, in Water and the Food Source (*Chlamydomonas reinhardtii*). Ph.D. Thesis, The University of Wisconsin-Milwaukee, Milwaukee, WI, USA, 1984.
47. Ingersoll, C.G.; Dwyer, F.J.; May, T.W. Toxicity of Inorganic and Organic Selenium to Daphnia Magna (Cladocera) and Chironomus Riparius (Diptera). *Environ. Toxicol. Chem.* **1990**, *9*, 1171–1181. [CrossRef]
48. Kumar, C.M.V.; Karthick, V.; Inbakandan, D.; Kumar, V.G.; Rene, E.R.; Dhas, T.S.; Ravi, M.; Sowmiya, P.; Das, C.G.A. Effect of Selenium Nanoparticles Induced Toxicity on the Marine Diatom Chaetoceros Gracilis. *Process Saf. Environ. Prot.* **2022**, *163*, 200–209. [CrossRef]
49. Moges, F.D.; Hamdi, H.; Al-Barty, A.; Zaid, A.A.; Sundaray, M.; Parashar, S.K.S.; Gubale, A.G.; Das, B. Effects of Selenium Nanoparticle on the Growth Performance and Nutritional Quality in Nile Tilapia, Oreochromis Niloticus. *PLoS ONE* **2022**, *17*, e0268348. [CrossRef]
50. Ibrahim, A.M.; Mohamed, F.; Al-Quraishy, S.; Abdel-Baki, A.A.S.; Abdel-Tawab, H. Green Synthesis of Cerium Oxide / Moringa Oleifera Seed Extract Nano-Composite and Its Molluscicidsal Activities against Biomophalaria Alexanderina. *J. King Saud Univ.—Sci.* **2021**, *33*, 101368. [CrossRef]
51. Verma, R.; Vijayalakshmy, K.; Chaudhiry, V. Detrimental Impacts of Heavy Metals on Animal Reproduction: A Review. *J. Entomol. Zool. Stud.* **2018**, *6*, 27–30.
52. Fahmy, S.R.; Abdel-Ghaffar, F.; Bakry, F.A.; Sayed, D.A. Ecotoxicological Effect of Sublethal Exposure to Zinc Oxide Nanoparticles on Freshwater Snail Biomphalaria Alexandrina. *Arch. Environ. Contam. Toxicol.* **2014**, *67*, 192–202. [CrossRef]
53. Khorrami, M.B.; Sadeghnia, H.R.; Pasdar, A.; Ghayour-Mobarhan, M.; Riahi-Zanjani, B.; Hashemzadeh, A.; Zare, M.; Darroudi, M. Antioxidant and Toxicity Studies of Biosynthesized Cerium Oxide Nanoparticles in Rats. *Int. J. Nanomed.* **2019**, *14*, 2915–2926. [CrossRef] [PubMed]
54. Ibrahim, A.M.; Sayed, D.A. Toxicological Impact of Oxyfluorfen 24% Herbicide on the Reproductive System, Antioxidant Enzymes, and Endocrine Disruption of Biomphalaria Alexandrina (Ehrenberg, 1831) Snails. *Environ. Sci. Pollut. Res.* **2019**, *26*, 7960–7968. [CrossRef] [PubMed]
55. Oliveira-Filho, E.C.; de Freitas Muniz, D.H.; de Carvalho, E.L.; Cáceres-Velez, P.R.; Fascineli, M.L.; Azevedo, R.B.; Grisolia, C.K. Effects of AgNPs on the Snail Biomphalaria Glabrata: Survival, Reproduction and Silver Accumulation. *Toxics* **2019**, *7*, 12. [CrossRef]
56. Hasheesh, W.S.; Marie, M.A.S.; El-Deeb, F.A.A.; Sayed, S.S.M. Impact of Asparagus Densiflours and Oreopanax Guatemalensis Plants and Difenoconazole Fungicide on Biochemical Parameters of Biomaphalaria Alexandrina Snails. *Austral. J. Basic Appl. Sci.* **2011**, *5*, 366–378.
57. Ibrahim, A.M.; Bakry, F.A. Assessment of the Molluscicidal Impact of Extracted Chlorophyllin on Some Biochemical Parameters in the Nervous Tissue and Histological Changes in Biomphalaria Alexandrina and Lymnaea Natalensis Snails. *Invertebr. Neurosci.* **2019**, *19*, 7. [CrossRef]
58. Zarai, Z.; Boulais, N.; Karray, A.; Misery, L.; Bezzine, S.; Rebai, T.; Gargouri, Y.; Mejdoub, H. Immunohistochemical Localization of Hepatopancreatic Phospholipase A2 in Hexaplex Trunculus Digestive Cells. *Lipids Health Dis.* **2011**, *10*, 91. [CrossRef]
59. Elwasify, A.Y.H.; Ghanem, M.H.; El-Bamby, M.M.M.; Ali, F.A.F. Effect of Bioaccumulation and Biosedimentation of Some Heavy Metals on Histological Features in the Cichlid Fish, Tilapia Zillii Inhabiting Lake Qarun, Egypt. *Egypt. J. Aquat. Biol. Fish.* **2021**, *25*, 695–711. [CrossRef]
60. Harris, L.; O'Byrne-Ring, N.; Lambkin, H. Characterisation of Cell Types in Abalone (*Haliotis* Spp.) Tissues Using Immunohistochemical Techniques. *Aquaculture* **2006**, *261*, 1413–1421. [CrossRef]
61. Nakopoulou, L.; Janinis, J.; Panagos, G.; Comin, G.; Davaris, P. The Immunohistochemical Expression of Proliferating Cell Nuclear Antigen (PCNA/Cyclin) in Malignant and Benign Epithelial Ovarian Neoplasms and Correlation with Prognosis. *Eur. J. Cancer* **1993**, *29*, 1599–1601. [CrossRef] [PubMed]
62. Ohta, Y.; Ichimura, K. Proliferation Markers, Proliferating Cell Nuclear Antigen, Ki67, 5-Bromo-2′-Deoxyuridine, and Cyclin D1 in Mouse Olfactory Epithelium. *Ann. Otol. Rhinol. Laryngol.* **2000**, *109*, 1046–1048. [CrossRef] [PubMed]
63. Alao, J.P. The Regulation of Cyclin D1 Degradation: Roles in Cancer Development and the Potential for Therapeutic Invention. *Mol. Cancer* **2007**, *6*, 24. [CrossRef] [PubMed]
64. Bartkova, J.; Lukas, J.; Strauss, M.; Bartek, J. Cell Cycle-Related Variation and Tissue-Restricted Expression of Human Cyclin D1 Protein. *J. Pathol.* **1994**, *172*, 237–245. [CrossRef] [PubMed]
65. Tousson, E.; Ali, E.M.M.; Moustafa, A.H.A.; Moselhey, S.S.; El-said, K.S. Proliferating Cell Nuclear Antigen as A Biomarker for Thioacetamide Induced Hepatotoxicity of Rat Liver. *Am. J. Zool. Res.* **2014**, *2*, 51–54.
66. Torlakovic, E.; Nielsen, S.; Vyberg, M. Antibody Selection in Immunohistochemical Detection of Cyclin D1 in Mantle Cell Lymphoma. *Am. J. Clin. Pathol.* **2005**, *124*, 782–789. [CrossRef]

67. Yousef, A.A.A.-H.; EI-Kassas, N.B. Ultrastructure and Histopathological Effects of Some Plant Extracts on Digestive Gland of Biomphalaria Alexandrina and Bulinus Truncatus. *J. Basic Appl. Zool.* **2013**, *66*, 27–33. [CrossRef]
68. Hamed, S.S.; Abdelmeguied, N.E.; Essaway, A.E.; Radwan, M.A.; Hegazy, A.E. Histological and Ultrastructural Changes Induced by Two Carbamate Molluscicides on the Digestive Gland of Eobania Vermiculata. *J. Biol. Sci.* **2007**, *7*, 1017–1037. [CrossRef]
69. Zhou, W.; Tempel, W.; Shah, A.; Chen, L.; Liu, Z.-J.; Lee, D.; Lin, D.; Chang, S.-H.; Brereton, P.S.; Izumi, M.; et al. *1XI9: Alanine Aminotransferase from Pyrococcus Furiosus Pfu-1397077-001*; Southeast Collaboratory for Structural Genomics (SECSG): Athens, GA, USA, 2004.
70. Almo, S.C.; Smith, D.L.; Danishefsky, A.T.; Ringe, D. The Structural Basis for the Altered Substrate Specificity of the R292D Active Site Mutant of Aspartate Aminotransferase from *E. coli*. *Protein Eng. Des. Sel.* **1994**, *7*, 405–412. [CrossRef] [PubMed]
71. Wang, H.L.; Chu, C.H.; Tsai, S.J.; Yang, R.J. Aspartate Aminotransferase and Alanine Aminotransferase Detection on Paper-Based Analytical Devices with Inkjet Printer-Sprayed Reagents. *Micromachines* **2016**, *7*, 9. [CrossRef] [PubMed]

microorganisms

MDPI

Article

Biosynthesis of Novel Tellurium Nanorods by *Gayadomonas* sp. TNPM15 Isolated from Mangrove Sediments and Assessment of Their Impact on Spore Germination and Ultrastructure of Phytopathogenic Fungi

Mohamed N. Abd El-Ghany [1,*], Salwa A. Hamdi [2], Shereen M. Korany [3], Reham M. Elbaz [4,5] and Mohamed G. Farahat [1,6,*]

1 Botany and Microbiology Department, Faculty of Science, Cairo University, Giza 12613, Egypt
2 Zoology Department, Faculty of Science, Cairo University, Giza 12613, Egypt
3 Department of Biology, College of Science, Princess Nourah bint Abdulrahman University, P.O. Box 84428, Riyadh 11671, Saudi Arabia
4 Botany and Microbiology Department, Faculty of Science, Helwan University, Cairo 11795, Egypt
5 Department of Biology, College of Science, University of Bisha, P.O. Box 551, Bisha 61922, Saudi Arabia
6 Biotechnology Department, Faculty of Nanotechnology for Postgraduate Studies, Sheikh Zayed Branch Campus, Cairo University, Sheikh Zayed City 12588, Egypt
* Correspondence: mabdelghany@sci.cu.edu.eg or dr.mohamed.naguib@cu.edu.eg (M.N.A.E.-G.); farahat@cu.edu.eg (M.G.F.)

Abstract: The biosynthesis of nanoparticles using green technology is emerging as a cost-efficient, eco-friendly and risk-free strategy in nanotechnology. Recently, tellurium nanoparticles (TeNPs) have attracted growing attention due to their unique properties in biomedicine, electronics, and other industrial applications. The current investigation addresses the green synthesis of TeNPs using a newly isolated mangrove-associated bacterium, *Gayadomonas* sp. TNPM15, and their impact on the phytopathogenic fungi *Fusarium oxysporum* and *Alternaria alternata*. The biogenic TeNPs were characterized using transmission electron microscopy (TEM), scanning electron microscopy (SEM), energy-dispersive X-ray spectroscopy (EDX), X-ray diffraction (XRD), Raman spectroscopy and Fourier transform infrared (FTIR). The results of TEM revealed the intracellular biosynthesis of rod-shaped nanostructures with a diameter range from 15 to 23 nm and different lengths reaching up to 243 nm. Furthermore, the successful formation of tellurium nanorods was verified by SEM-EDX, and the XRD pattern revealed their crystallinity. In addition, the FTIR spectrum provided evidence for the presence of proteinaceous capping agents. The bioinspired TeNPs exhibited obvious inhibitory effect on the spores of both investigated phytopathogens accomplished with prominent ultrastructure alternations, as evidenced by TEM observations. The biogenic TeNPs impeded spore germination of *F. oxysporum* and *A. alternata* completely at 48.1 and 27.6 μg/mL, respectively. Furthermore, an increase in DNA and protein leakage was observed upon exposure of fungal spores to the biogenic TeNPs, indicating the disruption of membrane permeability and integrity. Besides their potent influence on fungal spores, the biogenic TeNPs demonstrated remarkable inhibitory effects on the production of various plant cell wall-degrading enzymes. Moreover, the cytotoxicity investigations revealed the biocompatibility of the as-prepared biogenic TeNPs and their low toxicity against the human skin fibroblast (HSF) cell line. The biogenic TeNPs showed no significant cytotoxic effect towards HSF cells at concentrations up to 80 μg/mL, with a half-maximal inhibitory concentration (IC_{50}) value of 125 μg/mL. The present work spotlights the antifungal potential of the biogenic TeNPs produced by marine bacterium against phytopathogenic fungi as a promising candidate to combat fungal infections.

Keywords: tellurium nanoparticles; green synthesis; biogenic nanoparticles; phytopathogenic fungi; antifungal activity

check for updates

Citation: Abd El-Ghany, M.N.; Hamdi, S.A.; Korany, S.M.; Elbaz, R.M.; Farahat, M.G. Biosynthesis of Novel Tellurium Nanorods by *Gayadomonas* sp. TNPM15 Isolated from Mangrove Sediments and Assessment of Their Impact on Spore Germination and Ultrastructure of Phytopathogenic Fungi. *Microorganisms* **2023**, *11*, 558. https://doi.org/10.3390/microorganisms11030558

Academic Editor: Gustavo Dominguez-Bernal

Received: 14 January 2023
Revised: 17 February 2023
Accepted: 20 February 2023
Published: 22 February 2023

1. Introduction

The antimicrobial activity of various nanomaterials has gained increasing interest owing to their diverse medical, agricultural and environmental applications. Undoubtedly, nanoparticles have extraordinary advantages compared to bulk materials due to their larger surface-to-volume ratios and higher spatial confinement [1]. In this trend, extensive attempts have been employed to fabricate metal, metal oxide and metalloid nanoparticles as alternative antimicrobial agents, such as silver, gold, copper, zinc, selenium, and tellurium nanoparticles [2–11]. Recently, tellurium nanoparticles (TeNPs) have drawn significant attention due to their antibacterial and antifungal activity, besides biocompatibility [12]. Tellurium is a toxic rare element metalloid that belongs to chalcogens, with intermediate properties between metals and non-metals [13]. In a recent approach, the employment of the high efficiency of some tellurite-resistant microorganisms to biomineralize the metalloid oxyanions to less toxic elemental TeNPs resulted in a reduction in tellurium toxicity and increased its biocompatibility [14].

Green nanotechnology employs natural biological resources such as bacteria, fungi, algae and plants to achieve an eco-friendly cost-effective fabrication of nanoparticles [15–17]. The ability of these organisms to act as nanofactories for the biosynthesis nanoparticles makes them an excellent green alternative to chemical synthesis, avoiding the use of harsh and toxic chemicals. In this regard, many TeNPs-producing microorganisms have been isolated from diverse environments and have been reported to synthesize various shapes of TeNPs, such as a pillar, needle, spherical, cube and rod-shaped nanoparticles [18–21]. It has been reported that tellurite-biomineralizing bacteria produce black precipitates within and/or outside the cells as a common feature for intracellular and extracellular production of TeNPs [22–24]. In extracellular biosynthesis, the added tellurite salts are reduced into TeNPs in the culture broth, whereas, in the intracellular process, the biosynthesis of TeNPs occurs inside the cells, followed by cell lysis to be recovered and purified [25–28]. It has been believed that mangrove sediments contain numerous microbial groups that play vital roles and environmental processes in the mangrove ecosystem. Indeed, mangrove sediments-derived microbes have been supposed as a rich reservoir of natural product diversity and contribute in biomineralization and biotransformation of minerals [29–36]. Though tellurite-reducing bacteria were isolated from various terrestrial and marine habitats [37–41], their isolation from mangrove ecosystems is relatively unexplored.

Today, extensive losses of crops at the global level are caused by fungal infections, especially with the continuous emergence of fungicide resistance. It has been thought that phytopathogenic fungi are responsible for losing up to 20% of the global crop yields yearly, and about 10% of crops are destroyed due to post-harvest fungal infections [42]. The phytopathogen *Fusarium oxysporum* is a ubiquitous soilborne fungal pathogen that ranks 5th among the top 10 fungal plant pathogens, and was reported to cause devastating vascular wilt in more than 100 plant species [43]. Besides being a phytopathogen, *F. oxysporum* has also emerged as a human pathogen [44–46]. Additionally, *Alternaria alternata* is an important fungal pathogen that has a wide host range, causing considerable losses in the yield of many crops [47–50]. Like *F. oxysporum*, the fungal pathogen *A. alternata* has been documented as a causal agent of human infections [51–53]. The antimicrobial activity of TeNPs against human bacterial pathogens is well-documented. However, little data unveiling the antifungal activity of TeNPs against phytopathogens is available. In a recent study, citric juice-mediated TeNPs exhibited antibacterial effects against multidrug-resistant *Escherichia coli* and methicillin-resistant *Staphylococcus aureus* [54]. TeNP-induced antibacterial activity has been attributed to superoxide-mediated oxidative stress, leading to cytoplasmic thiol oxidation and inactivation of the iron–sulfur center-containing enzymes as well as the peroxidation of the membrane lipids, causing cell death [55,56]. In addition, biogenic TeNPs produced by *Ochrobactrum* sp. MPV1 exerted antibacterial and biofilm eradication activity against *Escherichia coli* JM109, *Pseudomonas aeruginosa* PAO1, and *Staphylococcus aureus* ATCC 25923 that was associated with the production of reactive oxygen species [57]. On the other hand, a limited number of published articles demonstrate the antifungal activity of TeNPs against yeasts and filamentous fungi. The antifungal activity of biogenic TeNPs against *Candida albicans* ATCC14053 is assumed to be due to the inhibition of squalene monooxygenase, which is a rate step-limiting enzyme in the ergosterol and cholesterol biosynthesis pathway involved in the synthesis of ergosterol, cholesterol, and phytosterols [58]. Herein, we report for the first time on the facile biosynthesis and characterization of TeNPs produced as intracellular nanorods by *Gayadomonas* sp. TNPM15 isolated from mangrove sediments and evaluate their impact on *F. oxysporum*, and *A. alternata*.

2. Materials and Methods

2.1. Sample Collection

Mangrove sediments were collected from Safaga Island, the Red Sea, Egypt (26°45′55.6″ N 33°57′38.3″ E) using sterile disposable wooden spatulas during low tides. The collected samples were placed in sterile Whirl-Pak® sample bags (Nasco, Kellogg, ID, USA) and transported to the laboratory on ice.

2.2. Isolation of TeNPs-Producing Bacteria

Mangrove sediment samples were suspended in sterile seawater and vigorously vortexed for 15 min, and then serially diluted in sterile seawater. Thereafter, 100 µL of each dilution were plated onto tryptone yeast extract seawater (TYS) agar (tryptone 10 g, yeast extract 5 and agar 15 g/L natural seawater) plates supplemented with 1 mM potassium tellurite. After 24–72 h of incubation at 30 °C, black-colored colonies were picked as presumptive TeNPs-producing bacteria. The selected colonies were re-streaked on the same medium until discrete colonies were obtained. To ensure that the color was not an organic bacterial pigment, the selected colonies were re-streaked on TYS agar plates without potassium tellurite and monitored for black pigment production during the incubation period (24–72 h) at 30 °C. Furthermore, the formation of intracellular tellurium nanostructures was confirmed by transmission electron microscopy (JEM-1400, JEOL, Tokyo, Japan) operating at 80 KV, after negative staining of bacterial cells with 1% phosphotungstic acid (*w*/*v*).

2.3. Bacterial Identification

A promising TeNPs-producing strain designated TNPM15 was identified by amplification and sequencing of its 16S rRNA gene. Typically, the genomic DNA was extracted using a PureLink DNA Purification Kit (Thermo Scientific, Waltham, MA, USA) according to the manufacturer's protocol. Genomic DNA was used as a template for the amplification of the 16S rRNA gene by polymerase chain reaction (PCR), using the universal primers 27F (5′-AGAGTTTGATCMTGGCTCAG-3′) and 1492R (5′-GGYTACCTTGTTACGACTT-3′). The PCR reaction was carried out by mixing 5 µL 10X Dream Taq buffer, 50 ng genomic DNA template, 0.4 µM of each primer, 0.2 mM of each dNTP, and one Unit Dream Taq DNA polymerase (Thermo Scientific, MA, USA), and, finally, nuclease-free water was added to increase the volume up to 50 µL. The amplification program contained an initial denaturation step at 95 °C for 4 min, followed by 30 cycles of denaturation at 95 °C for 40 s, annealing at 55 °C for 40 s, extension at 72 °C for 1.5 min and a final extension at 72 °C for 8 min in a ProFlex thermocycler (Applied Biosystems, Foster City, CA, USA). The amplified PCR products were purified using a GeneJET PCR purification kit (Thermo Scientific, MA, USA) according to the manufacturer's protocol.

The nucleotide sequence of the purified PCR product was sequenced in both directions using the 27F and 1492R primers using an ABI 3730xl sequence analyzer (Applied Biosystems, USA). Subsequently, CodonCode aligner software v. 7.1.2 was used to assemble the forward and reverse DNA sequence reads. The sequence similarity search was performed at the National Center for Biotechnology Information (NCBI) server using BLASTN (http://www.ncbi.nlm.nih.gov/blast; accessed on 20 December 2022) and the EzTaxon-e server database [59]. The phylogenetic tree in the context of 16S rRNA gene sequence was constructed using the Neighbor-Joining method of the MEGA-X software (version 10.0.5) with a bootstrap analysis based on 1000 replicates.

2.4. Biosynthesis of Tellurium Nanoparticles

The biosynthesis of TeNPs was achieved by culturing the strain TNPM15 in TYS broth amended with 1 mM potassium tellurite at 30 °C for 24 h, with continuous shaking (180 rpm). Afterwards, the bacterial cells were harvested by centrifugation at 10,000× *g* for 20 min, followed by washing the pellet twice with sterile saline solution. The pellets were then resuspended in a lysis buffer containing 10 mM Tris-HCl (pH 8.3), 1 mM EDTA, 100 mM NaCl, 1% SDS (*w*/*v*), and 0.5% Triton X100 (*v*/*v*); and the bacterial cells were disrupted by sonication for 5 min at 100 W. Thereafter, the released TeNPs were purified by sequential centrifugation at 12,000× *g* for 15 min and washing with 1.5 M Tris-HCl buffer (pH 8.3) containing 0.5% SDS (*w*/*v*). Then, the purified TeNPs were washed three times with deionized water and, finally, the pellet was resuspended in ultrapure water (18.2 Mcm @25 °C) and sterilized by filtration through 0.45 µm syringe filters.

2.5. Characterization of Tellurium Nanoparticles

The size and shape of biogenic TeNPs were determined using transmission electron microscopy. In brief, a drop of the diluted suspension of the biogenic TeNP was loaded on a carbon-coated copper grid and air-dried, and observed using a TEM (JEM-1400, JEOL, Japan) operating at 80 KV. Scanning electron microscopy (SEM) analysis was performed using a Quattro S instrument (Thermo Scientific, MA, USA) equipped with energy-dispersive X-ray spectroscopy (EDX) to determine the chemical composition of the as-prepared nanostructures. The X-ray diffraction (XRD) was investigated by a D8 Discover diffractometer (Bruker, Karlsruhe, Germany) using Cu Kα radiation as a light source with a current of 30 mA and an applied voltage of 40 kV. The 2θ angles ranged from 20 to 60° with a scan speed of 0.5°/min. Raman spectra of the biogenic TeNPs were recorded using a LabRAM HR Evolution Raman spectrometer (HORIBA Ltd., Kyoto, Japan) equipped with an air-cooled open electrode 1024 × 256-pixel CCD detector. The TeNPs were placed as thin films on a glass slide and dried in air at ambient temperature. Normal Raman spectra were acquired using 532 nm excitation (spectral range 50–300 cm^{-1}) with an acquisition interval of 20 s. Fourier-transform infrared (FTIR) spectroscopy was used to determine the functional groups. The scanning spectra were obtained in the range of 4000 to 400 cm^{-1} using a Nicolet 6700 FT-IR spectrometer (Thermo Scientific, MA, USA).

2.6. Fungal Phytopathogens

The antifungal effect of the biogenic TeNPs was investigated against two phytopathogenic fungi, *Fusarium oxysporum* AUMC 10313 and *Alternaria alternata* AUMC 3882, provided from the Assiut University Mycology Center (AUMC), Assiut, Egypt.

2.7. Germination Test

The inhibitory effect of the biogenic TeNPs on spore germination was evaluated in vitro. Briefly, *F. oxysporum* and *A. alternata* were cultured on potato dextrose agar (PDA) plates (Condalab, Madrid, Spain) at 28 °C for 7 days. Afterwards, spores were harvested, suspended in sterile saline solution (NaCl, 0.9%) and filtered through sterile muslin to remove mycelia. The spore count was adjusted to 2 × 10^6 spores/mL using a hemocytometer. Then, 50 µL of spore suspension was added to each well of a 96-well microtiter plate containing 100 µL potato dextrose broth (PDB) (Condalab, Madrid, Spain) with different TeNPs concentrations. The fungicide mancozeb (Santa Cruz Biotechnology, Dallas, TX, USA) was added at various concentrations (10, 20, 30, 40 and 50 µg/mL) instead of the TeNPs as a positive control, and sterile saline solution was used as a negative control. The plates were incubated at 28 °C for 16 h for germination. Subsequently, 100 spores were analyzed with a DM500 optical microscope (Leica, Heerbrugg, Switzerland) and the percentage of germinated spores was calculated. A spore was considered germinated if the germ tube length is equal to or longer than the spore length. The minimum inhibitory concentration (MIC) was determined as the lowest concentration of biogenic TeNPs that completely inhibits spore germination. The MIC values were expressed as μg/mL and calculated from the regression analysis curves.

2.8. Spore Morphology and Ultrastructure Analysis

The probable morphological changes and ultrastructural damage of fungal spores exposed to TeNPs were observed by using TEM. In brief, spore suspensions of *F. oxysporum* and *A. alternata* were prepared as above and exposed to TeNPs at the half concentration of their corresponding MIC values. Control groups were prepared using the same procedure, but TeNPs were not incorporated. After incubation at 28 °C for 3 h, the spores were collected by centrifugation at 6000× *g* for 10 min, washed twice with sterile saline and fixed with 3% (*v*/*v*) glutaraldehyde for 4 h overnight, then post fixed with 1% (*w*/*v*) osmium tetroxide at 4 °C for 3 h. Then, samples were sequentially dehydrated using a graded ethanol series (20 to 100%). Subsequently, the dehydrated spores were placed into an embedding plate, then permeated with pure propylene oxide, then 100% Spurr. Finally, ultrathin sections were prepared with a microtome and observed at 80 kV using a JEM-1400 transmission electron microscope (JEOL, Japan). Ultrastructural alterations of the spores were compared with untreated controls to assess the effect of TeNPs against spores of *F. oxysporum* and *A. alternata*.

2.9. Leakage of Proteins and DNA

To determine the influence of TeNPs on membrane integrity, the leakage of proteins and DNA from exposed phytopathogens was assessed. In brief, spore suspension of *F. oxysporum* and *A. alternata* (10^6 spore/mL) were incubated for 24 h at 28 °C with TeNPs at the half MIC values. Untreated spore suspension sets (without TeNPs) served as controls. At the end of incubation period, cultures were centrifuged at 10,000 rpm for 15 min at 4 °C and the supernatants were collected to measure the extracellular proteins and DNA. Subsequently, leakage of proteins and DNA upon exposure to the biogenic TeNPs was investigated following the method described by Khalil et al. [60].

2.10. Impact of TeNPs on Plant Cell Wall-Degrading Enzymes (CWDEs)

The potential inhibitory effect of TeNPs on the CWDEs of *F. oxysporum* and *A. alternata* was investigated. We focused on the influence of TeNPs on essential hydrolases associated with their pathogenicity as cellulase, pectinase and xylanase. To determine the activities of the CWDEs, each phytopathogen was inoculated into an induction broth [(g/L): $NaNO_3$ 2; K_2HPO_4 1; KCl 0.5; $MgSO_4$ $7H_2O$ 0.5; $FeSO_4$ $7H_2O$ 0.01] amended with 1% carboxymethyl cellulose (CMC), pectin or xylan for cellulase, pectinase and xylanase, respectively. All treated media were supplemented with TeNPs at the half concentration of their corresponding MIC values. Control groups were prepared using the same procedure, but TeNPs were not incorporated. After incubation at 28 °C for 7 days, cellulases, pectinases and xylanases were determined by measuring the amounts of reducing sugar released from 1% CMC, pectic acid and xylan by 3,5-dinitrosalicylic acid (DNSA) method using glucose, D-galacturonic acid and xylose as a standard, respectively [61]. Finally, the enzymatic activities were compared with those of untreated controls.

2.11. Cytotoxicity Assay

The Human Skin Fibroblast (HSF) cell line, obtained from Nawah Scientific Inc. (Mokatam, Cairo, Egypt), was used for the cytotoxicity investigations. The Dulbecco's Modified Eagle medium (DMEM) supplemented with 10% heat-inactivated fetal bovine serum (Gibco; Thermo Fisher Scientific, Loughborough, UK) and antibiotics (100 μg/mL streptomycin and 100 units/mL penicillin) was used for culturing and maintenance of the cell line. The HSF cells were cultured in a water-jacketed incubator (BINDER GmbH, Tuttlingen, Germany) at 37 °C under 5% CO_2 in a high humidity atmosphere.

The cytotoxicity of TeNPs against HSF cells was assessed using the sulforhodamine B (SRB) assay. Briefly, aliquots of 100 µL cell suspension (5×10^3 cells) were added to a 96-well microtitre plate, and various concentrations of TeNPs were added and incubated for 72 h. In another set, untreated HSF cells served as control. Afterwards, the media was replaced with 150 µL of 10% trichloroacetic acid (TCA) solution and incubated at 4 °C for 1 h. Subsequently, the TCA solution was discarded and the plates were gently washed five times with distilled water and allowed to dry. Aliquots of 70 µL SRB solution (0.4% *w*/*v* in 1% glacial acetic acid) were added and incubated in the dark at room temperature for 10 min. Thereafter, plates were washed thrice with 1% acetic acid and allowed to air-dry again. Then, the protein-bound SRB stain was dissolved in 150 µL of Tris buffer (10 mM, pH 10.5) and the absorbance was measured at 540 nm using Epoch2 microplate spectrophotometer (BioTek, Santa Clara, CA, USA).

2.12. Statistical Analysis

The measured data were subjected to the analysis of variance (ANOVA) appropriate to the design. The significant differences between treatments were compared with the critical difference at 5% level of probability by the Duncan's test using IBM SPSS software version 22.

3. Results

3.1. Isolation of TeNPs-Producing Bacteria

Several strains isolated from the mangrove sediments were able to grow in the presence of 1 mM potassium tellurite. The reduction of Te (IV) to $Te^{(0)}$ by bacterial isolates was detected by the black-colored precipitate in the culture, indicating the reduction of tellurite to black $Te^{(0)}$ precipitate. Black precipitates were not detected in cultures grown on the same medium without emendation, confirming the production of elemental tellurium rather than organic pigmentation. Furthermore, TEM analysis confirmed the production of intracellular rod-shaped nanostructures by a bacterial strain designated TNPM15 exhibiting a characteristic black color in the presence of tellurite (Figure 1). The selected strain TNPM15 was identified as *Gayadomonas* sp. based on its 16S rRNA gene sequence (accession number OP020452). BLAST results revealed that the strain TNPM15 shares 98.92% identity with *Gayadomonas joobiniege* strain G7 (accession number NR_133751.1) and 96.32% with *Catenovulum agarivorans* YM01 (accession number NR_117273.1). The Neighbor-Joining phylogenetic tree in the context of 16S rRNA gene sequence was constructed (Figure 2).

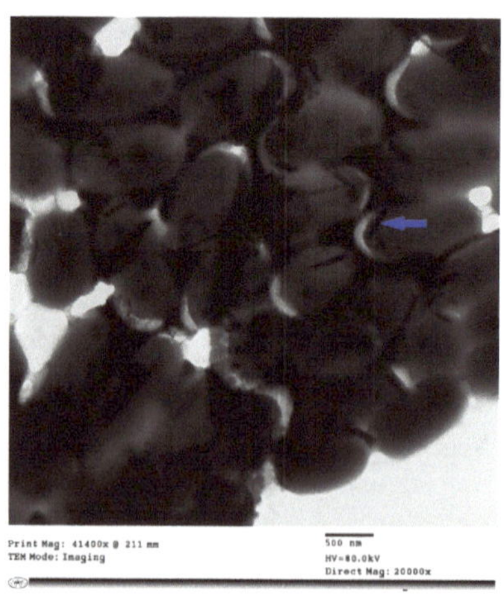

Figure 1. TEM micrograph showing intracellular formation of nanorods by *Gayadomonas* sp. TNPM15. The blue arrow indicates the formation of intracellular rod-shaped nanostructure.

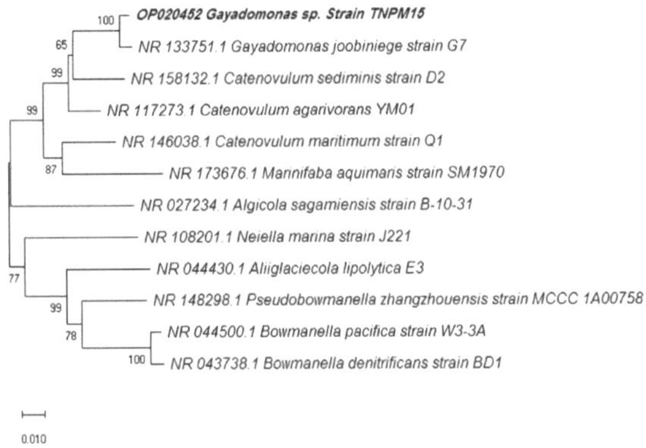

Figure 2. Phylogenetic tree showing the relationships between *Gayadomonas* sp. TNPM15 and the most closely related species.

3.2. Biosynthesis and Characterization of TeNPs

Biogenic TeNPs were produced by exploiting the tellurite-reduction capability of *Gayadomonas* sp. strain TNPM15 cultured in TYS broth supplemented with potassium tellurite. The produced intracellular TeNPs were recovered and resuspended in ultrapure water after purification steps. Electron microscopy was employed to clarify the size and shape of the purified nanoparticles. The TEM demonstrated that the purified TeNPs appeared as nanorods with a diameter range from 15 to 23 nm and different lengths reaching up to 243 nm (Figure 3). Moreover, SEM observation clearly confirmed the formation of rod-shaped nanostructures (Figure 4). Based on the results of elemental analysis by EDX, we evidenced that the black biogenic nanostructures were mainly composed of tellurium (Figure 5). Furthermore, the XRD spectrum revealed the crystalline nature of the produced TeNPs (Figure 6). Additionally, Raman spectra appeared in the whole lower-frequency region, showing characteristic vibration peaks attributed to tellurium at 121.3 and 140.3 cm^{-1}, affirming the formation of TeNPs (Figure 7). In addition, the FTIR spectrum provided insights into the functional groups incorporated into the surface of the as-prepared biogenic tellurium nanostructures (Figure 8).

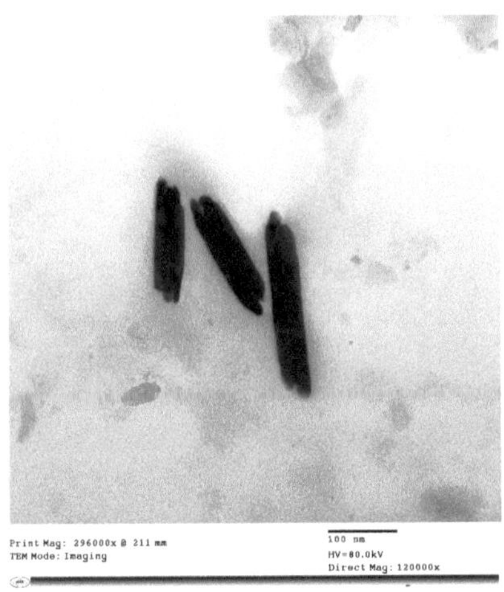

Figure 3. Transmission electron micrograph showing the rod-shaped biogenic TeNPs produced by *Gayadomonas* sp. TNPM15.

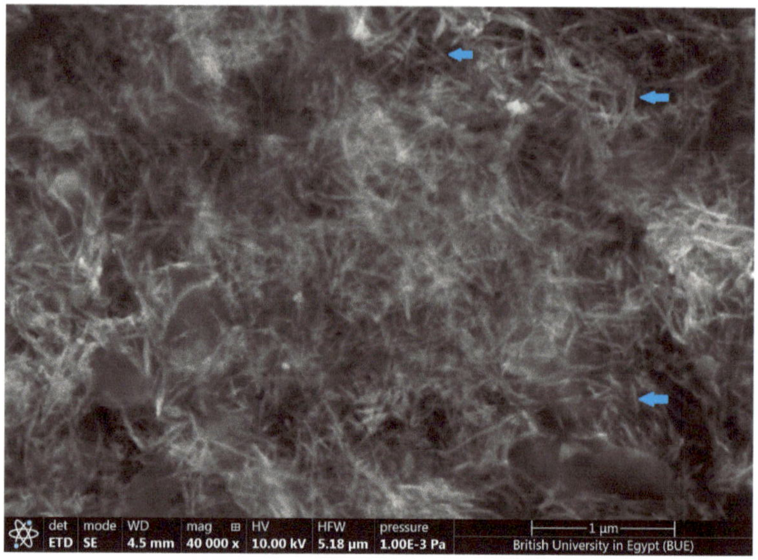

Figure 4. Scanning electron micrograph showing the rod-shaped biogenic TeNPs produced by *Gayadomonas* sp. TNPM15. The blue arrows indicate the formation of rod-shaped nanostructures.

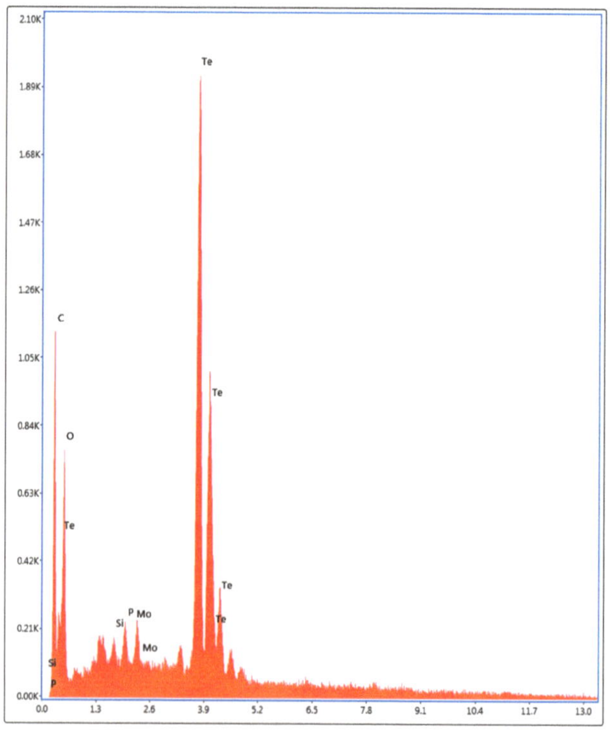

Figure 5. EDX spectrum of the biogenic TeNPs produced by *Gayadomonas* sp. TNPM15.

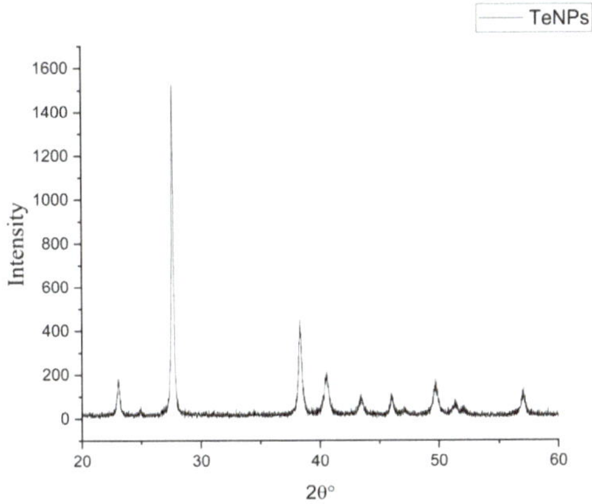

Figure 6. XRD spectrum of the biogenic TeNPs produced by *Gayadomonas* sp. TNPM15.

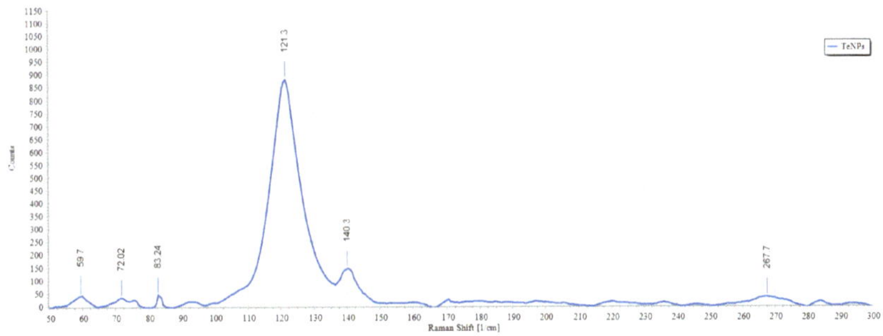

Figure 7. RAMAN spectrum of the biogenic TeNPs produced by *Gayadomonas* sp. TNPM15.

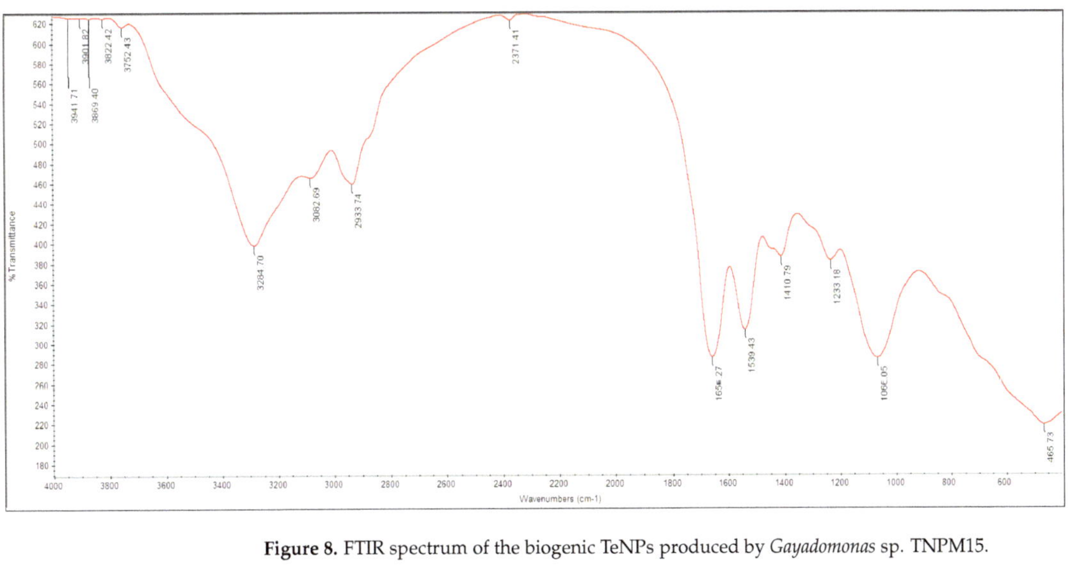

Figure 8. FTIR spectrum of the biogenic TeNPs produced by *Gayadomonas* sp. TNPM15.

3.3. Germination Test

The antifungal activity of the biosynthesized TeNPs was assessed against the phytopathogenic fungi *F. oxysporum* and *A. alternata*. Our results proved the ability of the biogenic TeNPs to inhibit fungal spore germination. The MIC values which completely inhibited spore germination of *F. oxysporum* and *A. alternata* were statistically deduced from the linear regression model fit (Figure 9). In the case of *F. oxysporum*, the deduced MIC were 48.1 and 38.0 μg/mL for TeNPs and mancozeb, respectively. On the other hand, TeNPs and mancozeb had MIC values of 27.6 and 30.5 μg/mL against *A. alternata*, respectively.

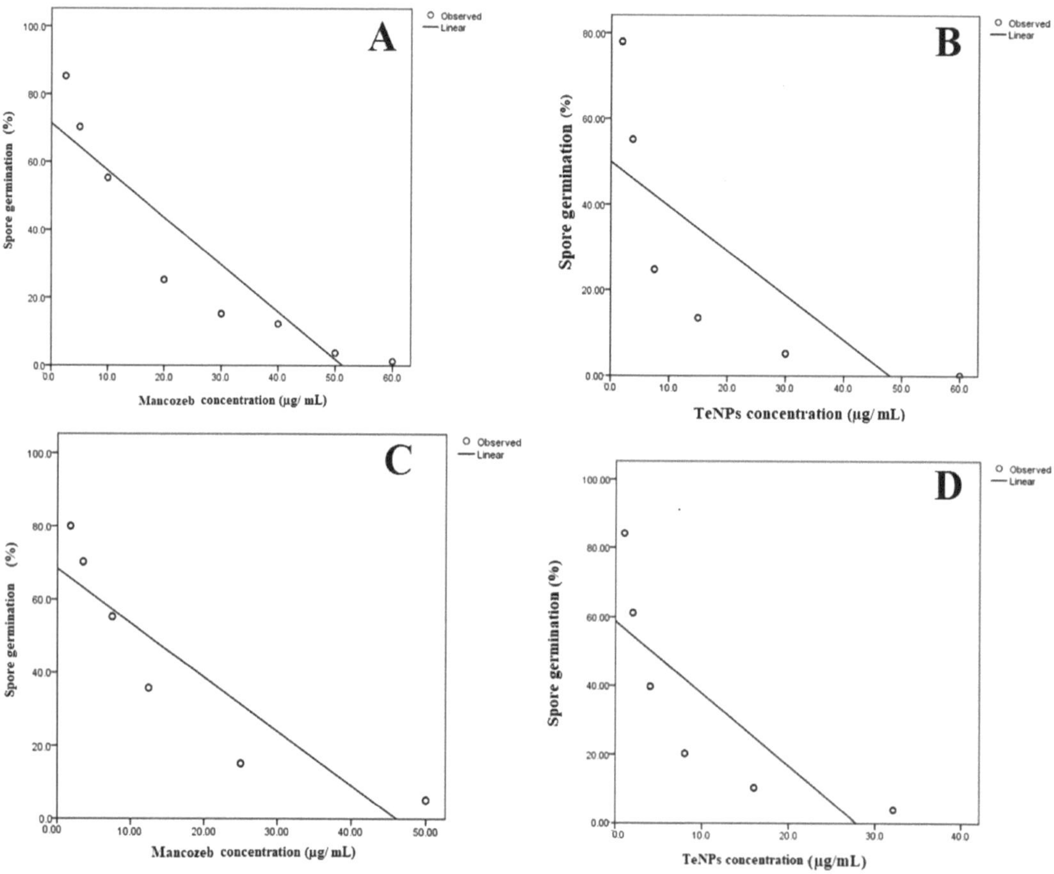

Figure 9. Linear regression model fit of spore germination. (**A**) *F. oxysporum* treated with mancozeb, (**B**) *F. oxysporum* treated with TeNPs, (**C**) *A. alternata* treated with mancozeb, (**D**) *A. alternata* treated with mancozeb.

3.4. Spore Morphology and Ultrastructure Analysis

The ultrastructural analysis of *F. oxysporum* and *A. alternata* spores was investigated by electron microscopy. Generally, spores in control experiments appeared intact with well-defined ultrastructure and smooth cell walls, whereas those exposed to biogenic TeNP exhibited prominent ultrastructure alternations (Figure 10). TEM observations indicated that the treated fungal spores seem to be deformed with ruptured walls in both investigated phytopathogenic fungi. Results indicated more deformation in the case of *A. alternata* than in *F. oxysporum*.

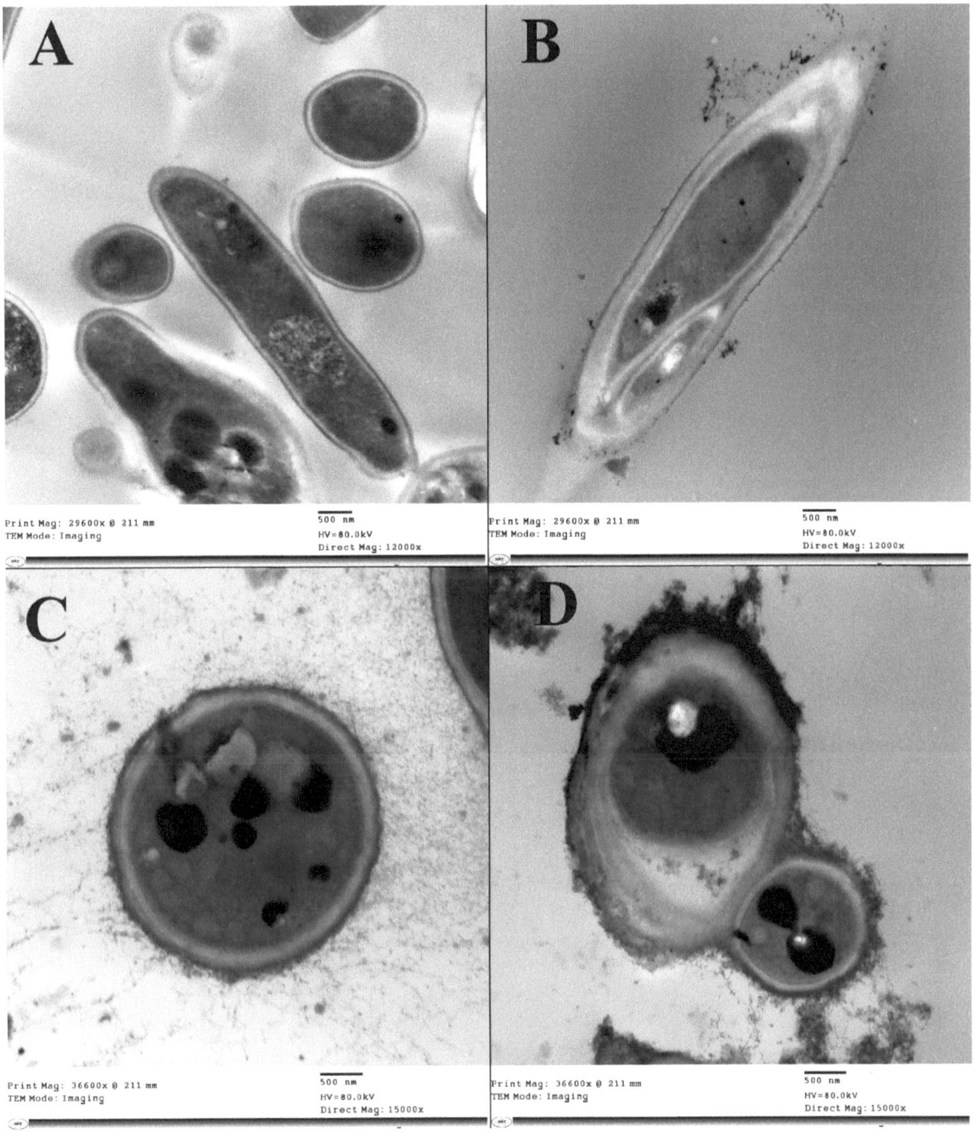

Figure 10. TEM micrographs showing the ultrastructure alternations in fungal spores upon exposure to the biogenic TeNP produced by *Gayadomonas* sp. TNPM15. (**A**) Untreated *F. oxysporum* (Control), (**B**) TeNPs-treated *F. oxysporum*, (**C**) Untreated *A. alternata* (Control), (**D**) TeNPs-treated *A. alternata.*

3.5. Leakage of Proteins and DNA

To gain further insight into the antifungal activity of TeNPs, we studied the probable leakage of proteins and DNA in TeNPs-treated spores compared with untreated control. The results revealed a significant increase in extracellular protein and DNA levels, indicating increased permeability of spore membrane and leakage of an intracellular substance (Figure 11). Likewise, TeNPs were more effective against *A. alternata* than *F. oxysporum* in the disruption of spore membranes.

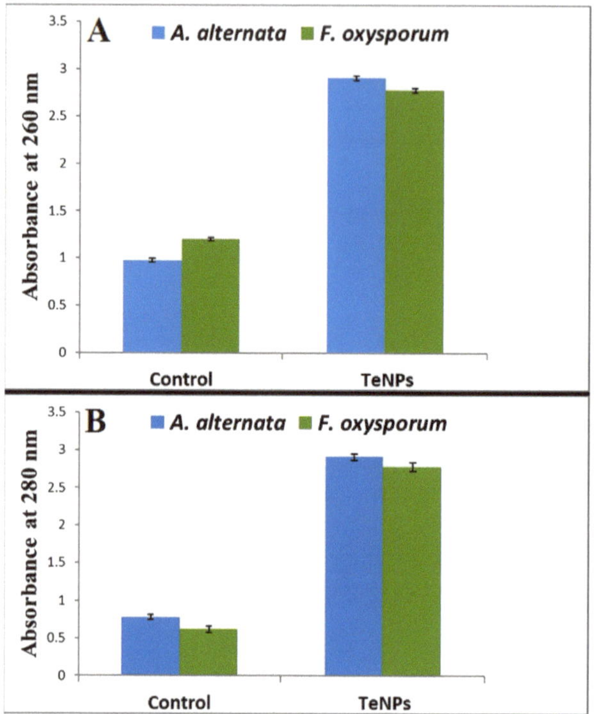

Figure 11. Effect of the biogenic TeNPs produced by *Gayadomonas* sp. TNPM15. on the leakage of proteins (**A**) and DNA (**B**) from spore suspension of *F. oxysporum* and *A. alternata*.

3.6. Impact of TeNPs on Plant Cell Wall-Degrading Enzymes

Application of the TeNPs to the growth medium at half of their MIC values resulted in significant inhibition of cellulase, pectinase, and xylanase activities in both investigated phytopathogenic fungi (Table 1). The results demonstrated about 78.2 and 84.9% inhibition of cellulase activity in treated *F. oxysporum* AUMC 10313 and *A. alternata* AUMC 3882, respectively, whereas the biogenic TeNPs reduced their pectinase activity by 51.3 and 59.3%, respectively. On the other hand, TeNPs exerted a relatively tenuous inhibitory effect against xylanase activity compared with that against cellulase and pectinase.

Table 1. Impact of the biogenic TeNPs produced by *Gayadomonas* sp. TNPM15 on the plant cell wall-degrading enzymes of *F. oxysporum* AUMC 10313 and *A. alternata* AUMC 3882.

Phytopathogen	Relative Enzyme Activity (%)		
	Cellulase	Pectinase	Xylanase
F. oxysporum AUMC 10313	21.8 ± 2.4	48.7 ± 1.6	72.1 ± 1.1
A. alternata AUMC 3882	15.1 ± 2.7	40.7 ± 1.9	70.7 ± 1.3

3.7. Cytotoxicity Assay

In the present study, the cytotoxicity and biocompatibility of the bacterial-mediated biogenic TeNPs were tested, in vitro, against HSF cell lines. Indeed, the SRB assay indicated that the biogenic TeNPs showed no significant cytotoxic effect towards HSF cells at concentrations up to 80 μg/mL. The results revealed that the half-maximal inhibitory concentration (IC_{50}) value of the biofabricated TeNPs against HSF cells was 125 μg/mL.

4. Discussion

The exploitation of the biomineralization potential of some tellurite-resistant bacteria in the bioconversion of chalcogen oxyanions is considered a valuable route to develop green strategies to fabricate unique nanoscale materials. In this investigation, the tellurite-resistant marine bacterium, *Gayadomonas* sp. TNPM15, was isolated from mangrove sediments on a tellurite-containing culture media. The appearance of the black color was the initial indication of the mineralization of tellurite to elemental tellurium, and TEM observations confirmed the formation of intracellular TeNPs. Mangrove has been reported as a highly productive ecosystem with noteworthy microbial diversity [62,63]. Likewise, the tellurite-resistant *Raoultella* sp. strain WAY isolated from wastewater reduced the Te(IV) forming elemental tellurium as black colonies exhibited the potentiality for the green synthesis of tellurium nanorods [64]. Previous studies have reported the isolation of tellurite-reducing bacteria that produce TeNPs from various natural habitats [65–68].

In our study, the biogenic TeNPs were produced as intracellular nanorods. These intracellular nanostructures were recovered following the lysis of bacterial cells by sonication. The size and shape of the purified bioinspired nanostructures were determined by TEM and SEM, which confirmed the production of nanorods. Furthermore, EDX analysis revealed that the prepared nanorods were composed mainly of tellurium, in which organic compounds were incorporated due to the biosynthesis process. The presence of organic matter, including proteins in the biogenic TeNPs, was evidenced by FTIR analysis. Bands at 1656 cm^{-1} and 1539 cm^{-1} are mainly associated with a C-O stretching peptide bonds from amide I and N-H bending and C-N stretching vibration from amide II, respectively [69–71]. The band at 2933 cm^{-1} donates the C-H asymmetric stretching of alkanes [72]. In addition, the peak at 1053 cm^{-1} could correspond to the C–N stretching vibration of the amine, while the band at 1410 cm^{-1} corresponds to the N–H bending of primary amides due to carbonyl stretch in proteins [73]. It has been suggested that proteins may be involved in the biosynthesis of various metal nanoparticles as reducing and capping agents that stabilize the biogenic nanoparticles. In agreement with our study, the FTIR spectrum of silver nanoparticles by *Penicillium* sp. revealed bands at 1644, 1538, and 3290 cm^{-1} correspond to the binding vibrations of amide I and amide II band of protein, respectively [74]. These results are in harmony with previous investigations reporting the presence of organics such as proteins on the surface of various biogenic nanoparticles as capping and stabilizing agents [19,54,75,76]. Furthermore, the nanoparticles were characterized by XRD, and their pattern confirmed that *Gayadomonas* sp. TNPM15 produced crystalline nanoparticles of elemental tellurium nanoscale hexagonal crystal structure [77,78]. In addition, Raman spectra revealed the appearance of the specific peaks attributed to the A1 and E phonon modes of vibration for tellurium [79,80]. The biosynthesis of tellurium nanorods by *Gayadomonas* sp. TNPM15 supports the findings of previous works that demonstrated the biosynthesis of rod-shaped TeNPs by various microorganisms [2,81–83]. Our findings agree with most previous literature reporting the intracellular biosynthesis of TeNPs [25,26,75,84]. On the other hand, the extracellular biosynthesis of TeNPs is infrequently reported using certain bacterial species such as *Rhodobacter capsulatus* and *Shewanella oneidensis* [22,85].

The findings of the current study illustrate the potent antifungal effect of the biogenic TeNPs-produced *Gayadomonas* sp. TNPM15 against two important fungal phytopathogens. Practically, the biogenic tellurium nanorods inhibited the spore germination of both investigated fungi and were evidenced to cause significant malformation and ultrastructural alternation on fungal spores by electron microscopy. Due to the rupture of cell walls and disintegration of membranes, the spores lost their intracellular content. This disintegration of the membrane was affirmed by increasing in the leakage of DNA and proteins. These findings concord with previous reports, suggesting that the antimicrobial effect of various nanoparticles may be attributed to the disintegration of membranes and leakage of the intracellular content, as well as the rupture of cell walls [2,86–91]. Interestingly, the biogenic TeNPs produced by *Gayadomonas* sp. TNPM15 were more effective than mancozeb against *A. alternata*, and their antifungal activity is more or less comparable to the mancozeb (positive control) against *F. oxysporum*. Mancozeb is a broad-spectrum fungicide that is used worldwide for fungal disease control. For decades, it has been an invaluable tool to farmers all over the world faced with the challenge of controlling fungal pests in their crops [92]. Likewise, several previous studies used mancozeb as a standard positive control for evaluating the antifungal efficacy of various biological and chemical antifungal agents [69,93,94]. In addition to their direct antifungal effects, TeNPs caused a remarkable inhibition of the plant cell wall-degrading enzymes. This inhibitory potential against cellulase, pectinase, and xylanase could result in the attenuation of the virulence of the exposed phytopathogens. Cellulose, hemicellulose, lignin, and pectin are the most common polysaccharides in plant cell walls. Thus phytopathogens secrete several extracellular enzymes responsible for cell wall degradation such as cellulase, pectinase, and xylanase to facilitate their invasion, resulting in adverse effects on the infected plants [95]. Accordingly, attempts to inhibit the cell wall, degrading enzymes secreted by phytopathogens, may restrict and attenuate their invasion, colonization, and expansion. Regarding the potential cytotoxicity, the cytotoxic effects of the biogenic TeNPs produced by *Gayadomonas* sp. TNPM15 were assessed in vitro against HSF cells. It is worth mentioning that TeNPs exhibited excellent biocompatibility towards HSF cells with no apparent toxicity at concentrations up to 80 μg/mL. By comparing the effective antifungal concentrations of TeNPs produced by *Gayadomonas* sp. TNPM15 with the IC_{50} value (125 μg/mL) against HSF cells, it can be concluded that the biogenic TeNPs produced by *Gayadomonas* sp. TNPM15 could be used as a safe non-toxic antifungal agent against phytopathogenic fungi.

5. Conclusions

The antifungal potential of the biogenic rod-shaped TeNPs produced by *Gayadomonas* sp. TNPM15 provides a promising solution to combat fungal infections such as root rot and black spot diseases caused by *Fusarium* and *Alternaria* spp., respectively. The biosynthesized TeNPs induce prominent ultrastructure alternations such as deformation and damage in the cell wall. Furthermore, the antifungal potential of the biogenic TeNPs is attributed to their inhibitory effect of fungal spore germination and disruption of membrane permeability and integrity, as evidenced by TEM and increased leakage of the intracellular components, in addition to their inhibitory effects on plant cell wall-degrading enzymes. The cytotoxicity assessments revealed their good biocompatibility and low toxicity toward HSF cells. We concluded that the biogenic tellurium nanorods produced by *Gayadomonas* sp. TNPM15 are a potential candidate as a robust antifungal with non-toxic doses.

Author Contributions: Conceptualization, M.N.A.E.-G. and M.G.F.; Methodology, M.N.A.E.-G. and M.G.F.; Software, M.G.F.; Formal analysis, M.N.A.E.-G. and M.G.F.; Investigation, M.G.F.; Resources, S.A.H., S.M.K., R.M.E. and M.G.F.; Writing—original draft, M.N.A.E.-G.; Writing – review & editing, M.G.F.; Visualization, M.G.F.; Funding acquisition, S.A.H., S.M.K. and R.M.E. All authors have read and agreed to the published version of the manuscript.

Funding: Research was funded by Princess Nourah bint Abdulrahman University Researchers Supporting Project number (PNURSP2023R214), Princess Nourah bint Abdulrahman University, Riyadh, Saudi Arabia.

Data Availability Statement: The essential data supporting the reported results are contained in this study. All other data are available on request from the corresponding authors.

Acknowledgments: The authors acknowledge the Princess Nourah bint Abdulrahman University Researchers Supporting Project number (PNURSP2023R214), Princess Nourah bint Abdulrahman University, Riyadh, Saudi Arabia.

Conflicts of Interest: The authors declare no conflict of interest.

References

1. Paras; Yadav, K.; Kumar, P.; Teja, D.R.; Chakraborty, S.; Chakraborty, M.; Mohapatra, S.S.; Sahoo, A.; Chou, M.M.C.; Liang, C.-T.; et al. A Review on Low-Dimensional Nanomaterials: Nanofabrication, Characterization and Applications. *Nanomatererials* **2022**, *13*, 160. [CrossRef]
2. Tang, A.; Ren, Q.; Wu, Y.; Wu, C.; Cheng, Y. Investigation into the Antibacterial Mechanism of Biogenic Tellurium Nanoparticles and Precursor Tellurite. *Int. J. Mol. Sci.* **2022**, *23*, 11697. [CrossRef]
3. Domínguez-Álvarez, E.; Rácz, B.; Marć, M.A.; Nasim, M.J.; Szemerédi, N.; Viktorová, J.; Jacob, C.; Spengler, G. Selenium and Tellurium in the Development of Novel Small Molecules and Nanoparticles as Cancer Multidrug Resistance Reversal Agents. *Drug Resist. Updat.* **2022**, *63*, 100844. [CrossRef]
4. Shahbaz, M.; Fatima, N.; Mashwani, Z.U.R.; Akram, A.; ul Haq, E.; Mehak, A.; Abasi, F.; Ajmal, M.; Yousaf, T.; Raja, N.I.; et al. Effect of Phytosynthesized Selenium and Cerium Oxide Nanoparticles on Wheat (*Triticum aestivum* L.) against Stripe Rust Disease. *Molecules* **2022**, *27*, 8149. [CrossRef]
5. Salem, M.F.; Abd-Elraoof, W.A.; Tayel, A.A.; Alzuaibr, F.M.; Abonama, O.M. Antifungal Application of Biosynthesized Selenium Nanoparticles with Pomegranate Peels and Nanochitosan as Edible Coatings for Citrus Green Mold Protection. *J. Nanobiotechnol.* **2022**, *20*, 182. [CrossRef]
6. Malandrakis, A.A.; Kavroulakis, N.; Chrysikopoulos, C.V. Zinc Nanoparticles: Mode of Action and Efficacy against Boscalid-Resistant *Alternaria alternata* Isolates. *Sci. Total Environ.* **2022**, *829*, 154638. [CrossRef]
7. Sahin, F.; Celik, N.; Ceylan, A.; Ruzi, M.; Onses, M.S. One-Step Green Fabrication of Antimicrobial Surfaces via in Situ Growth of Copper Oxide Nanoparticles. *ACS Omega* **2022**, *7*, 26504–26513. [CrossRef]
8. Ibarra-Laclette, E.; Blaz, J.; Pérez-Torres, C.A.; Villafán, E.; Lamelas, A.; Rosas-Saito, G.; Ibarra-Juárez, L.A.; García-ávila, C.D.J.; Martínez-Enriquez, A.I.; Pariona, N. Antifungal Effect of Copper Nanoparticles against *Fusarium kuroshium*, an Obligate Symbiont of *Euwallacea kuroshio* Ambrosia Beetle. *J. Fungi* **2022**, *8*, 347. [CrossRef]
9. Kamel, S.M.; Elgobashy, S.F.; Omara, R.I.; Derbalah, A.S.; Abdelfatah, M.; El-Shaer, A.; Al-Askar, A.A.; Abdelkhalek, A.; Abd-Elsalam, K.A.; Essa, T.; et al. Antifungal Activity of Copper Oxide Nanoparticles against Root Rot Disease in Cucumber. *J. Fungi* **2022**, *8*, 911. [CrossRef]
10. Da Silva, M.A.; García-Martínez, J.C.; Páez, P.L.; Paraje, M.G. Fungicidal and Antibiofilm Activities of Gold Nanoparticles on *Candida tropicalis*. *Nanomedicine* **2022**, *17*, 1663–1676. [CrossRef]
11. Soltani Nejad, M.; Samandari Najafabadi, N.; Aghighi, S.; Pakina, E.; Zargar, M. Evaluation of *Phoma* Sp. Biomass as an Endophytic Fungus for Synthesis of Extracellular Gold Nanoparticles with Antibacterial and Antifungal Properties. *Molecules* **2022**, *27*, 1181. [CrossRef]
12. Zambonino, M.C.; Quizhpe, E.M.; Jaramillo, F.E.; Rahman, A.; Vispo, N.S.; Jeffryes, C.; Dahoumane, S.A. Green Synthesis of Selenium and Tellurium Nanoparticles: Current Trends, Biological Properties and Biomedical Applications. *Int. J. Mol. Sci.* **2021**, *22*, 989. [CrossRef]
13. Ashraf, M.W.; Haider, S.I.; Solangi, A.R.; Memon, A.F. 6 Toxicity of tellurium and its compounds. In *Tellurium Chemistry*; Banik, B.K., Bajpai, S., Eds.; De Gruyter: Berlin, Germany; Boston, MA, USA, 2022. [CrossRef]
14. Espinosa-Ortiz, E.J.; Rene, E.R.; Guyot, F.; van Hullebusch, E.D.; Lens, P.N.L. Biomineralization of Tellurium and Selenium-Tellurium Nanoparticles by the White-Rot Fungus *Phanerochaete chrysosporium*. *Int. Biodeterior. Biodegrad.* **2017**, *124*, 258–266. [CrossRef]
15. Zwar, I.P.; Trotta, C.D.V.; Ziotti, A.B.S.; Neto, M.L.; Araújo, W.L.; de Melo, I.S.; Ottoni, C.A.; de Souza, A.O. Biosynthesis of Silver Nanoparticles Using Actinomycetes, Phytotoxicity on Rice Seeds, and Potential Application in the Biocontrol of Phytopathogens. *J. Basic Microbiol.* **2023**, *63*, 64–74. [CrossRef]
16. Nguyen, D.T.C.; Van Tran, T.; Nguyen, T.T.T.; Nguyen, D.H.; Alhassan, M.; Lee, T. New Frontiers of Invasive Plants for Biosynthesis of Nanoparticles towards Biomedical Applications: A Review. *Sci. Total Environ.* **2023**, *857*, 159278. [CrossRef]
17. Sivakumar, S.R.; Manimaran, K.; Govindasamy, M.; Alzahrani, F.M.; Alsaiari, N.S. Green Synthesis and Characterization of CuO Nanoparticles Using *Halymenia dilatata* Extract and Its Evaluation of Antimicrobial, Anticancer Activity. *Biomass Convers. Biorefinery* **2023**, *1*, 1–10. [CrossRef]
18. Liang, X.; Perez, M.A.M.J.; Nwoko, K.C.; Egbers, P.; Feldmann, J.; Csetenyi, L.; Gadd, G.M. Fungal Formation of Selenium and Tellurium Nanoparticles. *Appl. Microbiol. Biotechnol.* **2019**, *103*, 7241–7259. [CrossRef]
19. Abo Elsoud, M.M.; Al-Hagar, O.E.A.; Abdelkhalek, E.S.; Sidkey, N.M. Synthesis and Investigations on Tellurium Myconanoparticles. *Biotechnol. Rep.* **2018**, *18*, e00247. [CrossRef]
20. Ao, B.; He, F.; Lv, J.; Tu, J.; Tan, Z.; Jiang, H.; Shi, X.; Li, J.; Hou, J.; Hu, Y.; et al. Green Synthesis of Biogenetic Te(0) Nanoparticles by High Tellurite Tolerance Fungus *Mortierella* Sp. AB1 with Antibacterial Activity. *Front. Microbiol.* **2022**, *13*, 3973. [CrossRef]
21. Lin, Z.H.; Lee, C.H.; Chang, H.Y.; Chang, H.T. Antibacterial Activities of Tellurium Nanomaterials. *Chem. Asian J.* **2012**, *7*, 930–934. [CrossRef]
22. Borghese, R.; Baccolini, C.; Francia, F.; Sabatino, P.; Turner, R.J.; Zannoni, D. Reduction of Chalcogen Oxyanions and Generation of Nanoprecipitates by the Photosynthetic Bacterium *Rhodobacter capsulatus*. *J. Hazard. Mater.* **2014**, *269*, 24–30. [CrossRef]
23. Borghese, R.; Brucale, M.; Fortunato, G.; Lanzi, M.; Mezzi, A.; Valle, F.; Cavallini, M.; Zannoni, D. Extracellular Production of Tellurium Nanoparticles by the Photosynthetic Bacterium *Rhodobacter capsulatus*. *J. Hazard. Mater.* **2016**, *309*, 202–209. [CrossRef]

24. Beleneva, I.A.; Kharchenko, U.V.; Kukhlevsky, A.D.; Boroda, A.V.; Izotov, N.V.; Gnedenkov, A.S.; Egorkin, V.S. Biogenic Synthesis of Selenium and Tellurium Nanoparticles by Marine Bacteria and Their Biological Activity. *World J. Microbiol. Biotechnol.* **2022**, *38*, 188. [CrossRef]
25. Presentato, A.; Piacenza, E.; Darbandi, A.; Anikovskiy, M.; Cappelletti, M.; Zannoni, D.; Turner, R.J. Assembly, Growth and Conductive Properties of Tellurium Nanorods Produced by *Rhodococcus aetherivorans* BCP1. *Sci. Rep.* **2018**, *8*, 3923. [CrossRef]
26. Mirjani, R.; Faramarzi, M.A.; Sharifzadeh, M.; Setayesh, N.; Khoshayand, M.R.; Shahverdi, A.R. Biosynthesis of Tellurium Nanoparticles by *Lactobacillus plantarum* and the Effect of Nanoparticle-Enriched Probiotics on the Lipid Profiles of Mice. *IET Nanobiotechnol.* **2015**, *9*, 300–305. [CrossRef]
27. Ghosh, S.; Sarkar, B.; Kaushik, A.; Mostafavi, E. Nanobiotechnological Prospects of Probiotic Microflora: Synthesis, Mechanism, and Applications. *Sci. Total Environ.* **2022**, *838*, 156212. [CrossRef]
28. Ramos-Ruiz, A.; Field, J.A.; Wilkening, J.V.; Sierra-Alvarez, R. Recovery of Elemental Tellurium Nanoparticles by the Reduction of Tellurium Oxyanions in a Methanogenic Microbial Consortium. *Environ. Sci. Technol.* **2016**, *50*, 1492–1500. [CrossRef]
29. Ancheeva, E.; Daletos, G.; Proksch, P. Lead Compounds from Mangrove-Associated Microorganisms. *Mar. Drugs* **2018**, *16*, 319. [CrossRef]
30. Debasmita, D.; Sourav, G.; Srimoyee, B. A Review of Metal Resistance Mechanisms by Mangrove Bacteria. *Res. J. Biotechnol.* **2022**, *17*, 209–215. [CrossRef]
31. Alongi, D.M.; Pfitzner, J.; Trott, L.A.; Tirendi, F.; Dixon, P.; Klumpp, D.W. Rapid Sediment Accumulation and Microbial Mineralization in Forests of the Mangrove *Kandelia candel* in the Jiulongjiang Estuary, China. *Estuar. Coast. Shelf Sci.* **2005**, *63*, 605–618. [CrossRef]
32. Holguin, G.; Vazquez, P.; Bashan, Y. The Role of Sediment Microorganisms in the Productivity, Conservation, and Rehabilitation of Mangrove Ecosystems: An Overview. *Biol. Fertil. Soils* **2001**, *33*, 265–278. [CrossRef]
33. Zhu, D.H.; Song, Q.L.; Nie, F.H.; Wei, W.; Chen, M.M.; Zhang, M.; Lin, H.Y.; Kang, D.J.; Chen, Z.B.; Hay, A.G.; et al. Effects of Environmental and Spatial Variables on Bacteria in Zhanjiang Mangrove Sediments. *Curr. Microbiol.* **2022**, *79*, 97. [CrossRef]
34. Li, K.; Chen, S.; Pang, X.; Cai, J.; Zhang, X.; Liu, Y.; Zhu, Y.; Zhou, X. Natural Products from Mangrove Sediments-Derived Microbes: Structural Diversity, Bioactivities, Biosynthesis, and Total Synthesis. *Eur. J. Med. Chem.* **2022**, *230*, 114117. [CrossRef]
35. Gonzalez-Acosta, B.; Bashan, Y.; Hernandez-Saavedra, N.Y.; Ascencio, F.; De La Cruz-Agüero, G. Seasonal Seawater Temperature as the Major Determinant for Populations of Culturable Bacteria in the Sediments of an Intact Mangrove in an Arid Region. *FEMS Microbiol. Ecol.* **2006**, *55*, 311–321. [CrossRef]
36. Mishra, R.R.; Prajapati, S.; Das, J.; Dangar, T.K.; Das, N.; Thatoi, H. Reduction of Selenite to Red Elemental Selenium by Moderately Halotolerant *Bacillus megaterium* Strains Isolated from Bhitarkanika Mangrove Soil and Characterization of Reduced Product. *Chemosphere* **2011**, *84*, 1231–1237. [CrossRef]
37. Farias, P.; Francisco, R.; Morais, P.V. Potential of Tellurite Resistance in Heterotrophic Bacteria from Mining Environments. *iScience* **2022**, *25*, 104566. [CrossRef]
38. Wang, Z.; Bu, Y.; Zhao, Y.; Zhang, Z.; Liu, L.; Zhou, H. Morphology-Tunable Tellurium Nanomaterials Produced by the Tellurite-Reducing Bacterium *Lysinibacillus* Sp. ZYM-1. *Environ. Sci. Pollut. Res.* **2018**, *25*, 20756–20768. [CrossRef]
39. Horiike, T.; Otsuka, O.; Tanaka, Y.; Terahara, T.; Imada, C.; Yamashita, M. Diversity of Salt-Tolerant Tellurate-Reducing Bacteria in a Marine Environment. *J. Gen. Appl. Microbiol.* **2019**, *65*, 246–253. [CrossRef]
40. Wang, Z.; Yi, X.; Liu, Y.; Zhou, H. Complete Genome Sequence of a Tellurate Reducing Bacteria *Sporosarcina* Sp. Te-1 Isolated from Bohai Sea. *Mar. Genom.* **2021**, *60*, 100888. [CrossRef]
41. Soda, S.; Ma, W.; Kuroda, M.; Nishikawa, H.; Zhang, Y.; Ike, M. Characterization of Moderately Halotolerant Selenate- and Tellurite-Reducing Bacteria Isolated from Brackish Areas in Osaka. *Biosci. Biotechnol. Biochem.* **2018**, *82*, 173–181. [CrossRef]
42. Davies, C.R.; Wohlgemuth, F.; Young, T.; Violet, J.; Dickinson, M.; Sanders, J.W.; Vallieres, C.; Avery, S.V. Evolving Challenges and Strategies for Fungal Control in the Food Supply Chain. *Fungal Biol. Rev.* **2021**, *36*, 15–26. [CrossRef]
43. Husaini, A.M.; Sakina, A.; Cambay, S.R. Host-Pathogen Interaction in *Fusarium oxysporum* Infections: Where Do We Stand? *Mol. Plant-Microbe Interact.* **2018**, *31*, 889–898. [CrossRef]
44. Mallik, S.K.; Shahi, N.; Joshi, N.; Pant, K.; Kala, K.; Chandra, S.; Sarma, D. The Emergence of Zoonotic *Fusarium oxysporum* Infection in Captive-Reared Fingerlings of Golden Mahseer, *Tor putitora* (Hamilton, 1822) from the Central Himalayan Region of India. *Transbound. Emerg. Dis.* **2020**, *67*, 555–563. [CrossRef]
45. Arasaki, R.; Tanaka, S.; Okawa, K.; Tanaka, Y.; Inoue, T.; Kobayashi, S.; Ito, A.; Maruyama-Inoue, M.; Yamaguchi, T.; Muraosa, Y.; et al. Endophthalmitis Outbreak Caused by *Fusarium oxysporum* after Cataract Surgery. *Am. J. Ophthalmol. Case Rep.* **2022**, *26*, 101397. [CrossRef]
46. Badali, H.; Cañete-Gibas, C.; Patterson, H.; Sanders, C.; Mermella, B.; Garcia, V.; Mele, J.; Fan, H.; Wiederhold, N.P. In Vitro Activity of Olorofim against Clinical Isolates of the *Fusarium oxysporum* and *Fusarium solani* Species Complexes. *Mycoses* **2021**, *64*, 748–752. [CrossRef]
47. Martinko, K.; Ivanković, S.; Lazarević, B.; Đermić, E.; Đermić, D. Control of Early Blight Fungus (*Alternaria alternata*) in Tomato by Boric and Phenylboronic Acid. *Antibiotics* **2022**, *11*, 320. [CrossRef]
48. Kgatle, M.G.; Flett, B.; Truter, M.; Aveling, T.A.S. Control of Alternaria Leaf Blight Caused by *Alternaria alternata* on Sunflower Using Fungicides and *Bacillus amyloliquefaciens*. *Crop Prot.* **2020**, *132*, 105146. [CrossRef]

49. Ghosh, R.; Barman, S.; Khatun, J.; Mandal, N.C. Biological Control of *Alternaria alternata* Causing Leaf Spot Disease of Aloe Vera Using Two Strains of Rhizobacteria. *Biol. Control* **2016**, *97*, 102–108. [CrossRef]
50. De Mers, M. *Alternaria alternata* as Endophyte and Pathogen. *Microbiology* **2022**, *168*, 001153. [CrossRef]
51. Alastruey-izquierdo, A.; Cuesta, I.; Ros, L.; Mellado, E.; Rodriguez-tudela, J.L. Antifungal Susceptibility Profile of Clinical *Alternaria* spp. Identified by Molecular Methods. *J. Antimicrob. Chemother.* **2011**, *66*, 2585–2587. [CrossRef]
52. Kim, G.; Yoo, S.J.; Yoo, J.R.; Seo, K.B. The First Case Report of Thorn-Induced *Alternaria alternata* Infection of the Hand in an Immunocompetent Host. *BMC Infect. Dis.* **2022**, *22*, 304. [CrossRef]
53. Hattab, Z.; Ben Lasfar, N.; Abid, M.; Bellazreg, F.; Fathallah, A.; Hachfi, W.; Letaief, A. *Alternaria alternata* Infection Causing Rhinosinusitis and Orbital Involvement in an Immunocompetent Patient. *New Microbes New Infect.* **2019**, *32*, 100561. [CrossRef]
54. Medina Cruz, D.; Tien-Street, W.; Zhang, B.; Huang, X.; Vernet Crua, A.; Nieto-Argüello, A.; Cholula-Díaz, J.L.; Martínez, L.; Huttel, Y.; González, M.U.; et al. Citric Juice-Mediated Synthesis of Tellurium Nanoparticles with Antimicrobial and Anticancer Properties. *Green Chem.* **2019**, *21*, 1982–1998. [CrossRef]
55. Shah, V.; Medina-Cruz, D.; Vernet-Crua, A.; Truong, L.B.; Sotelo, E.; Mostafavi, E.; González, M.U.; García-Martín, J.M.; Cholula-Díaz, J.L.; Webster, T.J. Pepper-Mediated Green Synthesis of Selenium and Tellurium Nanoparticles with Antibacterial and Anticancer Potential. *J. Funct. Biomater.* **2022**, *14*, 24. [CrossRef]
56. Morena, A.G.; Bassegoda, A.; Hoyo, J.; Tzanov, T. Hybrid Tellurium-Lignin Nanoparticles with Enhanced Antibacterial Properties. *ACS Appl. Mater. Interfaces* **2021**, *13*, 14885–14893. [CrossRef]
57. Zonaro, E.; Lampis, S.; Turner, R.J.; Junaid, S.; Vallini, G. Biogenic Selenium and Tellurium Nanoparticles Synthesized by Environmental Microbial Isolates Efficaciously Inhibit Bacterial Planktonic Cultures and Biofilms. *Front. Microbiol.* **2015**, *6*, 584. [CrossRef]
58. Zare, B.; Sepehrizadeh, Z.; Faramarzi, M.A.; Soltany-Rezaee-Rad, M.; Rezaie, S.; Shahverdi, A.R. Antifungal Activity of Biogenic Tellurium Nanoparticles against *Candida albicans* and Its Effects on *Squalene monooxygenase* Gene Expression. *Biotechnol. Appl. Biochem.* **2014**, *61*, 395–400. [CrossRef]
59. Yoon, S.H.; Ha, S.M.; Kwon, S.; Lim, J.; Kim, Y.; Seo, H.; Chun, J. Introducing EzBioCloud: A Taxonomically United Database of 16S RRNA Gene Sequences and Whole-Genome Assemblies. *Int. J. Syst. Evol. Microbiol.* **2017**, *67*, 1613–1617. [CrossRef]
60. Khalil, N.M.; Abd El-Ghany, M.N.; Rodríguez-Couto, S. Antifungal and Anti-Mycotoxin Efficacy of Biogenic Silver Nanoparticles Produced by *Fusarium chlamydosporum* and *Penicillium chrysogenum* at Non-Cytotoxic Doses. *Chemosphere* **2019**, *218*, 477–486. [CrossRef]
61. Miller, G.L. Use of Dinitrosalicylic Acid Reagent for Determination of Reducing Sugar. *Anal. Chem.* **1959**, *31*, 426–428. [CrossRef]
62. Liang, J.-B.; Chen, Y.-Q.; Lan, C.-Y.; Tam, N.F.Y.; Zan, Q.-J.; Huang, L.-N. Recovery of Novel Bacterial Diversity from Mangrove Sediment. *Mar. Biol.* **2007**, *150*, 739–747. [CrossRef]
63. Farahat, M.G. Enhancement of β-Cyclodextrin Production and Fabrication of Edible Antimicrobial Films Incorporated with Clove Essential Oil/β-Cyclodextrin Inclusion Complex. *Microbiol. Biotechnol. Lett* **2020**, *48*, 12–23. [CrossRef]
64. Nguyen, V.K.; Choi, W.; Ha, Y.; Gu, Y.; Lee, C.; Park, J.; Jang, G.; Shin, C.; Cho, S. Microbial Tellurite Reduction and Production of Elemental Tellurium Nanoparticles by Novel Bacteria Isolated from Wastewater. *J. Ind. Eng. Chem.* **2019**, *78*, 246–256. [CrossRef]
65. Baesman, S.M.; Stolz, J.F.; Kulp, T.R.; Oremland, R.S. Enrichment and Isolation of *Bacillus beveridgei* sp. Nov., a Facultative Anaerobic Haloalkaliphile from Mono Lake, California, That Respires Oxyanions of Tellurium, Selenium, and Arsenic. *Extremophiles* **2009**, *13*, 695–705. [CrossRef]
66. Lloyd-Jones, G.; Williamson, W.M.; Slootweg, T. The Te-Assay: A Black and White Method for Environmental Sample Pre-Screening Exploiting Tellurite Reduction. *J. Microbiol. Methods* **2006**, *67*, 549–556. [CrossRef]
67. Chien, C.C.; Jiang, M.H.; Tsai, M.R.; Chien, C.C. Isolation and Characterization of an Environmental Cadmium- and Tellurite-Resistant *Pseudomonas* Strain. *Environ. Toxicol. Chem.* **2011**, *30*, 2202–2207. [CrossRef]
68. Amoozegar, M.A.; Ashengroph, M.; Malekzadeh, F.; Reza Razavi, M.; Naddaf, S.; Kabiri, M. Isolation and Initial Characterization of the Tellurite Reducing Moderately Halophilic Bacterium, *Salinicoccus* Sp. Strain QW6. *Microbiol. Res.* **2008**, *163*, 456–465. [CrossRef]
69. Gutierrez-Monsalve, J.A.; Mosquera, S.; González-Jaramillo, L.M.; Mira, J.J.; Villegas-Escobar, V. Effective Control of Black Sigatoka Disease Using a Microbial Fungicide Based on *Bacillus subtilis* EA-CB0015 Culture. *Biol. Control* **2015**, *87*, 39–46. [CrossRef]
70. Araya-Castro, K.; Chao, T.C.; Durán-Vinet, B.; Cisternas, C.; Ciudad, G.; Rubilar, O. Green Synthesis of Copper Oxide Nanoparticles Using Protein Fractions from an Aqueous Extract of Brown Algae *Macrocystis pyrifera*. *Processes* **2020**, *9*, 78. [CrossRef]
71. Labeeb, M.; Badr, A.; Haroun, S.A.; Mattar, M.Z.; El-kholy, A.S. Ultrastructural and Molecular Implications of Ecofriendly Made Silver Nanoparticles Treatments in Pea (*Pisum sativum* L.). *J. Genet. Eng. Biotechnol.* **2022**, *20*, 5. [CrossRef]
72. Rizwana, H.; Alwhibi, M.S.; Al-Judaie, R.A.; Aldehaish, H.A.; Alsaggabi, N.S. Sunlight-Mediated Green Synthesis of Silver Nanoparticles Using the Berries of *Ribes rubrum* (Red Currants): Characterisation and Evaluation of Their Antifungal and Antibacterial Activities. *Molecules* **2022**, *27*, 2186. [CrossRef]
73. Barabadi, H.; Kobarfard, F.; Vahidi, H. Biosynthesis and Characterization of Biogenic Tellurium Nanoparticles by Using *Penicillium chrysogenum* PTCC 5031: A Novel Approach in Gold Biotechnology. *Iran. J. Pharm. Res. IJPR* **2018**, *17*, 87.

74. Singh, D.; Rathod, V.; Ninganagouda, S.; Hiremath, J.; Singh, A.K.; Mathew, J. Optimization and Characterization of Silver Nanoparticle by Endophytic Fungi *Penicillium* sp. Isolated from *Curcuma longa* (Turmeric) and Application Studies against MDR *E. coli* and *S. aureus*. *Bioinorg. Chem. Appl.* **2014**, *2014*, 408021. [CrossRef]
75. Zonaro, E.; Piacenza, E.; Presentato, A.; Monti, F.; Dell'Anna, R.; Lampis, S.; Vallini, G. *Ochrobactrum* sp. MPV1 from a Dump of Roasted Pyrites Can Be Exploited as Bacterial Catalyst for the Biogenesis of Selenium and Tellurium Nanoparticles. *Microb. Cell Fact.* **2017**, *16*, 215. [CrossRef]
76. Arab, F.; Mousavi-Kamazani, M.; Salavati-Niasari, M. Facile Sonochemical Synthesis of Tellurium and Tellurium Dioxide Nanoparticles: Reducing Te(IV) to Te via Ultrasonic Irradiation in Methanol. *Ultrason. Sonochem.* **2017**, *37*, 335–343. [CrossRef]
77. Song, J.M.; Lin, Y.Z.; Zhan, Y.J.; Tian, Y.C.; Liu, G.; Yu, S.H. Superlong High-Quality Tellurium Nanotubes: Synthesis, Characterization, and Optical Property. *Cryst. Growth Des.* **2008**, *8*, 1902–1908. [CrossRef]
78. Zare, B.; Faramarzi, M.A.; Sepehrizadeh, Z.; Shakibaie, M.; Rezaie, S.; Shahverdi, A.R. Biosynthesis and Recovery of Rod-Shaped Tellurium Nanoparticles and Their Bactericidal Activities. *Mater. Res. Bull.* **2012**, *47*, 3719–3725. [CrossRef]
79. Liu, J.W.; Chen, F.; Zhang, M.; Qi, H.; Zhang, C.L.; Yu, S.H. Rapid Microwave-Assisted Synthesis of Uniform Ultralong Te Nanowires, Optical Property, and Chemical Stability. *Langmuir* **2010**, *26*, 11372–11377. [CrossRef]
80. Manikandan, M.; Dhanuskodi, S.; Maheswari, N.; Muralidharan, G.; Revathi, C.; Rajendra Kumar, R.T.; Mohan Rao, G. High Performance Supercapacitor and Non-Enzymatic Hydrogen Peroxide Sensor Based on Tellurium Nanoparticles. *Sens. Bio-Sensing Res.* **2017**, *13*, 40–48. [CrossRef]
81. Najimi, S.; Shakibaie, M.; Jafari, E.; Ameri, A.; Rahimi, N.; Forootanfar, H.; Yazdanpanah, M.; Rahimi, H.R. Acute and Subacute Toxicities of Biogenic Tellurium Nanorods in Mice. *Regul. Toxicol. Pharmacol.* **2017**, *90*, 222–230. [CrossRef]
82. Alonso-Fernandes, E.; Fernández-Llamosas, H.; Cano, I.; Serrano-Pelejero, C.; Castro, L.; Díaz, E.; Carmona, M. Enhancing Tellurite and Selenite Bioconversions by Overexpressing a Methyltransferase from *Aromatoleum* sp. CIB. *Microb. Biotechnol.* **2022**, 1–16. [CrossRef]
83. Gomez-Gomez, B.; Corte-Rodríguez, M.; Perez-Corona, M.T.; Bettmer, J.; Montes-Bayón, M.; Madrid, Y. Combined Single Cell and Single Particle ICP-TQ-MS Analysis to Quantitatively Evaluate the Uptake and Biotransformation of Tellurium Nanoparticles in Bacteria. *Anal. Chim. Acta* **2020**, *1128*, 116–128. [CrossRef]
84. Srivastava, P.; Nikhil, E.V.R.; Bragança, J.M.; Kowshik, M. Anti-Bacterial TeNPs Biosynthesized by Haloarcheaon *Halococcus salifodinae* BK3. *Extremophiles* **2015**, *19*, 875–884. [CrossRef]
85. Mohanty, A.; Kathawala, M.H.; Zhang, J.; Chen, W.N.; Loo, J.S.C.; Kjelleberg, S.; Yang, L.; Cao, B. Biogenic Tellurium Nanorods as a Novel Antivirulence Agent Inhibiting Pyoverdine Production in *Pseudomonas aeruginosa*. *Biotechnol. Bioeng.* **2014**, *111*, 858–865. [CrossRef]
86. Elakraa, A.A.; Salem, S.S.; El-Sayyad, G.S.; Attia, M.S. Cefotaxime Incorporated Bimetallic Silver-Selenium Nanoparticles: Promising Antimicrobial Synergism, Antibiofilm Activity, and Bacterial Membrane Leakage Reaction Mechanism. *RSC Adv.* **2022**, *12*, 26603–26619. [CrossRef]
87. Parveen, A.; Yalagatti, M.S.; Abbaraju, V.; Deshpande, R. Emphasized Mechanistic Antimicrobial Study of Biofunctionalized Silver Nanoparticles on Model *Proteus mirabilis*. *J. Drug Deliv.* **2018**, *2018*, 3850139. [CrossRef]
88. Balakumaran, M.D.; Ramachandran, R.; Kalaichelvan, P.T. Exploitation of Endophytic Fungus, *Guignardia mangiferae* for Extracellular Synthesis of Silver Nanoparticles and Their in Vitro Biological Activities. *Microbiol. Res.* **2015**, *178*, 9–17. [CrossRef]
89. Li, L.; Pan, H.; Deng, L.; Qian, G.; Wang, Z.; Li, W.; Zhong, C. The Antifungal Activity and Mechanism of Silver Nanoparticles against Four Pathogens Causing Kiwifruit Post-Harvest Rot. *Front. Microbiol.* **2022**, *13*, 3418. [CrossRef]
90. Zhou, L.; Zhao, X.; Li, M.; Lu, Y.; Ai, C.; Jiang, C.; Liu, Y.; Pan, Z.; Shi, J. Antifungal Activity of Silver Nanoparticles Synthesized by Iturin against *Candida albicans* in Vitro and in Vivo. *Appl. Microbiol. Biotechnol.* **2021**, *105*, 3759–3770. [CrossRef]
91. Balakumaran, M.D.; Ramachandran, R.; Balashanmugam, P.; Mukeshkumar, D.J.; Kalaichelvan, P.T. Mycosynthesis of Silver and Gold Nanoparticles: Optimization, Characterization and Antimicrobial Activity against Human Pathogens. *Microbiol. Res.* **2016**, *182*, 8–20. [CrossRef]
92. Gullino, M.L.; Tinivella, F.; Garibaldi, A.; Kemmitt, G.M.; Bacci, L.; Sheppard, B. Mancozeb: Past, Present, and Future. *Plant Dis.* **2010**, *94*, 1076–1087. [CrossRef]
93. Dania, V.O.; Olaleye, E.A. Chemical Characterization of Essential Oil Constituents of Three Selected Botanicals and Their Antimicrobial Activity against Postharvest Rot Pathogens of Tomato (*Solanum lycopersicum* L.). *Arch. Phytopathol. Plant Prot.* **2022**, *55*, 564–582. [CrossRef]
94. Sun, Z.; Yang, L.M.; Han, M.; Han, Z.M.; Yang, L.; Cheng, L.; Yang, X.; Lv, Z. liang Biological Control Ginseng Grey Mold and Plant Colonization by Antagonistic Bacteria Isolated from Rhizospheric Soil of *Panax ginseng* Meyer. *Biol. Control* **2019**, *138*, 104048. [CrossRef]
95. Lorrai, R.; Ferrari, S. Host Cell Wall Damage during Pathogen Infection: Mechanisms of Perception and Role in Plant-Pathogen Interactions. *Plants* **2021**, *10*, 399. [CrossRef]

microorganisms

MDPI

Article

Biofilm Inhibitory Activity of Actinomycete-Synthesized AgNPs with Low Cytotoxic Effect: Experimental and In Silico Study

Sabah A. AboElmaaty [1], Ali A. Shati [2], Mohammad Y. Alfaifi [2], Serag Eldin I. Elbehairi [2,3], Norhan S. Sheraba [4], Mervat G. Hassan [1], Mona Shaban E. M. Badawy [5], Ahmed Ghareeb [6], Ahmed A. Hamed [7,*] and Ebtsam Z. Gabr [1]

1 Botany and Microbiology Department, Faculty of Science, Benha University, Benha 13511, Egypt
2 Department of Biology, Faculty of Science, King Khalid University, Abha 9004, Saudi Arabia
3 Cell Culture Lab, Egyptian Organization for Biological Products and Vaccines (VACSERA Holding Company), Giza 12511, Egypt
4 VACSERA, The Holding Company for Biological Products and Vaccines, Giza 12511, Egypt
5 Department of Microbiology and Immunology, Faculty of Pharmacy (Girls), Al-Azhar University, Cairo 11884, Egypt
6 Botany and Microbiology Department, Faculty of Science, Cairo University, Giza 12613, Egypt
7 Microbial Chemistry Department, National Research Center, 33 El-Buhouth Street, Giza 12622, Egypt
* Correspondence: ahmedshalbio@gmail.com

Abstract: The emergence of resistance by biofilm-forming bacteria has reached alarming and dangerous levels that threaten human civilization. The current study sought to investigate the antibiofilm potential of green-synthesized silver nanoparticles, mediated by a new *Streptomyces* strain. Zeta potential, transmission electron microscopy (TEM), and UV-Vis spectroscopy were used to analyze the biosynthesized AgNPs. Results revealed that silver nanoparticles had a size of (5.55 and 45.00 nm) nm and a spherical shape, with surface plasmon resonance (SPR) absorption at 400–460 nm in the UV-vis spectra establishing the formation of *Streptomyces*-Ag-NPs. The biosynthesized AgNPs showed a pronounced antibacterial efficacy against *Escherichia coli*, *Pseudomonas aeruginosa*, *Bacillus subtilis*, and *Staphylococcus aureus*. Moreover, the obtained *Streptomyces*-AgNPs exerted biofilm inhibition activity against nosocomial hospital-resistant bacteria, including *Bacillus subtilis*, *Staphylococcus aureus*, and *Escherichia coli*. The mechanism of biogenic AgNPs antibacterial action was visualized using TEM, which indicated the AgNPs accumulation and disruption of bacterial cell membrane function. Additionally, a molecular docking study was conducted to evaluate the binding mode of AgNPs with an Escherichia coli outer membrane. Furthermore, the cytotoxic profile of the AgNPs was evaluated toward three cell lines (MCF-7, HepG2 & HCT 116), and the low cytotoxic effects of the obtained nanoparticles indicated their possible medical application with low risks to human health.

Keywords: biogenic AgNPs; *streptomyces*; antibifilm

Citation: AboElmaaty, S.A.; Shati, A.A.; Alfaifi, M.Y.; Elbehairi, S.E.I.; Sheraba, N.S.; Hassan, M.G.; Badawy, M.S.E.M.; Ghareeb, A.; Hamed, A.A.; Gabr, E.Z. Biofilm Inhibitory Activity of Actinomycete-Synthesized AgNPs with Low Cytotoxic Effect: Experimental and In Silico Study. *Microorganisms* **2023**, *11*, 102. https://doi.org/10.3390/microorganisms11010102

Academic Editor: Kamel Abd-Elsalam

Received: 30 October 2022
Revised: 27 December 2022
Accepted: 27 December 2022
Published: 30 December 2022

1. Introduction

Antimicrobial resistance (AMR) is one of the greatest public health challenges that currently threaten the successful prevention and treatment of a rising variety of bacterial, parasitic, viral, and fungal infections [1]. Nosocomial pathogenic bacteria became a significant obstacle to medical remediation after the emergence of antibiotic-resistant bacteria, with new resistance mechanisms emerging as a result of the abuse of antibiotics. [2–4]. One of these mechanisms is the development of bacterial biofilms, in which bacteria employ a special form of communication known as quorum sensing (QS). This enables other bacteria to react and cooperate to create bacterial biofilm aggregations [5,6]. This biofilm is made up of different macromolecules known as extracellular polymeric substances (EPS), types of tertiary biofilm structure that allow surface adherence and cell-to-cell interaction [6,7]. This matrix protects dangerous bacteria against potential antibiotic damage and attacks from the human immune system, making it one of the most serious challenges to medication and infectious disease prevention [7–9].

Despite many attempts that have been conducted to eradicate biofilm formation, there is still a definite need to discover an efficient agent to control biofilm formation by bacteria [10]. In hospitals and other healthcare facilities, the rate of nosocomial infections caused by antibiotic-resistant pathogens has increased [11]. The three most prevalent bacterial pathogens that cause nosocomial infections are *E. coli*, *Klebsiella* sp., and *S. aureus* [12–15].

Nanostructured silver is one of the noblest metals that has recently been used in several applications, starting from agriculture and ranging to medical use [16]. It has been estimated that the global world demand for nanomaterials ranges from 0.3 million tons to 1.6 million tons, including the Asian region, which accounts for approx. "(34%)", followed by North America (about 31%) and Europe (about 30%), respectively [16–18]. Silver nanoparticles (AgNPs) are structures with unique physical, chemical, and biological features of less than 100 nm [19–21]. AgNPs have been recently proven to be very good candidates as antimicrobial agent against some serious superbugs including *E. coli*, *S. typhi*, *S. epidermidis*, and *S. aureus*.

In this work, we examined a new actinomycete strain for its ability to perform biosynthesis of an eco-friendly silver nanoparticle (AgNPs) from a cell supernatant of *Streptomyces* sp. ASM23 (CS-AgNPs) and cell filtrate (CF-AgNPs). The biosynthesized AgNPs exhibited a remarkable biofilm inhibitory activity toward some tested nosocomial-resistant bacteria, including *B. subtilis*, *S. aureus*, and *E. coli*. Furthermore, we reported the in vitro cytotoxicity of the *streptomyces*-biosynthesized AgNPs against three cell lines (MCF-7, HepG2 & HCT 116), and the obtained results showed that the cell filtrate AgNPs (CF-AgNPs) have low cytotoxic activity. This suggests the possible use of biosynthesized nanometal in the medical field. Moreover, we studied the mechanism of inhibitory activity in silico aside from microscopic visualization.

2. Materials and Methods

2.1. Chemicals

All reagents were of analytical grade and purchased from Sigma–Aldrich (Burlington, MA, USA) and Himedia (Mumbai, India).

2.2. Microbial Strains and Cell Cultures

Bacillus subtilis ATCC6633, *Escherichia coli* ATCC25955, *Staphylococcus aureus* ATCC 33591, *Pseudomonas aeruginosa* ATCC10145, *Candida albicans* ATCC10231, and *Aspergillus niger* ATCC 6275 were obtained from the Microbiology and Immunology Department, Faculty of Medicine, Al-Azhar University, Egypt. Human breast adenocarcinoma (MCF-7), hepatocellular carcinoma (HepG-2), and colorectal adenocarcinoma (HCT 116) cell lines were obtained from the American Type Culture Collection (ATCC), USA. Cells were cultured in RPMI-1640 supplemented with streptomycin (100 μg/mL), and then with penicillin (100 units/mL) and heat-inactivated fetal bovine serum (10% *v*/*v*) in a humidified, 5% (*v*/*v*) CO_2 atmosphere at 37 °C.

2.3. Sample Collection and Isolation of the Bacterial Strain

Soil samples were collected from the Dakahlia Governorate, Egypt. Each sample was placed in clean, sterile bottles stored at 4 °C. Via intensive vortexing, the soil samples were suspended in sterile distilled water, and serial dilutions were made in sterile distilled water [18]. A volume of 0.1 mL of appropriate dilution was added to the Petri plate containing a starch nitrate agar media. Incubation was performed at 30 °C, and pH 7 was observed for seven days until the growth of actinomycetes [22,23]. A well-isolated formed colony was picked and purified by streaking on a starch nitrate agar medium.

2.4. Phenotypic Characterization of the Isolated Strain

Cultural characteristics studies of the ASM23 isolate have been studied according to the guidelines set out in the International *Streptomyces* Project. After its growth on various (ISP) media, the isolate was observed for its morphological features, pigmentation, and aerial mycelium. The microscopic examination was carried out to visualize the spore chain morphology of the bacterial isolate on starch nitrate agar plates [24,25]. The electron micrograph was captured by a Transmission Electron Microscope (TEM) at the National Research Centre (NRC), Egypt. Moreover, the utilization of various carbon sources was also studied.

2.5. Molecular Identification of Isolated Strain

The genetic identification of the isolated actinomycete strain was carried out through 16S rRNA gene sequencing. DNA extraction was performed according to the manufacturer's instructions using the DNeasy Blood & Tissue Kit (QIAGEN). Two universal primers (27F-5′-AGAGTTTGATCCTGGCTCAG-3′ and 1492R 5′GGTTACCTTGTTACGACTT-3′) were used to conduct 50 μL PCR amplification reactions. The purified PCR products were then sent to the Macrogen Company, South Korea, for sequencing [26]. The sequence was then investigated for homology and similarity using online NCBI BLAST search tools to compare similar known sequences in the NCBI database. The phylogenetic trees have been built using MEGA-X software.

2.6. Actinomycetes Cell-Free Extract Preparation

The strain *Streptomyces* sp. ASM23 was cultivated in yeast malt extract broth with the following composition g/L: yeast extract, 4.0; malt extract, 10.0; dextrose, 4.0. The broth had a pH of 7.2, and the strain was incubated at 32 °C in a rotary shaker at 150 rpm for six days. After incubation, the bacterial cell pellets were separated from the culture supernatant by centrifuging the whole content at 10,000 rpm for 15 min. To prepare the cell filtrate, 1 g of the collected pellets was re-suspended in 50 mL sterile distilled water, incubated for 48 h at 28 °C under shaking conditions (150 rpm), and centrifuged at 10,000 rpm for 15 min [24].

2.7. Green Synthesis of Silver Nanoparticles Using Actinomycetes Extracts

AgNPs biosynthesis was carried out by mixing an equal volume of bacterial supernatants and cell filtrate at a final concentration (1.0 mM). The mixture was incubated in a dark rotary shaker (150 rpm) for 3 days at 32 °C. The control experiment was conducted to determine the role of bacteria in the biosynthesis of AgNPs by mixing uninoculated culture media with the solution of $AgNO_3$. After incubation with actinomycetes supernatant and cell filtrate, the reduction of silver ions in the incubated solution was followed by a color change from colorless to yellowish-brown. The UV–Vis spectra of the incubated solution at different time intervals were measured using (SPECTROstar Nano absorbance plate reader—BMG LABTECH, Offenburg, Germany) with an increment of 2 nm and with a slot width of 5 nm in the 200 to 900 nm range [25–28].

2.8. Transmission Electron Microscopy (TEM)

Transmission electron microscopy (TEM) was used to measure the size and morphology of the synthesized AgNPs. A volume of 2 μL of AgNPs solution was mounted on carbon coated copper grids to prepare the sample. After adding the sample, the thin film formed was air-dried and observed using a Philips 10 Technai with an accelerating voltage of about 180 keV and with a wavelength (λ) of 0.0251 Å at the National Research Center, Egypt.

2.9. Studies of X-ray Diffraction (XRD)

The X-ray diffraction (XRD) pattern of the generated green AgNPs was measured using a PAN analytical X'pert PRO X-ray diffractometer (Philips, Eindhoven, The Netherlands) at the National Research Center, Egypt, with Cu Kα1 radiation. It was determined that there was an operating voltage and tube current of around 40 kV and 30 mA, respectively. After the material was drop-coated onto a glass substrate, the X-ray diffraction patterns were collected at 2θ from 10° to 80° at a scanning speed of 0.02°/min.

2.10. Zeta Potential

Zeta potential measurements are done using a high-performance two-angle Zetasizer Nano ZS device at the National Research Center, Egypt.

2.11. Determination of the Antimicrobial Activity of (AgNPs)

The antimicrobial activity of the biosynthesized AgNPs was measured using the agar well diffusion method. To measure the antimicrobial activity, a set of pathogenic test bacteria comprising Gram-positive bacteria (*Staphylococcus aureus* and *Bacillus subtilis*), Gram-negative bacteria (*Pseudomonas aeruginosa* and *Escherichia coli*), yeast (*Candida albicans*), and the fungus (*Aspergillus niger*) were selected. Both bacterial and yeast microbes were grown on a nutrient agar medium: this was composed of 10 g L^{-1} peptone, 3 g L^{-1} beef extract, and 20 g L^{-1} agar. The pH was adjusted to 7.2. On the other hand, the fungal strain was cultivated on a potato dextrose agar medium (PDA) of the following ingredients: 4 g L^{-1} potato extract, 20 g L^{-1} dextrose, and 20 g L^{-1} agar (pH 6). The culture of each microorganism was diluted with sterile distilled water to concentrations ranging from 10^7 to 10^8 CFU/mL. After preparing the inoculated agar plates, wells of 5 mm diameter were punched into the agar medium and filled with 50 µL of conc. and 50 µg/mL of the biosynthesized AgNPs and allowed to diffuse at 5 °C for 2 h. The plates were incubated for 24 hrs at 37 °C for bacteria and 48h at 30 °C for fungi and yeast. The inhibition zone diameter was measured after the incubation period [29]. Ciprofloxacin (5 µg/disc) and nystatin (10 µg/disc) were used as positive controls.

2.12. Monitoring Microbial Growth Curve Dynamics for E. coli in the Presence of Silver Nanoparticles Solutions

The most successful treatment was subjected to further investigations of the effect of different AgNPs concentrations along the *E. coli* growth curve. The antibacterial effectiveness of different concentrations of filtrate AgNPs' solution (0, 20, 40, 60 and 80, 100 µg/mL) was tested to examine its activity on the bacterial growth curve for *E. coli* bacterial colonies. These were inoculated with nutrient broth media till reaching optical culture density (OD600) 0.25–0.35. Different concentrations of AgNPs' solution (0, 20, 40, 60, 80, and 100 µg/mL) were added to the inoculated cultures. The cultures were incubated at 37 °C at 180 rpm. Bacterial densities were followed by determining the optical density (OD) at 600 nm after 0, 24, 48, 72, and 96 hrs [30].

2.13. In Vitro Antibiofilm Activity

The formed AgNPs' inhibitory biofilm activity was evaluated by applying a microtiter plate (MTP) assay to four resistant bacteria obtained from Al-Azhar university hospitals in Egypt (*Escherichia coli*, *Pseudomonas aeruginosa*, *Staphylococcus aureus*, and *Bacillus subtilis*). Briefly, each well was filled with LB broth (180 µL). It was then inoculated with pathogenic bacterial suspension (10 µL), 10 µL of (50 µg/mL) AgNPs was added along with the control (without the test sample) and the solution was incubated statically at 37 °C for 24 h. After the incubation period, the floating bacterial content was carefully removed and well washed with 200 µL of phosphate buffer saline (PBS) pH 7.2, left to dry, and stained with 200 µL/well of crystal violet solution (0.1%, *w*/*v*) for 10 min. The plate wells were washed with 200 µL/well-distilled water three times and kept for drying. To solubilize the dye, 150 µL/well of 95% ethanol was added, and the optical density at 570 nm was measured by (SPECTROstar Nano absorbance plate reader (BMG LABTECH)). All isolates were duplicated at least four times [31].

2.14. In Vitro Cytotoxic Activity

The cytotoxicity of the biosynthesized AgNPs from *Streptomyces* sp. ASM23 culture supernatant and cell filtrate was assessed against human tumor cells (MCF-7, HepG-2 & HCT 116) using the Sulphorhodamine B assay (SRB). Additionally, 90% confluence growing cells were trypsinized and cultured for 24 h before treatment with the biosynthesized AgNPs in a 96-well tissue culture plate (3000 cells/well). Cells were exposed to the six different concentrations of each compound (0.01, 0.1, 1, 10, and 1000 μg/mL); untreated cells (control) were added. The cells were incubated with the concentrations for 72 h and subsequently fixed with TCA (10% *w*/*v*) for 1 h at 4 °C. After washing, cells were stained with 0.4% (*w*/*v*) SRB solution for 10 min in a dark place. The excess stain was washed with 1% (*v*/*v*) glacial acetic acid. The SRB-stained cells were dissolved with Tris-HCl after drying overnight, and the color intensity in the microplate reader was measured at 540 nm. The relation between percentages of cell line viability and concentrations of the compounds has been analyzed to determine the IC_{50} (drug dose that decreases survival to 50%) using SigmaPlot 12.0 software.

2.15. Molecular Docking

The outer membrane proteins (OMPs) of Gram-negative bacteria play an important role in mediating antibacterial resistance and bacterial virulence, thus affecting the pathogenic ability of the bacteria. The outer membrane proteins and OMPX (1QJ8) crystal structure were obtained from the PDB database. The three-dimensional structure of this protein was taken as the target for molecular docking, with Ag using the PatchDock server [32]. The SDF file of silver was downloaded from the PubChem database and converted into a PDB file using online SMILES translator (https://www.cactus.nci.nih.gov/translate/, accessed on 12 October 2021). The PDB files of the respective protein and silver were uploaded under the receptor–ligand interaction mode, with default values and an RMSD cutoff value of <4.0 Å. Based on the scoring, the top-ranked solution was selected for interpretation. The amino acid residues of the proteins that are within 4 Å were analyzed for their molecular interaction with silver.

2.16. Statistical Analysis

Data were analyzed using the one-way analysis for variance (ANOVA) method and carried out by the SPSS program based on the mean of the triplicate results.

3. Results

3.1. Collection and Isolation of Bacterial Isolate

A single colony of strain ASM23 was isolated using serial dilution techniques from the soil collected from Mansoura. The isolate was coded, retained at 4 °C on starch nitrate agar slants, and deposited at the Culture Collection Center, Department of Microbial Chemistry, National Research Center, Egypt.

3.2. Phenotypic Studies of Strain ASM23

Morphological properties of the isolated strain ASM23 on different ISP media (ISP2-ISP5) showed that the strain has very close characteristics to members of the genus *Streptomyces* [24,25]. The aerial mycelium of the strain ASM23 is mainly whitish-grey on different ISP media, with no diffusible pigments (Figure 1a). Microscopic examination revealed that the spore of the ASM23 is short with a smooth spore surface (Figure 1b). On the other hand, the strain utilizes D-glucose, sucrose, D-Fructose, and l-arabinose as carbon sources.

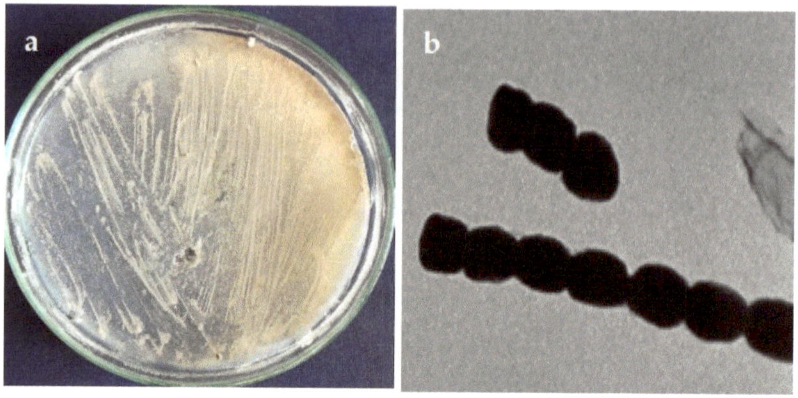

Figure 1. (**a**) Aerial and substrate mycelium morphology; (**b**) spore chain morphology under transmission electron micrography.

3.3. Molecular Identification of Selected Actinomycetes

Sequencing of 16S rDNA has been carried out to identify the ASM23 isolate. The DNA was isolated, amplified using PCR, sequenced, and aligned using the NCBI Online BLAST tool to examine similarity scores with other identified strains. The results confirmed a high similarity rating of 99.72% homology of isolate ASM23 with *Streptomyces* sp. The partial 16S rDNA sequence of the *Streptomyces* isolate was deposited in the GenBank with the MN689775 accession number under the title *Streptomyces* sp. ASM23. The phylogenetic trees have been built by MEGA-X and a maximum likelihood method (Figure 2) [33].

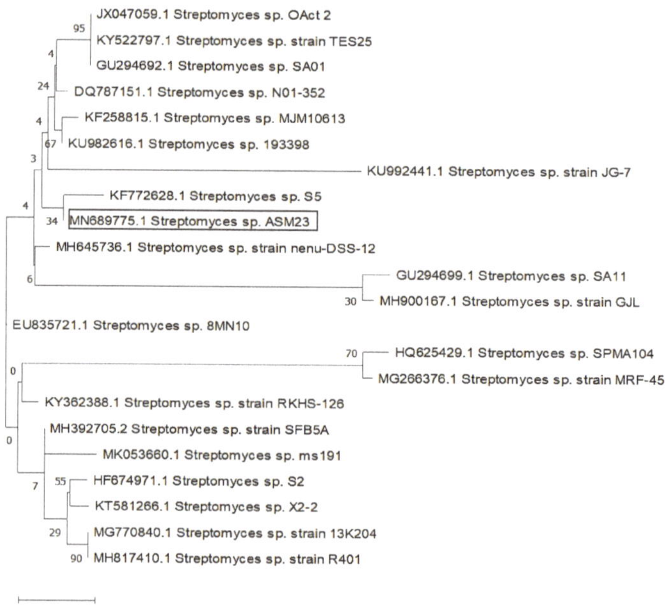

Figure 2. Evolutionary relationships of taxa related to *Streptomyces* sp. ASM23 is based on a 16S rDNA sequence using the maximum likelihood method.

3.4. Green Biosynthesis of AgNPs

The green biosynthesis of silver nanoparticles AgNPs was performed using *Streptomyces* sp. ASM23 supernatant (CS) and cell filtrate (CF). The synthesis was performed under dark conditions at 32 °C. A solution of a brownish color forms after the addition of AgNO3 aqueous solution (1 mM) to the *Streptomyces* sp. ASM23 supernatant and cell filtrate, implying metal ions reduction and the formation of AgNPs (Figure 3b,c). AgNPs have light yellow to brown colors due to the vibration of surface plasm in the particles confirmed by spectra for UV-Vis absorption (Figure 3a). The biosynthesized AgNPs in the presence of actinomycete cell filtrate exhibited a characteristic surface plasmon band (SPR) at a wavelength ranging from 400 to 460 nm, indicating the formation of AgNPs at different time intervals. These findings follow early reports by Składanowski et al. [34] and Al-Dhabi et al. [35].

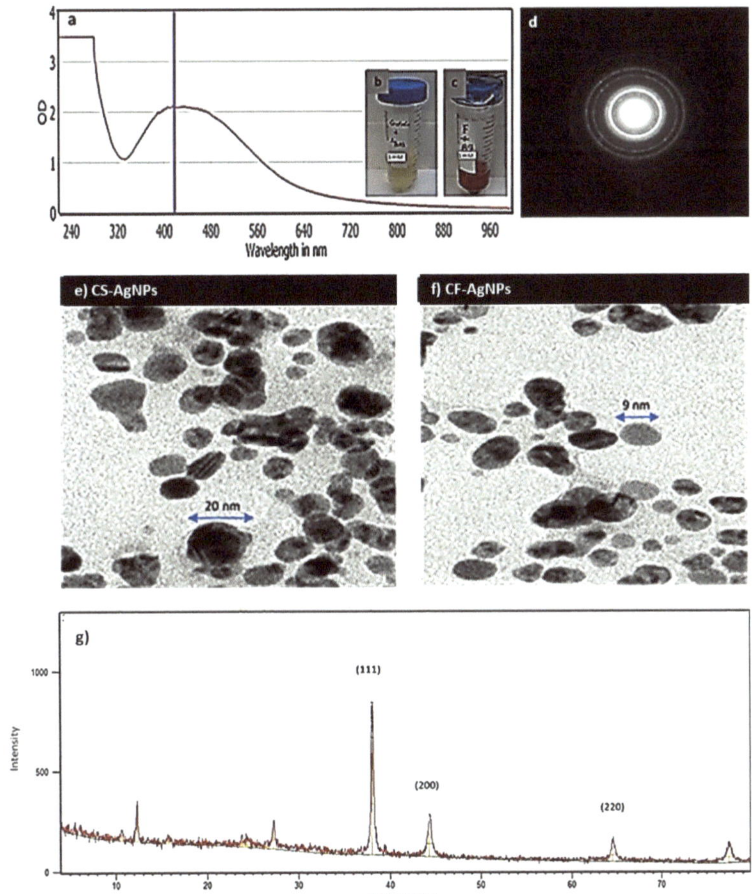

Figure 3. (**a**) Surface plasmon absorption bands (SPR) of AgNPs formed in the presence of *Streptomyces* sp. ASM23 cell filtrate (CF-AgNPs). (**b**) The color change due to the reduction of metal ions and (**c**) the formation of AgNPs represented by (**a**) Control+ AgNO3 and (**b**) *Streptomyces* sp. ASM23 cell filtrate+ $AgNO_3$. (**d**) Selected area (electron) diffraction (SAED) confirms the crystalization of nanoparticles. (**e**) Transmission electron microscope micrographs for (**f**) AgNPs biosynthesized by *Streptomyces* sp. ASM23 cell filtrate (CF-AgNPs) and cell supernatant (CS-AgNPs). (**g**) X-ray diffraction (XRD) pattern of silver nanoparticles.

3.5. Characterization of Silver Nanoparticles

Dynamic light scattering (DLS) and zeta potential measurements were used to investigate the biosynthesized AgNPs' surface and colloidal stability. The results indicated that the particle size (i.e., hydrodynamic, HD diameter) increased when adding the cell supernatant and cell filtrate of *Streptomyces* sp. ASM23 to the aqueous solution of $AgNO_3$. This results from an Ag^+ ion being reduced in Ag nuclei and forming large particles. Moreover, through *Streptomyces* sp. ASM23, the zeta potential for the AgNPs formed by the aqueous solution of the $AgNO_3$ samples was −27.6 and −30.5 mV, respectively. The results showed that the prepared AgNPs tend to attract particles electrostatically in the solution. The particle size and shape have been demonstrated via transmission electron microscope (TEM) measurements for two prepared samples (CS-AgNPs and CF-AgNPs), as shown in (Figure 3e,f). The average particle size for silver nanoparticles biosynthesized by the *Streptomyces* sp. The particle sizes ASM23 cell-free filtrate and supernatant range from about ~5.55 to 35 nm, with spherical and cubic shapes (Figure 3d–f). The X-ray diffraction (XRD) pattern of silver nanoparticles (NPs) was measured and presented in Figure 3g. Peaks of XRD are assigned to diffraction from the (111), (200), and (220) planes of silver.

3.6. Antimicrobial Activity of Biosynthesized AgNPs

The antimicrobial activity of the synthesized silver nanoparticles has been measured toward different human pathogens using the agar well diffusion method. In general, the AgNPs, prepared by the cell filtrate of *Streptomyces* sp. ASM23, showed more antimicrobial activity than the supernatant solutions. The results showed that *E. coli* and *S. aureus* were the most susceptible bacteria to the antibacterial effects of AgNPs, prepared by *Streptomyces* sp. ASM23 cell filtrate (CF-AgNPs), with inhibition zone diameters of 19.00 and 16.00 mm, respectively. On the other hand, the formed CF-AgNPs displayed moderate antimicrobial activity toward *P. aeruginosa* and *B. subtilis*. Additionally, the AgNPs of both filtrate and supernatant showed good antibacterial activity against *S. aureus* while exhibiting low activity against *C. albicans* and no response was detected against *A. niger* (Table 1).

Table 1. Antimicrobial activity of AgNPs produced by *Streptomyces* sp. ASM23 cell supernatant (CS-AgNPs) and cell filtrate (CF-AgNPs).

Test Microbes	Zone of Inhibition (mm)			
	S-AgNPs	**CF-AgNPs**	**Cip**	**Nys**
P. aeruginosa	11.00	14.00	18.00	-
B. subtilis	10.00	13.00	19.00	-
E. coli	15.00	19.00	18.00	-
S. aureus	8.00	16.00	-	
C. albicans	7.05	9.00	-	18
A. niger	-	-	-	15

All data are expressed as the mean of three reads, Cip: ciprofloxacin, Nys: Nystatin.

3.7. Effect of Different Concentrations of AgNPs on E. coli Growth Curve Dynamics

The effect of CF-AgNPs, biosynthesized by the filtrate solution on the growth curve dynamics of *E. coli*, has been investigated. The bacterial growth curve slope dwindled perpetually with elevating nanoparticle concentration. The highest inhibition activity has been detected at an 80 µg/mL concentration, which started with a dwarfed long period. At 80 mg/mL of AgNPs, the most marked effect is easily visualized in the late logarithmic phase and during the bacterial stationary growth phase (Figure 4).

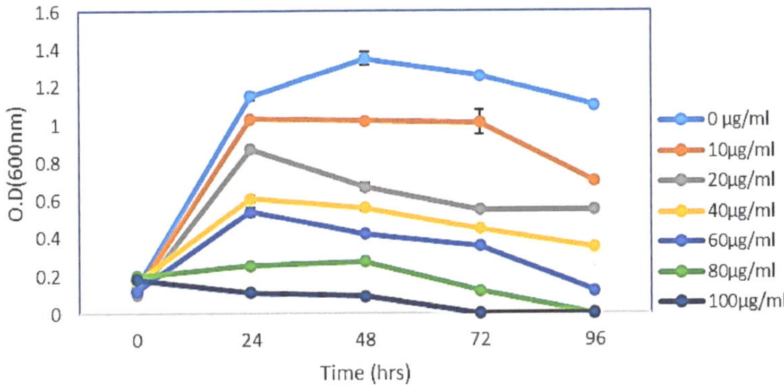

Figure 4. Growth curve dynamics of *E. coli* with CF-AgNPs solution. Each experiment was carried out in triplicate and the results are presented as mean values ± standard deviations (SD).

3.8. Antibacterial Mechanism of Action

The antibacterial mechanism of the AgNPs has been visualized using transmission electron microscopy (TEM). The *S. aureus* was treated with CF-AgNPs at a final concentration, of 50 µg mL^{-1}, and incubated at 37 °C overnight. The TEM micrograph displayed the accumulation of CF-AgNPs on the *S. aureus* cell membrane, broken membranes of the treated *S. aureus*, and the appearance of CF-AgNPs inside the bacterial cells (Figure 5a,b).

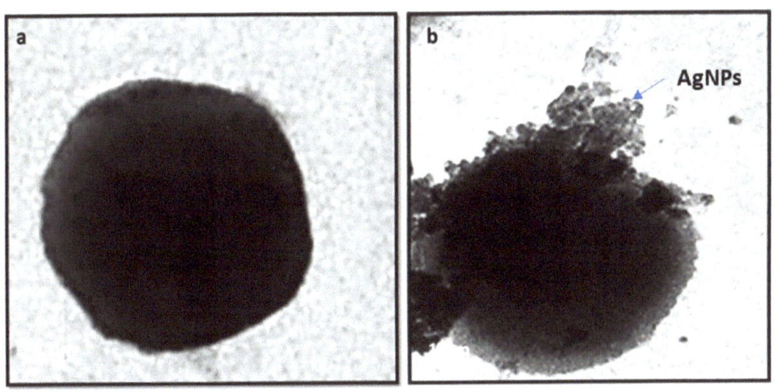

Figure 5. Accumulation of AgNPs on the *S. aureus* outer cell membrane (**a**) Bacteria with no treatment and (**b**) Bacteria treated with AgNPs.

3.9. Antibiofilm Activity of Biosynthesized AgNPs

The AgNPs, synthesized by the *Streptomyces* sp. ASM23 cell filtrate (CF-AgNPs), displayed a potent biofilm inhibition activity against *E. coli* and *B. subtilis*, with biofilm inhibitory ratios of 74.20 and 70.52%. At the same time, the AgNPs synthesized by *Streptomyces* sp. ASM23 cell filtrate (CF-AgNPs) only displayed a moderate antibiofilm activity toward *S. aureus*, up to 50.30%. On the other hand, *P. aeruginosa* displayed an inadequate response, with a biofilm formation inhibition rate up to 20.00% (Figure 6). Statistical analysis of the biofilm data, based on mean of triplicate and standard deviations (SD), showed that the p value is lower than 0.05. This indictaed that the result is statistically significant.

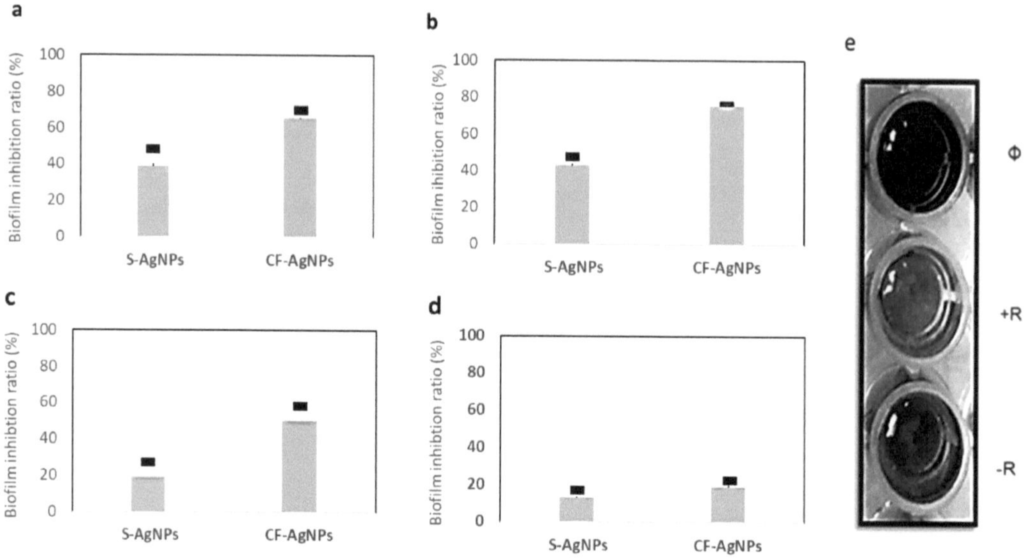

Figure 6. The biofilm inhibition ratio of AgNPs prepared by *Streptomyces* sp. ASM23 cell supernatant (S) and cell filtrate (CF) against four clinical microbes (**a**) *B. subtilis*, (**b**) *E. coli* (**c**) *S. aureus*, and (**d**) *P. aeruginosa*. The bars on the graph represent the mean of triplicate, and SD. (**e**) Microtiter plate, (Ø) blank, (+R) positive result, and (−R) negative result.

3.10. In Vitro Cytotoxic Activity

The sulforhodamine B (SRB) assay was used to assess the cytotoxicity of the green CF-AgNPs and CS-AgNPs towards MCF-7, HepG2, and HCT 116 tumor cell lines over a concentration range from 0.01 to 1000 µg/mL (Table 2 and Figures 7 and 8). Tested compounds showed a variety of cytotoxicity profiles against solid tumor cells. Biosynthesized AgNPs from the cell supernatant of *Streptomyces* spp ASM23 showed a potent cytotoxic profile in all tested cell lines (MCF-7, HepG2 & HCT 116), with IC_{50}s of 27.8, 18, and 73.2, respectively, more than biosynthesized AgNPs from the cell filtrate of *Streptomyces* spp ASM23. The toxicity of biosynthesized AgNPs from the cell supernatant of *Streptomyces* spp ASM23 on HCT 116 was weaker compared to MCF-7 and HepG2 cells, whereas biosynthesized AgNPs from cell filtrate of *Streptomyces* spp ASM23 have shown promising toxicity against MCF-7 cells with IC_{50} 65.1, relevant to HepG2 and HCT 116 cells with IC_{50}s 82.9 and 81.9, respectively, shown in (Table 2, Figures 7 and 8).

Table 2. The IC_{50} (µg/mL) of the biosynthesized AgNPs from the cell supernatant (CS-AgNPs) and Cell filtrate (CF-AgNPs) of *Streptomyces* sp. ASM23 against different tumor cell lines.

Sample	IC50 (µg/mL)		
	MCF-7	**HepG2**	**HCT 116**
CS-AgNPs	27.8 ± 2.3	18 ± 0.6	73.2 ± 1.5
CF-AgNPs	65.1 ± 2.4	82.9 ± 1.5	81.9 ± 1.4

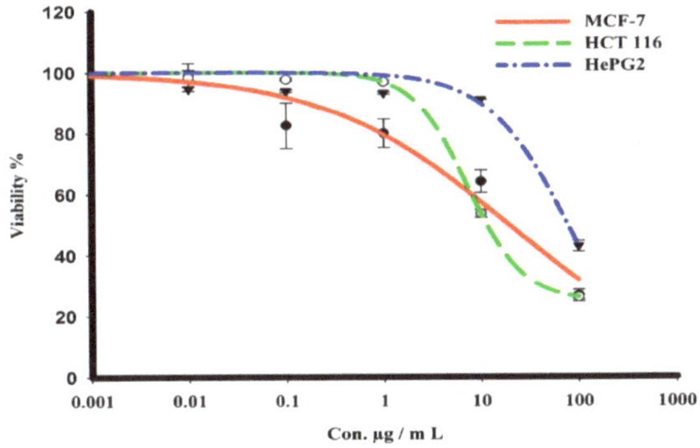

Figure 7. The dose–response curve of biosynthesized AgNPs from the cell supernatant of *Streptomyces* sp. ASM23 on the cytotoxicity in MCF-7, HePG2, and HCT 116 cell lines. Cells were exposed to biosynthesized AgNPs from the cell supernatant of *Streptomyces* sp. ASM23, with different concentrations of compounds for 72 h. Cell viability was detected by SRB stain.

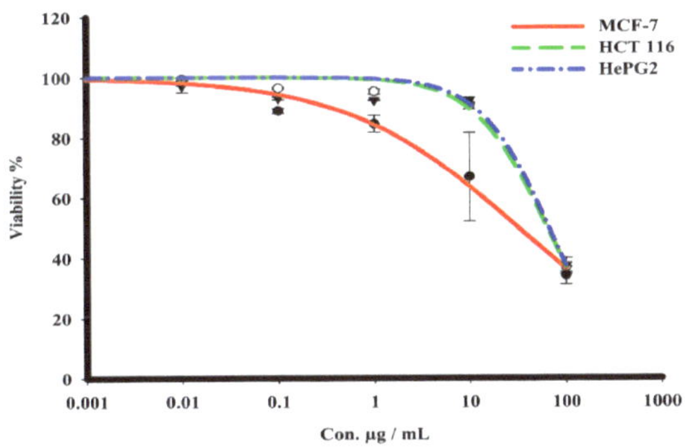

Figure 8. The dose–response curve of biosynthesized AgNPs from cell filtrate of *Streptomyces* sp. ASM23 on the cytotoxicity in MCF-7, HePG2, and HCT 116 cell lines. Cells were exposed to biosynthesized AgNPs from cell filtrate of *Streptomyces* sp. ASM23 with different concentrations of compounds for 72 h. Cell viability was detected by SRB stain.

3.11. Molecular Docking Determine the AgNP's Binding Capacity to E. coli

The antibacterial effect was visualized and confirmed using a TEM micrograph. This showed the accumulation of formed Ag nanoparticles on bacterial cell walls that consist of peptidoglycan; these work in turn as an essential protective barrier for bacteria and encapsulate the cytoplasmic membrane of both Gram-negative and Gram-positive bacterial cells. To assess and predict the antibacterial activity of silver nanoparticles, molecular docking was performed using the PatchDock server for the three-dimensional structures of *E. coli* outer membrane proteins, the OMPX (1QJ8) crystal structure. The amino acid residues that interact with Ag, along with the type of interactions, are presented in Figure 9. Ag interacted with OMPX (1QJ8) through Phe148, which may lead to disruption of the *E. coli* outer membrane.

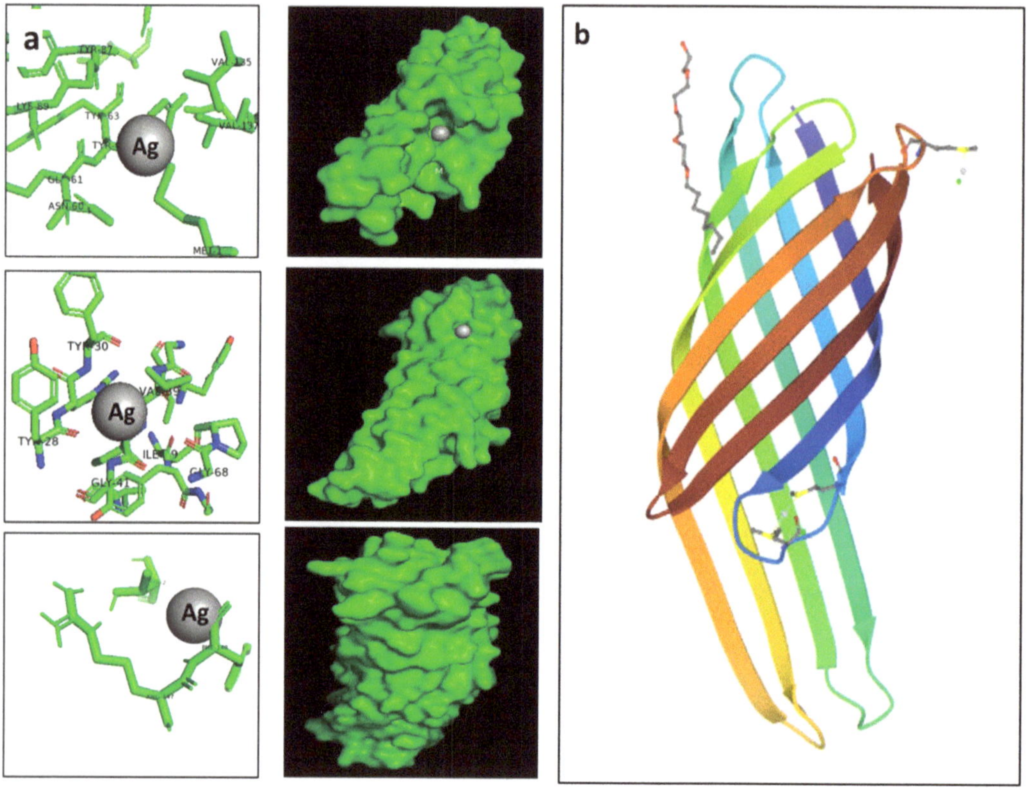

Figure 9. (**a**) showed the possible binding modes of AgNPs to *E. coli* outer membrane protein, while (**b**) 3D structure of *E. coli* outer membrane protein 1QJ8. Molecular docking was performed to identify the interaction of synthesized nanoparticles with amino acid residues. The docking model shows an interaction of Ag with 1QJ8 through Phe146.

4. Discussion

Despite the pioneer achievements in organic and synthetic chemistry to develop effective antimicrobial agents, new candidates with novel action mechanisms are still urgently needed to counter the ongoing increase in antibiotic resistance due to biofilm formation. AgNPs have unique physical and optical properties that make them suitable for a wide range of potential applications such as antimicrobial agents, coating of biomedical devices, and drug nano-delivery approaches [19]. Recently, AgNPs' antibacterial effects have been the subject of current controversy for the inhibition of most common nosocomial pathogens that cause various clinical diseases such as *E. coli*, *K. pneumonia* and *S. aureus* [36].

The in vitro microbicidal activity of the AgNPs was previously tested against diverse pathogenic fungi such as *F. solani*, *A. niger*, *A. parasiticus*, and *C. albicans*, and different pathogenic bacteria such as *B. megaterium*, *Bacillus subtilis*, *Proteus vulgaris* and, *Escherichia coli*. Chudasama et al. and Ramgopal et al. [6,37] have stated that AgNPs displayed inhibitory activity toward *S. aureus* (Gram-positive) and *E. coli* (Gram-negative). Prakash et al. [38] have also highlighted a similar result for the antibacterial effects of the AgNPs on Gram-negative bacteria (*E. coli*) and Gram-positive (*S. pyogene*). Conversely, Dipak and Sankar et al. [39] have shown the highest and the lowest zone of inhibition formation against *E. coli* and *Staphylococcus* sp., respectively. The obtained results showed that AgNPs, prepared by *Streptomyces* sp. ASM23 cell filtrate, have antibacterial activity against *E. coli* and *S. aureus*, while it showed moderate antibacterial activity against *B. subtilis*. Several studies have reported the possible antimicrobial mechanism of silver nanoparticles and reported the AgNPs' primary mechanism of action.

Therefore these particles can penetrate the cell membrane, or attach themselves according to their size to the bacterial surface [7]. Moreover, the bactericidal efficacy is further increased by reducing the size of the particle [40]. There have been many arguments to clarify why microbial growth inhibition by AgNPs should occur; the most convincing is free radical formation, a process which was supported in the spectrum of AgNPs by the peak appearance at 336.33 in the electron spin resonance (ESR) spectrum of AgNPs [41]. The generation of free radicals primarily disrupts the microbial membrane lipids structure, dissociate it, and hinders microbial growth [42]. In one study, the release of silver cations from AgNPs was attributed to the high antibacterial activity [43]. These Ag^+ ions are charged positively and far less than neutral AgNPs. They can easily interact with the S or P and N electron-rich biomolecules in the bacterial cell walls. Some researchers have hypothesized that a key to microorganism growth inhibition is the interaction between the positive AgNPs' charge and the negative charge on the cell membrane [44–46]. The nanoparticles form pits within the microbial cell wall; they then accumulate and eventually permeate into the bacterial cells [41,47]. In the presented work, analysis by TEM micrograph proved the accumulation of AgNPs, both on the bacterial cell membrane and inside the bacterial cells. Some reports declared that the AgNPs adhere primarily to the bacterial surface and alter membrane properties, but that they may cause DNA damage inside the bacterial cell [48]. McQuillan and colleagues stated that cell membrane dissolution is the primary mechanism for the action of silver nanoparticles [49]. Several reports suggested that silver nanoparticles' mechanism of inhibition started through the binding and interacting of silver with thiol-containing proteins in the cell wall and influencing their functions [50].

Computational study, carried out through omlecular docking, was performed using a PatchDock server for the three-dimensional structures of *E. coli* outer membrane proteins with an OMPX (1QJ8) crystal structure. The results showed that the biosynthesized AgNPs bind to 1QJ8 through Phe146, an occurrence which may be the cause of disrupting the *E. coli* outer membrane. Additionally, the antibiofilm activity of the formed silver nanoparticles against a set of hospital nosocomial bacterial pathogens such as *B. subtilis* and *E. coli* indicated its possible use in the medical field. Moreover, the low cytotoxic effect of AgNPs, synthesized via a filtrate of *Streptomyces* sp. ASM23 against all tested cell lines (MCF-7, HepG2 & HCT 116), makes them promising and possible candidates for application in the biomedical field. Additionally, they have a strong safety profile.

5. Conclusions

The spread of antimicrobial resistance has made it imperative to look for an alternative source for an adequate drug to combat multi-drug-resistant bacteria. The obtained results concluded that actinomycetes could be an effective, cheap, environmentally friendly, and secure strategy for the green biosynthesis of nanoparticles. Additionally, silver nanoparticles showed pronounced antibiofilm activity against a set of hospital nosocomial tests pathogenic such as *B. subtilis* and *E. coli*, indicating their possible applicability in the medical field. Moreover, AgNPs synthesized via filtrates of *Streptomyces* sp. ASM23 showed a low cytotoxic profile in all tested cell lines (MCF-7, HepG2 & HCT 116), reflecting their possible use in the biomedical field and their high safety profile. The obtained results highlighted the possible use of green AgNPs as an anti-biofilm agent with an efficient effect on hospital nosocomial bacterial pathogens.

Author Contributions: Conceptualization, S.E.I.E., A.A.S., A.A.H. and M.Y.A.; methodology, S.A.A., A.A.H., S.E.I.E., A.A.S., M.S.E.M.B., N.S.S., E.Z.G., A.G., M.G.H. and M.Y.A.; software, N.S.S.; validation, M.Y.A. and S.E.I.E.; formal analysis, N.S.S., A.A.H. and A.G.; investigation, A.A.H., M.Y.A. and A.A.S.; resources, M.Y.A., A.A.H., A.A.S. and S.E.I.E.; data curation, S.A.A.; writing—original, A.A.H., M.G.H. and A.G.; draft preparation, A.A.H., E.Z.G. and A.G.; writing—review and editing, A.A.H. and A.G.; visualization, A.A.S. and M.Y.A.; supervision, M.Y.A., A.A.H. and A.G.; project administration, S.E.I.E. and M.Y.A.; funding acquisition, M.Y.A. and S.E.I.E. All authors have read and agreed to the published version of the manuscript.

Funding: The authors extend their appreciation to the deanship of scientific research at King Khalid University for funding this work through a research group project under grant number (R.G.P. 2/78/43).

Data Availability Statement: The data presented in this study are available on request from the corresponding author.

Acknowledgments: The authors extend their appreciation to the Deanship of Scientific Research at King Khalid University for funding this work.

Conflicts of Interest: The authors declare no competing interests.

References

1. Mohammed, A.B.A.; Mohamed, A.; El-Naggar, N.E.-A.; Mahrous, H.; Nasr, G.M.; Abdella, A.; Ahmed, R.H.; Irmak, S.; Elsayed, M.S.A.; Selim, S.; et al. Antioxidant and Antibacterial Activities of Silver Nanoparticles Biosynthesized by Moringa oleifera through Response Surface Methodology. *J. Nanomater.* **2022**, *2022*, 1–15. [CrossRef]
2. Alhadrami, H.A.; Hamed, A.A.; Hassan, H.M.; Belbahri, L.; Rateb, M.E.; Sayed, A.M. Flavonoids as Potential anti-MRSA Agents through Modulation of PBP2a: A Computational and Experimental Study. *Antibiotics* **2020**, *9*, 562. [CrossRef] [PubMed]
3. Hifnawy, M.S.; Hassan, H.M.; Mohammed, R.; Fouda, M.M.; Sayed, A.M.; Hamed, A.A.; Abouzid, S.F.; Rateb, M.E.; Alhadrami, H.A.; Abdelmohsen, U.R. Induction of Antibacterial Metabolites by Co-Cultivation of Two Red-Sea-Sponge-Associated Actinomycetes *Micromonospora* sp. UR56 and *Actinokinespora* sp. EG49. *Mar. Drugs* **2020**, *18*, 243. [CrossRef] [PubMed]
4. Soliman, M.; Qari, S.H.; Abu-Elsaoud, A.; El-Esawi, M.; Alhaithloul, H.; Elkelish, A. Rapid green synthesis of silver nanoparticles from blue gum augment growth and performance of maize, fenugreek, and onion by modulating plants cellular antioxidant machinery and genes expression. *Acta Physiol. Plant.* **2020**, *42*, 1–16. [CrossRef]
5. Iravani, S.; Korbekandi, H.; Mirmohammadi, S.V.; Zolfaghari, B. Synthesis of silver nanoparticles: Chemical, physical and biological methods. *Res. Pharm. Sci.* **2014**, *9*, 385–406.
6. Ramgopal, M.; Saisushma, C.; Atti, I.H.; Alhasin, A.M. A Facile Green Synthesis of Silver Nanoparticles using Soap Nuts. *Res. J. Microbiol.* **2011**, *6*, 432–438. [CrossRef]
7. Rai, M.; Yadav, A.; Gade, A. Silver nanoparticles as a new generation of antimicrobials. *Biotechnol. Adv.* **2009**, *27*, 76–83. [CrossRef]
8. Donlan, R.M. Biofilms: Microbial Life on Surfaces. *Emerg. Infect. Dis.* **2002**, *8*, 881–890. [CrossRef]
9. Flemming, H.-C. EPS—Then and Now. *Microorganisms* **2016**, *4*, 41. [CrossRef]
10. Labena, A.; Hegazy, M.A.; Kamel, W.M.; Elkelish, A.; Hozzein, W.N. Enhancement of A Cationic Surfactant by Capping Nanoparticles: Synthesis, Characterization and Multiple Applications. *Molecules* **2020**, *25*, 2007. [CrossRef]
11. Salem, S.S.; El-Belely, E.F.; Niedbała, G.; Alnoman, M.M.; Hassan, S.E.-D.; Eid, A.M.; Shaheen, T.I.; Elkelish, A.; Fouda, A. Bactericidal and In-Vitro Cytotoxic Efficacy of Silver Nanoparticles (Ag-NPs) Fabricated by Endophytic Actinomycetes and Their Use as Coating for the Textile Fabrics. *Nanomaterials* **2020**, *10*, 2082. [CrossRef]
12. Hamed, A.A.; Soldatou, S.; Qader, M.M.; Arjunan, S.; Miranda, K.J.; Casolari, F.; Pavesi, C.; Diyaolu, O.A.; Thissera, B.; Eshelli, M.; et al. Screening Fungal Endophytes Derived from Under-Explored Egyptian Marine Habitats for Antimicrobial and Antioxidant Properties in Factionalised Textiles. *Microorganisms* **2020**, *8*, 1617. [CrossRef]
13. Hathout, A.; El-Nekeety, A.; Hamed, A.; Sabry, B.; Abdel-Aziz, M.; Ghareeb, M.; Aly, S. Novel Egyptian bacterial strains exhibiting antimicrobial and antiaflatoxigenic activity. *J. Appl. Pharm. Sci.* **2016**, *6*, 1–10. [CrossRef]
14. Kinross, P.; Petersen, A.; Skov, R.; Van Hauwermeiren, E.; Pantosti, A.; Laurent, F.; Voss, A.; Kluytmans, J.; Struelens, M.J.; Heuer, O.; et al. Livestock-associated meticillin-resistant Staphylococcus aureus (MRSA) among human MRSA isolates, European Union/European Economic Area countries, 2013. *Eurosurveillance* **2017**, *22*, 16-00696. [CrossRef]
15. Palomino, J.C.; Martin, A. Drug Resistance Mechanisms in Mycobacterium tuberculosis. *Antibiotics* **2014**, *3*, 317–340. [CrossRef]
16. Bhattacharya, D.; Gupta, R.K. Nanotechnology and Potential of Microorganisms. *Crit. Rev. Biotechnol.* **2005**, *25*, 199–204. [CrossRef]
17. Khatoon, N.; Mazumder, J.A.; Sardar, M. Biotechnological Applications of Green Synthesized Silver Nanoparticles. *J. Nanosci. Curr. Res.* **2017**, *2*, 1–8. [CrossRef]
18. Zhang, X.-F.; Liu, Z.-G.; Shen, W.; Gurunathan, S. Silver Nanoparticles: Synthesis, Characterization, Properties, Applications, and Therapeutic Approaches. *Int. J. Mol. Sci.* **2016**, *17*, 1534. [CrossRef]

19. Abdel-Azeem, A.; Nada, A.A.; O'Donovan, A.; Kumar Thakur, V.; Elkelish, A. Mycogenic Silver Nanoparticles from Endophytic Trichoderma atroviride with Antimicrobial Activity. *J. Renew. Mater.* **2020**, *8*, 171–185. [CrossRef]
20. Calderón-Jiménez, B.; Johnson, M.E.; Montoro Bustos, A.R.; Murphy, K.E.; Winchester, M.R.; Vega Baudrit, J.R. Silver Nanoparticles: Technological Advances, Societal Impacts, and Metrological Challenges. *Front. Chem.* **2017**, *5*, 6. [CrossRef]
21. Pulit-Prociak, J.; Banach, M. Silver nanoparticles—A material of the future ... ? *Open Chem.* **2016**, *14*, 76–91. [CrossRef]
22. Carlson, C.; Hussain, S.M.; Schrand, A.M.; Braydich-Stolle, L.K.; Hess, K.L.; Jones, R.L.; Schlager, J.J. Unique Cellular Interaction of Silver Nanoparticles: Size-Dependent Generation of Reactive Oxygen Species. *J. Phys. Chem. B* **2008**, *112*, 13608–13619. [CrossRef] [PubMed]
23. Liu, W.; Wu, Y.; Wang, C.; Li, H.C.; Wang, T.; Liao, C.Y.; Cui, L.; Zhou, Q.F.; Yan, B.; Jiang, G.B. Impact of silver nanoparticles on human cells: Effect of particle size. *Nanotoxicology* **2010**, *4*, 319–330. [CrossRef] [PubMed]
24. Hamed, A.A.; Abdel-Aziz, M.S.; Fadel, M.; Ghali, M.F. Antimicrobial, antidermatophytic, and cytotoxic activities from *Streptomyces* sp. MER4 isolated from Egyptian local environment. *Bull. Natl. Res. Cent.* **2018**, *42*, 22. [CrossRef]
25. Abdel-Aziz, M.S.; Hathout, A.S.; A El-Neleety, A.; A Hamed, A.; A Sabry, B.; E Aly, S.; Abdel-Wahhab, M.A. Molecular identification of actinomycetes with antimicrobial, antioxidant and anticancer properties. *Comun. Sci.* **2019**, *10*, 218–231. [CrossRef]
26. Wang, X.; Ji, Z.; Chang, C.H.; Zhang, H.; Wang, M.; Liao, Y.-P.; Lin, S.; Meng, H.; Li, R.; Sun, B.; et al. Use of Coated Silver Nanoparticles to Understand the Relationship of Particle Dissolution and Bioavailability to Cell and Lung Toxicological Potential. *Small* **2013**, *10*, 385–398. [CrossRef]
27. Li, N.; Xia, T.; Nel, A.E. The role of oxidative stress in ambient particulate matter-induced lung diseases and its implications in the toxicity of engineered nanoparticles. *Free. Radic. Biol. Med.* **2008**, *44*, 1689–1699. [CrossRef]
28. Han, X.; Gelein, R.; Corson, N.; Wade-Mercer, P.; Jiang, J.; Biswas, P.; Finkelstein, J.N.; Elder, A.; Oberdörster, G. Validation of an LDH assay for assessing nanoparticle toxicity. *Toxicology* **2011**, *287*, 99–104. [CrossRef]
29. Gliga, A.R.; Skoglund, S.; Wallinder, I.O.; Fadeel, B.; Karlsson, H.L. Size-dependent cytotoxicity of silver nanoparticles in human lung cells: The role of cellular uptake, agglomeration and Ag release. *Part. Fibre Toxicol.* **2014**, *11*, 11. [CrossRef]
30. Kaiser, J.-P.; Roesslein, M.; Diener, L.; Wichser, A.; Nowack, B.; Wick, P. Cytotoxic effects of nanosilver are highly dependent on the chloride concentration and the presence of organic compounds in the cell culture media. *J. Nanobiotechnology* **2017**, *15*, 1–11. [CrossRef]
31. Kaiser, T.D.L.; Pereira, E.M.; dos Santos, K.R.N.; Maciel, E.L.N.; Schuenck, R.P.; Nunes, A.P.F. Modification of the Congo red agar method to detect biofilm production by Staphylococcus epidermidis. *Diagn. Microbiol. Infect. Dis.* **2013**, *75*, 235–239. [CrossRef]
32. Schneidman-Duhovny, D.; Inbar, Y.; Nussinov, R.; Wolfson, H.J. PatchDock and SymmDock: Servers for rigid and symmetric docking. *Nucleic Acids Res.* **2005**, *33*, W363–W367. [CrossRef]
33. Felsenstein, J. Confidence limits on phylogenies: An approach using the bootstrap. *Evolution* **1985**, *39*, 783–791. [CrossRef]
34. Składanowski, M.; Wypij, M.; Laskowski, D.; Golińska, P.; Dahm, H.; Rai, M. Silver and gold nanoparticles synthesized from *Streptomyces* sp. isolated from acid forest soil with special reference to its antibacterial activity against pathogens. *J. Clust. Sci.* **2016**, *28*, 59–79. [CrossRef]
35. Al-Dhabi, N.A.; Ghilan, A.-K.M.; Arasu, M.V. Characterization of Silver Nanomaterials Derived from Marine *Streptomyces* sp. Al-Dhabi-87 and Its In Vitro Application against Multidrug Resistant and Extended-Spectrum Beta-Lactamase Clinical Pathogens. *Nanomaterials* **2018**, *8*, 279. [CrossRef]
36. Lakhundi, S.; Zhang, K. Methicillin-Resistant Staphylococcus aureus: Molecular Characterization, Evolution, and Epidemiology. *Clin. Microbiol. Rev.* **2018**, *31*, e00020-18. [CrossRef]
37. Chudasama, B.; Vala, A.K.; Andhariya, N.; Mehta, R.V.; Upadhyay, R.V. Highly bacterial resistant silver nanoparticles: Synthesis and antibacterial activities. *J. Nanoparticle Res.* **2010**, *12*, 1677–1685. [CrossRef]
38. Prakash, A.; Sharma, S.; Ahmad, N.; Ghosh, A.; Sinha, P. Synthesis of Agnps By Bacillus Cereus Bacteria and Their Antimicrobial Potential. *J. Biomater. Nanobiotechnology* **2011**, *2*, 155–161. [CrossRef]
39. Paul, D.; Sinha, S.N. Extracellular Synthesis of Silver Nanoparticles Using *Pseudomonas Aeruginosa* KUPSB12 and Its Antibacterial Activity. *Jordan J. Biol. Sci.* **2014**, *7*, 245–250. [CrossRef]
40. Agnihotri, S.; Mukherji, S.; Mukherji, S. Size-controlled silver nanoparticles synthesized over the range 5–100 nm using the same protocol and their antibacterial efficacy. *RSC Adv.* **2013**, *4*, 3974–3983. [CrossRef]
41. Kim, J.S.; Kuk, E.; Yu, K.N.; Kim, J.-H.; Park, S.J.; Lee, H.J.; Kim, S.H.; Park, Y.K.; Park, Y.H.; Hwang, C.-Y.; et al. Antimicrobial effects of silver nanoparticles. *Nanomed. Nanotechnol. Biol. Med.* **2007**, *3*, 95–101, Erratum in *Nanomed. Nanotechnol. Biol. Med.* **2014**, *10*, e1119. [CrossRef] [PubMed]
42. Mendis, E.; Rajapakse, N.; Byun, H.-G.; Kim, S.-K. Investigation of jumbo squid (Dosidicus gigas) skin gelatin peptides for their in vitro antioxidant effects. *Life Sci.* **2005**, *77*, 2166–2178. [CrossRef] [PubMed]
43. Priyadarshini, S.; Gopinath, V.; Priyadharsshini, N.M.; MubarakAli, D.; Velusamy, P. Synthesis of anisotropic silver nanoparticles using novel strain, Bacillus flexus and its biomedical application. *Colloids Surfaces B Biointerfaces* **2013**, *102*, 232–237. [CrossRef] [PubMed]
44. Dibrov, P.; Dzioba, J.; Gosink, K.K.; Hase, C.C. Chemiosmotic Mechanism of Antimicrobial Activity of Ag^+ in Vibrio cholerae. *Antimicrob. Agents Chemother.* **2002**, *46*, 2668–2670. [CrossRef] [PubMed]

45. Hamouda, T.; Myc, A.; Donovan, B.; Shih, A.Y.; Reuter, J.D.; Baker, J.R. A novel surfactant nanoemulsion with a unique non-irritant topical antimicrobial activity against bacteria, enveloped viruses and fungi. *Microbiol. Res.* **2001**, *156*, 1–7. [CrossRef]
46. Sondi, I.; Salopek-Sondi, B. Silver nanoparticles as antimicrobial agent: A case study on *E. coli* as a model for Gram-negative bacteria. *J. Colloid Interface Sci.* **2004**, *275*, 177–182. [CrossRef]
47. Amro, N.A.; Kotra, L.P.; Wadu-Mesthrige, K.; Bulychev, A.; Mobashery, S.; Liu, G.-Y. High-Resolution Atomic Force Microscopy Studies of the *Escherichia coli* Outer Membrane: Structural Basis for Permeability. *Langmuir* **2000**, *16*, 2789–2796. [CrossRef]
48. Durán, N.; Durán, M.; de Jesus, M.B.; Seabra, A.B.; Fávaro, W.J.; Nakazato, G. Silver nanoparticles: A new view on mechanistic aspects on antimicrobial activity. *Nanomedicine* **2016**, *12*, 789–799. [CrossRef]
49. McQuillan, J.S.; Infante, H.G.; Stokes, E.; Shaw, A.M. Silver nanoparticle enhanced silver ion stress response in *Escherichia coli* K12. *Nanotoxicology* **2011**, *6*, 857–866. [CrossRef]
50. Holt, K.B.; Bard, A.J. Interaction of Silver(I) Ions with the Respiratory Chain of *Escherichia coli*: An Electrochemical and Scanning Electrochemical Microscopy Study of the Antimicrobial Mechanism of Micromolar Ag^+. *Biochemistry* **2005**, *44*, 13214–13223. [CrossRef]

microorganisms

Article

Living and Regenerative Material Encapsulating Self-Assembled *Shewanella oneidensis*-CdS Hybrids for Photocatalytic Biodegradation of Organic Dyes

Mingyue Tao [1], Chenyang Jin [1], Hongfei Lu [1], Kai Jin [1], Lin Yu [2], Jinliang Liu [1], Jing Zhang [1], Xiaohui Zhu [1,*] and Yihan Wu [1,*]

1 Department of Chemical and Environmental Engineering, Shanghai University, Shanghai 200433, China
2 Medical School, Shanghai University, Shanghai 200433, China
* Correspondence: xhzhu@shu.edu.cn (X.Z.); yihanw@shu.edu.cn (Y.W.)

Abstract: Reductive biodegradation by microorganisms has been widely explored for detoxifying recalcitrant contaminants; however, the biodegradation capacity of microbes is limited by the energy level of the released electrons. Here, we developed a method to self-assemble *Shewanella oneidensis*-CdS nanoparticle hybrids with significantly improved reductive biodegradation capacity and constructed a living material by encapsulating the hybrids in hydrogels. The material confines the nano-bacteria hybrids and protects them from environmental stress, thus improving their recyclability and long-term stability (degradation capacity unhindered after 4 weeks). The developed living materials exhibited efficient photocatalytic biodegradation of various organic dyes including azo and nitroso dyes. This study highlights the feasibility and benefits of constructing self-assembled nano-bacteria hybrids for bioremediation and sets the stage for the development of novel living materials from nano-bacteria hybrids.

Keywords: nano-bacteria hybrid; hydrogel; biodegradation; *Shewanella oneidensis*; living material

Citation: Tao, M.; Jin, C.; Lu, H.; Jin, K.; Yu, L.; Liu, J.; Zhang, J.; Zhu, X.; Wu, Y. Living and Regenerative Material Encapsulating Self-Assembled *Shewanella oneidensis*-CdS Hybrids for Photocatalytic Biodegradation of Organic Dyes. *Microorganisms* **2022**, *10*, 2501. https://doi.org/10.3390/microorganisms10122501

Academic Editor: Kamel Abd-Elsalam

Received: 14 November 2022
Accepted: 14 December 2022
Published: 16 December 2022

1. Introduction

Microbial biodegradation has been extensively investigated as a simple, cost-effective, and environment-friendly strategy for detoxifying pollutants. Microbes have been demonstrated to degrade a variety of chemicals, such as small-molecule organic dyes and synthetic plastics [1,2]. Under aerobic conditions, organic contaminants can be oxidized by microorganisms and utilized as carbon or energy sources through aerobic respiration [3]. In addition to oxidization, microbial reductive biodegradation has been explored to biotransform recalcitrant pollutants such as halogenated chemicals and azo dyes [4]. During reductive biodegradation, microbes can reduce contaminants through their reductases. For instance, azo-reductases in bacteria and fungi have been characterized and recognized as enzymes contributing to the decolorization of azo dyes [5].

Because enzymes have substrate specificity, the application of enzyme-based biodegradation is limited to certain compounds with specific functional groups. A more general approach can be developed by harnessing the reductive degradation capacity of electrochemically active bacteria (EAB). These EAB are capable of respiring using extracellular electron acceptors through extracellular electron transfer. Chemicals such as azo dyes can capture electrons released from EAB and can be reduced without catalysis by azo reductases [6]. However, the reductive biodegradation capacity of bacterial cells is limited by the energy levels of the electrons released from the EAB. Strategies to aid electron transfer or enhance the energy level of electrons from the EAB can improve the reductive degradation performance [7].

Attempts have been made to engineer bacterial cells by constructing nano-bio hybrids. For example, nano-bacteria hybrids constructed from non-photosynthetic bacteria can harvest solar energy and produce acetic acids from CO_2. Nano-bio hybrids have recently been applied in diverse fields including catalysis, energy storage, and electronics. Here, we report an approach for the *in situ* self-assembly of *S. oneidensis*-CdS nanoparticle hybrids under pseudo anaerobic conditions, which can be directly employed for the photocatalytic degradation of organic dyes (Figure 1). *Shewanella oneidensis* is a Gram-negative facultative anaerobe with a sophisticated respiratory network that uses diverse electron acceptors [8,9]. Owing to its prominent capability for extracellular electron generation and transfer, *S. oneidensis* has been explored for a wide range of applications including biodegradation [10–16]. The biodegradation performance of the hybrids was significantly improved compared to that of *S. oneidensis* without engineering. The application of nano-bio hybrids is primarily limited by the fragility and susceptibility of hybrid cells [17]. To address the challenges, we constructed an engineered living material (ELM) [18,19] based on the *S. oneidensis*-CdS hybrids (Figure 1). The effectiveness of the ELMs for the photocatalytic degradation of trypan blue (containing azo bonds) and naphthol green B (NGB; containing nitroso groups) was demonstrated. The material was self-sufficient with nutrients to support encapsulated live hybrid bacterial cells. These cells could survive the construction and storage processes and maintain their ability to photocatalytically degrade organic pollutants. This functional living material improved the stability of the nano-bacteria hybrids and prevented their escape. Outside ELMs, nano-bacteria hybrids cannot survive beyond 3 days, whereas inside ELMs, their viability and photocatalytic degradation capacity remained unhindered after 4 weeks. The ELMs also possessed regenerative abilities; that is, successive regeneration of new ELMs could be performed using one parent ELM containing nano-bacteria hybrids as the seed. Our study highlights the feasibility and benefits of constructing self-assembled nano-bacteria hybrids for bioremediation and sets the stage for the development of novel living materials from nano-bacteria hybrids.

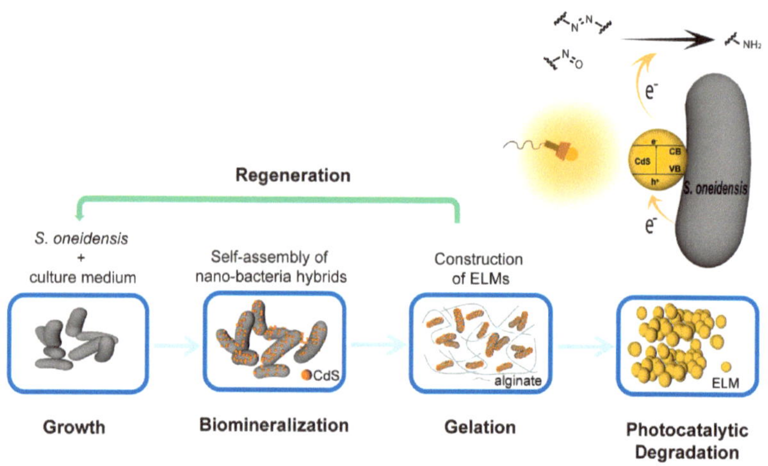

Figure 1. Schematic showing the workflow to construct engineered living materials (ELMs) with encapsulated nano-bacteria hybrids. The CdS nanoparticles biomineralized by *Shewanella oneidensis* are deposited on the surface of cells to self-assemble nano-bacteria hybrids. The hybrids are encapsulated in an alginate hydrogel containing nutrients to form ELMs. Nano-bacteria hybrids in the ELMs can remove contaminants via photocatalytic reductive degradation.

2. Materials and Methods

2.1. Materials

Shewanella oneidensis MR-1 (JCM31522) was obtained from the Typical Culture Conservation Center of China. Luria Bertani (LB) broth and phosphate-buffered saline (PBS) powder were purchased from Solarbio. Sodium alginate, lactic acid, urea, piperazine-1,4-bis (2-ethanesulfonic acid) monosodium salt (PIPES), cadmium chloride, and sodium thiosulfate were purchased from Shanghai Titan Scientific. Modified Wolfe's mineral solution (1.5 g/L nitrilotriacetic acid, 3 g/L $MgSO_4$•$7H_2O$, 0.5 g/L $MnSO_4$•H_2O, 1 g/L NaCl, 0.1 g/L $FeSO_4$•$7H_2O$, 0.1 g/L $CoSO_4$•$7H_2O$, 0.1 g/L $CaCl_2$•$2H_2O$, 0.1 g/L $ZnSO_4$•$7H_2O$, 0.01 g/L AlK $(SO_4)_2$, 0.01 g/L H_3BO_3, 0.01 g/L Na_2MoO_4•$2H_2O$, 0.01 g/L $CuSO_4$•$5H_2O$, and 0.01 g/L $NiCl_2$•$6H_2O$, pH = 7) and Wolfe's vitamin solution (20 mg/L biotin, 20 mg/L folic acid, 100 mg/L pyridoxine hydrochloride, 50 mg/L thiamine hydrochloride, 50 mg/L riboflavin, 50 mg/L nicotinic acid, 50 mg/L calcium D- (+)-pantothenate, 1 mg/L vitamin B12, 50 mg/L p-aminobenzoic acid and 50 mg/L thioctic acid) were purchased from Biofeng. Lactic acid was filtered through a 0.22 μm filter membrane before use.

2.2. Methods

2.2.1. Self-Assembly of Nano-Bacteria Hybrids

A single colony of *S. oneidensis* grown on LB agar was used to inoculate 5 mL of LB medium in a 14 mL culture tube. The culture was incubated at 30 °C and 250 rpm for 12 h. A fresh LB medium solution (5 mL) in a 14 mL culture tube was inoculated with 50 μL of the 12-h culture and cultured at 30 °C and 250 rpm. After 24-h incubation, the bacterial cells were collected by centrifugation at 4500× *g* for 10 min. The cell pellet was washed three times with 10 mL of sterile PBS solution and then 10 mL of mineral medium (0.3 g NaOH, 1.5 g NH_4Cl, 0.1 g KCl, 5.85 g NaCl, 15.12 g PIPES, 1 mL Wolfe's vitamin solution, 1 mL modified Wolfe's mineral solution, and 9 mM lactic acid dissolved in 1 L of ultrapure water). After washing, the cell pellet was resuspended in a mineral medium to obtain a solution with a cell density of $OD_{600\ nm}$ = 0.5. $Na_2S_2O_3$ (final concentration, 2 mM) and lactic acid (final concentration, 9 mM) were added to the cell suspension. Capped culture tubes filled with the as-prepared cell suspension (microaerobic conditions) were incubated at 30 °C for 24 h. $CdCl_2$ was added at a final concentration of 1 mM and incubated at 30 °C for an additional 48 h.

2.2.2. TEM/EDS/SAED Analysis of Nano-Bacteria Hybrids

After biosynthesis, the nano-bacteria hybrids were collected from 0.01 mL of solution and washed three times with PBS. A small drop of the resulting diluent was placed on a carbon grid and heated to dehydrate. The as-prepared sample was analyzed using a HITACHI-HT7820 TEM and JEM-2010F HRTEM.

2.2.3. Characterization of CdS Isolated from Nano-Bacteria Hybrids

The nano-bacteria hybrids from 420 mL of the biosynthesis culture were washed with water and resuspended in 20 mL of water. The mixture was then processed using an ultrasonic homogenizer at a power density of 350 W for 2 h. After homogenization, the pellet was collected and heated at 30 °C for 1 h in a solution containing 50 mM Tris-HCl, 50 mM NaCl, and 8 M urea (50 mL; pH 7.5). The precipitate was spun down from the resulting mixture and heated at 30 °C for 1 h in a solution containing 50 mM Tris-HCl, 50 mM NaCl, and 0.2% Tween-20 (50 mL, pH 7.5). The collected precipitate was washed three times each with PBS (50 mL), ethanol (20 mL), and cyclohexane (20 mL). After thorough washing, the isolated CdS nanoparticles were freeze-dried for 12 h. The fluorescence spectrum (excitation wavelength of 350 nm) was measured on a spectrofluorometer (FS5) using a suspension of isolated CdS in a mineral solution.

2.2.4. Degradation of Trypan Blue or NGB by Nano-Bacteria Hybrids

Nano-bacteria hybrids from 56 mL of biosynthesis culture were washed three times with mineral medium (20 mL) and added to a mineral medium solution (50 mL) containing 20 mg/L trypan blue or 100 mg/L NGB. The resulting mixture was irradiated in a capped 50 mL falcon tube with a tungsten filament lamp (60 W) at a power density of 12.9 mW/cm^2. The control group samples (isolated CdS or *S. oneidensis* cells) were washed and resuspended using the same procedure. The non-irradiated control groups were wrapped in tin foil. The concentrations of trypan blue and NGB over time were determined based on absorbance at 583 nm and 714 nm, respectively.

2.2.5. Preparation of ELMs

Nano-bacteria hybrids collected from 56 mL of biosynthesis culture were washed three times with LB (10 mL) and then resuspended in 5 mL of LB. The resuspension was mixed with 2 mL sterilized sodium alginate solution (0.1 g/mL). The as-prepared mixture was added dropwise (1 drop every three seconds) to a $CaCl_2$ (0.1 M) solution using a syringe. The hydrogel spheres encapsulating the nano-bacteria hybrids were washed three times with a PBS solution (60 mL). For the control groups containing unmodified *S. oneidensis* cells, 56 mL of cell culture without $CdCl_2$ was used. For the control groups containing only isolated CdS nanoparticles, CdS nanoparticles were isolated from 56 mL of nano-bacteria hybrid solution following the procedure described above and resuspended in 5 mL of LB to be mixed with the sodium alginate solution.

2.2.6. Degradation of Trypan Blue or NGB with ELMs

ELM spheres with encapsulated nano-bacteria hybrids were added to 50 mL of a mineral medium solution containing 20 mg/L trypan blue or 100 mg/L NGB. The samples were irradiated at room temperature with a tungsten filament lamp (60 W, 250–3000 nm) at a power density of 12.9 mW/cm^2 in capped 50 mL falcon tubes. The non-irradiated control groups were wrapped in tin foil. The concentrations of trypan blue and NGB over time were determined based on absorbance at 583 nm and 714 nm, respectively. In addition to the mineral medium solution, the same degradation procedure was performed in PBS and water.

2.2.7. Recycling and Recharging of ELMs

After degradation, ELM spheres were collected by filtration, washed three times with 20 mL PBS, and then incubated in LB (20 mL) at 30 °C for 24 h. After incubation, the spheres were collected by filtration and washed three times with 20 mL mineral medium solution. The degradation capacity of recycled ELM spheres was assayed according to the procedure described above.

2.2.8. Assessment of Cell Viability in ELMs

The ELMs were incubated in 20 mg/L trypan blue aqueous solution for 24 h with or without irradiation. The ELMs incubated in water without trypan blue were used as control groups. After 24 h of incubation, 10 hydrogel spheres were washed three times with 60 mL of PBS and subsequently incubated with 2 mL of EDTA solution (0.1 M EDTA, 0.2 M K_2HPO_4, pH 6.9). A dilution of the resulting solution was plated on an LB agar plate and incubated at 30 °C for 24 h to count the colonies. The resulting solution was subjected to fluorescence microscopy, flow cytometry, and TEM analyses. The samples were stained with a mixture of SYTO 9 and PI for 25 min and analyzed using a confocal laser scanning microscope (OLYMPUS FV3000) and a flow cytometer (CytoFLEX). For TEM analysis, 10 µL of the sample was dropped onto a carbon grid and dried. After recycling, the same analyses were performed on the recovered ELMs.

2.2.9. Biosafety Assessment of ELMs

After 12 h of degradation, the ELMs were removed from the solution via filtration. Then, 0.1 mL of the residual solution was plated on an LB agar plate and incubated at 30 °C. After more than 24 h, the plates were inspected to determine whether any colonies had formed. The same procedure was performed for the samples after a 36-h degradation process.

2.2.10. Stability Assessment of ELMs

The ELM spheres were stored at 4 °C, −20 °C, and −80 °C. The degradation capacity of the samples was assayed after 14 and 28 days to evaluate the stability of the ELMs under different storage conditions.

2.2.11. Regeneration of ELMs

After 24 h of degradation, the ELM spheres were crushed after washing three times with water (10 mL). The crushed spheres were incubated in 5 mL of LB medium at 30 °C and 250 rpm in a 14 mL culture tube. After 12 h, 10 μL of the suspension was plated on an LB agar plate and incubated at 30 °C. After 24-h incubation, a colony was picked from the plate and inoculated into 5 mL of LB medium. After incubation at 30 °C and 250 rpm for 24 h, the resulting cell culture was used for the biosynthesis of nano-bacteria hybrids and regeneration of ELMs. This procedure was repeated for 10 generations.

3. Results and Discussion

3.1. *Self-Assembly and Characterization of Nano-Bacteria Hybrids*

As *S. oneidensis* is a facultative anaerobic bacterium, it can consume oxygen via aerobic respiration. We speculated that *S. oneidensis* could deplete dissolved oxygen molecules to generate a pseudo anaerobic environment [20] and use $Na_2S_2O_3$ as the electron acceptor producing S^{2-} to form CdS with Cd^{2+}. Thus, a strict anoxic culturing process is not required. *Shewanella oneidensis* was cultured in capped culture vials supplemented with $Na_2S_2O_3$ as the electron acceptor. The headspace volume was maintained at 5% of the vials and was not exchanged to eliminate oxygen. Upon addition of $CdCl_2$ to the *S. oneidensis* culture, the culture solution gradually turned yellow, indicating CdS nanocrystal formation (Figure S1).

Self-biomineralized CdS nanoparticles were deposited on the surface of the *S. oneidensis* cells (Figure 2a). The formation of such nano-bacteria hybrids did not significantly alter the morphology of *S. oneidensis* cells, suggesting that the integrity of cell membranes was preserved, which is critical for electron generation and transfer. To determine the chemical composition and morphology of the biosynthesized CdS nanoparticles, they were isolated from the nano-bacteria hybrids. The SEM-EDS results demonstrated that the ratio of Cd to S atoms in the crystals was nearly 1:1, suggesting that the chemical composition was close to that of stoichiometric CdS (Figure S2). According to the high-resolution TEM results, the d-spacing of the isolated CdS nanoparticles was 3.3 Å which aligned with the (111) plane [21] of cubic CdS (Figure 2b). The electron diffraction ring pattern agreed with that of CdS in the face-centered cubic structure (Figure 2c) [22]. The fluorescence spectrum excited at 350 nm showed an emission peak at 450 nm, which is comparable to that of the chemically synthesized CdS (Figure S3) [23].

Collectively, these results unambiguously suggest that *S. oneidensis*-CdS nano-bacteria hybrids were constructed by supplementing the microaerobic *S. oneidensis* cultures with $Na_2S_2O_3$ and $CdCl_2$. The CdS nanoparticles biosynthesized and deposited on *S. oneidensis* cells had face-centered cubic structures.

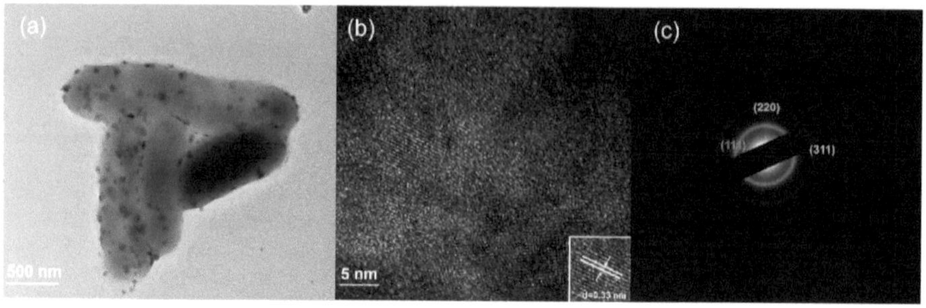

Figure 2. Characterization of the CdS-*Shewanella oneidensis* hybrids self-assembled under pseudo anaerobic conditions: (**a**) TEM image of nano-bacteria hybrids showing self-biomineralized CdS nanocrystals deposited on the cell surface; (**b**) Electron diffraction pattern of the CdS nanocrystals in nano-bacteria hybrids from HRTEM results. The d-spacing of the CdS nanoparticles is 3.3 Å which aligns with the (111) plane of face-centered cubic CdS; (**c**) SAED analysis of the CdS nanocrystals in nano-bacteria hybrids. The indices are assigned to the diffraction rings in accordance with the face-centered cubic lattice of CdS.

3.2. Photocatalytic Degradation of Azo Dyes by Nano-Bacteria Hybrids

The semiconductor CdS nanoparticles in the nano-bacteria hybrids can accept electrons from *S. oneidensis* cells. By absorbing light, CdS nanoparticles can excite the accepted electrons to a higher energy level for better reductive degradation. To determine whether the CdS nanoparticles can promote reductive degradation by *S. oneidensis*, the degradation performance of nano-bacteria hybrids for trypan blue (organic dye with azo groups) was compared to that of *S. oneidensis* without engineering. Because oxygen competes with contaminants as electron acceptors, reductive biodegradation is commonly performed under rigorous anaerobic conditions to prevent inhibition by oxygen [24]. Because *S. oneidensis* is a facultative anaerobe capable of both anaerobic and aerobic respiration [25,26], we speculated that the nano-bacteria hybrids could perform reductive biodegradation under microaerobic conditions as the oxygen would be consumed. Compared to other microbes explored for bioremediation (e.g., *Geobacter sulfurreducens*), the system is more robust and easier to handle, without the need for rigorous anaerobic operation.

The photocatalytic degradation of azo dyes by the nano-bacteria hybrids was assayed under microaerobic conditions. Nano-bacteria hybrid and wild-type *S. oneidensis* cells with the same number of colony-forming units were incubated in a mineral medium solution containing 20 mg/L trypan blue (simulating wastewater containing dyes and heavy metals). The reduction in the azo bond to the amine group in trypan blue results in a decrease in the characteristic absorption at 583 nm. The absorbance of the reaction mixture at 583 nm was therefore used to determine the degradation efficiency of each sample. The reaction mixture was sampled and measured every 2 h (Figure 3a). The degradation efficiency of the nano-bacteria hybrids was 33-fold higher than that of *S. oneidensis* cells (Figure 3b, Figure S4). After 12 h of treatment, the sample treated with nano-bacteria hybrids under light irradiation turned almost colorless, whereas the other control samples remained blue.

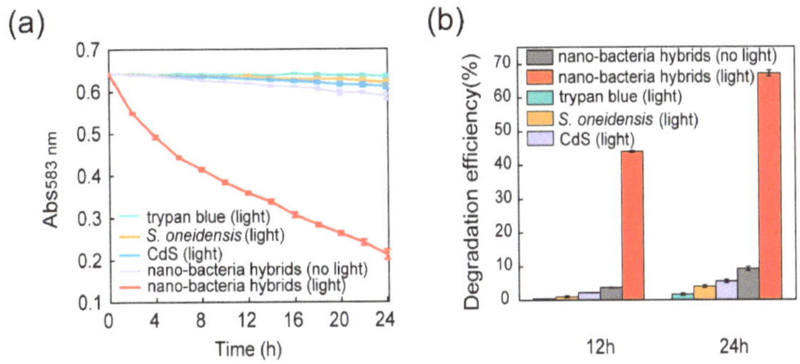

Figure 3. Photocatalytic degradation of trypan blue by nano-bacteria hybrids: (**a**) Change in concentration of trypan blue after treatment with nano-bacteria hybrids and controls. Nano-bacteria hybrids were incubated in a mineral medium solution containing 20 mg/L trypan blue and irradiated by a tungsten filament lamp. The absorbance of the reaction mixture at 583 nm was used to determine the change in concentration of trypan blue. Control groups included trypan blue only (with light), *Shewanella oneidensis* only (with light), CdS only (with light), and nano-bacteria hybrids (without light). The photocatalytic degradation performance of nano-bacteria hybrids was substantially improved compared to that of wild-type *S. oneidensis* cells; (**b**) The concentration of trypan blue after 12 h and 24 h of treatment was compared with the initial concentration to assess trypan blue degradation efficiency. The concentration was determined by comparing it to a trypan blue calibration curve generated based on $Abs_{583\,nm}$. The photocatalytic degradation efficiency of nano-bacteria hybrids was 33 times better than that of wild-type *S. oneidensis* cells. The data are presented as mean ± s.d ($n = 2$).

Without light irradiation, the degradation efficiency of nano-bacteria hybrids was similar to that of wild-type *S. oneidensis*, which is the baseline degradation from anaerobic respiration (biodegradation of less than 5% of dyes after 24 h) [27]. With light irradiation, the degradation efficiency of nano-bacteria hybrids was improved 7-fold. This improvement depends on photoexcitation, suggesting that the semiconductor CdS nanoparticles played a role [28]. For the abiotic sample containing only CdS nanoparticles but no *S. oneidensis*, no significant degradation was observed, even when photoexcited, suggesting that *S. oneidensis* cells provide electrons to fill the holes in the CdS nanoparticle's valence band to sustain reductive degradation. Taken together, these results suggest a degradation mechanism for the self-assembled nano-bacteria hybrids. The CdS nanoparticles generated excited electrons to reduce dyes under illumination. The paired holes in the valence band received electrons from the anaerobic microbial respiration of *S. oneidensis* to maintain the cycle. This mechanism is consistent with the recognized electron transfer process between *S. oneidensis* and abiotic materials [29,30].

3.3. Engineered Living Material with Encapsulated Nano-Bacteria Hybrids

The application of nano-bio hybrids is primarily limited by the fragility and susceptibility of hybrid cells, thus we attempted to transform the as-prepared nano-bacteria hybrids into ELMs to simplify the operation and enhance performance. The ELMs comprised functional live cells and scaffolding matrices, thus integrating the properties (e.g., mechanical properties and porosity) of conventional materials and the distinct functionality of live cells (e.g., responsiveness and regeneration) [31].

The self-assembled nano-bacteria hybrids were washed with PBS and resuspended in an LB medium. The suspension was mixed with sodium alginate powder, followed by injection into a $CaCl_2$ aqueous solution to form ELM spheres. The photocatalytic degradation performance of trypan blue by ELMs was evaluated (Figure 4a). For comparison with the results for nano-bacteria hybrids not encapsulated, the colony-forming units in the ELMs used were kept consistent with those in Section 3.2. After 24 h of incubation with a mineral medium solution containing trypan blue, ELMs containing nano-bacteria hybrids removed 73.9% of the dye compounds under light irradiation (Figure 4d, Figure S5). Without photoexcitation, only 19.3% of trypan blue could be removed by the ELMs, which is similar to the performance of alginate hydrogels containing *S. oneidensis* cells without engineering (Figure 4d). The slight increase in removal rate compared to the *S. onediensis* cells not encapsulated is attributed to the adsorption of dyes by alginate hydrogels. Taken together, the fabrication process of ELMs does not harm the activity of *S. oneidensis* or the photocatalytic degradation capacity of the nano-bacteria hybrids.

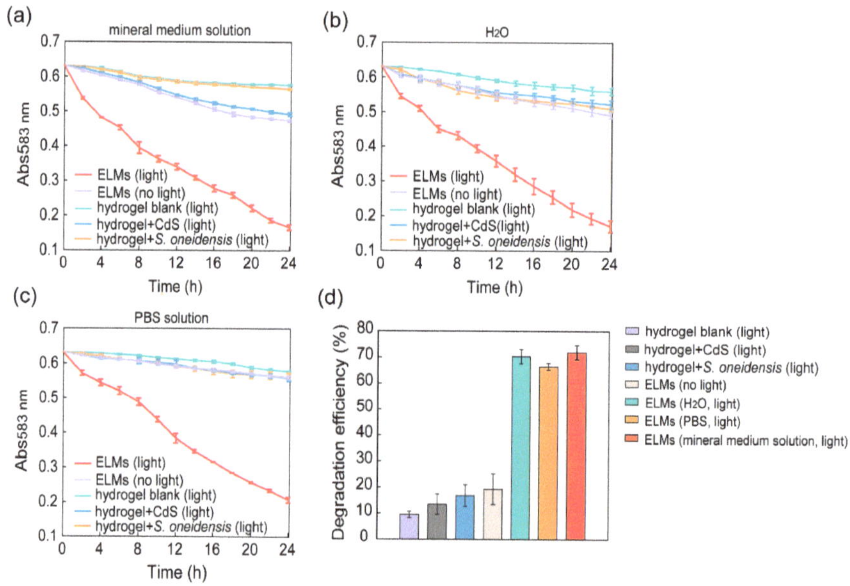

Figure 4. Photocatalytic degradation performance of engineered living materials (ELMs) with encapsulated nano-bacteria hybrids: (**a**) Change in concentration change of trypan blue after treatment with ELMs and controls in mineral medium solution. The ELMs were incubated in a mineral medium solution containing 20 mg/L trypan blue and irradiated by a tungsten filament lamp. Control groups included hydrogels (light), *Shewanella oneidensis* encapsulated in hydrogels (light), CdS encapsulated in hydrogels (light), and ELMs (no light). The photocatalytic degradation performance of ELMs was significantly improved compared to that of wild-type *S. oneidensis* cells encapsulated in hydrogels; (**b**) Change in concentration of trypan blue after treatment with ELMs and controls in water; (**c**) Change in concentration of trypan blue after treatment with ELMs and controls in PBS solution; (**d**) The concentration of trypan blue after 24 h of treatment was compared with the initial concentration to assess the trypan blue degradation efficiency. The concentration was determined by comparing it to a trypan blue calibration curve generated based on $Abs_{583\ nm}$. The photocatalytic degradation efficiency of ELMs was not affected by the media (i.e., simulated wastewater, water, and PBS solution). The data are presented as mean ± s.d (n = 2).

The bioremediation by *S. oneidensis* requires carbon sources to support cell survival [30,32]. As ELMs can contain nutrients, we speculated that photocatalytic degradation by ELMs can be performed in media without carbon sources. The degradation performance of the ELMs was assessed in PBS and water (Figure 4b,c). The ELMs degraded 74.4% and 67.6% of trypan blue in water and PBS solutions, respectively, after 24 h of treatment (Figure 4d, Figure S5). These results demonstrated that ELMs are self-sustaining and can support the survival of nano-bacteria hybrids as well as their ability to biodegrade in solutions without nutrients.

Recent reports have shown that biopolymer scaffolds of ELMs do not interfere with the biosensing and biosynthesis capabilities of encapsulated microbial cells [33,34]. However, there have been few reports examining ELMs containing nano-bacteria hybrids, and little is known about whether biopolymer scaffolds affect the light energy conversion of nano-bacteria hybrids. Our results illustrate the feasibility of fabricating ELMs from nano-bacteria hybrids and demonstrate that nano-bacteria hybrids can continue to perform photocatalytic degradation when encapsulated in biopolymers. Alginate was chosen as the biopolymer scaffold for ELMs because it is a widely used polymer for cell immobilization [35–39]. However, other hydrogel matrices, such as gelatin and bacterial cellulose, can be tested in the future to determine whether nano-bacterial hybrids can survive and perform photocatalytic degradation in other biopolymer scaffolds.

3.4. Recycling and Reuse of Encapsulated Nano-Bacteria Hybrids

The ELMs contain live *S. oneidensis* cells and the nutrients within ELMs support cell survival. The viability of nano-bacteria hybrid cells in ELMs was assessed after degradation treatment. After the degradation of trypan blue, the ELM spheres were crushed and incubated in an LB medium. Colonies formed on LB agar by plating the media confirmed the presence of live cells. Cell identity was confirmed using 16S rDNA sequencing. These results confirmed that there were nano-bacteria hybrid cells that remained alive throughout the treatment process. Because ELMs can preserve the viability of nano-bacteria hybrids, we attempted to recycle and reuse the materials after degradation treatment. After degrading trypan blue, ELM spheres were collected by spin-down and added to fresh trypan blue solution (Figure 5a). The recycled ELMs exhibited photocatalytic degradation capability, but their efficiency was reduced to 40.0% (Figure 5b and Figure S6). The cells were extracted from recycled ELMs to analyze their viability. Live cells were observed in the recycled ELM, but the ratio of live to dead cells decreased to 1:5 (Figure S7). To restore the viability of the encapsulated bacterial cells, the recycled ELM spheres were soaked in fresh LB medium for 24 h to allow cell growth. After incubation, the photocatalytic degradation capacity of trypan blue by ELM spheres was measured again. Our results demonstrate that incubation with LB media can rehabilitate recycled ELMs. The recycled ELMs treated with fresh LB media exhibited a photocatalytic degradation efficiency comparable to that of freshly prepared ELMs (Figure 5b). Incubation with LB media increased the percentage of live *S. oneidensis* cells in ELMs from 18.7% to 67.3% (Figure 6 and Figures S7–S10).

Collectively, these results suggest that the nano-bacteria hybrids can survive in ELMs throughout the degradation treatment, but their viability decreases, probably owing to the toxicity of trypan blue. The recycled ELMs contained fewer live bacterial cells; therefore, their degradation performance was hindered. Recharging used ELMs via incubation with a fresh culture medium could restore their photodegradation capacity by increasing the number of live bacterial cells. Although bacterial cells were observed to have nanoparticles deposited on their surfaces in recharged ELMs (Figure S10), the interaction between the nanoparticles and newly born bacteria needs further investigation.

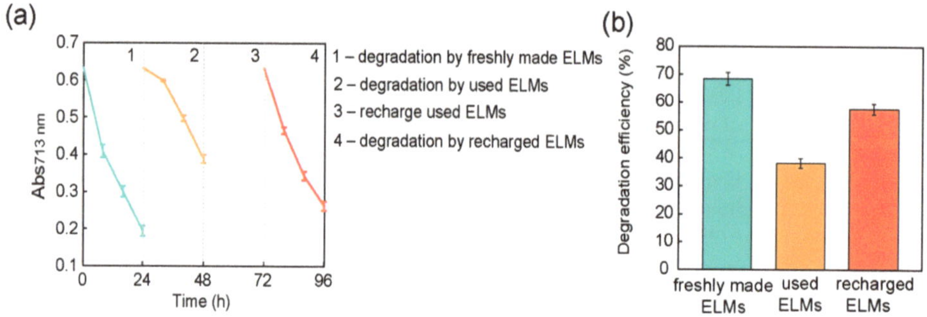

Figure 5. Recycle and recharge engineered living materials (ELMs) after the photocatalytic degradation process: (**a**) Recycling and reuse cycle of ELMs. After degrading trypan blue for 24 h, the ELMs were collected and used to degrade fresh trypan blue solution. The photocatalytic degradation efficiency was lower than that of freshly prepared ELMs. The recycled ELMs were incubated in Luria Bertani (LB) medium solution. After a 24-h incubation, the recharged ELMs were used for the photocatalytic degradation of trypan blue; the observed efficiency was comparable to that of freshly prepared ELMs; (**b**) Degradation efficiency of freshly prepared ELMs, used ELMs, and recharged ELMs after 24 h of treatment. The photocatalytic degradation efficiency of ELMs decreased after incubation with trypan blue but recovered after incubation with LB medium solution. The data are presented as mean ± s.d (n = 2).

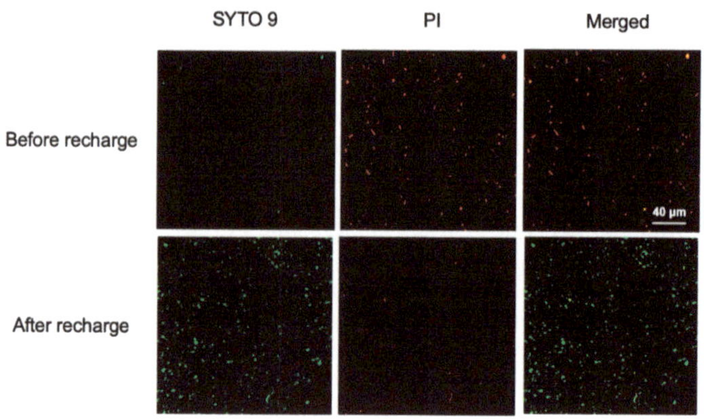

Figure 6. Cell viability in recharged engineered living materials (ELMs). After photocatalytic degradation of trypan blue, spheres were incubated in a culture medium solution. The recharged ELMs spheres were washed with PBS solution and dissolved using EDTA solution. The released cells were co-stained with SYTO 9 and PI and subsequently analyzed under a confocal laser scanning microscope. The majority of the hybrid cells in the recharged ELMs were alive.

3.5. Safety of Encapsulated Nano-Bacteria Hybrids

Biosafety is always a concern that must be addressed when microorganisms are involved in environmental applications. Although no genetic modification was performed on our nano-bacteria hybrids, the escape risk of the bacterial cells was carefully assessed. The solution was sampled after 12 h or 48 h of degradation and plated on LB agar. No bacterial colonies were observed on the inoculated plates after incubation at 30 °C for more than 24 h, confirming that no detectable leakage of bacterial cells occurred during the degradation process (Figure S11). The ELMs confined the nano-bacteria hybrids and prevented potential biosafety risks associated with live cell escape but allowed the exchange of organic small molecules for degradation.

A commonly overlooked problem in the application of nano-bacteria hybrids is that nanoparticles can peel off from cells. Nanoparticles, especially those containing heavy metals, have nanotoxicity. We observed that by encapsulating nano-bacteria hybrids in ELMs, not only were bacterial cells confined, but the leakage of nanoparticles was also significantly attenuated (Figure S12). The ELM-treated solution contained 15 times less cadmium than the solution treated with nano-bacteria hybrids not encapsulated in hydrogels. Taken together, ELMs can improve the biosafety of applications involving nano-bacteria hybrids and have the potential to reduce the risk of nanotoxicity associated with nanoparticles.

3.6. Stability of Encapsulated Nano-Bacteria Hybrids

The stability and required storage conditions of a material affect the ease of industrial application. Without additional protection, nano-bacterial hybrids did not survive beyond 3 days. The prepared ELMs contained alginate hydrogels, which are expected to protect nano-bacteria hybrids from environmental stress. To assess the stability of ELMs, ELM spheres containing nano-bacteria hybrids were stored at 4 °C, −20 °C, and −80 °C without additional protection. After 14 and 28 days, the photocatalytic degradation capability of the stored ELM spheres was analyzed. The ELMs stored under different conditions demonstrated similar degradation capacities for trypan blue. The degradation efficiency after storage was comparable to that of freshly prepared ELMs (Figure 7 and Figure S13). These results show that the fabricated ELMs can be stably stored over the long term, with the photocatalytic degradation capacity of nano-bacteria hybrids well preserved.

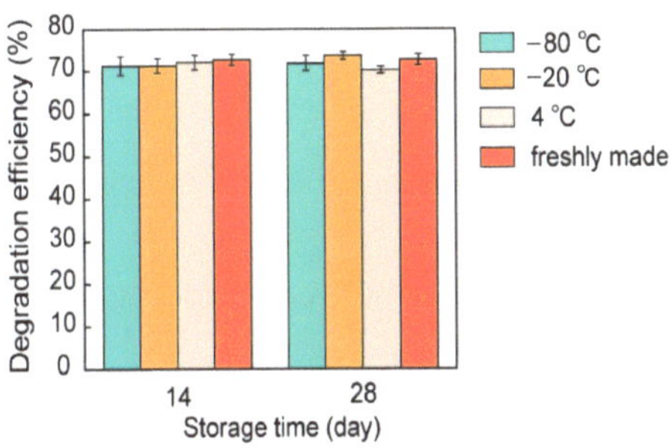

Figure 7. Assessment of the photocatalytic degradation performance of engineered living materials (ELMs) stored under different conditions. The concentration of trypan blue after 24 h of treatment was compared with the initial concentration to assess the trypan blue degradation efficiency. The concentration was determined by comparing it to a trypan blue calibration curve generated based on $Abs_{583\ nm}$. The ELMs could retain their photocatalytic degradation capacity for up to 28 days when stored at temperatures lower than 4 °C. The data are presented as mean ± s.d (n = 2).

3.7. Regeneration of ELMs

There has been growing interest in the development of regenerative ELMs [40–42]. The observation that nano-bacteria hybrids remain alive in ELMs suggests that recycled ELMs can be utilized to regenerate successive generations of ELMs. To assess the regenerative potential of the ELMs containing nano-bacteria hybrids, the spheres were crushed and incubated in LB medium solution after degradation treatment. Such a solution containing live bacterial cells was used to regenerate bacterial cultures for the biosynthesis of nano-bacteria hybrids. The as-prepared nano-bacteria hybrids were encapsulated in alginate hydrogels to regenerate ELMs. The regeneration process was repeated for 10 generations. For each generation of ELMs, the photocatalytic degradation capacity for trypan blue was measured and compared with that of the parental generation, which showed no significant decline (Figure 8 and Figure S14). The nano-bacteria hybrids in our ELMs could be extracted and used to regenerate more ELMs. The regenerated ELMs exhibited photocatalytic degradation performance that was comparable to that of the parental generation, suggesting that nano-bacteria hybrids in ELMs did not experience degeneration. These results highlight the sustainability of the proposed materials. The regenerative property makes it practical to construct ELMs in the field, which is useful when the ELMs are damaged or shipment of large amounts of ELMs is impractical.

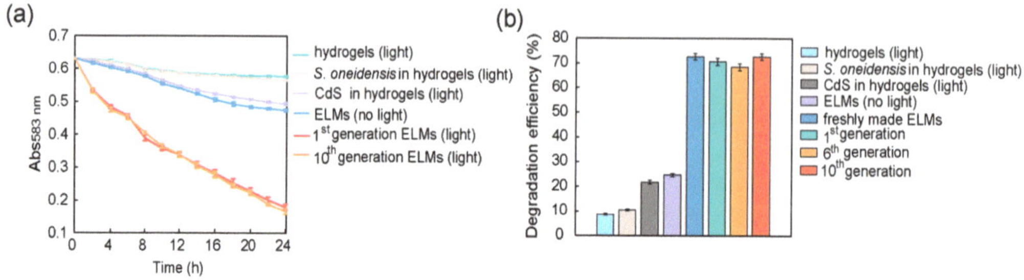

Figure 8. Assessment of the regenerative properties of engineered living materials (ELMs): (**a**) Change in concentration of trypan blue after treatment with the 1st and 10th generation of regenerated ELMs. Live cells were extracted from recycled ELMs and used to regenerate bacterial culture for the biosynthesis of nano-bacteria hybrids. The as-prepared nano-bacteria hybrids were encapsulated in alginate hydrogels to regenerate ELMs. The regeneration process was repeated for 10 generations; (**b**) The concentration of trypan blue after 24 h of treatment was compared with the initial concentration to assess the trypan blue degradation efficiency. The concentration was determined by comparing it to a trypan blue calibration curve generated based on $Abs_{583\ nm}$. Regenerated ELMs exhibited photocatalytic degradation efficiency comparable to that of freshly prepared ELMs. No significant difference was observed between the degradation performance of different generations of ELMs. The data are presented as mean ± s.d (n = 2).

3.8. Degradation of Other Dyes by Nano-Bacteria Hybrids and ELMs

Real wastewater has a complex composition; thus, techniques that can be applied to various contaminants are appealing. The ELM with nano-bacteria hybrids would be practically more useful if it could be applied to the reductive degradation of other types of contaminants as well. Because *S. oneidensis* is capable of biodegrading various organic pollutants other than azo dyes, such as sulfonamide and nitroso compounds [43,44], we tested the degradation capacity of ELMs for NGB. Naphthol green B contains nitroso groups, which can be converted into amino groups by reductive degradation. The degradation efficiency of NGB by *S. oneidensis*-CdS nano-bacteria hybrids and ELMs was assessed. The degradation of NGB led to a decrease in the absorption at 713 nm, which was used to measure its concentration. Treatment of 100 mg/L NGB solution with nano-bacteria hybrids for 24 h under light irradiation consumed more than 60% of the contaminants, which was a marked improvement compared to that with only *S. oneidensis* cells or the no-light control group (Figure S15). After encapsulating nano-bacteria hybrids in alginate hydrogels, the photocatalytic degradation efficiency of NGB by ELMs was not affected (Figure S15). The degradation efficiency of NGB by the ELMs was not hindered after long-term storage (Figure S16). The degradation performance of regenerated ELMs toward NGB was also evaluated and was found to be comparable to that of freshly prepared ELMs (Figure S17). These results demonstrate that ELMs constructed from *S. oneidensis*-CdS nano-bacteria hybrids have the potential to biodegrade recalcitrant chemicals with various functional groups.

4. Conclusions

We developed a method to self-assemble *S. oneidensis* with bacterially biosynthesized CdS nanoparticles under pseudo anaerobic conditions. The self-assembled nano-bacteria hybrids can be directly applied to degradation without the need to isolate and purify the nanoparticles. To improve the resistance of the material to environmental stress, nano-bacteria hybrids were encapsulated in ELMs using an alginate matrix. Our results suggest that immobilization with alginate hydrogels preserves the photocatalytic properties of nano-bacteria hybrids. More importantly, ELMs could enhance the viability of nano-bacteria hybrids. Outside ELMs, nano-bacteria hybrids cannot survive beyond 3 days, whereas inside ELMs, their viability and photocatalytic degradation capacity remained unhindered after 4 weeks.

The unique properties of ELMs in this study (i.e., recyclable, stable, and regenerative) make nano-bacteria hybrids a step closer to becoming wastewater treatment materials. To further explore these possibilities, the degradation performance and stability of the ELMs will be examined in real wastewater containing diverse chemicals in future investigation.

Supplementary Materials: The following supporting information can be downloaded at: https://www.mdpi.com/article/10.3390/microorganisms10122501/s1, Figure S1: The color change of the solution after biosynthesis of CdS; Figure S2: EDS analysis of the CdS nanoparticles isolated from nano-bacteria hybrids; Figure S3: Fluorescence spectrum of the CdS nanoparticles isolated from nano-bacteria hybrids; Figure S4: Photocatalytic degradation efficiency of nano-bacteria hybrids towards trypan blue; Figure S5: Photocatalytic degradation performance of ELMs with encapsulated nano-bacteria hybrids in various media; Figure S6: The degradation efficiency of ELMs during the recycling and recharging process was monitored; Figure S7: Flow cytometry analysis of cells in ELMs before recharge; Figure S8: Flow cytometry analysis of cells in ELMs after recharge; Figure S9: Cell viability in ELMs assessed by plate counting before and after recharge; Figure S10: TEM image of nano-bacteria hybrids extracted from ELMs; Figure S11: Biosafety assessment of ELMs; Figure S12: Assessment of nanoparticle leakage from nano-bacteria hybrids and ELMs; Figure S13: Photocatalytic degradation performance of trypan blue by ELMs stored under different conditions; Figure S14: Photocatalytic degradation performance of trypan blue by regenerated ELMs; Figure S15: Photocatalytic degradation of naphthol green B by nano-bacteria hybrids and ELMs; Figure S16: Photocatalytic degradation performance of NGB by ELMs stored under different conditions; Figure S17: Photocatalytic degradation performance of NGB by regenerated ELMs.

Author Contributions: Conceptualization, X.Z. and Y.W.; Data curation, M.T., C.J., H.L. and L.Y.; Formal analysis, M.T.; Investigation, M.T.; Project administration, J.L., X.Z. and Y.W.; Supervision, J.Z. and X.Z.; Vali-dation, C.J., H.L., K.J., L.Y. and J.L.; Writing, review & editing, K.J. and Y.W. All authors have read and agreed to the published version of the manuscript.

Funding: We received the financial support from Shanghai Sailing Program (20YF1413900), National Natural Science Foundation of China (32000036), Innovative Research Team of High-Level Local Universities in Shanghai, and Shanghai University Sailing Program.

Institutional Review Board Statement: Not applicable.

Informed Consent Statement: Not applicable.

Data Availability Statement: The authors confirm that the data supporting the findings of this study are available within the article and its supplementary materials.

Conflicts of Interest: There are no conflicts to declare.

References

1. Varjani, S.; Rakholiya, P.; Ng, H.Y.; You, S.; Teixeira, J.A. Microbial Degradation of Dyes: An Overview. *Bioresour. Technol.* **2020**, *311*, 123728. [CrossRef] [PubMed]
2. Ru, J.; Huo, Y.; Yang, Y. Microbial Degradation and Valorization of Plastic Wastes. *Front. Microbiol.* **2020**, *11*, 1–20. [CrossRef] [PubMed]
3. Holliger, C.; Gaspard, S.; Glod, G.; Heijman, C.; Schumacher, W.; Schwarzenbach, R.P.; Vazquez, F. Contaminated Environments in the Subsurface and Bioremediation: Organic Contaminants. *FEMS Microbiol. Rev.* **1997**, *20*, 517–523. [CrossRef] [PubMed]
4. Kamal, I.M.; Abdeltawab, N.F.; Ragab, Y.M.; Farag, M.A.; Ramadan, M.A. Biodegradation, Decolorization, and Detoxification of Di-Azo Dye Direct Red 81 by Halotolerant, Alkali-Thermo-Tolerant Bacterial Mixed Cultures. *Microorganisms* **2022**, *10*, 994. [CrossRef] [PubMed]
5. Ramalho, P.A.; Cardoso, M.H.; Cavaco-Paulo, A.; Ramalho, M.T. Characterization of Azo Reduction Activity in a Novel Ascomycete Yeast Strain. *Appl. Environ. Microbiol.* **2004**, *70*, 2279–2288. [CrossRef]

6. Liu, Y.N.; Zhang, F.; Li, J.; Li, D.B.; Liu, D.F.; Li, W.W.; Yu, H.Q. Exclusive Extracellular Bioreduction of Methyl Orange by Azo Reductase-Free Geobacter Sulfurreducens. *Environ. Sci. Technol.* **2017**, *51*, 8616–8623. [CrossRef]
7. Yan, F.F.; He, Y.R.; Wu, C.; Cheng, Y.Y.; Li, W.W.; Yu, H.Q. Carbon Nanotubes Alter the Electron Flow Route and Enhance Nitrobenzene Reduction by Shewanella Oneidensis MR-1. *Environ. Sci. Technol. Lett.* **2013**, *1*, 128–132. [CrossRef]
8. Hirose, A.; Kasai, T.; Aoki, M.; Umemura, T.; Watanabe, K.; Kouzuma, A. Electrochemically Active Bacteria Sense Electrode Potentials for Regulating Catabolic Pathways. *Nat. Commun.* **2018**, *9*, 1–10. [CrossRef]
9. Sun, W.; Lin, Z.; Yu, Q.; Cheng, S.; Gao, H. Promoting Extracellular Electron Transfer of Shewanella Oneidensis MR-1 by Optimizing the Periplasmic Cytochrome c Network. *Front. Microbiol.* **2021**, *12*, 2963. [CrossRef]
10. Miran, W.; Long, X.; Huang, W.; Okamoto, A. Current Production Capability of Drug-Resistant Pathogen Enables Its Rapid Label-Free Detection Applicable to Wastewater-Based Epidemiology. *Microorganisms* **2022**, *10*, 472. [CrossRef]
11. Min, D.; Cheng, L.; Zhang, F.; Huang, X.N.; Li, D.B.; Liu, D.F.; Lau, T.C.; Mu, Y.; Yu, H.Q. Enhancing Extracellular Electron Transfer of Shewanella Oneidensis MR-1 through Coupling Improved Flavin Synthesis and Metal-Reducing Conduit for Pollutant Degradation. *Environ. Sci. Technol.* **2017**, *51*, 5082–5089. [CrossRef]
12. Ng, C.K.; Sivakumar, K.; Liu, X.; Madhaiyan, M.; Ji, L.; Yang, L.; Tang, C.; Song, H.; Kjelleberg, S.; Cao, B. Influence of Outer Membrane C-Type Cytochromes on Particle Size and Activity of Extracellular Nanoparticles Produced by Shewanella Oneidensis. *Biotechnol. Bioeng.* **2013**, *110*, 1831–1837. [CrossRef]
13. Khanal, A.; Ho, C.T.; Hur, H.-G.; Lee, J.-H. Comparative Study of Roxarsone Reduction by Shewanella Oneidensis MR-1 and Cellulomonas Sp. Strain Cellu-2a. *Appl. Sci.* **2022**, *12*, 1839. [CrossRef]
14. Xiao, X.; Ma, X.B.; Yuan, H.; Liu, P.C.; Lei, Y.B.; Xu, H.; Du, D.L.; Sun, J.F.; Feng, Y.J. Photocatalytic Properties of Zinc Sulfide Nanocrystals Biofabricated by Metal-Reducing Bacterium Shewanella Oneidensis MR-1. *J. Hazard. Mater.* **2015**, *288*, 134–139. [CrossRef]
15. Tan, W.; Wang, L.; Yu, H.; Zhang, H.; Zhang, X.; Jia, Y.; Li, T.; Dang, Q.; Cui, D.; Xi, B. Accelerated Microbial Reduction of Azo Dye by Using Biochar from Iron-Rich-Biomass Pyrolysis. *Materials* **2019**, *12*, 1079. [CrossRef]
16. Zhu, F.; Huang, Y.; Ni, H.; Tang, J.; Zhu, Q.; Long, Z.; Zou, L. Biogenic Iron Sulfide Functioning as Electron-Mediating Interface to Accelerate Dissimilatory Ferrihydrite Reduction by Shewanella Oneidensis MR-1. *Chemosphere* **2022**, *288*, 132661. [CrossRef]
17. Bouabidi, Z.B.; El-Naas, M.H.; Zhang, Z. Immobilization of Microbial Cells for the Biotreatment of Wastewater: A Review. *Environ. Chem. Lett.* **2019**, *17*, 241–257. [CrossRef]
18. Chen, A.Y.; Zhong, C.; Lu, T.K. Engineering Living Functional Materials. *ACS Synth. Biol.* **2015**, *4*, 8–11. [CrossRef]
19. Nguyen, P.Q.; Courchesne, N.M.D.; Duraj-Thatte, A.; Praveschotinunt, P.; Joshi, N.S. Engineered living materials: Prospects and challenges for using biological systems to direct the assembly of smart materials. *Adv. Mater.* **2018**, *30*, 1704847. [CrossRef]
20. Fan, G.; Graham, A.J.; Kolli, J.; Lynd, N.A.; Keitz, B.K. Aerobic Radical Polymerization Mediated by Microbial Metabolism. *Nat. Chem.* **2020**, *12*, 638–646. [CrossRef]
21. Xiao, X.; Han, X.; Wang, L.G.; Long, F.; Ma, X.L.; Xu, C.C.; Ma, X.B.; Wang, C.X.; Liu, Z.Y. Anaerobically Photoreductive Degradation by CdS Nanocrystal: Biofabrication Process and Bioelectron-Driven Reaction Coupled with Shewanella Oneidensis MR-1. *Biochem. Eng. J.* **2020**, *154*, 107466. [CrossRef]
22. Bai, H.; Zhang, Z.; Guo, Y.; Jia, W. Biological Synthesis of Size-Controlled Cadmium Sulfide Nanoparticles Using Immobilized Rhodobacter Sphaeroides. *Nanoscale Res. Lett.* **2009**, *4*, 717–723. [CrossRef] [PubMed]
23. Wageh, S.; Maize, M.; Han, S.; Al-Ghamdi, A.A.; Fang, X. Effect of Solvent and Environmental Conditions on the Structural and Optical Properties of CdS Nanoparticles. *RSC Adv.* **2014**, *4*, 24110–24118. [CrossRef]
24. Cao, X.; Qi, Y.; Xu, C.; Yang, Y.; Wang, J. Transcriptome and Metabolome Responses of Shewanella Oneidensis MR-1 to Methyl Orange under Microaerophilic and Aerobic Conditions. *Appl. Microbiol. Biotechnol.* **2017**, *101*, 3463–3472. [CrossRef] [PubMed]
25. Huang, S.; Tang, J.; Liu, X.; Dong, G.; Zhou, S. Fast Light-Driven Biodecolorization by a Geobacter Sulfurreducens-CdS Biohybrid. *ACS Sustain. Chem. Eng.* **2019**, *7*, 15427–15433. [CrossRef]
26. Anna Engel, C.E.; Vorländer, D.; Biedendieck, R.; Krull, R.; Dohnt, K. Quantification of Microaerobic Growth of Geobacter Sulfurreducens. *PLoS ONE* **2020**, *15*, 1–20. [CrossRef]
27. Richter, K.; Schicklberger, M.; Gescher, J. Dissimilatory Reduction of Extracellular Electron Acceptors in Anaerobic Respiration. *Appl. Environ. Microbiol.* **2012**, *78*, 913–921. [CrossRef]
28. Martins, M.; Toste, C.; Pereira, I.A.C. Enhanced Light-Driven Hydrogen Production by Self-Photosensitized Biohybrid Systems. *Angew. Chem. Int. Ed.* **2021**, *60*, 9055–9062. [CrossRef]
29. Graham, A.J.; Gibbs, S.L.; Saez Cabezas, C.A.; Wang, Y.; Green, A.M.; Milliron, D.J.; Keitz, B.K. In Situ Optical Quantification of Extracellular Electron Transfer Using Plasmonic Metal Oxide Nanocrystals. *ChemElectroChem* **2022**, *9*, e202101423. [CrossRef]
30. Yang, C.; Aslan, H.; Zhang, P.; Zhu, S.; Xiao, Y.; Chen, L.; Khan, N.; Boesen, T.; Wang, Y.; Liu, Y.; et al. Carbon Dots-Fed Shewanella Oneidensis MR-1 for Bioelectricity Enhancement. *Nat. Commun.* **2020**, *11*, 1379. [CrossRef]
31. Rodrigo-Navarro, A.; Sankaran, S.; Dalby, M.J.; del Campo, A.; Salmeron-Sanchez, M. Engineered Living Biomaterials. *Nat. Rev. Mater.* **2021**, *6*, 1175–1190. [CrossRef]
32. Zhao, C.; Li, Y.; Li, X.; Huang, H.; Zheng, G.; Chen, Y. Biological Removal of Sulfamethoxazole Enhanced by S. Oneidensis MR-1 via Promoting NADH Generation and Electron Transfer and Consumption. *J. Hazard. Mater.* **2022**, *426*, 127839. [CrossRef]
33. Tang, T.C.; An, B.; Huang, Y.; Vasikaran, S.; Wang, Y.; Jiang, X.; Lu, T.K.; Zhong, C. Materials Design by Synthetic Biology. *Nat. Rev. Mater.* **2021**, *6*, 332–350. [CrossRef]

34. Jin, K.; Jin, C.; Wu, Y. Synthetic Biology-Powered Microbial Co-Culture Strategy and Application of Bacterial Cellulose-Based Composite Materials. *Carbohydr. Polym.* **2022**, *283*, 119171. [CrossRef]
35. Samuel, J.; Pulimi, M.; Paul, M.L.; Maurya, A.; Chandrasekaran, N.; Mukherjee, A. Batch and Continuous Flow Studies of Adsorptive Removal of Cr(VI) by Adapted Bacterial Consortia Immobilized in Alginate Beads. *Bioresour. Technol.* **2013**, *128*, 423–430. [CrossRef]
36. Yan, F.F.; Wu, C.; Cheng, Y.Y.; He, Y.R.; Li, W.W.; Yu, H.Q. Carbon Nanotubes Promote Cr(VI) Reduction by Alginate-Immobilized Shewanella Oneidensis MR-1. *Biochem. Eng. J.* **2013**, *77*, 183–189. [CrossRef]
37. Pang, Y.; Zeng, G.M.; Tang, L.; Zhang, Y.; Liu, Y.Y.; Lei, X.X.; Wu, M.S.; Li, Z.; Liu, C. Cr(VI) Reduction by Pseudomonas Aeruginosa Immobilized in a Polyvinyl Alcohol/Sodium Alginate Matrix Containing Multi-Walled Carbon Nanotubes. *Bioresour. Technol.* **2011**, *102*, 10733–10736. [CrossRef]
38. Letnik, I.; Avrahami, R.; Rokem, J.S.; Greiner, A.; Zussman, E.; Greenblatt, C. Living Composites of Electrospun Yeast Cells for Bioremediation and Ethanol Production. *Biomacromolecules* **2015**, *16*, 3322–3328. [CrossRef]
39. Belkin, S.; Yagur-Kroll, S.; Kabessa, Y.; Korouma, V.; Septon, T.; Anati, Y.; Zohar-Perez, C.; Rabinovitz, Z.; Nussinovitch, A.; Agranat, A.J. Remote Detection of Buried Landmines Using a Bacterial Sensor. *Nat. Biotechnol.* **2017**, *35*, 308–310. [CrossRef]
40. Wang, J.; Mignon, A.; Snoeck, D.; Wiktor, V.; Van Vliergerghe, S.; Boon, N.; De Belie, N. Application of Modified-Alginate Encapsulated Carbonate Producing Bacteria in Concrete: A Promising Strategy for Crack Self-Healing. *Front. Microbiol.* **2015**, *6*, 1–14. [CrossRef]
41. Balasubramanian, S.; Yu, K.; Meyer, A.S.; Karana, E.; Aubin-Tam, M.E. Bioprinting of Regenerative Photosynthetic Living Materials. *Adv. Funct. Mater.* **2021**, *31*, 2011162. [CrossRef]
42. Heveran, C.M.; Williams, S.L.; Qiu, J.; Artier, J.; Hubler, M.H.; Cook, S.M.; Cameron, J.C.; Srubar, W.V. Biomineralization and Successive Regeneration of Engineered Living Building Materials. *Matter* **2020**, *2*, 481–494. [CrossRef]
43. Cai, P.J.; Xiao, X.; He, Y.R.; Li, W.W.; Yu, L.; Lam, M.H.W.; Yu, H.Q. Involvement of C-Type Cytochrome CymA in the Electron Transfer of Anaerobic Nitrobenzene Reduction by Shewanella Oneidensis MR-1. *Biochem. Eng. J.* **2012**, *68*, 227–230. [CrossRef]
44. Mao, F.; Liu, X.; Wu, K.; Zhou, C.; Si, Y. Biodegradation of Sulfonamides by Shewanella Oneidensis MR-1 and Shewanella Sp. Strain MR-4. *Biodegradation* **2018**, *29*, 129–140. [CrossRef]

microorganisms

Article

Rhizobium pusense-Mediated Selenium Nanoparticles–Antibiotics Combinations against *Acanthamoeba* sp.

Pradnya B. Nikam [1], Jitendra D. Salunkhe [1], Kiran R. Marathe [1], Mousa A. Alghuthaymi [2], Kamel A. Abd-Elsalam [3,*] and Satish V. Patil [1,*]

1 Department of Biochemistry, School of Life Sciences, Kavayitri Bahinabai Chaudhari North Maharashtra University, Jalgaon 425001, India
2 Biology Department, Science and Humanities College, Shaqra University, Alquwayiyah 11971, Saudi Arabia
3 Plant Pathology Research Institute, Agricultural Research Center, Giza 12619, Egypt
* Correspondence: kamelabdelsalam@gmail.com (K.A.A.-E.); svpatil@nmu.ac.in or satish.patil7@gmail.com (S.V.P.); Tel.: +91-0257-2257421–25 (S.V.P.)

Abstract: Severe ocular infections by *Acanthamoeba* sp. lead to keratitis, resulting in irreversible vision loss in immune-compromised individuals. When a protozoal infection spreads to neural tissues, it causes granulomatous encephalitis, which can be fatal. Treatment often takes longer due to the transition of amoeba from trophozoites to cyst stages, cyst being the dormant form of *Acanthamoeba*. A prolonged use of therapeutic agents, such as ciprofloxacin (Cipro), results in severe side effects; thus, it is critical to improve the therapeutic efficacy of these widely used antibiotics, possibly by limiting the drug-sensitive protozoal-phase transition to cyst formation. Owing to the biomedical potential of selenium nanoparticles (SeNPs), we evaluated the synergistic effects of ciprofloxacin and *Rhizobium pusense*–biogenic SeNPs combination. SeNPs synthesized using *Rhizobium pusense* isolated from root nodules were characterized using UV–Visible spectrophotometer, FT-IR, SEM with EDX, particle size analysis, and Zeta potential. The combination was observed to reduce the sub-lethal dose of Cipro, which may help reduce its side effects. The selenium and ciprofloxacin (SeNPs–Cipro) combination reduced the LC_{50} by 33.43%. The anti-protozoal efficacy of SeNPs–Cipro was found to transduce through decreased protozoal-cyst formations and the inhibition of the galactosidase and protease enzymes of trophozoites. Furthermore, high leakage of sugar, proteins, and amino acids during the SeNPs–Cipro treatment was one primary reason for killing the trophozoites. These experimental results may be helpful in the further pre-clinical evaluation of SeNPs–Cipro to combat protozoal infections. Future studies for combinations of SeNPs with other antibiotics need to be conducted to know the potential of SeNPs against antibiotic resistance in *Acanthamoeba*.

Keywords: *Acanthamoeba*; nano-selenium; trophozoites; cysts; ciprofloxacin; anti-protozoal

Citation: Nikam, P.B.; Salunkhe, J.D.; Marathe, K.R.; Alghuthaymi, M.A.; Abd-Elsalam, K.A.; Patil, S.V. *Rhizobium pusense*-Mediated Selenium Nanoparticles–Antibiotics Combinations against *Acanthamoeba* sp. *Microorganisms* **2022**, *10*, 2502. https://doi.org/10.3390/microorganisms10122502

Academic Editor: Juan Carlos Gutiérrez

Received: 25 October 2022
Accepted: 12 December 2022
Published: 16 December 2022

1. Introduction

Acanthamoeba is a eukaryotic organism found in freshwater and soil. Its cysts are resistant to ultraviolet light, heat, and harsh antibiotics. They feed on other microorganisms such as bacteria, yeast, fungi, and other protozoa. They are opportunistic pathogens and can cause severe problems in favorable conditions, such as *Acanthamoeba* granulomatous encephalitis and *Acanthamoeba* keratitis, caused by *Acanthamoeba castellanii*. *Acanthamoeba* has a two-stage life cycle, the vegetative growing trophozoite and the dormant cyst stage [1–3]. *Acanthamoeba* infections are treated with combinations of various antibiotics and chemotherapeutic agents such as fluconazole (200 mg), neomycin, paromomycin, chlorhexidine (0.02%), hexamidine (0.01%), amphotericin B (varying concentrations), chlorhexidine digluconate (0.02%), and many others in different dosages found to be effective against *Acanthamoeba* infections such as eye keratitis and encephalitis [4–8]. Additionally, high doses, prolonged treatment, poor absorption of drugs, and drug toxicity are major problems in the treatment of *Acanthamoeba* diseases [9]. In this background, there is a need to find effective and safe sources of amoebicidal molecules. In the current era of nanotechnology, various nanomaterials have been reported as therapeutic agents, including antibacterial, anti-fungal, anti-protozoal, etc. Although silver nanoparticles are already reported as anti-*Acanthamoeba* products [10–13], there is a need to screen other nanomaterials as anti-*Acanthamoeba* drugs due to their toxic effect on human cells [14,15]. In addition, ciprofloxacin, a chemotherapeutic agent, has been previously reported for its anti-*Acanthamoeba* potential. It has been proven to be a significant chemotherapeutic agent against *Acanthamoeba*, and its combinations with voriconazole and chlorohexidine have been found potent in controlling infections [16].

Selenium is recognized as one of the crucial trace elements for biological functioning and its deficiencies may lead to some disorders [17,18]. Various essential proteins, known as seleno-proteins, are also characterized for different roles in mammals, including humans. Nano-selenium, similar to other nanomaterials, has a variety of useful qualities that make it widely used in electronics, agriculture, and the medical field. Compared to bulk selenium, nano-selenium is more bioavailable, and lesstoxic. Selenium nanoparticles (SeNPs) have already been linked to a variety of biological activities, including those that are anti-diabetic, anti-bacterial, anti-fungal, anti-cancer, and anti-oxidative. SeNPs are well known for their drug-delivery capabilities in addition to their extensive therapeutic applications [19,20]. All these reports induced to test its anti-amoebicidal potential. The present study deals with the anti-*Acanthamoeba* capacity of biosynthesized selenium nanoparticles (SeNPs). SeNP combinations with ciprofloxacin were examined for their effectiveness and mode of action in killing both the cyst and trophozoite stages of the *Acanthamoeba* species.

2. Material and Methods

2.1. Materials

Sodium selenite was purchased from Hi-media (Mumbai) and stored at room temperature. The artificial substrate BAPNA (Benzoyl D-L-Arginine-paranitroanilide) and Triton X- 100 were procured from Sigma chemicals, USA. The standard antibiotic, ciprofloxacin, was purchased from SRL and stored at 4 °C. Ultra-pure grade water was used for all the experiments.

2.2. Sample Collection

The bacterial isolates for SeNPs synthesis were collected from the root nodules of black chickpeas (*Cicer arientinum*) from the farms in the Jalgaon district of Maharashtra state. These isolates were grown on yeast extract mannitol salt agar medium and later characterized morphologically, biochemically, and by 16S rRNA sequencing [21].

2.3. Synthesis of Selenium Nanoparticles (SeNPs)

The 24-hour-old culture was inoculated on a solid nutritional medium containing 2.5 g of yeast extract, 2.5 g of casein enzyme hydrolysate, 1 g of dextrose, and 5 g of sodium chloride per liter of pure distilled water. This medium was supplemented with sterile sodium selenite solution in distilled water (1 mg/mL) after sterilization of the medium. The selenium tolerance and nanoparticle synthesis were investigated by the potential growth and red color change of the colonies.

The concentration of sodium selenite for the synthesis of nanoparticles was optimized, ranging from 100 to 1000 µg/mL used for the liquid synthesis of nano-selenium by adding the loopful inoculum in the broth with the composition mentioned above. After incubation, the bacterial cells were pelleted out and transferred into the sterile sodium selenite solution to avoid medium-component interference in the biotransformation. The biotransformation was carried out at 37 °C temperature and under continuous shaking at 180 rpm. After complete transformation, the red-colored material was centrifuged (REMI C-24 BL, Mumbai (MH), India) at 18,785 g to remove the leftover selenite traces. The pellets were then subjected to a 10 min ultrasonic treatment, followed by repetitive centrifuge to remove cell debris. The supernatant was collected and used for additional bioassays and characterization. The lyophilized, red, evenly suspended supernatant was gathered and used for additional research and analysis.

2.4. Characterization of Selenium Nanoparticles

Different analytical approaches were used to analyze the biosynthesized selenium nanoparticles (SeNPs). One of the key stages to understanding the physiochemical nature of synthesized nanomaterials, which aids in researching their potential uses and toxicity, is their characterization [22]. The lyophilized pure SeNPs were used for further characterization.

The SeNPs synthesized using *Rhizobium* cells suspended in water were confirmed by UV–Vis spectroscopy using the principle of surface plasma resonance. After the incubation of 24 h, the synthesis of SeNPs was confirmed by UV–Vis spectroscopy by taking the spectra at 200–700 nm (Shimadzu 1601, Tokyo, Japan). The FT-IR (Fourier Transmission Infrared Spectroscopy) analysis of the synthesized nano-selenium was performed (Shimadzu FT-IR 8400, Tokyo, Japan) to investigate the role of *Rhizobium* cells in reduction and to cap around SeNPs. The SeNPs samples were scanned in the wavelength range of 400 to 4000 cm^{-1} after blending the dried samples with KBr.

Field Emission Gun-Scanning electron microscopy (JEOL JSM-7600F FEG-SEM, JEOL USA Inc., Peabody, MA, USA) was carried out to know the structural morphology of the particles. After FEG-SEM, to determine the elemental makeup of that particular region, the electron micrograph was subjected to EDAX (energy-dispersive spectroscopy, Billerica, MA, USA) mapping. By using a Zetasizer by Malvern instrument and a Dynamic Light Scattering instrument (Malvern Panalyticals, Malvern, UK), it was possible to estimate the hydrodynamic diameter and the total charge potential on the manufactured SeNPs.

2.5. Amoebicidal Effect

2.5.1. Test Organisms

The pathogenic strain of *A. castellanii* (ATCC 50492) was cultured in a PYG (peptone–yeast extract–glucose) medium (2% peptone, 0.2% yeast extract, and 1.8% glucose) at 30 °C. Further, 1 mL of the culture was centrifuged for 5 min at 2000 g (REMI RM 02 plus mini centrifuge, Mumbai, India). After discarding the supernatant, pellets were washed twice with phosphate buffer saline (PBS), followed by diluting it in a PYG medium to obtain its final concentration of 2×10^4 trophozoites per milliliter [10,23].

2.5.2. Axenization of *Acanthamoeba* Cysts

Acanthamoebae were axenized by harvesting cysts from the PYG plate cultures by adding 5 mL sterile saline in 2-week-old PYG *Acanthamoeba* seeded plates. This material was centrifuged at 13,149 g (REMI C24 BL centrifuge, Mumbai, India) for 10 min to collect the cysts from the pellet. The pellet was incubated with 3% HCl overnight to eliminate the bacteria. Subsequently, the cysts were washed 2–3 times with saline and again centrifuged at 13,149 g for 10 min to remove the remaining hydrochloric acid, further transferring it to liquid cultures in PYG medium. The culture was maintained in axenic conditions by transferring it to a fresh medium weekly [23].

2.5.3. Amoebicidal Assay

The amoebicidal activity was assessed by adding 100 µL of the culture of *A. castellanii* containing 50 trophozoites and 100 µL of each test (SeNPs) concentration at 5, 10 and 50 ppm in sterile polystyrene plates, and plates were sealed followed by incubation at 30 °C. The plates were monitored using an inverted microscope and counted in a Fuchs-Rosenthal counting chamber after 24 h. The trypan blue exclusion method was used to test the cyst viability [11,24]. The negative control consisted of sterile water in the trophozoite suspension. The experiments were performed in triplicate on two different days. The SeNP LC_{50} was first observed using the different test concentrations of SeNPs, i.e., concentrations of 1.56 to 50 ppm were assessed for significant amoebicidal activity. Chlorohexidine was used as a standard anti-*Acanthamoebal* drug (positive control), and its LC_{50} was determined by exposing an increasing concentration of ciprofloxacin solution, i.e., 10 to 100 μg/mL.

2.5.4. Combination Study of SeNPs (Se) and Ciprofloxacin (Cipro)

The following two types of combinations were designed for study:

1. Cipro-Se: The sub-lethal concentration, i.e., the LC_{30} of SeNPs, was combined with increasing ciprofloxacin concentrations, i.e., SeNPs were fixed with varying amounts of ciprofloxacin concentration;
2. Se-Cipro: The sub-lethal concentration of ciprofloxacin was kept constant, which was calculated previously, and the concentration of SeNPs varied from 1.56 to 50 ppm. Both the combinations were exposed to 50 trophozoites in triplicate for 24 h. The LC_{50} value for both combinations was determined by Trypan blue exclusion method.

2.5.5. Mode of Action of Ciprofloxacin and SeNPs Combinations

To determine the mechanism of action on trophozoites, 2×10^3/mL trophozoites in peptone–glucose broth were treated with the LC_{30} concentration of both combinations and diluted up to a final volume of 5 mL of phosphate buffer pH 7 (0.2M) and incubated for 30 min at 30 °C. After centrifugation, the supernatant was checked for various leakages such as sugar, amino acids, and protein. The untreated trophozoites were used as a control.

2.5.6. Preparations of *Amoeba* Cell Lysates

The freshly cultured 100 trophozoites from PYG agar were collected and added in 5 mL phosphate-buffered saline (pH 7.2) centrifuged and washed with phosphate buffer (pH 7.2). These trophozoites were exposed to sub-lethal concentrations of the SeNPs-Cipro combination and the control (without any treatment), and incubated for 2 h. The cysts were separated from the pellet by centrifuging this material at 13,149 g for 10 min. To get rid of the germs, the pellet was treated with 3% HCl overnight. The cysts were then transferred to liquid cultures in PYG medium after being washed with physiological saline two or three times. The cysts were then centrifuged once more at 13,149 g for 10 min to remove any remaining hydrochloric acid. By switching to a new medium every week, the culture was kept in axenic conditions.

2.5.7. Protease Inhibitor Assay (Trypsin)

Standard Protease (Trypsin) activity was determined by observing the hydrolysis of synthetic substrate BAPNA (Benzoyl D-L-Arginine-paranitroanilide) by measuring the product, i.e., the yellow-color product p-nitroanilidine, at 410 nm. Enzymatic activity is expressed as μM/mg protein/min. Enzyme inhibition as a function of time was determined by trypsin assay in the presence of 100 μL of LC_{30} concentration SeNPs and BAPNA as a substrate for determination of the trypsin inhibitory potential of SeNPs. Additionally, the protein content was quantified by the Bradford method [25].

2.5.8. Encystment Assay

Encystation assays were performed as per previous studies [26,27]. A total of 5×10^5 *Acanthamoeba* trophozoites were incubated with SeNPs and other combinations in a 24-well plate containing PBS and a medium having a composition of 5 mM $MgCl_2$ and 8% glucose. These plates were incubated at 30 °C for 72 h. To collect only cysts, 0.25% sodium dodecyl sulfate (SDS) was used. The leftover cysts were counted using a hemocytometer.

2.5.9. Excystation Assay

The excystation assay for *Acanthamoeba castellanii* cysts, was performed according to the method described earlier by Dudley et al. 2009 [28]. A total of 1×10^6 trophozoites were inoculated on non-nutrient agar plates and incubated at 30 °C for 14 days. After this, the cysts were added to 5 mL of PBS buffer saline (PBS) and collected by centrifugation at 3000× *g* for 10 min. These cysts were then suspended in 1 mL of PBS and preserved at 4°C for use in excystation assays after counting. The excystation was measured by incubating 1×10^5 *A. castellanii* cysts with SeNPs at 30 °C. This was observed under the microscope every 24 h for up to 72 h. The emergence of trophozoites was checked and counted by using a hemocytometer.

2.5.10. Toxicity Assay of SeNPs–Cipro Conjugates

The toxicity assays of SeNPs, ciprofloxacin and their conjugate were carried out by lactate dehydrogenase assay by exposing the keratenocytes cell line (HEKa, cell line, American Type Culture Collection, Manassas, VA, USA) to 200 μL cell suspension in each test well at its respective LC_{50} concentration at 24 h in a CO_2 incubator at 37 °C. This mixture was centrifuged at 400 *g* for 5 min and used for the assay. The supernatant was mixed with 2 μL of sodium lactate, 2 μL of tetrazolium salt and 50 μL of phosphate buffer, incubated at 20 min in the dark, and measured for absorbance at 490 nm. The negative control was PBS buffer, whereas the positive control was 1% triton X-100. All reactions were carried out in triplicate, and the percentage toxicity was calculated using the formula below [29]:

$$\%\ \text{Toxicity} = (\text{OD490 Test})\text{-}(\text{OD490 Negative control})/(\text{OD490 positive control})\text{-}(\text{OD490 Negative control}) \quad (1)$$

2.5.11. Statistical Test

The experiments were performed in triplicate and the data presented are in the form of mean ± standard deviation. One-way analysis of variance (ANOVA) was used to analyze the differences in the acquired data. Calculations were done using SPSS software v.20.0.20.0, Armonk, NY, USA. Statistically significant values were considered if $p \leq 0.05$.

3. Results

3.1. Synthesis and Characterization of Selenium Nanoparticles

The isolated bacteria from the root nodules were gram-negative and motile. They produced oxidase, catalase, and indole, whereas methyl red, Vogus-Prausker, and citrate utilization tested negative. The 16S rRNA gene sequencing confirmed that the bacteria were *Rhizobium pusense* with 100% similarity. The phylogenetic tree was designed by using the MEGA software v.11 (Figure 1). This isolate was able to tolerate about 1000 μg/mL sodium selenite. The transformation of selenite into red nano-selenium was observed in both solid and liquid medium after the incubation period. The color intensity increased with respect to the increase in incubation time. The SeNPs synthesized in the liquid medium were separated using centrifugation followed by sonication to separate them from the cellular material.

The red-colored suspended material was used for characterization using several methods, including UV–visible spectrophotometry, FTIR spectrophotometry, DLS, SEM, and TEM. The scanning of the red colloidal sample of the SeNPs using a UV–Visible spectrophotometer in the range 200–800 nm showed a sharp spectrum at 269 nm due to the Surface Plasmon Resonance mechanism [10] (Figure 2a). In Figure 2b, the FTIR spectrum gives away important functional groups responsible for the biotransformation of selenite into SeNPs. The intense broad peak at 3284 cm^{-1} belongs to the –OH stretch of the hydroxyl group, whereas the small outgrowths at 2959 and 2927 cm^{-1} are for the –C–H stretch of the methyl and methylene groups, respectively. The peak at 1658 cm^{-1} shows alkenyl C=C stretching. The aromatic nitro compounds can be detected because of the peak at 1536 cm^{-1}. Peaks at 1451 cm^{-1} are attributed to carbonate ions, and at 1402 cm^{-1}, to –OH bending because of the phenol or tertiary alcohol. The peak values 1233 cm^{-1} and 929 cm^{-1} refer to the P–O–P stretching of aromatic phosphates, while other peaks such as 1059 represent phosphate ions. The peak at 771 cm^{-1} represents skeletal C-C vibrations in methylene, and the one at 699 cm^{-1} represents C–H bending of the aromatic group. Therefore, we can say that these complex functional groups contribute to the synthesis and stabilization of the biogenic SeNPs [30,31].

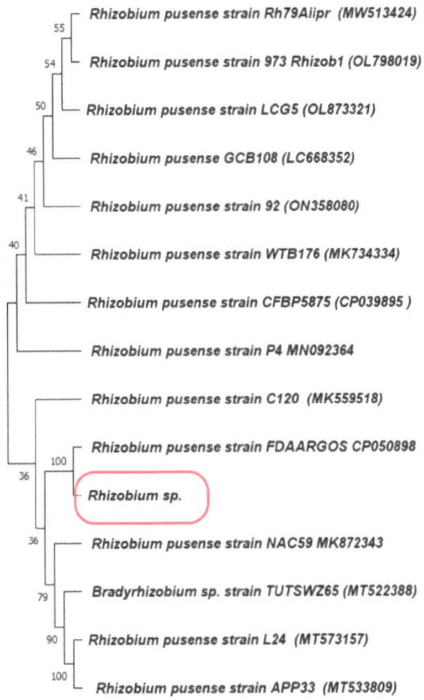

Figure 1. Phylogenetic tree of the bacterial isolate from the root nodules.

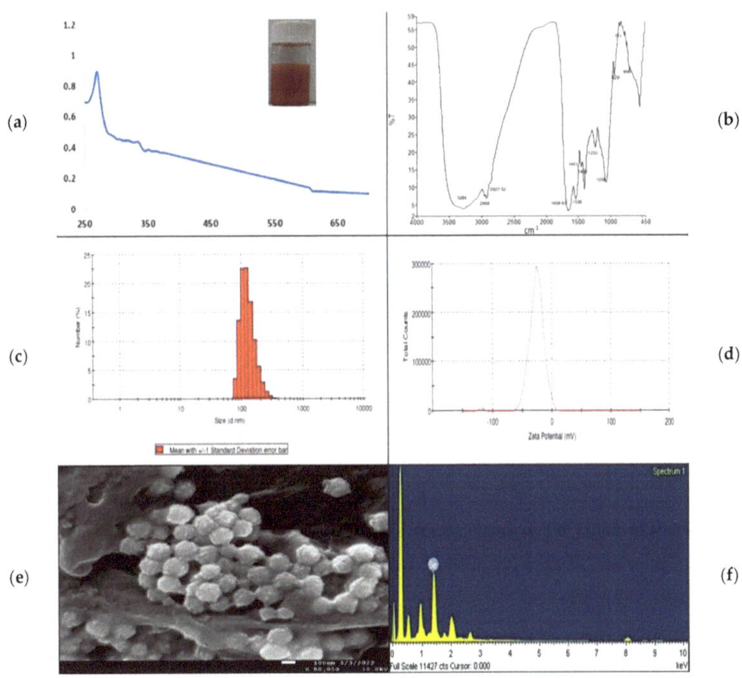

Figure 2. Characterization of selenium nanoparticles; (**a**) UV–Visible spectrum; (**b**) FTIR spectrum; (**c**) Particle size analysis graph; (**d**) Zeta potential; (**e**) FEG-SEM image; (**f**) EDAX spectrum.

The particle size of the biosynthesized SeNPs ranged from 78–190 nm, and the average particle size was found to be 172.8 nm, having a polydispersity index of 0.066 (Figure 2c), which corresponds to the homogeneity and fine dispersion with the most negligible aggregation of the nanoparticles [31,32]. Along with this, the zeta potential (Figure 2d) at −26.9 indicates good stability of the biogenic SeNPs. The FEG-SEM micrograph has the spherical shape of SeNPs with an average size between 90 and 170 nm (Figure 2e). The EDAX graph for the FEG-SEM image has the characteristic peak for selenium (Figure 2f). All these outcomes confirmed the biosynthesis of spherical and stable SeNPs.

3.2. *Anti-Acanthamoeba Activity*

Selenium nanoparticles have been extensively reported for their antibacterial, antifungal, anti-viral, and insecticidal applications [33,34]. For the first time, the anti-*Acanthamoeba* activity of these SeNPs has been studied by using the *Rhizobium* sp. For this, the lyophilized selenium nanomaterial was suspended in a phosphate buffer containing 0.01% starch, and different concentrations from 1.56 to 50 ppmwere tested against fifty trophozoites of *Acanthamoeba* and incubated for 48 h. It was observed that the SeNPs showed 50% mortality at a very high dose (LC_{50} at 72 ppm) against the tested trophozoites. As Se nanoparticles are well recognized as a vehicle for drug delivery [9], and ciprofloxacin has been reported as an effective drug for *Acanthamoeba* keratitis [20], the combination study of both components was designed. The conformation of the conjugate formation of SeNPs and ciprofloxacin was confirmed using a UV–Visible spectrophotometer (Shimadzu 1601, Tokyo, Japan). Standard ciprofloxacin had maximum absorbance at 271 nm [34], SeNPs at 269 nm, and the combination of SeNPsand ciprofloxacin had their maximum absorbance peak at 276 nm. This shifting of wavelength for the combination shows that SeNPs developed a conjugate with ciprofloxacin, which may be responsible for enhancing its activity. Such results have been reported in the case of gold nanoparticles conjugated with different antibiotics (Figure 3) [35].

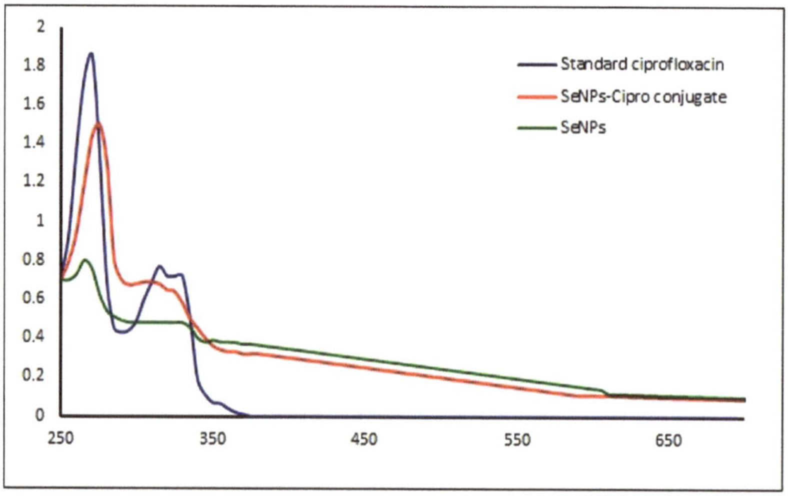

Figure 3. UV–Visible spectra of the combinations used for the experiment.

Primarily, the LC_{50} of ciprofloxacin against the trophozoites was determined and found to be 56.260 ppm after 48 h. Ciprofloxacin is a standard drug, and it showed a comparatively lower LC_{50} value than SeNPs. Hence, in the first combination, i.e., LC_{30} of SeNPs, the concentration was combined with different concentrations of ciprofloxacin and tested against the trophozoites. However, the Cipro–Se conjugation did not show any significant increase in amoebicidal activity, i.e., the LC_{50} was reduced by only 4.19%. In this background, the second combination, i.e., Se–Cipro, was tested against the trophozoites. The Se–Cipro combination showed a significant decrease in the LC_{50}.The LC_{50} was recorded at 37.452 ppm concentration of Se–Cipro for *Acanthamoeba*, reducing the dose by 34.5% (Table 1) compared to the ciprofloxacin treatment.

Table 1. Anti-*Acanthamoeba* activity of nano-selenium (Se), ciprofloxacin, and their conjugate.

Test Species	Test Products	$LC_{50} \pm$ SD (mg lit^{-1})	95% Fiducial Limits	$LC_{90} \pm$ SD (mg lit^{-1})	95% Fiducial Limits	Regression Equation
Acanthamoeba sp.	SeNPs	72.084 ± 10.87	56.593–106.58	115.058 ± 19.264	87.7664–176.491	Y = 0.423 + 0.149X
	Cipro	56.2608 ± 2.2078	51.7279–60.43	125.0330 ± 3.13	119.2960–131.69	Y = 9.27 + 0.274 X
	Cipro–Se	53.9078 ± 5.4802	45.4017–68.6011	94.1228 ± 11.056	77.1133–124.0217	Y = 1.55 + 0.302 X
	Se–Cipro	37.452 ± 3.31	32.196–46.153	94.122 ± 11.05	77.113–124.021	Y= 4.78 + 0.980 X
	Sodium selenite	120.143± 2.56	115.316- 125.444	205.1626± 6.389	193.82–219.166	Y= 8.53 + 0.960 X
	Chlorohexidine	24.580 ± 1.25	22.361- 27.36	56.070 ± 3.61	49.95–64.52	Y= 6.52 + 0.754 X

Level of significance: $p < 0.05$; Y—mortality rate (significant at $p < 0.05$ level); X—concentration (significant at $p < 0.05$ level); LC_{50}—lethal concentration that kills 50% of the exposed trophozoites; LC_{90}—lethal concentration that kills 90% of the exposed trophozoites; SE—standard error (all values are mean of four replicates); Cipro–Se—varied concentration of Cipro with constant sub-lethal concentration of Se; Se–Cipro—varied concentration of Se with constant sub-lethal concentration of Cipro; Chlorohexidine—a standard anti-*Acanthamoeba* drug (positive control). The LC_{50} and LC_{90} concentration units are given in mg lit^{-1}, which also mean as ppm.

3.3. The Cysticidal Effect of SeNPs and SeNP–Cipro (Se–Cipro)

SeNPs and Se-Cipro combinations tested for encystment and excystment effect on *Acanthamoebae* trophozoites showed significant mortality. Among all the given treatments, SeNPs allowed maximum cyst formation of the treated trophozoites (22.066% encystment), while Cipro–Se and ciprofloxacin allowed only 7% and 10% of encystment of the treated trophozoites, respectively (Table 2). The Se–Cipro, which proved to be an effective anti-*Acanthamoeba* preparation, also allowed less encystment (4.66%) of the treated trophozoites. The Se-Cipro treatment significantly inhibited the excystment process of the cysts and this inhibition was constant even after the addition of fresh culture medium into it.

Table 2. Encystment and excystment at LC_{50} concentration of nano-combinations and control.

Sr. No.	Concentrations ppm	Total No. of Cysts	No. Cyst After 24 hrs.	% Encystment	% Excystment
1	56.260 Cipro	50	5	10.66	Nil
2	72.084 Se	50	10	22.066	Nil
3	53.907 Cipro–Se	50	3	7.33	Nil
4	37.452 Se–Cipro	50	2	4.66	Nil
5	Control	50	45		39.33

The trypan blue exclusion assay was performed to check live and killed cysts after every treatment to the cysts using a 40X magnification of the light microscope. The live cysts exclude the stain due to their respiratory mechanism, whereas the killed cysts remain blue as they lose the ability to exclude the stain (Figure 4) [36].

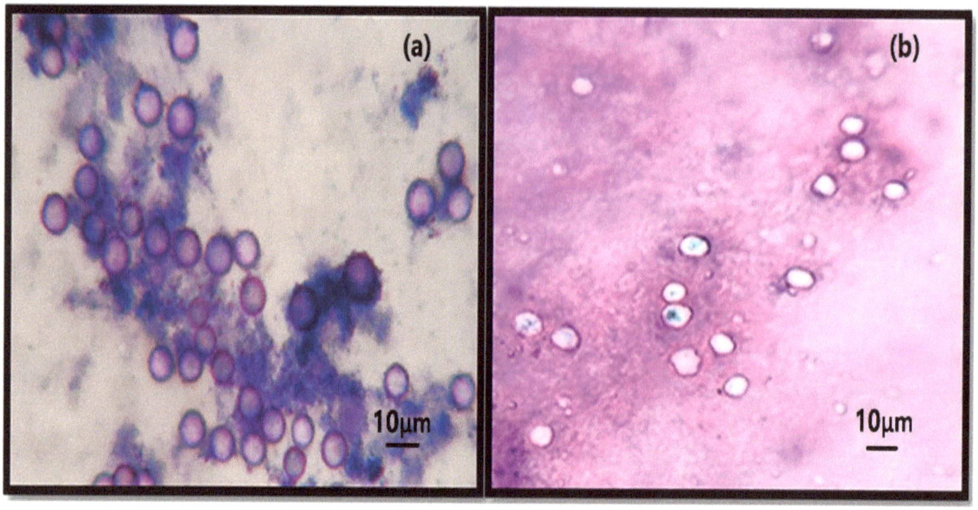

Figure 4. Cysticidal activity of SeNPs–Cipro conjugates: (**a**) Cyst treated with SeNPs–Cipro conjugate; (**b**) control cyst excludes trypan blue.

3.4. Mechanism of Anti-Acanthamoeba Activity of SeNPs

Similarly to earlier-reported drugs such as chlorohexidine [37,38], the SeNPs synthesized using the *Rhizobium* sp. and their combination with ciprofloxacin were found to have the same mode of action against the *Acanthamoeba* trophozoites, i.e., a significant leakage of essential proteins, enzymes, amino acids, and sugars. Microbial synthesized SeNPs were found to have significant anti-*Acanthamoeba* potential. As the membrane is the primary target for anti-protozoal drugs [39], the effect of selenium nanoparticles, standard antibiotics, and the combinations of both were studied on *Acanthamoeba* cells by analyzing the leakage assays. *Acanthamoeba* cells were treated with the LC_{30} of all the tested material for 3 h, and the broth was analyzed for the leakage of amino acids, proteins, and sugars. It was revealed that the Se–Cipro combination causes significant leakage of the protozoal membrane, indicating damage to cellular integrity as a potential mechanism of action of the combinations. There was significant protein, amino acid, and sugar leakage as vital nutrients compared to the control. There was no significant inhibition of enzyme action after the SeNPs treatment; however, Se–Cipro efficiently inhibited the proteases of *Acanthamoeba* lysates. As is the case with proteases, β-galactosidase is a vital enzyme of *Acanthamoeba* that plays an essential role in *Acanthamoeba*'s life cycle and pathogenesis. The Se–Cipro composite was also found to be a significant inhibitor of β-galactosidase of the *Acanthamoeba* cells, i.e., the lysates (Figure 5).

3.5. Toxicity Testing

The toxicity of all the compounds tested against the *Acanthamoeba* species was evaluated fortheir non-targeted toxicity against the keratinocytes cell line (HEKa, cell line). The study revealed that the LC_{50} value of the Se–Cipro nanoconjugate was non-toxic and the percentage toxicity calculated was 1.792%. The ciprofloxacin at its LC_{50} concentration showed a toxicity of 6.1%, but after conjugation, it was reduced. The SeNPs used were also found non-toxic at its LC_{50} concentrations. Overall, the study reveals the synthesized Se–Cipro conjugates were non-toxic (Figure 6).

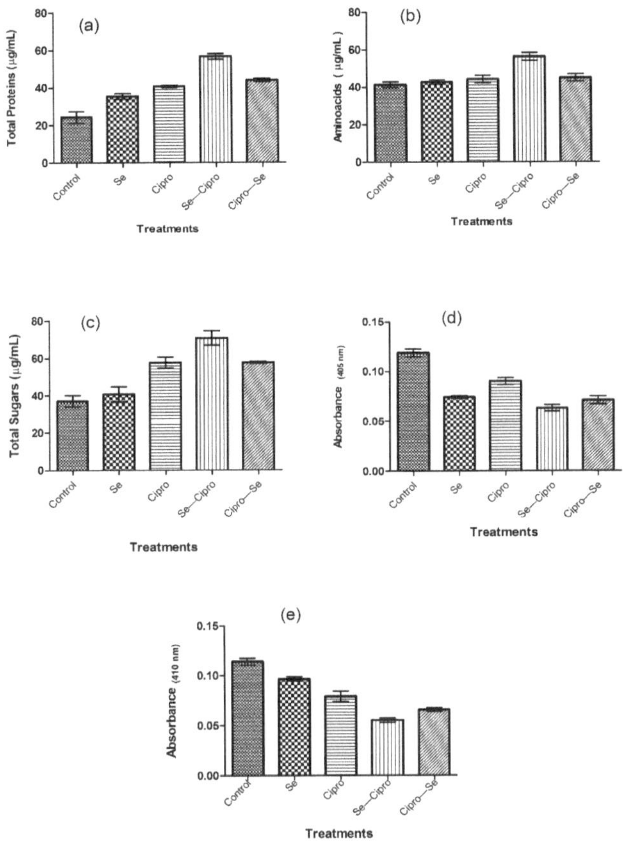

Figure 5. Graphs of bioassays after the different treatments of selenium nanoparticles (Se) and ciprofloxacin (Cipro) given to *Acanthamoeba* trophozoites; (**a**) Protein leakage; (**b**) Amino acid leakage; (**c**) Total sugar leakage; (**d**) Galactosidase inhibitions; (**e**) Protease inhibition.

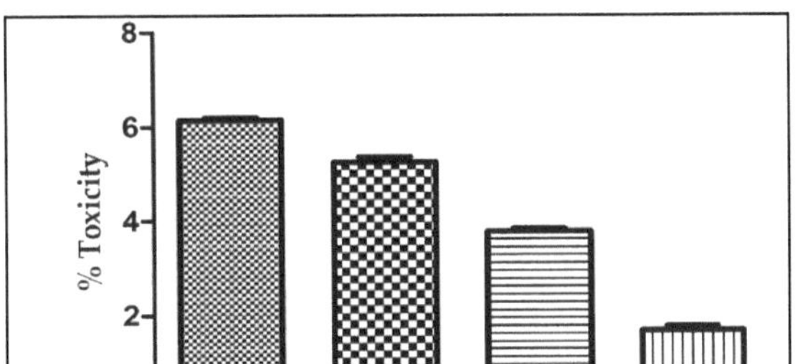

Figure 6. Graphical representation of the percentage toxicity of the different treatments given to the *Acanthamoeba* cells.

4. Discussion

The management of diseases caused by *Acanthamoeba*, particularly amoebic keratitis (AK), is very challenging for healthcare providers. The available therapeutic medications for the same have drawbacks such as prolonged treatment, and require the combination of multiple drug courses [40]. Hence, to combat ocular protozoa, it is essential to increase the efficacy of pre-existing antibiotics. Nanoparticle-based formulations have an effective drug-delivery system against the *Acanthamoeba* species due to their easy accessibility to the infected organ, the release of drugs at a high concentration to the target site, and the increase in drug solubility and stability [41]. Biocompatible SeNPs (selenium nanoparticles) are well known for their drug delivery to specific cells and their increasing drug efficacy [42,43]. To access an increase in the amoebicidal efficacy of the SeNPs–Cipro combination, we first aimed to biosynthesize SeNPs using the *Rhizobium* species isolated from the root nodules of *Cicer arientinum*. Therefore, first, we isolated gram-negative and motile bacteria from the root nodules. The biochemical test and 16S rRNA sequencing confirmed the isolate as *Rhizobium pusense*. This microbial isolate was able to produce dark-red colonies, which primarily indicates that the organism highly tolerates selenium and produces nano-selenium. The intense red-color formation indicates selenite was converted in the nanomaterial, similar to a previous report for the *Pseudomonas* sp. [44]. The bacterial isolate also transformed selenite into red nano-selenium in the liquid medium after an incubation of 24 h. It was observed that the accumulation of nanomaterial takes place both intra- and extracellularly; upon cell lysis, the SeNPs are released into the solution. SeNPs were separated using centrifugation followed by sonication. This red-colored suspended material showed a sharp spectrum at 269 nm when observed under a UV–Visible spectrophotometer. Earlier studies on synthesizing SeNPs using ethanolic extract of bee propolis and lysozyme-fabricated SeNPs have also remarked the spectrum of selenium nanoparticles at 265 nm [45,46]. The spectrum obtained is due to the surface plasmon resonance of the nanoparticles in the suspension. There are also some other shreds of evidence that manifest the SeNPs spectrum at different wavelengths. SeNPs synthesized using *Fusarium oxysporum* had the maximum absorbance at 217 nm [30]; *Emblica officinalis* fruit extract had the SeNPs spectrum at 270 nm [32]; SeNPs synthesized by the seed powder of *Mucuna pruriens* at 319 nm [31]; additionally, in many more reports, UV–Visible spectrum showed slight variations. The FTIR spectra of *Rhizobium*-mediated synthesized SeNPs confirm the presence of complex functional groups that contribute to the synthesis and stabilization of the biogenic SeNPs [30,47]. The FEG-SEM micrograph of selenium nanoparticles has a spherical shape with an average size between 90 and 170 nm, which also confirmed the biosynthesis of SeNPs.

Currently available multipurpose contact-lens disinfection systems are not fully effective against *Acanthamoeba* trophozoites and cysts. It was observed that even after treatment with strong amoebicidal drugs such as chlorohexidine and polyhexamethylene biguanide, *Acanthamoeba* infections persist as active infections. Hence, there is an urge to increase the disinfecting activity of these systems to prevent *Acanthamoeba keratitis* infections [48]. Niyyati et al. (2018) also advocated using gold and silver nanoparticles as lens disinfectants to avoid *Acanthamoeba* infections [49].

SeNPs have been mostly reported as potential antibacterial, antifungal, anti-viral and insecticidal agents [20,33]. The anti-*Acanthamoebal* activity of various nanoparticles such as silver, gold, cobalt, and zinc have been extensively studied [41], but there are no reports on selenium nanoparticles. However, the present study provides the basis for further applications of selenium nanoconjugates as anti-*Acanthamoebal* agents. In the present study, we have tested antibiotics and selenium-nanoparticle conjugates against the human pathogen *Acanthamoeba castellanii*. It was observed that there is a significant decrease in the LC_{50} value of the Se–Cipro nanoconjugates (34.5%) compared to ciprofloxacin [50,51]. The hype in the anti-*Acanthamoeba* activity of antibiotics and nano-selenium conjugates compared to basic compounds is similar to our previous reports of Salunkhe et al. (2022), who advocated that the naringenin–nano-silver conjugates reduced the LC_{50} of naringenin by

50.56% [52]. Similarly, Padzik et al. (2018) also observed that the conjugated nanomaterial increases the activity against *Acanthamoeba* keratitis [53]. The rise in anti-*Acanthamoebal* potential tannic acid–silver nanoparticle conjugates (AgTANPs) may be due to their high absorption in amoebic cells. Aqeel et al. (2016) reported that gold nanomaterial conjugates with biguanides, such as chlorhexidine digluconate, and other drugs are more effective against the *Acanthamoeba* sp. than the standard drug [54]. The two major modes of action for the Se–Cipro conjugate tested were found to be the leakage of membrane and the inhibition of essential enzymes such us trypsin in the amoebic cells. It was observed that the *Acanthamoeba* cells treated with the LC_{30} for 3 h by the Se–Cipro conjugate caused the leakage of amino acids, proteins, and sugars. Hence, it was revealed that the damage in cellular integrity is an indication of the potential mechanism of the drug and SeNPs combinations as compared to standard ciprofloxacin. Xuguang et al. (2003) advocated the same mode of action, i.e., disruptions of the *Acanthamoebal* membrane by the major available chemotherapeutic drugs [55]. Beside these, Se–Cipro significantly inhibits crude enzymes such as the proteases of *Acanthamoeba* lysates. In addition to the proteases, Se–Cipro significantly inhibited β-galactosidase, which is a vital enzyme required for completing the life cycles of *Acanthamoeba* [56]. Many reports suggest that enzymatic activities, including cytolytic mechanisms and vacuole digestions, may play a significant role in the life cycle and metabolism of trophozoites. Amoebae produce a variety of proteases that can participate in the damage of the corneal tissue. Amoebic proteolytic enzymes include serine proteases [57,58], contact-dependent metalloproteases [59], elastases [60], cysteine proteases [57], and mannose-mediated adhesion-induced cytotoxic proteases [61].

All these tested materials were checked for their toxicity on human keratinocytes by LDH assay. The results show that the positive control, i.e., 1% triton X-100, shows 95% toxicity, ciprofloxacin shows 6.1%, whereas the cells without any treatment (negative control) had 2% toxicity, and the Se–Cipro shows 1.79% toxicity. A lower level of LDH corresponds to less cellular disruption [29] and, hence, it indicates that the Se–Cipro combination has negligible toxicity on human cells.

5. Conclusions

The *Acanthamoeba* sp. is well known for its opportunistic pathogenesis in human diseases such as an eye infection i.e. Acanthamoeba keratitisand Acanthamoeba granulomatous encephalitis. Although the development of various drugs and supportive therapies is available, *Acanthamoeba* infections are still a major public health concern. Currently available drugs are not fully effective against the trophozoite and cyst forms of *Acanthamoeba*. Although various drugs, nanomaterials, and their conjugates, e.g., cobalt, tannic acid, silver, etc., have been reported for their anti-*Acanthamoeba* activity, there are some toxicity concerns regarding these metal conjugates, due to the generation of highly reactive oxygen species [48,49]. For this reason, the current selenium–antibiotic combination is a safe formulation, as selenium is well known for its antioxidant potential. In an epitome, we biosynthesized NPs using *Rhizobium pusense* isolated from the rhizosphere. The biosynthesized SeNPs were observed to have anti-*Acanthamoeba* activity. The sub-lethal concentration of SeNPs enhanced the anti-protozoal activity of ciprofloxacin. The combination also effectively limited the *Acanthamoeba* phase maturation to dormancy, which can increase the protozoal sensitivity towards antibiotics. The anti-protozoal action of the combo transduces through the destabilization of the protozoal membrane and the inhibition of essential enzymes. The toxicity assay, i.e., LDH, proved that the LC_{50} and LD_{90} concentrations of selenium and Se–Cipro conjugate are comparatively non-toxic, indicating the proposed conjugate may be a future anti-*Acanthamoeba* drug. The experimental observations of the study demand further preclinical studies for using SeNPs and their combination with antibiotics as potential amoebicidal agents.

6. Future Perspectives

The studies conducted were limited to using combinations of ciprofloxacin with biogenic SeNPs against the protozoan *Acanthamoeba*. More investigations on the combinational studies of SeNPs with other antibiotics, natural products, and other nanomaterials to combat antibiotic-resistance in the *Acanthamoeba* species need to be carried out in the future.

Author Contributions: S.V.P. supervised, conceived, and designed the experiments. P.B.N. performed the experiments and wrote the manuscript. J.D.S. helped in data collection and generated pictures. K.R.M. gave scientific validation and edited the manuscript. M.A.A. edited the text. K.A.A.-E. revised, supervised, and edited the text. All authors have read and agreed to the published version of the manuscript.

Funding: This research was funded by University Grants Commission, New Delhi, under the CSIR–UGC NET–JRF Scheme to P.B.N [UGC Ref. No. 816 (CSIR–UGC NET Dec 2018)} and the APC was funded by Kamel A. Abd-Elsalam, Agricultural research center (Egypt).

Institutional Review Board Statement: Not applicable.

Informed Consent Statement: Not applicable.

Data Availability Statement: Not applicable.

Conflicts of Interest: The authors declare no conflict of interest.

References

1. Trabelsi, H.; Dendana, F.; Sellami, A.; Sellami, H.; Cheikhrouhou, F.; Neji, S.; Makni, F.; Ayadi, A. Pathogenic free-living amoebae: Epidemiology and clinical review. *Pathol. Biol.* **2012**, *60*, 399–405. [CrossRef] [PubMed]
2. Marciano-Cabral, F.; Cabral, G. *Acanthamoeba* spp. as agents of disease in humans. *Clin. Microbiol. Rev.* **2003**, *16*, 273–307. [CrossRef] [PubMed]
3. Siddiqui, R.; Khan, N.A. Biology and pathogenesis of *Acanthamoeba*. *Parasites Vectors* **2012**, *5*, 1–13. [CrossRef] [PubMed]
4. Gautom, R.K.; Fritsche, T.R.; Lindquist, T.D.; Tasman, W.; Jaeger, E.A. Acanthamoeba keratitis. *Duane's Found. Clin. Ophthalmol.* **1997**, *2*, 1–15.
5. El-Sayed, N.M.; Ismail, K.A.; Ahmed, S.A.E.G.; Hetta, M.H. In vitro amoebicidal activity of ethanol extracts of *Arachis hypogaea* L., *Curcuma longa* L. and *Pancratium maritimum* L. on *Acanthamoeba castellanii* cysts. *Parasitol. Res.* **2012**, *110*, 1985–1992. [CrossRef]
6. Khan, N.A. Pathogenesis of *Acanthamoeba* infections. *Microb. Pathog.* **2003**, *34*, 277–285. [CrossRef]
7. Martín-Navarro, C.M.; Lorenzo-Morales, J.; Cabrera-Serra, M.G.; Rancel, F.; Coronado-Alvarez, N.M.; Pinero, J.E.; Valladares, B. The potential pathogenicity of chlorhexidine-sensitive *Acanthamoeba* strains isolated from contact lens cases from asymptomatic individuals in Tenerife, Canary Islands, Spain. *J. Med. Microbiol.* **2008**, *57*, 1399–1404. [CrossRef]
8. Schuster, F.L.; Visvesvara, G.S. Opportunistic amoebae: Challenges in prophylaxis and treatment. *Drug Resist. Updat.* **2004**, *7*, 41–51. [CrossRef]
9. Leitsch, D.; Köhsler, M.; Marchetti-Deschmann, M.; Deutsch, A.; Allmaier, G.; Duchêne, M.; Walochnik, J. Major role for cysteine proteases during the early phase of *Acanthamoeba castellanii encystment*. *Eukaryot. Cell.* **2010**, *9*, 611–618. [CrossRef]
10. Borase, H.P.; Patil, C.D.; Sauter, I.P.; Rott, M.B.; Patil, S.V. Amoebicidal activity of phytosynthesized silver nanoparticles and their in vitro cytotoxicity to human cells. *FEMS Microbiollett.* **2013**, *345*, 127–131. [CrossRef]
11. Anwar, A.; Mungroo, M.R.; Anwar, A.; Sullivan, W.J.; Khan, N.A.; Siddiqui, R. Repositioning of guanabenz in conjugation with gold and silver nanoparticles against pathogenic amoebae *Acanthamoeba castellanii* and *Naegleriafowleri*. *ACS Infect. Dis.* **2019**, *5*, 2039–2046. [CrossRef] [PubMed]
12. Hendiger, E.B.; Padzik, M.; Żochowska, A.; Baltaza, W.; Olędzka, G.; Zyskowska, D.; Bluszcz, J.; Jarzynka, S.; Chomicz, L.; Grodzik, M.; et al. Tannic acid-modified silver nanoparticles enhance the anti-Acanthamoeba activity of three multipurpose contact lens solutions without increasing their cytotoxicity. *Parasites Vectors* **2020**, *13*, 1–8. [CrossRef] [PubMed]
13. Hendiger, E.B.; Padzik, M.; Sifaoui, I.; Reyes-Batlle, M.; López-Arencibia, A.; Zyskowska, D.; Grodzik, M.; Pietruczuk-Padzik, A.; Hendiger, J.; Olędzka, G.; et al. Silver Nanoparticles Conjugated with Contact Lens Solutions May Reduce the Risk of Acanthamoeba Keratitis. *Pathogens* **2021**, *10*, 583. [CrossRef] [PubMed]
14. Liao, C.; Li, Y.; Tjong, S.C. Bactericidal and cytotoxic properties of silver nanoparticles. *Int. J. Mol. Sci.* **2019**, *20*, 449. [CrossRef]
15. Ferdous, Z.; Nemmar, A. Health impact of silver nanoparticles: A review of the bio distribution and toxicity following various routes of exposure. *Int. J. Mol. Sci.* **2020**, *21*, 2375. [CrossRef]
16. Ortillés, A.; Belloc, J.; Rubio, E.; Fernández, M.T.; Benito, M.; Cristóbal, J.A.; Calvo, B.; Goñi, P.P. In-vitro development of an effective treatment for Acanthamoeba keratitis. *Int. J. Antimicrob. Agents* **2017**, *50*, 325–333. [CrossRef]
17. Eszenyi, P.; Sztrik, A.; Babka, B.; Prokisch, J. Elemental, nano-sized (100–500 nm) selenium production by probiotic lactic acid bacteria. *Int. J. Biosci. Biochem. Bioinform.* **2011**, *1*, 148. [CrossRef]

18. Xu, C.; Guo, Y.; Qiao, L.; Ma, L.; Cheng, Y.; Roman, A. Biogenic synthesis of novel functionalized selenium nanoparticles by *Lactobacillus casei* ATCC 393 and its protective effects on intestinal barrier dysfunction caused by enterotoxigenic *Escherichia coli* K88. *Front. Microbiol.* **2018**, *9*, 1129. [CrossRef]
19. Maiyo, F.; Singh, M. Selenium nanoparticles: Potential in cancer gene and drug delivery. *Nanomedicine* **2017**, *12*, 1075–1089. [CrossRef]
20. Khurana, A.; Tekula, S.; Saifi, M.A.; Venkatesh, P.; Godugu, C. Therapeutic applications of selenium nanoparticles. *Biomed. Pharmacother.* **2019**, *111*, 802–812. [CrossRef]
21. Basaglia, M.; Toffanin, A.; Baldan, E.; Bottegal, M.; Shapleigh, J.P.; Casella, S. Selenite-reducing capacity of the copper-containing nitrite reductase of *Rhizobium sullae*. *FEMS Microbiollett.* **2007**, *269*, 124–130. [CrossRef] [PubMed]
22. Al Thaher, Y.; Chandrasekaran, B.; Panchu, S.J. The Importance of Nano-materials Characterization Techniques. In *Integrative Nanomedicine for New Therapies*; Engineering Materials; Krishnan, A., Chuturgoon, A., Eds.; Springer: Cham, Switzerland, 2020; pp. 19–37.
23. Makdoumi, K.; Bäckman, A.; Mortensen, J.; Magnuson, A.; Crafoord, S. Comparison of UVA-and UVA/riboflavin-induced growth inhibition of *Acanthamoeba castellanii*. *Graefes. Arch. Clin. Exp. Ophthalmol.* **2013**, *251*, 509–514. [CrossRef] [PubMed]
24. Lakhundi, S.; Khan, N.A.; Siddiqui, R. Inefficacy of marketed contact lens disinfection solutions against keratitis-causing *Acanthamoeba castellanii* belonging to the T4 genotype. *Exp. Parasitol.* **2014**, *141*, 122–128. [CrossRef] [PubMed]
25. Bradford, M.M. A rapid and sensitive method for the quantitation of microgram quantities of protein utilizing the principle of protein-dye binding. *Anal Biochem.* **1976**, *72*, 248–254. [CrossRef]
26. Anwar, A.; Siddiqui, R.; Hussain, M.A.; Ahmed, D.; Shah, M.R.; Khan, N.A. Silver nanoparticle conjugation affects anti-acanthamoebic activities of amphotericin B, nystatin, and fluconazole. *Parasitol. Res.* **2018**, *117*, 265–271. [CrossRef]
27. Anwar, A.; Ting, E.L.S.; Anwar, A.; ulAin, N.; Faizi, S.; Shah, M.R.; Khan, N.A.; Siddiqui, R. Antiamoebic activity of plant-based natural products and their conjugated silver nanoparticles against Acanthamoeba castellanii (ATCC 50492). *AMB Express* **2020**, *10*, 1–10. [CrossRef]
28. Dudley, R.; Jarroll, E.L.; Khan, N.A. Carbohydrate analysis of *Acanthamoeba castellanii*. *Exp. Parasitol.* **2009**, *122*, 338–343. [CrossRef]
29. Vijayakumar, S.; Ganesan, S. In vitro cytotoxicity assay on gold nanoparticles with different stabilizing agents. *J. Nanomater.* **2012**, *2012*, 14. [CrossRef]
30. Asghari-Paskiabi, F.; Imani, M.; Razzaghi-Abyaneh, M.; Rafii-Tabar, H. *Fusariumoxysporum* a bio-factory for nano selenium compounds: Synthesis and characterization. *Sci. Iran* **2018**, *25*, 1857–1863.
31. Menon, S.; Agarwal, H.; Shanmugam, V.K. Catalytical degradation of industrial dyes using biosynthesized selenium nanoparticles and evaluating its antimicrobial activities. *Sustain. Environ. Res.* **2021**, *31*, 1–12. [CrossRef]
32. Gunti, L.; Dass, R.S.; Kalagatur, N.K. Phytofabrication of selenium nanoparticles from *Emblicaofficinalis* fruit extract and exploring its biopotential applications: Antioxidant, antimicrobial, and biocompatibility. *Front. Microbiol.* **2019**, *10*, 931. [CrossRef] [PubMed]
33. Murugesan, G.; Nagaraj, K.; Sunmathi, D.; Subramani, K. Methods involved in the synthesis of selenium nanoparticles and their different applications—A review. *Eur. J. Biomed.* **2019**, *6*, 189–194.
34. Rogachev, A.A.; Yarmolenko, M.A.; Rogachou, A.V.; Tapalski, D.V.; Liu, X.; Gorbachev, D.L. Morphology and structure of antibacterial nanocomposite organic–polymer and metal–polymer coatings deposited from active gas phase. *RSC Adv.* **2013**, *3*, 11226–11233. [CrossRef]
35. Saha, B.; Bhattacharya, J.; Mukherjee, A.; Ghosh, A.; Santra, C.; Dasgupta, A.K.; Karmakar, P. In vitro structural and functional 506 evaluation of gold nanoparticles conjugated antibiotics. *Nanoscale Res. Lett.* **2007**, *2*, 614–622. [CrossRef]
36. Strober, W. Trypan blue exclusion test of cell viability. *Curr. Protoc. Immunol.* **1997**, *21*, A–3B.
37. Khunkitti, W.; Hann, A.C.; Lloyd, D.; Furr, J.R.; Russell, A.D. Biguanide-induced changes in *Acanthamoeba castellanii*: An electron microscopic study. *J. Appl. Microbiol.* **1998**, *84*, 53–62. [CrossRef]
38. McDonnell, G.; Russell, A.D. Antiseptics and disinfectants: Activity, action, and resistance. *Clin. Microbiol. Rev.* **1999**, *12*, 147–179. [CrossRef]
39. Anwar, A.; Khan, N.A.; Siddiqui, R. Combating *Acanthamoeba* spp. cysts: What are the options? *Parasites Vectors* **2018**, *11*, 26. [CrossRef]
40. Lorenzo-Morales, J.; Martín-Navarro, C.M.; López-Arencibia, A.; Arnalich-Montiel, F.; Piñero, J.E.; Valladares, B. Acanthamoeba keratitis: An emerging disease gathering importance worldwide? *Trends. Parasitol.* **2013**, *29*, 181–187. [CrossRef]
41. Sharma, G.; Kalra, S.K.; Tejan, N.; Ghoshal, U. Nanoparticles based therapeutic efficacy against *Acanthamoeba*: Updates and future prospect. *Exp. Parasitol.* **2020**, *218*, 108008. [CrossRef]
42. Ferro, C.; Florindo, H.F.; Santos, H.A. Selenium nanoparticles for biomedical applications: From development and characterization to therapeutics. *Adv. Healthc. Mater.* **2021**, *10*, 2100598. [CrossRef] [PubMed]
43. Lin, W.; Zhang, J.; Xu, J.F.; Pi, J. The advancing of selenium nanoparticles against infectious diseases. *Front. Pharmacol.* **2021**, *12*, 1971. [CrossRef] [PubMed]
44. Avendaño, R.; Chaves, N.; Fuentes, P.; Sánchez, E.; Jiménez, J.I.; Chavarría Vargas, M. Production of selenium nanoparticles in Pseudomonas putida KT2440. *Sci. Rep.* **2016**, *6*, 37155. [CrossRef] [PubMed]
45. Shubharani, R.; Mahesh, M.; Yogananda Murthy, V. Biosynthesis and characterization, antioxidant and antimicrobial activities of selenium nanoparticles from ethanol extract of Bee Propolis. *J. Nanomed. Nanotechnol.* **2019**, *10*, 2.

46. Vahdati, M.; Tohidi Moghadam, T. Synthesis and Characterization of Selenium Nanoparticles-Lysozyme Nanohybrid System with Synergistic Antibacterial Properties. *Sci. Rep.* **2020**, *10*, 510. [CrossRef]
47. Nandiyanto, A.B.D.; Oktiani, R.; Ragadhita, R. How to read and interpret FTIR spectroscope of organic material. *Indones. J. Sci. Technol.* **2019**, *4*, 97–118. [CrossRef]
48. Seal, D.V. Acanthamoeba keratitis update—Incidence, molecular epidemiology and new drugs for treatment. *Eye* **2003**, *17*, 893–905. [CrossRef]
49. Niyyati, M.; Sasani, R.; Mohebali, M.; Ghazikhansari, M.; Kargar, F.; Hajialilo, E.; Rezaeian, M. Anti-Acanthamoeba effects of silver and gold nanoparticles and contact lenses disinfection solutions. *Iran J. Parasitol.* **2018**, *13*, 180.
50. Abjani, F.; Khan, N.A.; Jung, S.Y.; Siddiqui, R. Status of the effectiveness of contact lens disinfectants in Malaysia against keratitis-causing pathogens. *Exp Parasitol.* **2017**, *183*, 187–193. [CrossRef]
51. Siddiqui, R.; Lakhundi, S.; Khan, N.A. Status of the effectiveness of contact lens solutions against keratitis-causing pathogens. *Cont. Lens Anterior Eye* **2015**, *38*, 34–38. [CrossRef]
52. Salunkhe, J.D.; Mohite, B.V.; Patil, S.V. Naringenin biosynthesis and fabrication of naringenin mediated nano silver conjugate for antimicrobial potential. *Nat. Prod. Res.* **2022**, 1–7. [CrossRef] [PubMed]
53. Padzik, M.; Hendiger, E.B.; Chomicz, L.; Grodzik, M.; Szmidt, M.; Grobelny, J.; Lorenzo-Morales, J. Tannic acid-modified silver nanoparticles as a novel therapeutic agent against Acanthamoeba. *Parasitol. Res.* **2018**, *117*, 3519–3525. [CrossRef] [PubMed]
54. Aqeel, Y.; Siddiqui, R.; Anwar, A.; Shah, M.R.; Khan, N.A. Nanoparticle conjugation enhances the antiacanthamoebic effects of chlorhexidine. *Antimicrob. Agents Chemother.* **2016**, *60*, 1283–1288. [CrossRef] [PubMed]
55. Xuguang, S.; Lin, C.; Yan, Z.; Zhiqun, W.; Ran, L.; Shiyun, L.; Xiuying, J. Acanthamoeba keratitis as a complication of orthokeratology. *Am. J. Ophthalmol.* **2003**, *136*, 1159–1161. [CrossRef] [PubMed]
56. Henrissat, B.; Callebaut, I.; Fabrega, S.; Lehn, P.; Mornon, J.P.; Davies, G. Conserved catalytic machinery and the prediction of a common fold for several families of glycosyl hydrolases. *Proc. Nat. Acad. Sci. USA* **1995**, *92*, 7090–7094. [CrossRef]
57. Hadas, E.; Mazur, T. Proteolytic enzymes of pathogenic and non-pathogenic strains of *Acanthamoeba* spp. *Trop. Med. Parasitol. Off. Organ Dtsch. Trop. Ges. Dtsch. Ges. Tech. (GTZ)* **1993**, *44*, 197–200.
58. Mitro, K.; Bhagavathiammai, A.; Zhou, O.M.; Bobbett, G.; McKerrow, J.H.; Chokshi, R.; Chokshi, B.; James, E.R. Partial characterization of the proteolytic secretions of *Acanthamoeba polyphaga. Exp. Parasitol.* **1994**, *78*, 377–385. [CrossRef]
59. Khan, N.A.; Jarroll, E.L.; Panjwani, N.; Cao, Z.; Paget, T.A. Proteases as markers for differentiation of pathogenic and non-pathogenic species of *Acanthamoeba. J. Clin. Microbiol.* **2000**, *38*, 2858–2861. [CrossRef] [PubMed]
60. Ferrante, A.; Bates, E.J. Elastase in the pathogenic free-living amoebae *Naegleria* and *Acanthamoeba* spp. *Infect. Immun.* **1998**, *56*, 3320–3321. [CrossRef]
61. Leher, H.; Silvany, R.; Alizadeh, H.; Huang, J.; Niederkorn, J.Y. Mannose induces the release of cytopathic factors from *Acanthamoeba castellanii. Infect. Immun.* **1998**, *66*, 5–10. [CrossRef]

microorganisms

MDPI

Article

Effect of Biogenic Silver Nanoparticles on the Quorum-Sensing System of *Pseudomonas aeruginosa* PAO1 and PA14

Erika Kushikawa Saeki [1,*], Heloísa Moreira Martins [2], Larissa Ciappina de Camargo [2], Laís Anversa [3], Eliandro Reis Tavares [4], Sueli Fumie Yamada-Ogatta [4], Lucy Megumi Yamauchi Lioni [4], Renata Katsuko Takayama Kobayashi [2] and Gerson Nakazato [2]

1 Regional Laboratory Center, Adolfo Lutz Institute, Presidente Prudente 19013-050, SP, Brazil
2 Laboratory of Basic and Applied Bacteriology, Department of Microbiology, Biological Sciences Center, State University of Londrina, Londrina 86057-970, PR, Brazil
3 Regional Laboratory Center, Adolfo Lutz Institute, Bauru 17015-110, SP, Brazil
4 Laboratory of Molecular Biology of Microorganisms, Department of Microbiology, Biological Sciences Center, State University of Londrina, Londrina 86057-970, PR, Brazil
* Correspondence: erikaksaeki@gmail.com

Abstract: The increase in multidrug-resistant microorganisms represents a global threat requiring the development novel strategies to fight bacterial infection. This study aimed to assess the effect of silver nanoparticles (bio-AgNPs) on bacterial growth, biofilm formation, production of virulence factors, and expression of genes related to the quorum-sensing (QS) system of *P. aeruginosa* PAO1 and PA14. Biofilm formation and virulence assays were performed with bio-AgNPs. RT-qPCR was carried out to determine the effect of bio-AgNPs on the QS regulatory genes *lasI, lasR, rhlI, rhlR, pqsA*, and *mvfR*. Bio-AgNPs had an MIC value of 62.50 μM, for both strains. Phenotypic and genotypic assays were carried out using sub-MIC values. Experimental results showed that treatment with sub-MICs of bio-AgNPs reduced ($p < 0.05$) the motility and rhamnolipids and elastase production in *P. aeruginosa* PAO1. In PA14, bio-AgNPs stimulated swarming and twitching motilities as well as biofilm formation and elastase and pyocyanin production. Bio-AgNP treatment increased ($p < 0.05$) the expression of QS genes in PAO1 and PA14. Despite the different phenotypic behaviors in both strains, both showed an increase in the expression of QS genes. Demonstrating that the bio-AgNPs acted in the induction of regulation. The possible mechanism underlying the action of bio-AgNPs involves the induction of the rhl and/or pqs system of PAO1 and of the las and/or pqs system of PA14. These results suggest that exposure to low concentrations of bio-AgNPs may promote the expression of QS regulatory genes in *P. aeruginosa*, consequently inducing the production of virulence factors such as elastase, pyocyanin, and biofilms.

Keywords: AgNPs; antivirulence; gene regulation; quorum quenching

Citation: Saeki, E.K.; Martins, H.M.; Camargo, L.C.d.; Anversa, L.; Tavares, E.R.; Yamada-Ogatta, S.F.; Lioni, L.M.Y.; Kobayashi, R.K.T.; Nakazato, G. Effect of Biogenic Silver Nanoparticles on the Quorum-Sensing System of *Pseudomonas aeruginosa* PAO1 and PA14. *Microorganisms* **2022**, *10*, 1755. https://doi.org/10.3390/microorganisms10091755

Academic Editors: Kamel A. Abd-Elsalam and Alexander I. Netrusov

Received: 21 July 2022
Accepted: 24 August 2022
Published: 30 August 2022

1. Introduction

Pseudomonas aeruginosa is a Gram-negative, mobile, aerobic, non-spore-forming bacterium. Due to its versatile metabolism, *P. aeruginosa* can grow in aquatic environments and in soil, causing infections in plants and animals and a variety of diseases in humans, particularly in immunocompromised individuals [1–3].

The World Health Organization recognized the difficulty in treating *P. aeruginosa* infections and included the bacterium in its priority list of pathogens for which new drugs are urgently required, particularly for the treatment of carbapenem-resistant isolates [4], which are considered a major threat to public health.

Compounds with antivirulence activity, that is, those that can control pathogens without exerting bacteriostatic or bactericidal action, have been proposed as potential therapeutic agents [5,6]. The quorum-sensing (QS) system can be affected by attenuation of QS communication, disruption of receptor proteins, degradation of autoinducing signals, or inhibition of the synthesis of signaling molecules [7,8]. The QS system is a regulatory center of virulence factors and has been widely studied as a target for antivirulence therapy [6,9].

QS is a mechanism of cellular communication between bacteria mediated by the secretion of extracellular signaling molecules (autoinducers). The concentration of autoinducers depends on microbial density. In general, the system consists of an enzyme that catalyzes the synthesis of signaling molecules and a receptor that binds to the signaling molecule and induces the expression of genes responsible for various physiological mechanisms, such as cell density and virulence factor production [10,11]. Chemical communication through the QS system is a central feature of bacterial life, allowing bacteria to remain in a community and defining collective behaviors [12].

The QS system of *P. aeruginosa* consists of three hierarchically organized systems, namely las, rhl, and PQS [13–15]. Due to the importance of QS to bacterial pathogenicity, many studies have focused on QS interference. Metal nanoparticles have been identified as potent agents against QS [16].

Studies on the QS system of *P. aeruginosa* and its regulation are of great importance to better understand the response of bacteria to metal nanoparticles. For this, it is necessary to identify the mechanisms involved in bacterial pathogenicity. This study aimed to investigate the effect of biogenic silver nanoparticles (bio-AgNPs) on *P. aeruginosa* growth, virulence factors, biofilm formation, and QS.

2. Materials and Methods

2.1. Bacterial Strains and Culture Conditions

The reference strains *P. aeruginosa* PAO1 and UCBPP-PA14 (PA14) were kindly provided by Regina Lúcia Baldini (University of São Paulo) and Laurence G. Rahme (Harvard University). Stock cultures were kept at −80 °C in Luria–Bertani (LB) broth (Neogen, Lansing, MI, USA) containing 20% (v/v) glycerol (Synth, Diadema, SP, Brazil).

2.2. Bio-AgNPs

Bio-AgNPs were synthesized by reduction of silver nitrate catalyzed by a cell-free enzyme preparation from fungus *Fusarium oxysporum* (strain 551). The fungal inoculum was obtained from the Laboratory of Molecular Genetics, University of São Paulo, Piracicaba, São Paulo State, Brazil. *F. oxysporum* was grown for 7 days at 28 °C on 0.5% w/v yeast extract (Neogen, USA), 2% w/v malt extract (Neogen, USA), 2% w/v agar (Acumedia, Lansing, MI, USA), and distilled water. Then, the fungal biomass (0.1 g/mL) was mixed with sterile distilled water and incubated at 28 °C for 72 h under stirring (150 rpm). Next, the cell-free filtrate was mixed with 0.01 M silver nitrate ($AgNO_3$, Sigma–Aldrich, Steinheim am Albuch, Germany) and incubated for 15 days at 28 °C in the dark. Finally, Bio-AgNPs were washed with distilled water, centrifuged at 27,000× g and 4 °C for 30 min, and incubated in an ultrasonic bath for 30 min. Washing steps were repeated three times [17].

These bio-AgNPs have been synthesized, quantified, and characterized by our research group and these results have been previously published [18]. According to Scandorieiro et al. [18], the nanoparticle used in this study exhibited a plasmonic absorption band in the visible region of the spectrum (near 420 nm), spherical shape, and an average size and zeta potential of 73.1 ± 0.5 nm and −24.2 ± 2.1 mV, respectively.

2.3. Minimum Inhibitory Concentration (MIC)

The MIC of bio-AgNPs for *P. aeruginosa* PAO1 and PA14 was determined by the microdilution method in 96-well plates [19]. For that, bio-AgNps were diluted in Mueller–Hinton broth (MHB, Difco, Sparks, MD, USA) to concentrations ranging from 7.81 to 1000 µM. The plates were incubated at 37 °C for 24 h. Positive (*P. aeruginosa* grown in the absence of bio-AgNPs) and negative (bio-AgNPs or MHB alone) controls were performed. All experiments were conducted in triplicate and repeated at least three times.

2.4. Growth Curve Analysis

The effect of bio-AgNPs on the growth kinetics of *P. aeruginosa* PAO1 and PA14 was assessed by the plating method, as described by Jorgensen [20]. Bacterial inoculum was cultivated in LB broth (Neogen, Lansing, MI, USA) with bio-AgNPs at concentrations of 1/2, 1/4, and 1/8 MIC at 37 °C for 48 h. A control growth curve was constructed by incubating bacterial strains in the absence of bio-AgNPs. Bacterial growth was analyzed at six incubation times (0, 4, 8, 24, 32, and 48 h).

2.5. Assessment of Virulence Factors

2.5.1. Biofilm Formation

Biofilm formation capacity was analyzed in 96-well polystyrene plates using the crystal violet method described by Ramos-Vivas et al. [21], with some modifications (Supplementary Table S1). Bacterial isolates were cultured on LB agar (Neogen, Lansing, MI, USA) at 37 °C for 24 h. Then, 180 µL of LB broth (Neogen, Lansing, MI, USA) and 20 µL of *P. aeruginosa* inoculum (initial concentration of 1.5×10^6 CFU/mL) were added to each well and incubated at 37 °C for 24 h in the presence (1/2, 1/4, and 1/8 MIC) and absence (control) of bio-AgNPs. The supernatant of each well was discarded, and the cell layer was washed three times with phosphate buffer solution (PBS, pH 7.2), fixed with 250 µL of absolute methanol PA (Merck, Darmstadt, Germany) for 10 min, and stained with a 1.0% *w*/*v* aqueous solution of crystal violet (Merck, Darmstadt, Germany) for 15 min. Subsequently, the dye solution was discarded, and each well was washed three times with ultrapure water and treated with 250 µL of 33% *v*/*v* glacial acetic acid (Merck, Darmstadt, Germany). Assays were conducted in the presence (1/2, 1/4, and 1/8 MIC) and absence of bio-AgNPs. Absorbance readings were taken at 620 nm on a spectrophotometer (Multiskan FC, Thermo Scientific, Waltham, MA, USA). Experiments were repeated four times per isolate, and results are presented as mean and standard deviation.

2.5.2. Swarming Motility

P. aeruginosa isolates were grown in LB broth (Neogen, Lansing, MI, USA) for 24 h at 30 °C. Ten microliters of a *P. aeruginosa* suspension containing 10^8 CFU/mL in the presence (1/2, 1/4, and 1/8 MIC) or absence (control) of subinhibitory concentrations of bio-AgNPs were inoculated at the center of swarming agar plates containing 1% *w*/*v* glucose (Synth, Diadema, SP, Brazil), 0.5% *w*/*v* peptone (Acumedia, Lansing, MI, USA), 0.2% *w*/*v* yeast extract (BD, Sparks, MD, USA), and 0.5% *w*/*v* agar (BD, Sparks, MD, USA). Plates were incubated without inversion for 24 h at 30 °C [22].

2.5.3. Swimming Motility

P. aeruginosa isolates were seeded on LB agar (Neogen, Lansing, MI, USA) at 37 °C for 24 h. Swimming agar plates containing 1.0% *w*/*v* tryptone (Acumedia, Lansing, MI, USA), 0.5% *w*/*v* sodium chloride (Merck, Darmstadt, Germany), and 0.3% *w*/*v* agar (BD, Sparks, MD, USA), previously equilibrated to room temperature, were inoculated on the surface with one colony in the presence (1/2, 1/4, and 1/8 MIC) or absence (control) of subinhibitory concentrations. Plates were incubated without inversion for 24 h at 30 °C [23].

2.5.4. Twitching Motility

P. aeruginosa isolates were seeded on LB agar (Neogen, Lansing, MI, USA) and incubated at 37 °C for 24 h. One colony of each isolate was inoculated in the presence (1/2, 1/4, and 1/8 MIC) or absence (control) of subinhibitory concentrations of bio-AgNPs at the bottom of twitching agar plates containing 1.0% *w*/*v* tryptone (Acumedia, Lansing, MI, USA), 0.5% *w*/*v* yeast extract (BD, Sparks, MD, USA), 1.0% *w*/*v* sodium chloride (Merck, Darmstadt, Germany), and 1.0% *w/v* agar (BD, Sparks, MD, USA). Plates were inverted and incubated at 37 °C for 24 h. Posteriorly, the agar was removed and stained with 2% *w*/*v* crystal violet (Laborclin, Pinhais, PR, Brazil) for 2 h [24]. The motility halo was measured to the nearest millimeter. As a negative control, each isolate was inoculated in tryptone soy agar (BD, Sparks, MD, USA) under the same conditions.

2.5.5. Rhamnolipids

P. aeruginosa strains were cultured in the presence (1/2, 1/4, and 1/8 MIC) or absence of bio-AgNPs in LB broth (Neogen, Lansing, MI, USA) at 37 °C for 24 h. Then, 10 µL of inoculum was placed at the center of a cetyltrimethylammonium bromide (CTAB) agar plate containing 0.09% *w/v* monobasic potassium phosphate (Synth, Diadema, SP, Brazil), 0.11% *w/v* bibasic sodium phosphate (Synth, Diadema, SP, Brazil), 0.25% *w/v* sodium nitrate (Synth, Diadema, SP, Brazil), 0.01% *w/v* calcium chloride (Synth, Diadema, SP, Brazil), 0.04% *w/v* magnesium sulfate (Synth, Diadema, SP, Brazil), 0.2% *w/v* CTAB (Sigma–Aldrich, Steinheim am Albuch, Germany), 0.005% *w/v* methylene blue (Synth, Diadema, SP, Brazil), 0.5% *w/v* glucose (Synth, Diadema, SP, Brazil), and 2.0% *w/v* agar (Acumedia, Lansing, MI, USA). Plates were incubated at 37 °C for 48 h [25]. Rhamnolipid production was determined by measuring the halo of blue precipitate surrounding colonies.

2.5.6. Alkaline Protease

P. aeruginosa strains were grown in LB broth (Neogen, Lansing, MI, USA) at 37 °C for 24 h in the presence (1/2, 1/4, and 1/8 MIC) or absence of bio-AgNPs. Then, 10 µL of supernatant from treated and untreated *P. aeruginosa* PAO1 and PA14 was added to a milk agar plate (pH 10.0) containing 1.0% *w/v* milk powder (Acumedia, Lansing, MI, USA), 0.1% *w/v* peptone (Acumedia, Lansing, MI, USA), 0.5% *w/v* NaCl (Synth, Diadema, SP, Brazil), and 2.0% *w/v* agar (Acumedia, Lansing, MI, USA) and was incubated at 37 °C for 24 h. Alkaline protease production was indicated by the formation of a clear halo around colonies. Halo diameters were measured and compared with those of the control [26].

2.5.7. Elastase B (LasB)

The elastin–Congo Red (ECR) method was used to investigate LasB activity [27]. *P. aeruginosa* was cultured in LB broth (Neogen, Lansing, MI, USA) at 37 °C for 24 h in the presence (1/2, 1/4, and 1/8 MIC) or absence of bio-AgNPs. After centrifugation (8000 rpm, 10 min), 500 µL of the supernatant was added to 500 µL of 100 mM Tris-HCl buffer, pH 7.5, containing 10 mg of ECR (Sigma-Aldrich, Steinheim am Albuch, Germany). The mixture was incubated at 37 °C for 6 h under stirring (120 rpm). After centrifugation (8000 rpm, 10 min), the absorbance was measured at 495 nm on a Thermo Scientific Multiskan GO spectrophotometer.

2.5.8. Pyocyanin

A liquid medium consisting of 2.0% *w/v* peptone (Neogen, Lansing, MI, USA), 0.14% *w/v* magnesium chloride (Synth, Diadema, SP, Brazil), and 1.0% *w/v* magnesium sulfate (Synth, Diadema, SP, Brazil) was used to assess pyocyanin production, as described by El-Mowafy et al. [28]. Briefly, 1 mL of a 0.5 McFarland-equivalent suspension of *P. aeruginosa* grown in LB broth (Neogen, Lansing, MI, USA) was inoculated in 30 mL of liquid medium in the presence (1/2, 1/4, or 1/8 MIC) or absence of bio-AgNPs. Cells were incubated at 37 °C for 24 h. Then, cultures were centrifuged, and 7.5 mL of the supernatant was transferred to a tube containing 4.5 mL of chloroform (Merck, Steinheim am Albuch, Germany). This mixture was vigorously homogenized on a vortex mixer for 20 s. The organic phase was collected (3 mL), acidified with 1.5 mL of 0.2 M hydrochloric acid, homogenized for 20 s, and centrifuged at 4600 rpm for 10 min. The absorbance of the resulting pink-colored solution was measured at 520 nm on a SPECORD S600 UV/VIS spectrophotometer (Analytica Jena, Germany). Pyocyanin concentrations were converted to µg/mL, by multiplying the optical density at 520 nm by 17,072 × 1.5.

2.6. RNA Extraction and Real-Time Polymerase Chain Reaction (RT-qPCR)

For analysis of the expression of QS-related genes, total RNA was extracted from *P. aeruginosa* PAO1 and PA14 cells grown in the absence (reference) or presence of bio-AgNPs (1/2 MIC). Extraction was performed using TRIzol reagent (Invitrogen, Waltham, MA, USA) and the RNeasy Mini kit (QIAGEN, Germantown, MA, EUA). DNA contamination was eliminated by using the RQ1-DNAse kit (Promega, USA). The quality of extracted RNA was verified by agarose gel electrophoresis and calculation of the A260/A280 ratio. Gene sequences were obtained from GenBank, and primers were designed and analyzed using Primer-BLAST and BioEdit Sequence Alignment Editor. Primer sequences are listed in Table 1. RT-qPCR was performed using the QuantiNova SYBR Green RT-PCR kit (QIAGEN, Germantown, MA, EUA) in a final volume of 20 µL, consisting of 10 µL of SYBR Green RT-PCR Master Mix, 0.2 µL of RT mix, 1 µL (20 µM) of each primer, 5 µL of bacterial RNA (50 ng/µL), and 2.8 µL of RNase-free water. Reactions were performed on a Rotor-Gene Q 2plex (QIAGEN, Germantown, MA, EUA) using the following steps: reverse transcription for 10 min at 50 °C, initial denaturation for 2 min at 95 °C, 40 cycles of 95 °C for 5 s, and hybridization and extension at 60 °C for 10 s. The *proC* gene was selected as an internal control and was used to normalize the expression of target genes.

Table 1. List of genes and their respective primers used in the RT-qPCR assay.

Target Gene	Primer	Sequence	Amplicon Size (bp)
proC	F	5′-GAG CAA CTG ATC GTC TCC ATC-3′	100
	R	5′-GGG TGT TGG GCA TGC AG-3′	
lasI	F	5′-GTA GGC GTG GAG AAG ATG ATG-3′	122
	R	5′-ATC TGG GTC TTG GCA TTG AG-3′	
lasR	F	5′-CTG TGG ATG CTC AAG GAC TAC-3′	111
	R	5′-CCA CTG CAA CAC TTC CTT CT-3′	
rhlI	F	5′-GTC GGT CTG GGA GCT TTC-3′	100
	R	5′-CAG GTA CCA GGC GCA TT-3′	
rhlR	F	5′-CTG TGG TGG GAC GGT TTG-3′	139
	R	5′-GGG TGA AGG GAA TCG TGT G-3′	
pqsA	F	5′-GCT GTA TTC GAT TCC CAA GAT G-3′	100
	R	5′-CCA GGT ATC GTC GAG CAG-3′	
mvfR	F	5′-GCT TCG CCT GAT CCC TTA C-3′	104
	R	5′-GCA GCA CCC GGA GAT TG-3′	

F: forward; R: reverse; pb: base pairs.

2.7. Statistical Analysis

Statistical analysis was performed using R Studio version 1.2.5001 (R, Boston, MA, USA), and graphs were constructed using GraphPad Prism version 8.4.2 (GraphPad Software Inc., San Diego, CA, USA). Values of $p < 0.05$ were considered significant. Data were subjected to analysis of variance followed by Tukey's test. For RT-qPCR data, randomization tests were conducted using REST software version 2.0.13 (Qiagen, Hilden, Germany) to identify statistically significant differences between groups.

3. Results

3.1. MIC of bio-AgNPs against *P. aeruginosa*

MIC values were determined by using bio-AgNPs at different concentrations (7.81 to 1000 µM). MIC values for *P. aeruginosa* PAO1 and PA14 were 62.5 µM. Bio-AgNPs were also tested at subinhibitory concentrations, 1/2 (31.25 µM), 1/4 (15.62 µM), and 1/8 MIC (7.81 µM), to evaluate whether treatments would have any effect on PAO1 and PA14 growth. As shown in Figure 1A,B, PAO1 and PA14 growth was not influenced ($p > 0.05$) by treatment with bio-AgNPs at subinhibitory concentrations after 24 and 48 h.

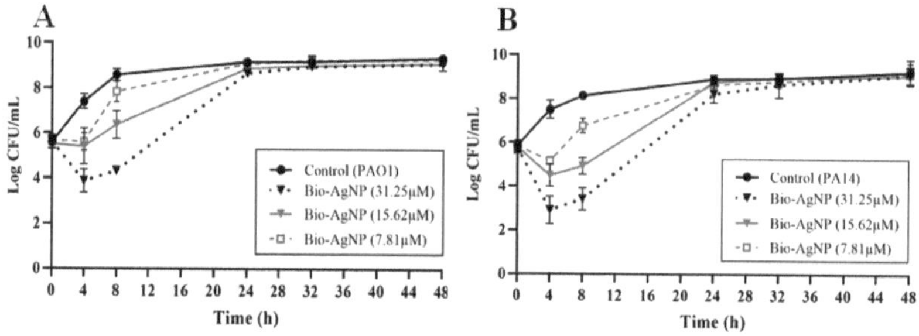

Figure 1. Growth curve of *Pseudomonas aeruginosa*. (**A**) PAO1 and (**B**) PA14 in the presence or absence (control) of subinhibitory concentrations (31.25 μM, 15.62 μM, and 7.81 μM) of biogenic silver nanoparticles (Bio-AgNP). Bacteria were inoculated at a density of 5×10^5 CFU/mL. Results are expressed as mean ± standard deviation.

3.2. *Effect of bio-AgNPs on P. aeruginosa Virulence Factors*

We investigated the effects of bio-AgNPs at subinhibitory concentrations (7.81 μM, 15.62 μM, and 31.25 μM) on QS-controlled virulence factors, such as biofilm formation, motility, and rhamnolipid, alkaline protease, LasB, and pyocyanin production.

3.2.1. Biofilm Formation

The crystal violet method revealed no significant change in biofilm formation in PAO1 after treatment with bio-AgNPs (Table 2). However, all three bio-AgNP treatments significantly increased ($p < 0.05$) biofilm formation in PA14, by 41.22% to 127.86%.

Table 2. Biofilm formation in *P. aeruginosa* after treatment with different subinhibitory concentrations of silver nanoparticles.

Strain	Compound	Concentration	Biofilm Formation [1]	Inhibition Rate (%) [2]	Increase Rate (%) [2]
PAO1	Controle	0 μM	0.260 ± 0.040	-	-
PAO1	bio-AgNP	7.81 μM	0.257 ± 0.021	1.31%	-
PAO1	bio-AgNP	15.62 μM	0.256 ± 0.051	1.73%	-
PAO1	bio-AgNP	31.25 μM	0.238 ± 0.068	8.74%	-
PA14	Controle	0 μM	0.052 ± 0.007 [c]	-	-
PA14	bio-AgNP	7.81 μM	0.118 ± 0.015 [a]	-	127.86%
PA14	bio-AgNP	15.62 μM	0.085 ± 0.020 [b]	-	64.09%
PA14	bio-AgNP	31.25 μM	0.073 ± 0.013 [b]	-	41.22%

[1] Values expressed in optical density (OD) 620 nm (mean ± standard deviation). [2] Inhibition rate or increase = (OD control group–OD experimental group)/OD control group. [a–c] Tukey's test with significant difference $p < 0.05$. bio-AgNP: biogenic silver nanoparticles.

3.2.2. Effect on Swarming, Swimming, and Twitching Motilities

In the present study, the effects of sub-MIC levels of bio-AgNPs on the swarming, swimming, and twitching motilities in *P. aeruginosa* were investigated by determining the diameter (mm) of the motility halo. As depicted in Figure 2A1, bio-AgNPs inhibited the swarming motility in PAO1. After treatment with 31.25 μM bio-AgNPs, the mean diameter of the motility halo was 45.67 mm, representing an inhibition rate of 22.45% compared with the untreated control (58.89 mm diameter). In PA14, bio-AgNPs induced a significant increase ($p < 0.05$) in the swarming motility (Figure 2A2). The diameter of the swarming motility halo increased from 13.89 mm (untreated control) to 45.00–45.89 mm, representing an increase of up to 230.40%.

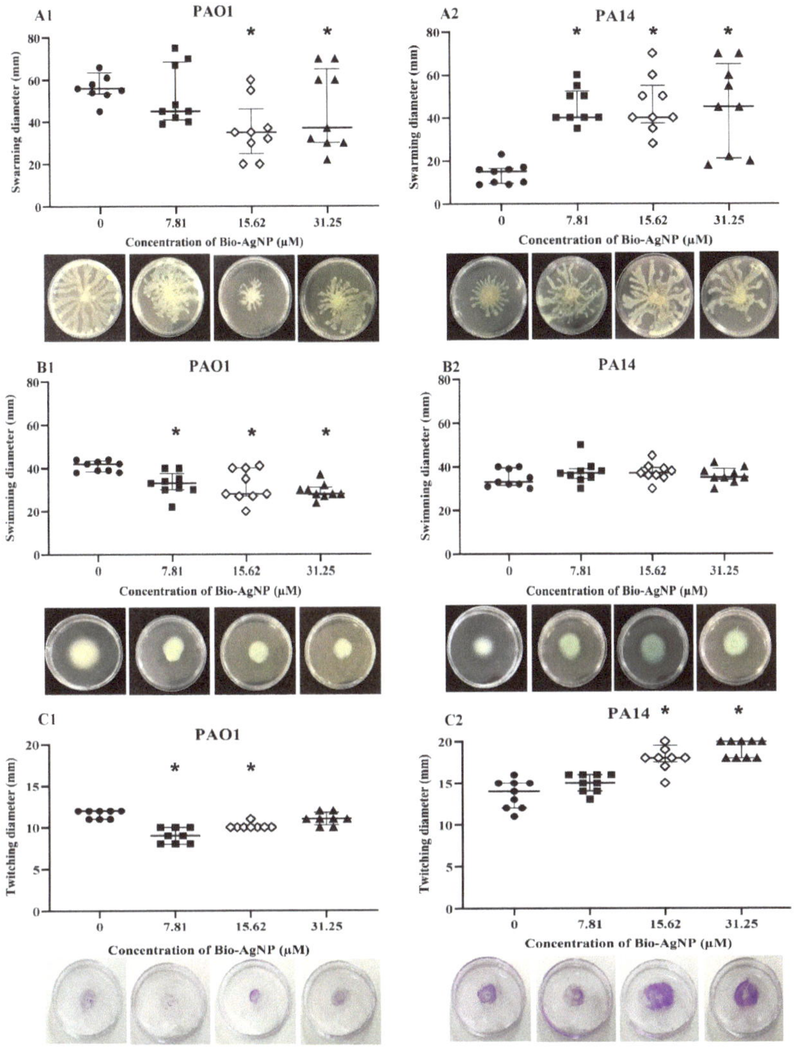

Figure 2. Effect of bio-AgNPs on swarming, swimming, and twitching motilities in *Pseudomonas aeruginosa* PAO1 and PA14. Swarming motility in (**A1**) PAO1 and (**A2**) PA14. Swimming motility in (**B1**) PAO1 and (**B2**) PA14. Twitching motility in (**C1**) PAO1 and (**C2**) PA14. Each data point represents the result of one treatment, and the mean of three replications is indicated by a black bar. Images below the graphs are representative photographs of PAO1 and PA14 motility halos. Asterisks indicate significant differences between means and the control at $p < 0.05$ by analysis of variance followed by Tukey's test.

Similarly, PAO1 swimming motility was significantly ($p < 0.05$) inhibited by bio-AgNP treatments (Figure 2B1). The control exhibited a motility halo of 41.00 mm in diameter, whereas treated bacteria had a swimming motility halo of 29.33 to 32.78 mm, with a reduction of 20.05% to 28.46%. In PA14, treatments did not influence the swimming motility (Figure 2B2).

PAO1 twitching motility was significantly inhibited ($p < 0.05$) by treatment with 7.81 and 15.62 µM bio-AgNPs, leading to reductions of 22.58% and 12.90%, respectively

(Figure 2C1). By contrast, in PA14, the twitching motility increased by 34.15% and 39.84% ($p < 0.05$) after treatment with 15.62 and 31.25 μM bio-AgNPs, respectively (Figure 2C2).

3.2.3. Rhamnolipids

All bio-AgNP treatments significantly reduced ($p < 0.05$) rhamnolipid production by PAO1 (Figure 3A1). However, treatments did not influence rhamnolipid production by PA14 (Figure 3A2).

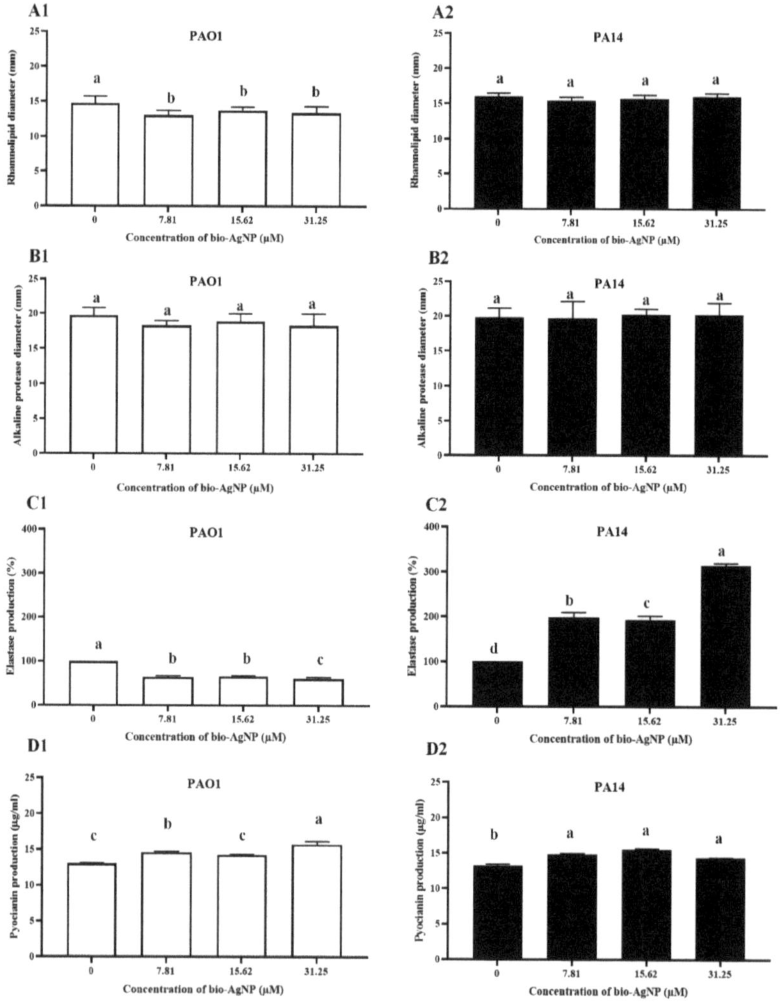

Figure 3. Effect of bio-AgNPs on virulence factors of *Pseudomonas aeruginosa* PAO1 and PA14. Rhamnolipid production in (**A1**) PAO1 and (**A2**) PA14. Alkaline protease production in (**B1**) PAO1 and (**B2**) PA14. Elastase B production in (**C1**) PAO1 and (**C2**) PA14. Pyocyanin production in (**D1**) PAO1 and (**D2**) PA14. Results are expressed as mean ± standard deviation (n = 3). a–c Tukey's honestly significant difference test at $p < 0.05$.

3.2.4. Alkaline Protease

Alkaline protease production was not affected by bio-AgNP treatments (Figure 3B1,B2). The mean halo diameters of PAO1 and PA14 were 19.67 and 19.78 mm, respectively, before treatment, and 18.33–18.89 and 19.67–20.22 mm, respectively, after treatment.

3.2.5. LasB

Figure 3C depicts the LasB activity of strains as a function of bio-AgNP concentration. In PAO1, LasB activity decreased ($p < 0.05$) by 33.79%, 30.27%, and 39.01% with exposure to bio-AgNPs at 7.81, 15.62, and 31.25 μM, respectively (Figure 3C1). By contrast, LasB activity increased ($p < 0.05$) by 99.18%, 93.76%, and 214.74% in PA14 with exposure to 7.81, 15.62, and 31.25 μM bio-AgNPs, respectively (Figure 3C2).

3.2.6. Pyocyanin

Pyocyanin production by *P. aeruginosa* in the presence of different concentrations of bio-AgNPs was determined spectrophotometrically at 520 nm. In both PAO1 and PA14, there was an increase in pyocyanin production with all treatments ($p < 0.05$). In the absence of bio-AgNPs, PAO1 and PA14 produced 13.51 and 14.74 μg/mL pyocyanin, respectively. PAO1 treated with 31.25 μM bio-AgNPs showed the highest pyocyanin production (16.16 μg/mL, representing an increase of 19.70%. In PA14, the highest pyocyanin production (16.39 μg/mL) was achieved by treatment with 31.25 μM bio-AgNPs, affording an increase of 11.21% (Figure 3D).

3.3. Gene Expression

The relative expression of the six regulatory genes of the QS system (*lasI, lasR, rhlI, rhlR, pqsA*, and *mvfR*) of *P. aeruginosa* PAO1 and PA14 was assessed after 24 h of treatment with 31.25 μM bio-AgNPs, as shown in Figure 4. Gene expression was higher in *P. aeruginosa* PAO1 treated with bio-AgNP than in the control ($p < 0.05$): the expression of *lasI, lasR, rhlI, rhlR, pqsA*, and *mvfR* increased by 2.0-, 1.1-, 2.1-, 1.4-, 2.1-, and 2.1-fold, respectively. In PA14, the expression of *lasI, lasR, rhlR, pqsA*, and *mvfR* increased ($p < 0.05$) by 3.3-, 1.4-, 1.7-, 3.3-, and 3.5-fold, respectively, with bio-AgNP treatment. *rhlI* expression, however, was downregulated, being 0.6-fold lower in treated bacteria ($p < 0.05$).

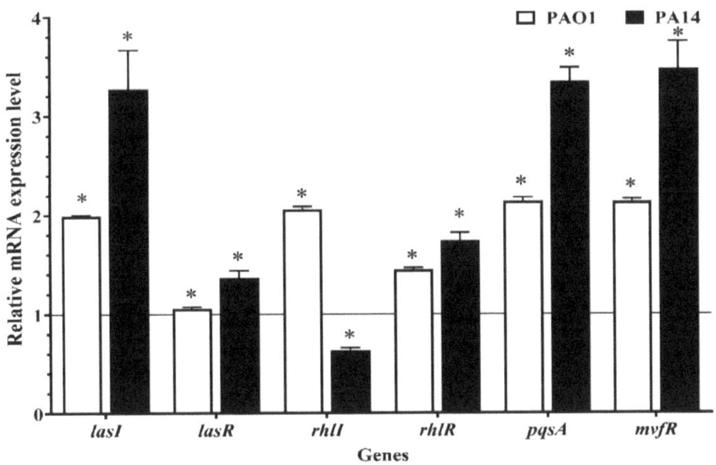

Figure 4. Relative expression of quorum-sensing regulatory genes in *Pseudomonas aeruginosa* treated with bio-AgNPs. The mean relative gene expression was normalized to that of the reference gene *proC*. * $p < 0.05$.

4. Discussion

Given the need for advances in the control of multidrug-resistant bacteria and antivirulence therapeutic strategies, we investigated the potential of bio-AgNPs as bacterial control agents. Research efforts have been directed toward the use of nanotechnology and nanoparticles to target the QS system and/or virulence of microorganisms such as *P. aeruginosa* [16,29,30]. One of the advantages of using bio-AgNPs is related to their low cytotoxicity. Scandorieiro et al. [31] showed that this biogenic silver nanoparticle did not show cytotoxicity at concentrations up to 97.22 μM in human RBC and HEp-2 cells. In the present study, we used concentrations up to 31.25 μM of bio-AgNPs.

In the current study, we identified significant differences in phenotypic characteristics between the evaluated reference strains. In PAO1, the swarming, swimming, and twitching motilities, as well as rhamnolipid and elastase production, were significantly reduced in the presence of bio-AgNPs at sub-MIC levels. However, in PA14, bio-AgNP treatment significantly enhanced swarming motility, twitching motility, biofilm formation, and elastase and pyocyanin production.

The reference strains exhibit different genotypic characteristics. *P. aeruginosa* PAO1 is a moderately virulent strain belonging to a relatively rare clonal group (ST 549). PA14, on the other hand, is highly virulent and belongs to the most common clonal group (ST 253). Furthermore, PA14 has two pathogenicity islands (PAPI-1 and PAPI-2), which are absent in PAO1 [32–34] The strains also differ in genome size (6.3 Mbp in PAO1 and 6.5 Mbp in PA14) and have unique genomic regions (54 regions in PAO1 and 58 regions in PA14) [35]. Such factors could explain the different responses of strains to bio-AgNP treatment.

As evidenced by the biofilm formation assay, untreated PAO1 showed greater capacity for biofilm formation than untreated PA14. The reduced capacity of PA14 for biofilm formation might be related to mutation of the *ladS* gene, which has a deleterious effect on biofilm production [33]. Kasetty et al. [36] also showed that in microfluidic biofilm culture conditions, PAO1 quickly outcompetes PA14 in density. Furthermore, PA14 exhibits a competitive fitness advantage when invading a preformed biofilm and is better able to tolerate starvation than PAO1 in the biofilm context.

Surprisingly, after bio-AgNP treatment, PA14 exhibited a significant increase in biofilm formation. The mechanism of biofilm formation differs between PAO1 and PA14, given that each strain uses a different exopolysaccharide as the predominant structural biofilm component. Whereas PAO1 uses the Wsp system to produce Psl, PA14 uses the Pil-Chp system for Pel production [37,38]. Biofilm formation is dependent on several factors, such as synthesis of exopolysaccharides Psl and Pel, alginates, extracellular DNA, and adhesins (flagella and type IV pili). These factors contribute to maturation, antibiotic resistance, and biofilm persistence [39]. Exopolysaccharides are responsible for bacterial adhesion, biofilm formation, and architecture stability [40]. Extracellular DNA contributes to twitching motility and the supply of nutrients to bacterial cells [41].

It is important to highlight that there is an indirect link between biofilm formation and the QS system, related to the control of the swarming and twitching motilities as well as rhamnolipid [42] and pyocyanin [43] production. *P. aeruginosa* displays three types of motilities (swarming, swimming, and twitching), which allow surface colonization and exploration of new environments [44]. The first step in biofilm formation is bacterial fixation to surfaces via polar flagella and production of adhesion proteins, such as type IV pili, thereby allowing cells to spread to surrounding areas [45].

In the present study, the significant increase in PA14 swarming motility, twitching motility, and pyocyanin production with bio-AgNP treatment contributed to biofilm formation. According to Persat et al. [46], virulence factors can be activated by structural elements (e.g., type IV pili). *P. aeruginosa* uses these structures for surface recognition and activation of the Chp chemosensory system, which regulates cAMP (second messenger) and induces virulence gene transcription. Therefore, the expressive increase in the swarming and twitching motilities in PA14, mediated by type IV pili, might have favored the increase in other virulence factors.

Ouyang et al. [30] treated *Pseudomonas putida* KT2440 with low concentrations of zinc nanoparticles (0.5–30 mg/L) and observed an increase in biofilm formation. Protein and sugar contents of biofilm also increased with treatment. Garuglieri et al. [47] reported that subinhibitory concentrations of AgNPs (0.01 μg/mL) might increase the swimming motility in *Escherichia coli*. In the study of Saeki et al. [48], *P. aeruginosa* isolates from clinical and environmental sources were found to exhibit increased swarming, swimming, and twitching motilities following treatment with bio-AgNPs. In such cases, bio-AgNPs possibly stimulated a response to stress induced by environmental conditions, as described by Villa et al. [49].

PAO1 showed a significant reduction in LasB production with bio-AgNP treatment, and the opposite was observed in PA14 ($p < 0.05$). LasB, also known as pseudolysin, is associated with vascular inflammation in *P. aeruginosa* infections [50]. In agreement with the present study, in which PA14 showed increased biofilm formation with bio-AgNP treatment, Yu et al. [51] demonstrated that LasB is crucial for biofilm formation. The authors found that ΔlasB mutant *P. aeruginosa* has reduced capacity for biofilm formation owing to a decrease in rhamnolipid synthesis.

The alkaline protease assay revealed no differences in protease production with bio-AgNP treatment in either PAO1 or PA14. Alkaline protease causes severe damage to host tissues by disruption of cytoskeleton structures and degradation of fibronectin and laminin, important components of the endothelium [50]. Both alkaline protease and LasB are able to inhibit neutrophil function, interfering with chemotaxis. As a result, bacteria gain an advantage by escaping phagocytes of the host defense system [52].

To further understand bio-AgNP at the molecular level and to lend support to the outcome of our in vitro results, we followed up with qRT-PCR analysis. *lasI, rhlI,* and *pqsA* expression is necessary for the synthesis of QS signaling molecules in *P. aeruginosa*. Thus, the transcriptional regulators lasR, rhlR, and mvfR were analyzed in this study. Although most PAO1 virulence factors decreased after bio-AgNP treatment, the respective genes were upregulated in both strains (PAO1 and PA14). Bio-AgNP-treated PAO1 bacteria exhibited a higher expression (upregulation) of the evaluated genes compared with the control ($p < 0.05$). In PA14, only rhlI was downregulated; the other genes were more highly expressed ($p < 0.05$) than in the control.

Typically, the common mechanisms underlying quorum-sensing interference include inhibition of signal biosynthesis, signal degradation, and interruption of the reception signal molecules [53]. Based on our results, the possible mechanism underlying the action of bio-AgNPs involves the induction of the rhl and/or pqs system of PAO1 and induction of the las and/or pqs system of PA14.

The las system is the master regulator of QS, as it induces the expression of both rhl and PQS pathways in *P. aeruginosa* [54]. The synthase proteins LasI and RhlI are responsible for the production of the autoinducers 3-oxo-C12-AHL and C4-AHL, respectively. When the concentration of such molecules reaches a given threshold, 3-oxo-C12-AHL binds to the receptor lasR, thereby inducing the expression of virulence factors, such as LasB, exotoxin, and proteases, and activating the rhlI/R system. The autoinducer C4-AHL binds to rhlR and controls motility and expression of genes that encode LasB, pyocyanin, and rhamnolipid. The third autoinducer, PQS, binds to the receptor mvfR, regulating pyocyanin expression and activating the rhlI system [14]. Therefore, the 2.0-fold increase in *pqsA* and *mvfR* gene expression in PAO1 and PA14 was probably associated with increased pyocyanin production resulting from bio-AgNP treatment; the 0.6-fold reduction in *rhlI* expression in PA14, however, did not lead to a decrease in virulence in phenotypic assays.

Liao et al. [55] demonstrated that the main mechanism of action of AgNPs against multidrug-resistant isolates of *P. aeruginosa* involves the imbalance of oxidation processes and impairment of the elimination of oxygen reactive species. Cugini et al. [56] proposed that compounds such as farnesol, which promote reactive oxygen species production, may enhance the expression of the PQS system by activating rhlR in *P. aeruginosa*.

The decrease in motility and rhamnolipid and LasB production in PAO1 might be explained by the effect of bio-AgNPs on *lasR* expression: expression increased 1.1-fold. Given the interconnection of QS systems and the fact that lasR controls the activation of connected cascades [57], it is suggested that the effect of bio-AgNP treatment on *lasR* expression was not sufficient to induce a phenotypic increase in virulence factors. By contrast, in PA14, the 1.4-fold increase in *lasR* expression resulted in increased motility and LasB production.

Our findings were similar to those of previous studies investigating the influence of AgNPs on the QS system of *P. aeruginosa*. Yang and Alvarez [58] observed that subinhibitory concentrations of AgNPs (21.6 and 108 μg/L) could induce the QS system (*lasI, lasR, rhlI,* and *rhlR* expression) in PAO1 and increase biofilm formation. Li et al. [29] studied several nanomaterials (e.g., silver, iron, zinc oxide, graphene) and found that AgNPs (100 μg/L) enhanced 3-oxo-C12-AHL synthesis, protease production, biofilm formation, and *lasR* expression in PAO1. Ouyang et al. [30], in assessing the effect of low concentrations of zinc nanoparticles on *P. putida*, observed an increase in the expression of QS regulatory genes with treatment.

Previous studies have found that AgNP treatment affected virulence factor production, and QS. Singh et al. [59] reported that AgNPs (28 nm) synthesized by *Rhizopus arrhizus* BRS-07 inhibited the expression of QS regulatory genes (*lasI, lasR, rhlI, rhlR,* and *fabH2*) in *P. aeruginosa* PAO1. Liu et al. [60] combined AgNPs with 4-nitropyridine N-oxide for inhibition of biofilm formation and QS gene expression (*lasI, lasR, rhlI, rhlR, pqsA,* and *pqsR*). In the current research, bio-AgNPs were synthesized by *F. oxysporum* and measured 73 nm in size. Liu et al. [60] used a combination of nanoparticles with 4-nitropyridine N-oxide, whereas we used bio-AgNPs only.

Given that QS inhibition has shown high potential as an antivirulence strategy [6,9], understanding the mechanisms of pathogenicity in *P. aeruginosa* is of extreme importance, to ensure that antivirulence compounds are applied efficiently. As highlighted by Mohanty et al. [61], the effects of metal nanoparticles on the QS system are highly dependent on bacterial species. Therefore, given the variability of bacterial responses to different metal compounds, this approach needs to be well-studied and scientifically proven with complementary molecular studies on reference strains and isolates obtained from different sources (environmental and clinical).

5. Conclusions

Our results suggest that bacterial exposure to low concentrations of bio-AgNPs may enhance the expression of QS regulatory genes in *P. aeruginosa*. This model indicates that the target compound (bio-AgNP) affected functional genes involved in the biofilms' formation and virulence production, mainly in PA14. This implies that the presence or accidental release of low concentrations of AgNPs may cause ecological imbalances, possibly leading to an increase in *P. aeruginosa* virulence. Thus, the new concept of antivirulence therapy needs to be carefully studied. Understanding bacterial virulence and pathogenicity mechanisms, which depend on a variety of extracellular and cell-associated factors, is essential for the development of potential targets for antivirulence therapy.

Supplementary Materials: The following supporting information can be downloaded at: https://www.mdpi.com/article/10.3390/microorganisms10091755/s1, Table S1: Comparison of biofilm formation methods.

Author Contributions: E.K.S. and H.M.M., conceptualization, methodology; E.K.S., H.M.M., L.C.d.C. and L.A., experiments and data curations; E.K.S. writing—original draft preparation; E.K.S., E.R.T., S.F.Y.-O., L.M.Y.L., R.K.T.K. and G.N., review and editing. All authors have read and agreed to the published version of the manuscript.

Funding: This research received no external funding.

Institutional Review Board Statement: Not applicable.

Informed Consent Statement: Not applicable.

Acknowledgments: The authors thank the Graduate Program in Microbiology of the State University of Londrina. The authors also thank Regina Lúcia Baldini (University of São Paulo) and Laurence G. Rahme (Harvard University) for providing the *P. aeruginosa* strains used in this study.

Conflicts of Interest: The authors declare no conflict of interest.

References

1. Tuon, F.F.; Dantas, L.R.; Suss, P.H.; Tasca Ribeiro, V.S. Pathogenesis of the *Pseudomonas aeruginosa* Biofilm: A Review. *Pathogens* **2022**, *11*, 300. [CrossRef] [PubMed]
2. Chadha, J.; Harjai, K.; Chhibber, S. Revisiting the virulence hallmarks of *Pseudomonas aeruginosa*: A chronicle through the perspective of quorum sensing. *Environ. Microbiol.* **2022**, *24*, 2630–2656. [CrossRef] [PubMed]
3. Cillóniz, C.; Dominedò, C.; Torres, A. Multidrug Resistant Gram-Negative Bacteria in Community-Acquired Pneumonia. *Crit. Care* **2019**, *23*, 79. [CrossRef] [PubMed]
4. Tacconelli, E.; Carrara, E.; Savoldi, A.; Harbarth, S.; Mendelson, M.; Monnet, D.L.; Pulcini, C.; Kahlmeter, G.; Kluytmans, J.; Carmeli, Y.; et al. Discovery, research, and development of new antibiotics: The WHO priority list of antibiotic-resistant bacteria and tuberculosis. *Lancet Infect. Dis.* **2018**, *18*, 318–327. [CrossRef]
5. Fleitas Martínez, O.; Cardoso, M.H.; Ribeiro, S.M.; Franco, O.L. Recent Advances in Anti-virulence Therapeutic Strategies with a Focus on Dismantling Bacterial Membrane Microdomains, Toxin Neutralization, Quorum-Sensing Interference and Biofilm Inhibition. *Front. Cell. Infect. Microbiol.* **2019**, *2*, 74. [CrossRef]
6. Piewngam, P.; Chiou, J.; Chatterjee, P.; Otto, M. Alternative approaches to treat bacterial infections: Targeting quorum-sensing. *Expert. Rev. Anti. Infect. Ther.* **2020**, *18*, 499–510. [CrossRef]
7. Li, Q.; Mao, S.; Wang, H.; Ye, X. The Molecular Architecture of *Pseudomonas aeruginosa* Quorum-Sensing Inhibitors. *Mar. Drugs* **2022**, *20*, 488. [CrossRef]
8. García-Contreras, R. Is Quorum Sensing interference a viable alternative to treat *Pseudomonas aeruginosa* infections? *Front. Microbiol.* **2016**, *14*, 1454. [CrossRef]
9. Defoirdt, T. Quorum-Sensing Systems as Targets for Antivirulence Therapy. *Trends Microbiol.* **2018**, *26*, 313–328. [CrossRef]
10. Castillo-Juárez, I.; Maeda, T.; Mandujano-Tinoco, E.A.; Tomás, M.; Pérez-Eretza, B.; García-Contreras, S.J.; Wood, T.K.; García-Contreras, R. Role of quorum sensing in bacterial infections. *World J. Clin. Cases* **2015**, *16*, 575–598. [CrossRef]
11. Goswami, J. Quorum Sensing by Super Bugs and their Resistance to Antibiotics: A Short Review. *Glob. J. Pharmaceu. Sci.* **2017**, *3*, 555614.
12. Papenfort, K.; Bassler, B.L. Quorum sensing signal-response systems in Gram-negative bacteria. *Nat. Rev. Microbiol.* **2016**, *11*, 576–588. [CrossRef]
13. Lee, J.; Wu, J.; Deng, Y.Y.; Wang, J.; Wang, C.; Wang, J.; Chang, C.; Dong, Y.; Williams, P.; Zhang, L.-H. A cell-cell communication signal integrates quorum sensing and stress response. *Nat. Chem. Biol.* **2013**, *9*, 339–343. [CrossRef]
14. Lee, J.; Zhang, L. The hierarchy quorum sensing network in *Pseudomonas aeruginosa*. *Protein Cell.* **2015**, *6*, 26–41. [PubMed]
15. Cornelis, P. Putting an end to the *Pseudomonas aeruginosa* IQS controversy. *Microbiologyopen* **2020**, *9*, e962. [CrossRef] [PubMed]
16. Ahmed, T.; Pattnaik, S.; Khan, M.B.; Ampasala, D.R.; Busi, S.; Sarma, V.V. Inhibition of quorum sensing-associated virulence factors and biofilm formation in *Pseudomonas aeruginosa* PAO1 by *Mycoleptodiscus indicus* PUTY1. *Braz. J. Microbiol.* **2020**, *51*, 467–487. [CrossRef] [PubMed]
17. Durán, N.; Marcato, P.D.; Alves, O.L.; Souza, G.I.H.; Esposito, E. Mechanistic aspects of biosynthesis of silver nanoparticles by several *Fusarium oxysporum* strains. *J. Nanobiotechnol.* **2005**, *3*, 8. [CrossRef] [PubMed]
18. Scandorieiro, S.; Rodrigues, B.C.D.; Nishio, E.K.; Panagio, L.A.; de Oliveira, A.G.; Durán, N.; Nakazato, G.; Kobayashi, R.K.T. Biogenic Silver Nanoparticles Strategically Combined With *Origanum vulgare* Derivatives: Antibacterial Mechanism of Action and Effect on Multidrug-Resistant Strains. *Front. Microbiol.* **2022**, *6*, 842600. [CrossRef]
19. CLSI-Clinical and Laboratory Standards Institute. Methods for Dilution Antimicrobial Susceptibility Tests for Bacteria That Grow Aerobically. In *Performance Standards for Antimicrobial Susceptibility Testing*, 10th ed.; Approved Standard; M07-A9; Clinical and Laboratory Standards Institute: Wayne, PA, USA, 2015.
20. Jorgensen, J.H. Antimicrobial susceptibility testing of bacteria that grow aerobically. *Infect. Dis. Clin. N. Am.* **1993**, *7*, 393–409. [CrossRef]
21. Ramos-Vivas, J.; Chapartegui-González, I.; Fernández-Martínez, M.; González-Rico, C.; Fortún, J.; Escudero, R.; Marco, F.; Linares, L.; Montejo, M.; Aranzamendi, M.; et al. Biofilm formation by multidrug resistant Enterobacteriaceae strains isolated from solid organ transplant recipients. *Sci. Rep.* **2019**, *9*, 8928. [CrossRef]
22. Norizan, S.N.; Yin, W.F.; Chan, K.G. Caffeine as a potential quorum sensing inhibitor. *Sensors* **2013**, *13*, 5117–5129. [PubMed]
23. Inoue, T.; Shingaki, R.; Fukui, K. Inhibition of swarming motility of *Pseudomonas aeruginosa* by branched-chain fatty acids. *FEMS Microbiol. Lett.* **2008**, *281*, 81–86. [CrossRef] [PubMed]
24. Otton, L.M.; Campos, M.S.; Meneghetti, K.L.; Corção, G. Influence of twitching and swarming motilities on biofilm formation in *Pseudomonas* strains. *Arch. Microbiol.* **2017**, *199*, 677–682. [CrossRef] [PubMed]
25. Siegmund, I.; Wagner, F. New method for detecting rhamnolipids excreted by *Pseudomonas* species during growth on mineral agar. *Biotechnol. Tech.* **1991**, *4*, 265–268. [CrossRef]
26. Aybey, A.; Demirkan, E. Inhibition of quorum sensing-controlled virulence factors in *Pseudomonas aeruginosa* by human serum paraoxonase. *J. Med. Microbiol.* **2016**, *65*, 106–113. [CrossRef]
27. Banerjee, M.; Moulick, S.; Bhattacharya, K.K.; Parai, D.; Chattopadhyay, S.; Mukherjee, S.K. Attenuation of *Pseudomonas aeruginosa* quorum sensing, virulence and biofilm formation by extracts of *Andrographis paniculata*. *Microb. Pathog.* **2017**, *113*, 85–93. [CrossRef]
28. El-Mowafy, S.A.; El Galil, K.H.A.; Habib, E.E.; Shaaban, M.I. Quorum sensing inhibitory activity of sub-inhibitory concentrations of β-lactams. *Afr. Health. Sci.* **2017**, *17*, 199–207. [CrossRef]

29. Li, N.; Wang, L.; Yan, H.; Wang, M.; Shen, D.; Yin, J.; Shentu, J. Effects of low-level engineered nanoparticles on the quorum sensing of *Pseudomonas aeruginosa* PAO1. *Environ. Sci. Pollut. Res. Int.* **2018**, *25*, 7049–7058. [CrossRef]
30. Ouyang, K.; Mortimer, M.; Holden, P.A.; Cai, P.; Wu, Y.; Gao, C.; Huang, Q. Towards a better understanding of *Pseudomonas putida* biofilm formation in the presence of ZnO nanoparticles (NPs): Role of NP concentration. *Environ. Int.* **2020**, *137*, 105485. [CrossRef]
31. Scandorieiro, S.; de Camargo, L.C.; Lancheros, C.A.; Yamada-Ogatta, S.F.; Nakamura, C.V.; de Oliveira, A.G.; Andrade, C.J.T.J.; Duran, N.; Nakazato, G.; Kobayashi, R.K.T. Synergistic and Additive Effect of Oregano Essential Oil and Biological Silver Nanoparticles against Multidrug-Resistant Bacterial Strains. *Front. Microbiol.* **2016**, *7*, 760. [CrossRef]
32. Harrison, E.M.; Carter, M.E.; Luck, S. Pathogenicity islands PAPI-1 and PAPI-2 contribute individually and synergistically to the virulence of *Pseudomonas aeruginosa* strain PA14. *Infect. Immun.* **2010**, *78*, 1437–1446. [CrossRef] [PubMed]
33. Mikkelsen, H.; McMullan, R.; Filloux, A. The *Pseudomonas aeruginosa* reference strain PA14 displays increased virulence due to a mutation in ladS. *PLoS ONE* **2011**, *6*, e29113. [CrossRef] [PubMed]
34. Sawa, T.; Momiyama, K.; Mihara, T.; Kainuma, A.; Kinoshita, M.; Moriyama, K. Molecular epidemiology of clinically high-risk *Pseudomonas aeruginosa* strains: Practical overview. *Microbiol. Immunol.* **2020**, *64*, 331–344. [CrossRef]
35. Lee, D.G.; Urbach, J.M.; Wu, G.; Liberati, N.T.; Feinbaum, R.L.; Miyata, S.; Diggins, L.T.; He, J.; Saucier, M.; Déziel, E.; et al. Genomic analysis reveals that *Pseudomonas aeruginosa* virulence is combinatorial. *Genome Biol.* **2006**, *7*, R90. [CrossRef]
36. Kasetty, S.; Katharios-Lanwermeyer, S.; O'Toole, G.A.; Nadell, C.D. Differential surface competition and biofilm invasion strategies of *Pseudomonas aeruginosa* PA14 and PA01. *J. Bacteriol.* **2021**, *203*, e0026521. [PubMed]
37. Colvin, K.M.; Gordon, V.D.; Murakami, K.; Borlee, B.R.; Wozniak, D.J.; Wong, G.C.; Parsek, M.R. The pel polysaccharide can serve a structural and protective role in the biofilm matrix of *Pseudomonas aeruginosa*. *PLoS Pathog.* **2011**, *7*, e1001264. [CrossRef] [PubMed]
38. Lee, C.K.; Vachier, J.; De Anda, J.; Zhao, K.; Baker, A.E.; Bennett, R.R.; Armbruster, C.R.; Lewis, K.A.; Tarnpol, R.L.; Lomba, C.J.; et al. Social cooperativity during surface attachment in early biofilms: A quantitative comparison between reversible attachment in *P. aeruginosa* PA14 and PAO1. *mBio* **2020**, *11*, e02644-19. [CrossRef]
39. Skariyachan, S.; Sridhar, V.S.; Packirisamy, S.; Kumargowda, S.T.; Challapilli, S.B. Recent perspectives on the molecular basis of biofilm formation by *Pseudomonas aeruginosa* and approaches for treatment and biofilm dispersal. *Folia Microbiol.* **2018**, *63*, 413–432. [CrossRef]
40. Thi, M.T.T.; Wibowo, D.; Rehm, B.H.A. *Pseudomonas aeruginosa* Biofilms. *Int. J. Mol. Sci.* **2020**, *21*, 8671. [CrossRef]
41. Brindhadevi, K.; Lewisoscar, F.; Mylonakis, E.; Shanmugam, S.; Verma, T.N.; Pugazhendhi, A. Biofilm and Quorum sensing mediated pathogenicity in *Pseudomonas aeruginosa*. *Process. Biochem.* **2020**, *96*, 49–57. [CrossRef]
42. Rasamiravaka, T.; Labtani, Q.; Duez, P.; El Jaziri, M. The Formation of Biofilms by *Pseudomonas aeruginosa*: A Review of the Natural and Synthetic Compounds Interfering with Control Mechanisms. *BioMed Res. Int.* **2015**, *2015*, 759348. [PubMed]
43. Das, T.; Kutty, S.K.; Tavallaie, R.; Ibugo, A.I.; Panchompoo, J.; Sehar, S.; Aldous, L.; Yeung, A.W.S.; Thomas, S.R.; Kumar, N.; et al. Phenazine virulence factor binding to extracellular DNA is important for *Pseudomonas aeruginosa* biofilm formation. *Sci. Rep.* **2015**, *5*, 8398. [PubMed]
44. Newman, J.W.; Floyd, R.V.; Fothergill, J.L. The contribution of *Pseudomonas aeruginosa* virulence factors and host factors in the establishment of urinary tract infections. *FEMS Microbiol. Lett.* **2017**, *364*, fnx124.
45. Rossi, E.; Paroni, M.; Landini, P. Biofilm and motility in response to environmental and host-related signals in Gram negative opportunistic pathogens. *J. Appl. Microbiol.* **2018**, *125*, 1587–1602.
46. Persat, A.; Inclan, Y.F.; Engel, J.N.; Stone, H.A.; Gitai, Z. Type IV pili mechanochemically regulate virulence factors in *Pseudomonas aeruginosa*. *PNAS* **2015**, *112*, 7563–7568.
47. Garuglieri, E.; Cattò, C.; Villa, F.; Zanchi, R.; Cappitelli, F. Effects of sublethal concentrations of silver nanoparticles on *Escherichia coli* and *Bacillus subtilis* under aerobic and anaerobic conditions. *Biointerphases* **2016**, *11*, 04B308.
48. Saeki, E.K.; Yamada, A.Y.; de Araujo, L.A.; Anversa, L.; Garcia, D.O.; Souza, R.L.B.; Martins, H.M.; Kobayashi, R.K.T.; Nakazato, G. Subinhibitory concentrations of biogenic silver nanoparticles affect motility and biofilm formation in *Pseudomonas aeruginosa*. *Front. Cell. Infect. Microbiol.* **2021**, *11*, 656984.
49. Villa, F.; Remelli, W.; Forlani, F.; Gambino, M.; Landini, P.; Cappitelli, F. Effects of chronic sub-lethal oxidative stress on biofilm formation by *Azotobacter vinelandii*. *Biofouling* **2012**, *28*, 823–833.
50. Galdino, A.C.M.; Branquinha, M.H.; Santos, A.L.S.; Viganor, L. Pseudomonas aeruginosa and Its Arsenal of Proteases: Weapons to Battle the Host. In *Pathophysiological Aspects of Proteases*; Chakraborti, S., Dhalla, N., Eds.; Springer: Singapore, 2017.
51. Yu, H.; He, X.; Xie, W.; Xiong, J.; Sheng, H.; Guo, S.; Huang, C.; Zhang, D.; Zhang, K. Elastase LasB of *Pseudomonas aeruginosa* promotes biofilm formation partly through rhamnolipid-mediated regulation. *Can. J. Microbiol.* **2014**, *60*, 227–235.
52. Hoge, R.; Pelzer, A.; Rosenau, F.; Wilhelm, S.; Duesseldorf, H. Weapons of a pathogen: Proteases and their role in virulence of *Pseudomonas aeruginosa*. *Curr. Res. Technol. Educ. Top. Appl. Microbiol. Microb. Biotechnol.* **2010**, *2*, 383–395.
53. Song, Y.; Cai, Z.H.; Lao, Y.M.; Jin, H.; Ying, K.-Z.; Lin, G.-H.; Zhou, J. Antibiofilm activity substances derived from coral symbiotic bacterial extract inhibit biofouling by the model strain *Pseudomonas aeruginosa* PAO1. *Microb. Biotechnol.* **2018**, *11*, 1090–1105. [PubMed]
54. Rathinam, P.; Vijay Kumar, H.S.; Viswanathan, P. Eugenol exhibits anti-virulence properties by competitively binding to quorum sensing receptors. *Biofouling* **2017**, *33*, 624–639.

55. Liao, S.; Zhang, Y.; Pan, X.; Zhu, F.; Jiang, C.; Liu, Q.; Cheng, Z.; Dai, G.; Wu, G.; Wang, L.; et al. Antibacterial activity and mechanism of silver nanoparticles against multidrug-resistant *Pseudomonas aeruginosa*. *J. Nanomed.* **2019**, *14*, 1469–1487.
56. Cugini, C.; Morales, D.K.; Hogan, D.A. Candida albicans-produced farnesol stimulates *Pseudomonas* quinolone signal production in LasR-defective *Pseudomonas aeruginosa* strains. *Microbiology* **2010**, *156*, 3096–3107. [PubMed]
57. Chuang, S.K.; Vrla, G.D.; Fröhlich, K.S.; Gitai, Z. Surface association sensitizes *Pseudomonas aeruginosa* to quorum sensing. *Nat. Commun.* **2019**, *10*, 4118.
58. Yang, Y.; Alvarez, P.J.J. Sublethal Concentrations of Silver Nanoparticles Stimulate Biofilm Development. *Environ. Sci. Technol. Lett.* **2015**, *2*, 221–226.
59. Singh, B.R.; Singh, B.N.; Singh, A.; Khan, W.; Naqvi, A.H.; Sigh, H.B. Mycofabricated biosilver nanoparticles interrupt *Pseudomonas aeruginosa* quorum sensing systems. *Sci. Rep.* **2015**, *5*, 13719. [PubMed]
60. Liu, L.; Li, J.; Zi, S.; Liu, F.; Deng, C.; Ao, X.; Zhang, P. AgNP combined with quorum sensing inhibitor increased the antibiofilm effect on *Pseudomonas aeruginosa*. *Appl. Microbiol. Biotechnol.* **2019**, *103*, 6195–6204.
61. Mohanty, A.; Tan, C.H.; Cao, B. Impacts of nanomaterials on bacterial quorum sensing: Differential effects on different signals. *Environ. Sci. Nano* **2016**, *3*, 351–356.

microorganisms

MDPI

Article

Assessing the Synergistic Activity of Clarithromycin and Therapeutic Oils Encapsulated in Sodium Alginate Based Floating Microbeads

Ikram Ullah Khan [1,*], Mehwish Shoukat [1], Muhammad Asif [2], Syed Haroon Khalid [1], Sajid Asghar [1], Muhammad Usman Munir [3], Muhammad Irfan [1], Akhtar Rasul [1], Sameer H. Qari [4], Alaa T. Qumsani [4], Mohamed M. Hassan [5], Maryam A. Alahdal [6], Muhammad Usman [7] and Zulqurnain Khan [8]

[1] Department of Pharmaceutics, Faculty of Pharmaceutical Sciences, Government College University Faisalabad, Faisalabad 38000, Pakistan; mesh.786@hotmail.com (M.S.); haroonkhalid80@gmail.com (S.H.K.); sajuhappa@gmail.com (S.A.); manipharma1@gmail.com (M.I.); dr.akhtarstar@gmail.com (A.R.)
[2] Department of Pharmacology, Faculty of Pharmacy, The Islamia University of Bahawalpur, Bahawalpur 63100, Pakistan; asif_pharmacist45@yahoo.com
[3] Department of Pharmaceutical Chemistry, College of Pharmacy, Jouf University Sakaka, Sakakah 72388, Saudi Arabia; mumunir@ju.edu.sa
[4] Department of Biology, Al-Jumum University College, Umm Al-Qura University, Makkah 24382, Saudi Arabia; shqari@uqu.edu.sa (S.H.Q.); atqumsani@uqu.edu.sa (A.T.Q.)
[5] Department of Biology, College of Science, Taif University, Taif 21944, Saudi Arabia; m.khyate@tu.edu.sa
[6] Biology Department, Faculty of Applied Sciences, Umm Al-Qura University, Makkah 21514, Saudi Arabia; mayahdal@uqu.edu.sa
[7] Department of Biochemistry and Biotechnology, MNS University of Agriculture, Multan 66000, Pakistan; muhammad.usman@mnsuam.edu.pk
[8] Institute of Plant Breeding and Biotechnology (IPBB), MNS University of Agriculture, Multan 66000, Pakistan; zulqurnain.khan@mnsuam.edu.pk
* Correspondence: ikramglt@gmail.com

Citation: Khan, I.U.; Shoukat, M.; Asif, M.; Khalid, S.H.; Asghar, S.; Munir, M.U.; Irfan, M.; Rasul, A.; Qari, S.H.; Qumsani, A.T.; et al. Assessing the Synergistic Activity of Clarithromycin and Therapeutic Oils Encapsulated in Sodium Alginate Based Floating Microbeads. *Microorganisms* **2022**, *10*, 1171. https://doi.org/10.3390/microorganisms10061171

Academic Editor: Juan Carlos Gutiérrez

Received: 24 March 2022
Accepted: 31 May 2022
Published: 7 June 2022

Abstract: We developed alginate-based floating microbeads of clarithromycin with therapeutic oils for the possible eradication of *Helicobacter pylori* (*H. pylori*) infections by enhancing the residence time of the carrier at the site of infection. In pursuit of this endeavor, the alginate was blended with hydroxy propyl methyl cellulose (HPMC) as an interpenetrating polymer to develop beads by ionotropic gelation using calcium carbonate as a gas generating agent. The developed microbeads remained buoyant under gastric conditions for 24 h. These microbeads initially swelled and afterwards decreased in size, possibly due to the erosion of the polymer. Furthermore, swelling was also affected by the type of encapsulated oil, i.e., swelling decreased with increasing concentrations of eucalyptus oil and increased with increasing concentrations of oleic acid. Antibacterial assays of the formulations showed significant antibacterial activity against *Staphylococcus aureus* (*S. aureus*) and *Escherichia coli* (*E. coli*); these assays also showed synergistic activity between clarithromycin and therapeutic oils as evident from the higher zone of inhibition of the microbeads as compared to the pure drug and oils. Scanning electron microscopy (SEM) images revealed a smoother surface for oleic acid containing the formulation as compared to eucalyptus oil containing the formulation. Differential scanning calorimetry (DSC) and thermogravimetric analysis (TGA) revealed the development of a stable formulation, while Fourier transform infrared spectrophotometry (FTIR) studies did not reveal any interaction between the polymers and the active ingredients. Optimized formulations (CLM3 and CLM6) were designed to release the drug in a controlled manner in gastric media by Fickian diffusion. Conclusively, the developed microbeads are a promising carrier to overcome the narrow therapeutic index and low bioavailability of clarithromycin, while the presence of therapeutic oils will produce synergistic effects with the drug to eradicate infection effectively.

Keywords: floating microbeads; clarithromycin; eucalyptus oil; oleic acid; sodium alginate

1. Introduction

Pharmaceutical scientists are continuously striving to achieve improved drug bioavailability after oral administration, as it is the most widely used and convenient route of administration. Conventional oral drug delivery systems are not able to overcome all the difficulties imposed by the gastrointestinal tract due to the various physiological constraints arising from the heterogeneity of the gastrointestinal tract. Various major obstacles faced during oral drug delivery include variable gastric emptying patterns, minimum drug residence, and negligible drug release from the carrier in a specific target area [1]. In recent times, gastroretentive drug delivery systems (GRDDS) have emerged as a promising approach to improve the bioavailability and effectiveness of active pharmaceutical ingredients (API) with an absorption window in the stomach or upper small intestine, or to promote localized effects in the stomach and duodenum [2,3]. GRDDS not only enhance the bioavailability, but also lower the required dose and minimize the gastrointestinal side effects. Candidates suitable for gastroretentive drug delivery systems include (a) drugs with a narrow therapeutic window in the stomach and small intestine, (b) drugs with a short half-life, (c) drugs that undergo instability and degradability in the lower gastrointestinal tract environment, (d) drugs that exhibit local activity in the upper part of the gastrointestinal tract (GIT), and (e) drugs with a low solubility at alkaline pH [4].

In GRDDS, various factors affect the gastric retention of the dosage forms including pharmaceutical technology factors (density of dosage form, size of dosage form), extrinsic factors (nature, posture, physical activity, sleep, frequency of food ingestion, body mass index, caloric content, and co-administration of drugs that affect the motility pattern) [5,6], and biological factors (age, illness, gender, and emotional state) [6,7]. The literature mentions several approaches to enhance the gastric residence time including muco- or bioadhesive systems, expandable systems, high-density systems, floating systems, magnetic systems, and superporous hydrogels [3].

Among the various GRDDS described in the literature, floating drug delivery systems (FDDS) are the most prominent. FDDS hold the capacity to float in and over the gastric contents owing to their low density and must be below 1.004 g/cm^3 [2,8]. FDDS are more suitable for localized treatments, such as those of local infections, peptic ulcers, gastritis, and Zollinger–Ellision syndrome, and are equally applicable for systemic applications. FDDS also provide protection to acid-labile and short half-life drugs. Furthermore, these systems are known to reduce the wastage of drugs and improve bioavailability by enhancing the residence time of drugs in the absorption window region of the GIT [8,9].

Clarithromycin (CLM) belongs to the macrolide class and inhibits bacterial protein synthesis. It belongs to BCS class II with a low solubility and high permeability. Its half-life is 3–4 h, and its bioavailability is 50% [10]. However, it is better absorbed from the stomach and also remains stable under acid conditions [11]. *H. pylori* resistance to various antibiotics including CLM is continuously rising [12,13], and thus there is need to develop new pharmaceutical strategies or carriers to overcome this resistance. This makes *H. pylori* infection a suitable candidate for GRDDS.

Low-density systems, such as FDDS, float on the stomach contents for a prolonged time while continuously releasing encapsulated contents. The composition of FDDS plays a key role in imparting floating characteristics and could comprise entrapped air or oil [14]. In the gas-generating system, various gas-generating agents such as calcium carbonate, sodium bicarbonate, tartaric acid, and citric acid are used. These gas-generating agents, upon contact with the gastric contents, produce carbon dioxide which is trapped in dosage form and keeps them buoyant in the gastric contents for a longer period and provides the slow release of the drug contents in the stomach [15]. Oils are trapped in carriers to impart a low density. Here, we develop microbeads with gas-generating agents, eucalyptus oil, and oleic acid. These oils may play multiple roles, i.e., (a) providing antibacterial activity against Gram-positive as well as Gram-negative bacteria [16] and (b) giving synergistic effects with clarithromycin. (c) Oil acts as a floating aid by creating multiple tiny pockets in the matrix to enhance buoyancy, and furthermore (d) helps to incorporate hydrophobic clarithromycin in microbeads. Thus, the developed floating microbeads containing clarithromycin and therapeutic oils will be a promising carrier for the treatment of *H. pylori*.

2. Material and Methods

2.1. Materials

Clarithromycin was received as a kind gift from Saffron Pharmaceutical Pvt (Ltd.), Faisalabad, Pakistan. Sodium alginate (CAS No 9005-38-3), hydroxy propyl methyl cellulose (HPMC) K15M (Viscosity = 15,000 cps), and calcium chloride (CAS No 10043-52-4) were obtained from Daejung Chemical and Metals Co., Ltd., Sinan, Gyeonggi, Korea. Calcium carbonate (CAS No 471-34-1) and acetic acid (CAS No 64-19-7) were obtained from E Merck, Darmstadt, Germany. Essential oils (eucalyptus oil (CAS 8000-48-4), oleic acid (CAS 112-80-1)) were obtained from Sigma Aldrich (St. Louis, MO, USA).

2.2. Preparation of Microbeads

Clarithromycin-loaded floating microbeads were prepared by an ionotropic gelation method. Initially, a 0.5% *w*/*v* solution of sodium alginate was prepared in distilled water. Afterward, 0.25% *w*/*v* HPMC K15M was added into the sodium alginate solution and stirred continuously by a magnetic stirrer. In the next step, 0.5% *w*/*v* calcium carbonate was added into this solution by continuous stirring. Clarithromycin was added to different concentrations of oils (0.2% to 0.5% *w*/*v*) and stirred to obtain a clear dispersion (Table 1). Later, the drug and oil mixture was added to the above polymeric solution and stirred continuously for 15 min. The final solution was added dropwise into a mixture containing 2% *w*/*v* calcium chloride and 10% *v*/*v* acetic acid solution. The polymeric solution was introduced into the cross linker solution at a constant flow rate of 10 mL/min by a needle of suitable gauge, using a syringe pump (ENMIND). After 30 min of curing, the microbeads were filtered, washed with distilled water, and air dried [11].

Table 1. Formulations of floating microbeads containing therapeutic oils and clarithromycin.

Code	Sodium Alginate (% *w*/*v*)	HPMC (% *w*/*v*)	$CaCO_3$ (% *w*/*v*)	Eucalyptus Oil (% *w*/*v*)	Oleic Acid (% *w*/*v*)	Clarithromycin (mg)
CLM 1	0.5	0.25	0.5	0.3	-	120
CLM 2	0.5	0.25	0.5	0.4	-	120
CLM 3	0.5	0.25	0.5	0.5	-	120
CLM 4	0.5	0.25	0.5	-	0.20	60
CLM 5	0.5	0.25	0.5	-	0.25	60
CLM 6	0.5	0.25	0.5	-	0.30	60
CLM 7	0.5	0.25	0.5	-	-	-

2.3. Characterization

2.3.1. Particle Size Analysis

The particle sizes of microbeads prepared with different gauge needles were determined using an ocular micrometer fitted in the eyepiece of an optical microscope and calibrated with a stage micrometer. Fifty microbeads were individually analyzed and the average reported [17].

2.3.2. Swelling Study

The swelling of the beads was checked by immersing the microbeads in 100 mL of 0.1 N HCl (pH 1.2). Afterwards, the beads were removed, dried, and weighed at different time intervals (0.5, 1, 2, 3, and 4 h). The dynamic weight changes of the microbeads were calculated as follows:

$$\%\ \text{weight change} = \text{Ws} - \text{Wi}/\text{Wi} \times 100$$

where "Ws" is the weight of the beads in the swollen state, and "Wi" is the initial weight of the beads [11].

2.3.3. In Vitro Buoyancy and Lag Time

The in vitro buoyancy was determined at different time intervals using fifty dry beads placed in 0.1 N HCL (pH 1.2) containing 0.02% *w*/*v* Tween 20. The temperature was maintained at 37 ± 1 °C and agitated at 100 rpm for 24 h using a USP Type II dissolution apparatus. The time taken by the beads to rise to the surface and float was taken as the lag time [11].

2.3.4. Density Measurements

The density of the microbeads was measured by filling 10 mL of a volumetric cylinder to the mark. The bulk density (Pb) of the beads was calculated by using the formula:

$$\text{Pb} = (\text{W2} - \text{W1})/\text{volume filled}$$

where "W2" is the weight of the cylinder and microbeads, and "W1" is the weight of the empty cylinder [18].

2.3.5. Drug Loading and Drug Entrapment Efficiency

A total of 25 mg of beads was placed in 20 mL of 0.1 N HCL and agitated at 100 rpm for 48 h. The drug loading and drug entrapment efficiency of all formulations were calculated by the following equations [11]:

$$\text{Drug loading } (\%\ w/w) = \text{amount of drug in sample (mg)}/\text{sample weight (mg)} \times 100$$

$$\text{DEE } (\%) = \%\text{ drug loading}/\%\text{ theoretical drug loading} \times 100$$

2.3.6. Microscopic Studies

A trinocular microscope (Accu-Scope 3001) fitted with a 5 megapixel camera was used to capture optical images of microbeads in dry and swollen state. The surface morphology and shape were assessed using a scanning electron microscope (SEM). The dry samples of blank, CLM3, and CLM6 were placed on aluminum stubs and coated with gold before taking SEM images [19].

2.3.7. Fourier Transform Infra-Red Spectroscopy (FTIR)

The FTIR spectra were recorded for the pure drug, blank, and drug-loaded microbeads to analyze the drug–polymer interactions. All samples were scanned in the range of 4000–400/cm [11].

2.3.8. Thermal Analysis

Thermogravimetric analysis (TGA) was used to assess the weight changes in a given sample as a function of heat. Differential scanning calorimetry (DSC) analysis was performed to determine any physical changes given a material under test with an increasing temperature over the time. For analysis, 5 mg of each sample was heated in a temperature range of 25 °C to 300 °C at a rate of 10 °C/min under a nitrogen atmosphere [11].

2.3.9. Powder X-ray Diffraction (XRD)

Powder X-ray diffraction was used for qualitative analysis of the crystalline and amorphous states of the pure drug and formulations. Pure drug, blank, CLM3, and CLM6 were scanned between 5° and 40° [11].

2.3.10. Antibacterial Assay

Antibacterial activity was evaluated against the Gram-positive and Gram-negative bacterial strains, namely *Staphylococcus aureus* (ATCC 6538P) and *Escherichia coli* (ATCC 25922), respectively. The agar well diffusion method was used to assess the antibacterial activity of the pure drug, oil, and developed formulations. Briefly, nutrient agar media was prepared and incubated for 2 h, then solidified in petri plates under a laminar airflow hood. Afterwards, the surface of the solidified nutrient media was streaked with a standard inoculum of microorganism. Bacterial cultures were grown overnight and adjusted to McFarland standard 1, which was equivalent to 3.0×10^8 CFU/mL of *S. aureus* and 3.5×10^8 CFU/mL of *E coli*. Then, a 5–8 mm hole was punched in the solidified media with a sterile cork borer and filled with the standard drug, pure oil, blank, and formulations (CLM1 to CLM6). These plates were incubated at 37 °C for 24 h, and zones of inhibition were measured [20].

2.3.11. In Vitro Drug Release and Release Kinetics

Drug release was assessed in a USP dissolution apparatus II using microbeads equivalent to 100 mg of drug. Each vessel contained 0.1 N HCl (pH 1.2) as a dissolution medium agitated at 100 rpm with the temperature maintained at 37 ± 1 °C. Samples (5 mL) were drawn at 0, 30, 60, 120, 180, 240, 300, 360, 420, 480, and 1440 min. Each sample was replaced with an equal volume of fresh dissolution media (0.1 N HCL). All samples were analyzed for drug content at 210 nm by using HPLC as described previously [21]. Release kinetics were assessed by the Korsemeyer–Peppas model [22].

3. Results and Discussion

3.1. Particle Size Analysis

The particle size of microbeads is an important parameter which regulates the amount of active ingredient to be encapsulated and determines the possible route of administration. The particle size of microbeads can be controlled by the quantity of polymer, the flow rate of pump, and by variating the needle gauge. Here, we kept the first two factors constant and only varied the needle gauge diameter, i.e., 18, 21, 23, and 25. It is evident from Figure 1 that as the needle size becomes narrow, the bead size also decreases. Amsden and Goosen prepared calcium alginate microbeads and obtained variable size beads by changing the gauge of needle. Smaller microbeads were formed using a higher gauge needle. A higher gauge needle has a small internal diameter that leads to a smaller cross-sectional surface area from which the drop releases, and hence the size of microbeads is reduced [23]. In another study, alginate microbeads were developed in the range of 200–800 μm by varying the alginate concentration, flow rate, air pressure, and needle diameter. They obtained smaller beads with a smaller needle diameter [24]. For further analysis, microbeads were generated with a needle of 21 gauge. This gauge was selected as it produced optimum size microbeads and did not face difficulty in transferring the viscous polymer dispersion via the needle into the cross-linking solution.

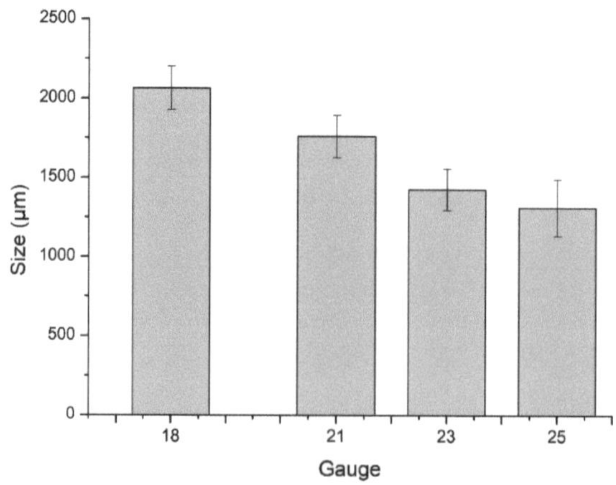

Figure 1. Microbead size variation with various syringe needle gauges. Error bars represent standard deviation (n = 50).

3.2. Swelling Studies

The swelling of all formulations was tested in simulated gastrointestinal media for four hours to assess the impact of formulation ingredients on swelling characteristics as it is one of the key parameters influencing the drug release pattern from polymeric beads [25,26]. It is evident that the swelling of the first three formulations, i.e., CLM1, CLM2, and CLM3, decreased with increasing concentrations of eucalyptus oil. In contrast, the swelling of the next three formulations, i.e., CLM4, CLM5, and CLM6, increased with increasing concentrations of oleic acid (derived from olive oil) as shown in Figure 2. This is probably due to the structural differences of the two oils. Eucalyptus oil contains three (aromatic ring, alkyl, and ether groups) lipophilic groups in its structure [27]. As the concentration of eucalyptus oil increases, the concentration of lipophilic moieties is also increased, thus impeding the penetration of fluid in microbeads. Hence, the swelling decreased as the concentration of eucalyptus oil increased. Our results are in agreement with previous studies where the presence of hydrophobic oils reduced the swelling behavior of carriers by hindering the interaction of polymeric chains with dissolution media [28,29].

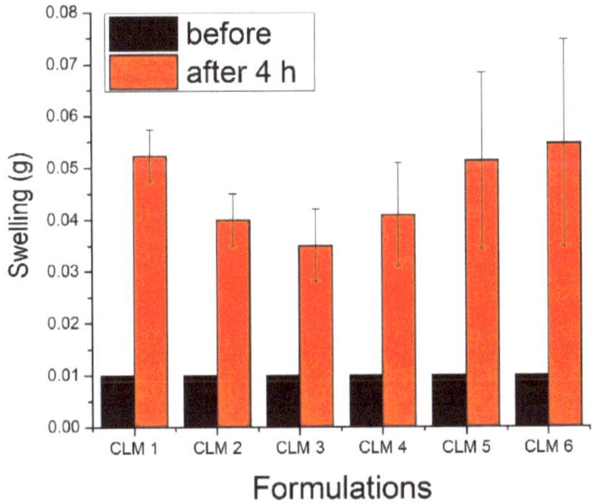

Figure 2. Swelling behavior of all formulations showing variation with respect to the concentration of oils. Error bars indicate standard deviation (n = 3).

In contrary to this, as the concentration of oleic acid increased, it's swelling also increased. Oleic acid is composed of long chain unsaturated fatty acid, which contains a hydrophilic group, i.e., carboxylic acid [30]. Thus, increasing the concentration of oleic acid, favors the penetration of fluid into the microbeads structure and hence increased swelling.

We also assessed the swelling of all formulations at different time intervals (i.e., 0 h, 0.5 h, 1 h, 2 h, 3 h, and 4 h) to evaluate the changes over the time. The results shows that initially weight of microbeads increased up to two hours and then it decreased (Figure 3).

Although, sodium alginate shows pH-dependent swelling, which is relatively higher at a higher pH owing to the presence of a –COOH group. At a low pH, this group remains partially unionized and develops hydrogen bonding between polymeric chains and solvent molecules, thus resulting in shrinkage and a low swelling profile at pH 1.2 [31]. The swelling of microbeads is also influenced by the presence of an interpenetrating polymer, i.e., HPMC K15, which hydrates rapidly upon contact with the dissolution media, thus leading to swelling of the microbeads. Similar observations have been reported for glimepiride-loaded microbeads developed with low-methoxy pectin and HPMC K15M. The presence of HPMC K15M imparts a rapid and higher swelling at pH 1.2 [32]. Maximum swelling was observed after 2 h, then it started to decrease—probably due to the erosion of the polymer as observed previously [33].

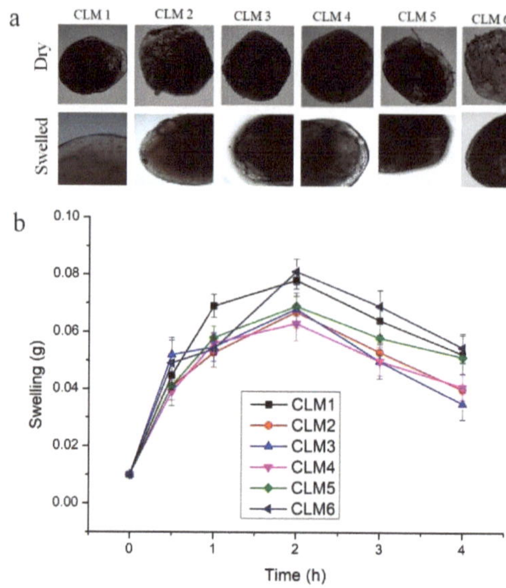

Figure 3. (**a**) Optical images of all formulations in dry and in swollen form after four hours in pH 1.2. (**b**) Swelling of all formulations at different time intervals. Error bars represent standard deviation (n = 3).

3.3. Buoyancy Studies

During buoyancy studies, all the microbeads started to float in less than one minute (Table 2) and continued to float for variable times. CLM1, CLM2, and CLM3 contain 0.3, 0.4, and 0.5% *w*/*v* of eucalyptus oil, respectively. As the concentration of oil was increased in the formulation, the in vitro buoyancy duration also increased. This is because oil not only imposes hydrophobic characteristics, but also aids in floating. Oil creates multiple tiny pockets in the polymer matrix to enhance buoyancy [11]. In CLM3, after 1 h 72% beads and after 4 h 60% beads were floating. At the end of 24 h, 6% beads were still floating, which shows excellent buoyancy (Table 2). A similar trend was observed with the formulations CLM4, CLM5, and CLM6 containing 0.2, 0.25, and 0.3% *w*/*v* oleic acid, respectively. As the concentration of oleic acid increased, the duration of floating of the microbeads also increased. In CLM6, which contains the maximum amount of oleic acid, after 1 h 66%, after 4 h 40%, and after 24 h only 4% beads were floating (Table 2). The slightly better floating duration of eucalyptus-oil-containing formulations was attributed to the higher concentration of oil as compared to the formulations containing oleic acid. In a previous study, Pornsak, Nartaya et al. assessed the floating behavior of oil-entrapped beads by using various types and concentrations of oils. They observed that as the concentration of oil was increased, its in vitro buoyancy also increased [34]. Clarithromycin has proven to be a very effective mono therapy for the treatment of *H. pylori*. However, its effectiveness is compromised due to the short dwelling time of clarithromycin in the stomach contents [35]. Increasing the residence time of the drug along with a co-agent could prove effective in the complete eradication of *H. pylori*.

3.4. Density Measurement

The results showed that CLM1, CLM2, and CLM3, which contain eucalyptus oil, have densities of less than one (Table 3). This reflects their suitability for floating drug delivery systems as this will aid in floating of the carriers. Similarly, the formulations containing oleic acid (CLM4, CLM5, and CLM6) have bulk densities of less than one, which promotes buoyancy (Table 3). Stops, Fell et al. determined the density of alginate-based microbeads.

The calculated densities of microbeads were less than 1 g/mL. The results therefore suggest that alginate beads float when placed in aqueous media (0.1 NHCl) [36].

Table 2. % Floating at different time intervals and floating lag time of all formulations.

Code	1 h	2 h	3 h	4 h	24 h	Floating Lag Time
CLM 1	60%	10%	6%	2%	-	<1 min
CLM 2	50%	24%	18%	8%	-	<1 min
CLM 3	72%	64%	64%	60%	6%	<1 min
CLM 4	52%	52%	52%	48%	-	<1 min
CLM 5	50%	48%	44%	44%	-	<1 min
CLM 6	66%	60%	40%	40%	4%	<1 min

Table 3. Physical characteristics of alginate microbeads.

Code	Bulk Density (g/cm^3)	% Drug Loading	% Entrapment Efficiency
CLM 1	0.33 ± 0.0006	2.51	26.10
CLM 2	0.32 ± 0.0009	2.76	28.71
CLM 3	0.29 ± 0.001	3.09	32.24
CLM 4	0.28 ± 0.033	2.85	29.78
CLM 5	0.31 ± 0.017	2.36	24.66
CLM 6	0.26 ± 0.03	2.04	21.26

3.5. *Drug Loading and Entrapment Efficiency*

Table 3 shows the drug loading and entrapment efficiency of all formulations containing clarithromycin with eucalyptus oil or oleic acid. It was observed that the entrapment efficiency increased with increasing concentrations of eucalyptus oil. This is because CLM3 contains the maximum amount of eucalyptus oil with the highest contents of lipophilic groups to entrap the lipophilic drug. In the case of the oleic acid formulation, the entrapment efficiency increased with decreasing concentrations of oil as shown in Table 3. The probable reason for this trend is the presence of hydrophilic groups in oleic acid. These groups repels drug molecules and reduce the entrapment efficiency. Many studies have reported variation in the entrapment efficiency of active ingredients in carriers using various lipids or oils. This is because drugs demonstrate a variable solubility in these lipids and oils [37,38].

3.6. *Antibacterial Assay*

Clarithromycin belongs to the macrolide antibiotic class and is effective against many Gram-positive and negative bacteria [10]. In this study, *S. aureus* and *E. coli* were taken as representative Gram-positive and Gram-negative bacteria, respectively. These strains were tested against the developed microbeads, the pure drug, and oils (eucalyptus oil and oleic acid) to evaluate the respective zones of inhibition. Overall, *S. aureus* was found to be more sensitive to these treatments as compared to *E. coli*, as evident from Figure 4. The probable answer lies behind the structural differences between the two strains. Gram-negative bacteria possess an additional hydrophilic lipo-polysaccharide layer, which resists the entry of hydrophobic constituents [27,29]. It has been observed that routine use of higher doses of antibiotics can lead to toxicity and the development of resistance. Pharmaceutical scientists are consistently trying to explore new and safe ways to overcome the above-mentioned hurdles to treat various infections. Therapeutic oils from natural sources possess potential antibacterial effects and can be safely combined with synthetic molecules for synergistic effects. Our results demonstrate synergistic activity between the drug and oils as evident from the higher zone of inhibition of the microbeads as compared to the pure drug and oils (Figure 4). Our results are supported by previous results where authors have reported synergistic activity between therapeutic oils and synthetic antibiotics [29,39–42]. From a technological point of view, the prepared beads showed better antimicrobial effects than clarithromycin. Similar observations have been reported previously for clarithromycin-loaded poly(d, l-lactide-co-glycolide) nanoparticles, which was attributed to the controlled release of the active ingredient and the better penetration of the nanoparticles into treated bacterial cells [43]. Based on the optimum physicochemical properties and strong antibacterial activity of CLM3 and CLM6, these two formulations were carried further for testing and analysis.

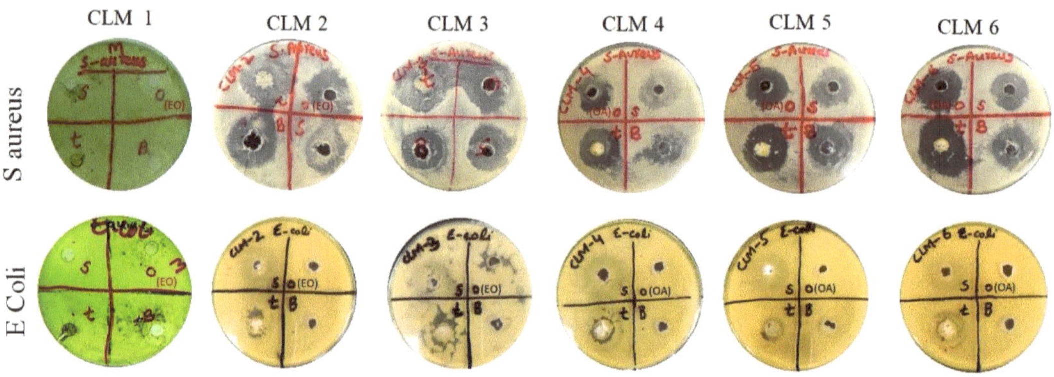

Figure 4. Antibacterial assays of all formulations against *S. aureus* and *E. coli* bacteria. In this figure, 't' represents microbeads, 's' represents the standard drug (clarithromycin), 'o' represents the tested oil (either eucalyptus oil (EO) or oleic acid (OA)), and 'B' represents the blank (0.1 N HCl).

3.7. Scanning Electron Microscopy (SEM)

SEM of the pure drug, CLM3, and CLM6 revealed detailed information about the external morphologies, as shown in Figure 5. Clarithromycin showed a crystalline nature as reported in the literature [11] and was further confirmed by XRD pattern of clarithromycin which show distinct crystalline peaks as shown in Figure 7. Under SEM, the beads appeared in a collapsed form, probably because of the removal of water during drying. Moreover, as a low concentration of polymer was used for the development of the microbeads, they have a loose network that collapsed on drying. Similar results have been reported for enalapril maleate-loaded floating particles based on sodium alginate. Particles observed to contain a low concentration of alginate collapsed after drying owing to the loose polymeric network [44]. Moreover, CLM6 microbeads showed less cracking and a comparatively smooth surface as compared to CLM3 (Figure 5). Similar results were reported by Bani-Jaber, Aideh et al. where under SEM, microbeads developed with oleic acid produced a smooth surface with few cracks [45].

3.8. Fourier Transform Infrared Spectroscopy

The FTIR spectra of clarithromycin, drug loaded, and blank hydrogel beads are shown in Figure 6. The pure drug shows an absorption band at 3471 cm^{-1}, 2977 cm^{-1}, 2360 cm^{-1}, 2341 cm^{-1}, 1691 cm^{-1}, and 1731 cm^{-1} which is due to the stretching of (O–H), (C–H), (C≡N), (C≡C), carbonyl groups (C=O), and lactone rings, respectively. Moreover, the absorbance bands of oxygen (ether linkage) were recorded at 1172 cm^{-1}, 1057 cm^{-1}, and 1008 cm^{-1} [19,46]. The characteristic peak at 1596 cm^{-1} indicates the stretching of a (COO-) group in sodium alginate, which was observed in the blank as well as both formulations [47]. Similarly, HPMC shows the absorption peak at 1050 cm^{-1}, which is due to stretching of the (C–O) group [48] and was observed in all formulations (Figure 6). The characteristic peak of the pure drug was not altered after encapsulation, indicating an absence of interaction between the drug and polymers.

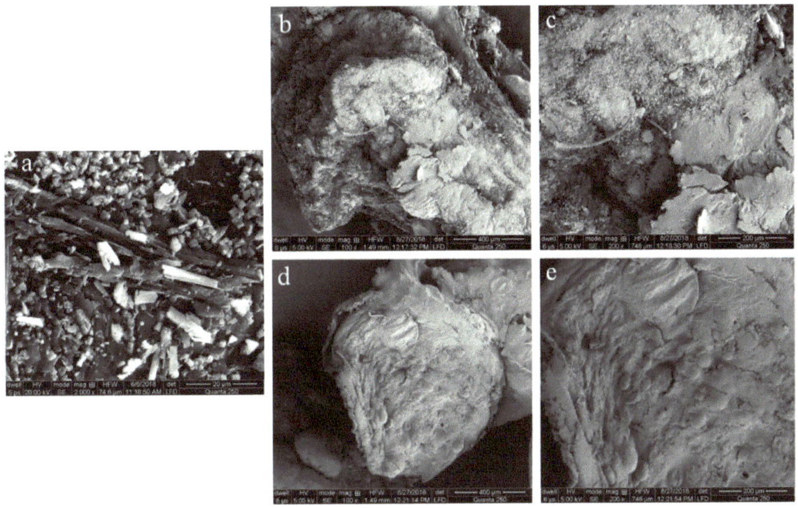

Figure 5. SEM microphotograph of (**a**) clarithromycin, (**b**,**c**) CLM3, and (**d**,**e**) CLM6.

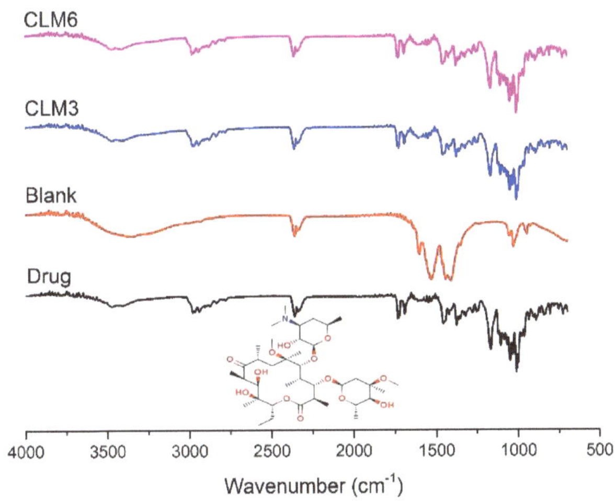

Figure 6. FTIR spectra of CLM3, CLM6, blank microbeads, and clarithromycin.

3.9. X-ray Diffraction (XRD)

X-ray diffraction describes the qualitative analysis of the crystalline phases and the state of crystallinity of the given material under test. The diffractograms of clarithromycin, blank, CLM3, and CLM6 are shown in Figure 7. The diffractogram of clarithromycin revealed the characteristic diffraction pattern for a crystalline material with intense reflections at Bragg's angle (2θ) of 11.41°, 13.69°, 15.11°, 17.23°, 20.38°, and 23.08°. This confirms the crystalline nature of clarithromycin as previously described in the literature [19]. The XRD results are also supported by our SEM analysis in which crystalline drug particles are clearly visible. In contrast, the XRD diffractogram of the blank, CLM3, and CLM6 showed a diffuse pattern indicating the amorphous nature of the carrier and the distribution of the drug at molecular level (Figure 7). Similar results have been previously reported where authors observed a significant reduction in clarithromycin peaks after encapsulation in muco-adhesive alginate beads [11].

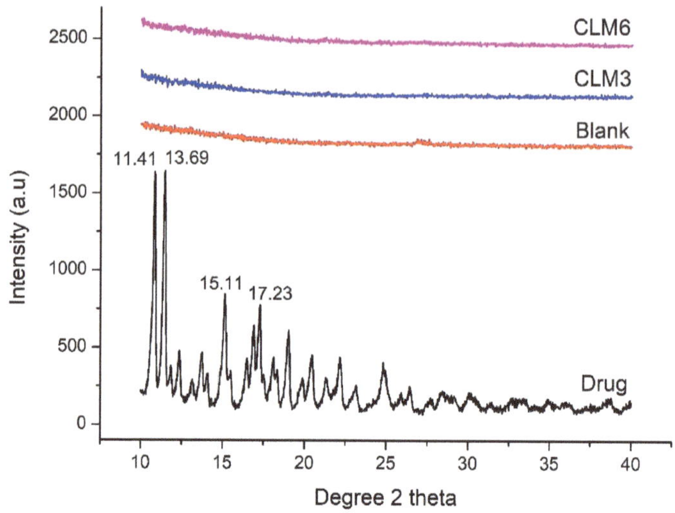

Figure 7. XRD diffraction pattern of CLM3, CLM6, blank microbeads, and clarithromycin.

3.10. Differential Scanning Calorimetry (DSC)

DSC analysis was conducted for clarithromycin, blank, CLM3, and CLM6 to assess their thermal behavior, as shown in Figure 8. The thermogram of the pure drug showed an endothermic peak at 220°, which corresponds to its melting point (Figure 8) and is in agreement with previously published studies about the melting point of clarithromycin [11,19]. Drug-loaded formulations did not show peaks of clarithromycin, which indicates that the formulation is stable and the drug is molecularly dispersed in the microbeads. Similar results have been reported for curcumin-loaded sodium alginate/montmorillonite polymeric composite beads, where curcumin peaks disappeared in the composite microbeads [49]. Furthermore, we did not observe degradation peaks for the blank and loaded microbeads within the test temperature and time.

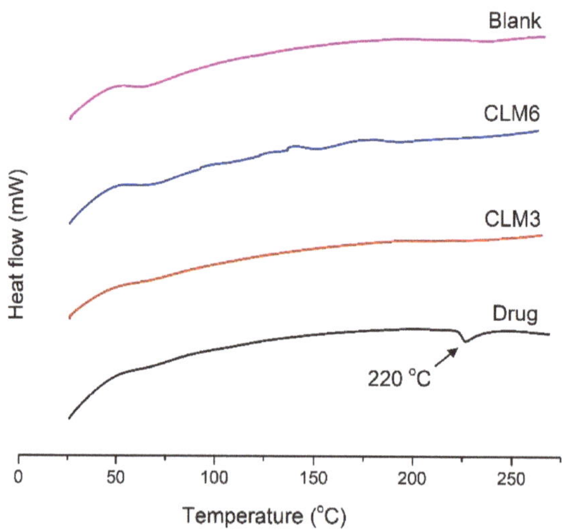

Figure 8. DSC thermogram of blank, CLM3 and CLM6 microbeads, and clarithromycin.

3.11. Thermogravimetric Analysis (TGA)

TGA was conducted on the pure drug and selected microbeads samples to assess the stability of samples by observing the mass losses of samples upon heating. The thermal characterization of clarithromycin, blank, CLM3, and CLM6 are shown in Figure 9. For pure clarithromycin, major thermal events started around 230 °C, representing the degradation of the drug above its melting point. We also observed abrupt mass loss for clarithromycin above 250 °C. Other authors have reported similar events for clarithromycin [50,51]. For blank and loaded microbeads, there was a mass loss between 50 and 200 °C, which was probably due to the loss of adsorbed water. No major degradation was observed in the tested temperature range, as the melting points of the polymers (sodium alginate and HPMC) are greater than 300 °C. Thus, we can conclude that the prepared microbeads are stable and the encapsulated active ingredients did not affect their stability. Our results are in agreement with previous results [52].

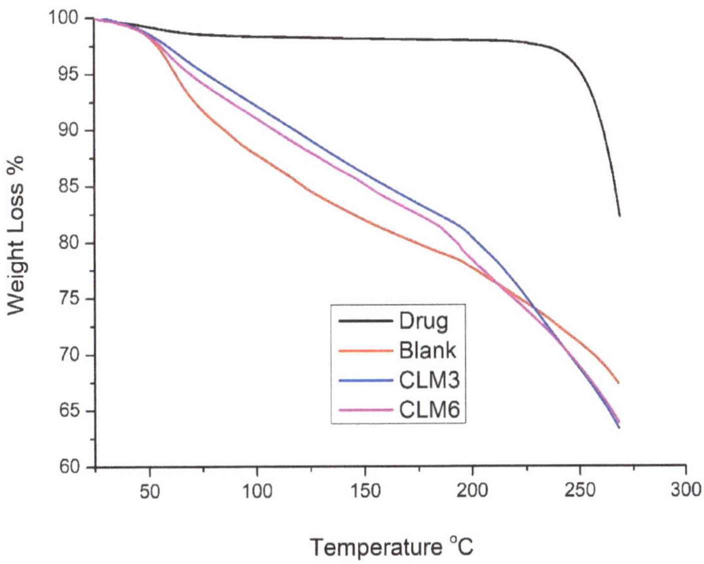

Figure 9. TGA graph of clarithromycin, blank, CLM3, and CLM6 formulations.

3.12. In Vitro Drug Release

Drug release behavior was studied for up to 24 h in simulated gastric fluid (0.1 N HCl having pH 1.2) for optimum formulation, i.e., of CLM3 and CLM6 (Figure 10). These formulations initially showed a rapid release that might be due to surface adhered drug particles or rapid release of drug present in the crona of microbeads owing to the solubility of clarithromycin in acidic pH. It is well known that during the drying of hydrogel, free drug molecules diffuse to the surface of the carrier. Upon contact with dissolution media, these molecules are rapidly released [29,53]. Afterwards, the drug is released in a sustained manner over 24 h. Similar results have been reported for clarithromycin-loaded floating microbeads that show a biphasic pattern with an initial rapid release followed by a slower release for up to six hours. After the incorporation of olive oil in microbeads, release was extended up to 12 h [11]. The controlled release of clarithromycin from microbeads in the presence of encapsulated oil could be due to the reduced porosity at the bead surface. Secondly, oil prevents rapid ingression of the dissolution media. Thirdly, HPMC K15 is a gelling agent and its presence as an interpenetrating polymer may also contribute to the slow release of the drug [11,53].

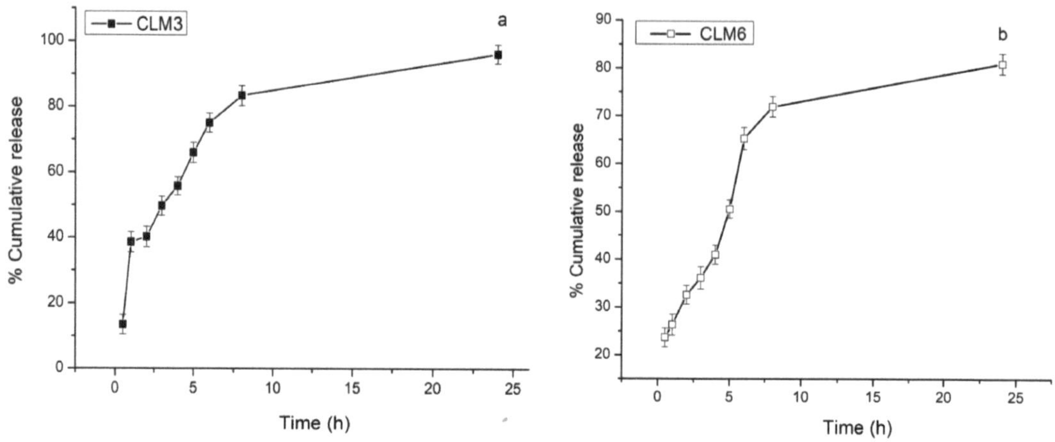

Figure 10. % Cumulative release of clarithromycin from (**a**) CLM3 and (**b**) CLM6.

The drug release mechanism from the optimized formulations was assessed by the Korsemeyer–Peppas model (Table 4). In this model, the diffusion coefficient "n" defines the involved drug transport mechanism. If n = 0.5, the drug is released through Fickian diffusion, n > 0.5 suggests non-Fickian transport, and n = 1 describes the non-Fickian case II (zero order) transport mechanism. The correlation coefficient (R^2) values for CLM3 and CLM6 were 0.95 and 0.99, indicating the fitness of the model used. The release exponent (n) value was 0.04 for CLM3 and 0.03 for CLM6. These values were less than 0.5, which represents Fickian diffusion [54].

Table 4. In vitro clarithromycin release kinetics from optimized microbeads.

Code	Korsemeyer–Peppas Model	
	Correlation Coefficients (R^2)	Release Exponent (n)
CLM 3	0.95	0.04
CLM 6	0.99	0.03

4. Conclusions

Clarithromycin-loaded alginate floating microbeads were successfully developed to extend the stay of the carrier in the stomach by an ionotropic gelation method. Beads modified with oil (eucalyptus oil and oleic acid) provided an excellent buoyancy and drug release profile. Swelling studies were performed on all formulations to assess the impact of formulation ingredients on the swelling characteristics. The swelling of formulations containing eucalyptus oil decreases as the concentration of eucalyptus oil increases because of the lipophilic moiety in their structure, while in the case of oleic acid formulations, the swelling increases as the concentration of oleic acid increases because of the hydrophilic groups. The physical parameters of all the formulations were within an acceptable range. Clarithromycin and therapeutic oils have shown antibacterial activity against both Gram-positive (*S. aureus*) and Gram-negative (*E. coli*) bacteria, which increases with increasing concentration. The drug and oils have shown synergistic effects against representative Gram-positive and Gram-negative bacteria. Under SEM, the microbeads appeared in a collapsed form because of the desorption of water upon drying. DSC and TGA indicated a stable formulation, while FTIR confirmed the absence of an interaction between the drug and polymers. Moreover, the optimized formulation released the drug in a controlled manner over twenty-four hours. Hence, our designed microbeads could not only curtail and alleviate the shortcomings of conventional drug delivery vehicles but will also be helpful for the targeted delivery of active ingredients to the *H. pylori* infected region; further studies are needed to evaluate this aspect. Based on these findings, it is concluded that our designed microbeads have synergistic anti-bacterial activities against the tested Gram-positive and negative bacterial strains, and thus can be used in the management of a wide variety of gastric disorders. Further studies to evaluate the efficacy of these microbeads against *H. pylori* using a series of in vitro and in vivo models are therefore warranted.

Author Contributions: Conceptualization, I.U.K. and S.A.; methodology, M.S., M.A. and M.U.M.; software, S.A., S.H.K. and M.U.M.; validation, S.H.Q., A.T.Q. and M.M.H.; formal analysis, S.H.K., M.I. and S.A.; investigation, M.S., M.A.A. and A.R.; resources, M.U., M.I. and A.R.; data curation, M.U., S.H.Q., A.T.Q. and M.M.H.; writing—original draft preparation, M.S.; writing—review and editing, I.U.K.; visualization, Z.K., M.U. and S.H.Q.; supervision, I.U.K., M.A.A. and S.H.K.; project administration, I.U.K.; funding acquisition, I.U.K. and M.U.M. All authors have read and agreed to the published version of the manuscript.

Funding: The authors extend their appreciation to Taif University for funding the current work via the Taif University Researchers Supporting Project number (TURSP-2020/59), Taif University, Taif, Saudi Arabia.

Institutional Review Board Statement: Not applicable.

Informed Consent Statement: Not applicable.

Data Availability Statement: Most of the data is presented in main article. Raw or processed data cannot be shared at this time due to technical or time limitations.

Acknowledgments: The authors are thankful to the Department of Pharmaceutics, Government College University Faisalabad for providing the necessary facilities for the execution of project.

Conflicts of Interest: The authors declare no conflict of interest.

References

1. Naseem, F.; Shah, S.U.; Rashid, S.A.; Farid, A.; Almehmadi, M.; Alghamdi, S. Metronidazole Based Floating Bioadhesive Drug Delivery System for Potential Eradication of *H. pylori*: Preparation and In Vitro Characterization. *Polymers* **2022**, *14*, 519. [CrossRef]
2. Lopes, C.M.; Bettencourt, C.; Rossi, A.; Buttini, F.; Barata, P. Overview on gastroretentive drug delivery systems for improving drug bioavailability. *Int. J. Pharm.* **2016**, *510*, 144–158. [CrossRef]
3. Kumar, M.; Kaushik, D. An Overview on Various Approaches and Recent Patents on Gastroretentive Drug Delivery Systems. *Recent Pat. Drug Deliv. Formul.* **2018**, *12*, 84–92. [CrossRef]
4. Vrettos, N.N.; Roberts, C.J. Gastroretentive Technologies in Tandem with Controlled-Release Strategies: A Potent Answer to Oral Drug Bioavailability and Patient Compliance Implications. *Pharmaceutics* **2021**, *13*, 1591. [CrossRef] [PubMed]
5. Klausner, E.A.; Lavy, E.; Friedman, M.; Hoffman, A. Expandable gastroretentive dosage forms. *J. Control. Release* **2003**, *90*, 143–162. [CrossRef]
6. Talukder, R.; Fassihi, R. Gastroretentive delivery systems: A mini review. *Drug Dev. Ind. Pharm.* **2004**, *30*, 1019–1028. [CrossRef]
7. Garg, R.; Gupta, G. Progress in controlled gastroretentive delivery systems. *Trop. J. Pharm. Res.* **2008**, *7*, 1055–1066. [CrossRef]
8. Namdev, A.; Jain, D. Floating Drug Delivery Systems: An Emerging Trend for the Treatment of Peptic Ulcer. *Curr. Drug Deliv.* **2019**, *16*, 874–886. [CrossRef]
9. Choudhary, S.; Jain, A.; Amin, M.C.I.M.; Mishra, V.; Agrawal, G.P.; Kesharwani, P. Stomach specific polymeric low density microballoons as a vector for extended delivery of rabeprazole and amoxicillin for treatment of peptic ulcer. *Colloids Surf. B Biointerfaces* **2016**, *141*, 268–277. [CrossRef]
10. Israr, M.; Pugliese, N.; Farid, A.; Ghazanfar, S.; Di Cerbo, A.; Muzammal, M.; Alamri, A.S.; Basheeruddin Asdaq, S.M.; Ahmad, A.; Khan, K.A. Preparation and Characterization of Controlled-Release Floating Bilayer Tablets of Esomeprazole and Clarithromycin. *Molecules* **2022**, *27*, 3242. [CrossRef]
11. Adebisi, A.O.; Laity, P.R.; Conway, B.R. Formulation and evaluation of floating mucoadhesive alginate beads for targeting *Helicobacter pylori*. *J. Pharm. Pharmacol.* **2015**, *67*, 511–524. [CrossRef]
12. Savoldi, A.; Carrara, E.; Graham, D.Y.; Conti, M.; Tacconelli, E. Prevalence of Antibiotic Resistance in Helicobacter pylori: A Systematic Review and Meta-analysis in World *Health Organ. Regions. Gastroenterol.* **2018**, *155*, 1372–1382.e1317. [CrossRef]
13. Fagoonee, S.; Pellicano, R. *Helicobacter pylori*: Molecular basis for colonization and survival in gastric environment and resistance to antibiotics. A short review. *Infect Dis.* **2019**, *51*, 399–408. [CrossRef]
14. Iglesias, N.; Galbis, E. In-Depth Study into Polymeric Materials in Low-Density Gastroretentive Formulations. *Pharmaceutics* **2020**, *12*, 636. [CrossRef]

15. Mandal, U.K.; Chatterjee, B.; Senjoti, F.G. Gastro-retentive drug delivery systems and their in vivo success: A recent update. *Asian J. Pharm. Sci.* **2016**, *11*, 575–584. [CrossRef]
16. Saporito, F.; Sandri, G.; Bonferoni, M.C.; Rossi, S.; Boselli, C.; Cornaglia, A.I.; Mannucci, B.; Grisoli, P.; Vigani, B.; Ferrari, F. Essential oil-loaded lipid nanoparticles for wound healing. *Int. J. Nanomed.* **2018**, *13*, 175–186. [CrossRef]
17. Rajinikanth, P.S.; Mishra, B. Stomach-site specific drug delivery system of clarithromycin for eradication of *Helicobacter pylori*. *Chem. Pharm. Bull.* **2009**, *57*, 1068–1075. [CrossRef]
18. Bai, Y.-X.; Li, Y.F. Preparation and characterization of crosslinked porous cellulose beads. *Carbohydr. Polym.* **2006**, *64*, 402–407. [CrossRef]
19. El Nashar, N.F.; Donia, A.A.; Mady, O.Y.; El Maghraby, G.M. Formulation of clarithromycin floating microspheres for eradication of Helicobacter pylori. *J. Drug Deliv. Sci. Technol.* **2017**, *41*, 213–221. [CrossRef]
20. Rajbhar, P.; Sahu, A.K.; Gautam, S.S.; Prasad, R.K.; Singh, V.; Nair, S.K. Formulation and Evaluation of Clarithromycin Co-Crystals Tablets Dosage Forms to Enhance the Bioavailability. *Pharma Innov.* **2016**, *5*, 5.
21. Morakul, B.; Suksiriworapong, J.; Chomnawang, M.T.; Langguth, P.; Junyaprasert, V.B. Dissolution enhancement and in vitro performance of clarithromycin nanocrystals produced by precipitation–lyophilization–homogenization method. *Eur. J. Pharm. Biopharm.* **2014**, *88*, 886–896. [CrossRef]
22. Peppas, N. Analysis of Fickian and non-Fickian drug release from polymers. *Pharm. Acta Helv.* **1985**, *60*, 110–111.
23. Amsden, B.; Goosen, M. An examination of factors affecting the size, distribution and release characteristics of polymer microbeads made using electrostatics. *J. Control. Release* **1997**, *43*, 183–196. [CrossRef]
24. Hadley, D.J.; Campbell, K.T.; Gabriel, M.H.; Silva, E.A. Open-source 3D printed air-jet for generating monodispersed alginate microhydrogels. *bioRxiv* **2019**, 804849. [CrossRef]
25. Nama, M.; Gonugunta, C.S.R.; Veerareddy, P.R. Formulation and evaluation of gastroretentive dosage forms of clarithromycin. *AAPS PharmSciTech* **2008**, *9*, 231. [CrossRef] [PubMed]
26. Azad, A.K.; Al-Mahmood, S.M.A.; Chatterjee, B.; Wan Sulaiman, W.M.A.; Elsayed, T.M.; Doolaanea, A.A. Encapsulation of black seed oil in alginate beads as a ph-sensitive carrier for intestine-targeted drug delivery: In vitro, in vivo and ex vivo study. *Pharmaceutics* **2020**, *12*, 219. [CrossRef]
27. Marzoug, H.N.B.; Romdhane, M.; Lebrihi, A.; Mathieu, F.; Couderc, F.; Abderraba, M.; Khouja, M.L.; Bouajila, J. Eucalyptus oleosa essential oils: Chemical composition and antimicrobial and antioxidant activities of the oils from different plant parts (stems, leaves, flowers and fruits). *Molecules* **2011**, *16*, 1695–1709. [CrossRef] [PubMed]
28. Li, A.; Khan, I.N.; Khan, I.U.; Yousaf, A.M.; Shahzad, Y. Gellan Gum-Based Bilayer Mucoadhesive Films Loaded with Moxifloxacin Hydrochloride and Clove Oil for Possible Treatment of Periodontitis. *Drug Des. Dev. Ther.* **2021**, *15*, 3937. [CrossRef]
29. Mahmood, H.; Khan, I.U.; Asif, M.; Khan, R.U.; Asghar, S.; Khalid, I.; Khalid, S.H.; Irfan, M.; Rehman, F. Shahzad, Y.; et al. In vitro and in vivo evaluation of gellan gum hydrogel films: Assessing the co impact of therapeutic oils and ofloxacin on wound healing. *Int. J. Biol. Macromol.* **2021**, *166*, 483–495. [CrossRef]
30. Wang, Q.; Geil, P.; Padua, G. Role of hydrophilic and hydrophobic interactions in structure development of zein films. *J. Polym. Environ.* **2004**, *12*, 197–202. [CrossRef]
31. Reddy, O.S.; Subha, M.C.S.; Jithendra, T.; Madhavi, C.; Rao, K.C. Fabrication and characterization of smart karaya gum/sodium alginate semi-IPN microbeads for controlled release of D-penicillamine drug. *Polym. Polym. Compos.* **2020**, *29*, 163–175. [CrossRef]
32. Patel, H.; Srinatha, A.; Sridhar, B.K. External Cross-linked Mucoadhesive Microbeads for Prolonged Drug Release: Development and In vitro Characterization. *Indian J. Pharm. Sci.* **2014**, *76*, 437–444. [PubMed]
33. Azad, A.K.; Al-Mahmood, S.M.A.; Kennedy, J.F.; Chatterjee, B.; Bera, H. Electro-hydrodynamic assisted synthesis of lecithin-stabilized peppermint oil-loaded alginate microbeads for intestinal drug delivery. *Int. J. Biol. Macromol.* **2021**, *185*, 861–875. [CrossRef]
34. Sriamornsak, P.; Thirawong, N.; Puttipipatkhachorn, S. Morphology and buoyancy of oil-entrapped calcium pectinate gel beads. *AAPS J.* **2004**, *6*, 65–71. [CrossRef]
35. Hardikar, S.; Bhosale, A. Formulation and evaluation of gastro retentive tablets of clarithromycin prepared by using novel polymer blend. *Bull. Fac. Pharm. Cairo Univ.* **2018**, *56*, 147–157. [CrossRef]
36. Stops, F.; Fell, J.T.; Collett, J.H.; Martini, L.G. Floating dosage forms to prolong gastro-retention—The characterisation of calcium alginate beads. *Int. J. Pharm.* **2008**, *350*, 301–311. [CrossRef]
37. Elhesaisy, N. l Swidan, S. Trazodone loaded lipid core poly (ε-caprolactone) nanocapsules: Development, characterization and in vivo antidepressant effect evaluation. *Sci. Rep.* **2020**, *10*, 1–10. [CrossRef]
38. Basso, J.; Mendes, M.; Cova, T.; Sousa, J.; Pais, A.; Fortuna, A.; Vitorino, R.; Vitorino, C. A Stepwise Framework for the Systematic Development of Lipid Nanoparticles. *Biomolecules* **2022**, *12*, 223. [CrossRef] [PubMed]
39. Rosato, A.; Vitali, C.; De Laurentis, N.; Armenise, D.; Milillo, M.A. Antibacterial effect of some essential oils administered alone or in combination with Norfloxacin. *Phytomedicine* **2007**, *14*, 727–732. [CrossRef]
40. Malik, T.; Singh, P.; Pant, S.; Chauhan, N.; Lohani, H. Potentiation of antimicrobial activity of ciprofloxacin by Pelargonium graveolens essential oil against selected uropathogens. *Phytother. Res.* **2011**, *25*, 1225–1228. [CrossRef]
41. Duarte, A.; Ferreira, S.; Silva, F.; Domingues, F.C. Synergistic activity of coriander oil and conventional antibiotics against *Acinetobacter baumannii*. *Phytomedicine* **2012**, *19*, 236–238. [CrossRef]

42. Lahmar, A.; Bedoui, A.; Mokdad-Bzeouich, I.; Dhaouifi, Z.; Kalboussi, Z.; Cheraif, I.; Ghedira, K.; Chekir-Ghedira, L. Reversal of resistance in bacteria underlies synergistic effect of essential oils with conventional antibiotics. *Microb. Pathog.* **2017**, *106*, 50–59. [CrossRef]
43. Öztürk, A.A.; Yenilmez, E.; Özarda, M.G. Clarithromycin-Loaded Poly (Lactic-co-glycolic Acid) (PLGA) Nanoparticles for Oral Administration: Effect of Polymer Molecular Weight and Surface Modification with Chitosan on Formulation, Nanoparticle Characterization and Antibacterial Effects. *Polymers* **2019**, *11*, 1632. [CrossRef]
44. Abbas, A.K.; Alhamdany, A.T. Floating Microspheres of Enalapril Maleate as a Developed Controlled Release Dosage Form: Investigation of the Effect of an Ionotropic Gelation Technique. *Turk. J. Pharm. Sci.* **2020**, *17*, 159–171. [CrossRef]
45. Bani-Jaber, A.; Aideh, K.; Hamdan, I.; Maraqa, R. Drug-loaded casein beads: Influence of different metal-types as cross-linkers and oleic acid as a plasticizer on some properties of the beads. *J. Drug Deliv. Sci. Technol.* **2009**, *19*, 125–131. [CrossRef]
46. Gattani, S.G.; Savaliya, P.J.; Belgamwar, V.S. Floating-mucoadhesive beads of clarithromycin for the treatment of *Helicobacter pylori* infection. *Chem. Pharm. Bull.* **2010**, *58*, 782–787. [CrossRef]
47. Rasel, M.A.T.; Hasan, M. Formulation and evaluation of floating alginate beads of diclofenac sodium. *Dhaka Univ. J. Pharm. Sci.* **2012**, *11*, 29–35. [CrossRef]
48. Van der Weerd, J.; Kazarian, S.G. Release of poorly soluble drugs from HPMC tablets studied by FTIR imaging and flow-through dissolution tests. *J. Pharm. Sci.* **2005**, *94*, 2096–2109. [CrossRef]
49. Reddy, O.S.; Subha, M.C.S.; Jithendra, T.; Madhavi, C.; Rao, K.C. Curcumin encapsulated dual cross linked sodium alginate/montmorillonite polymeric composite beads for controlled drug delivery. *J. Pharm. Anal.* **2021**, *11*, 191–199. [CrossRef]
50. Chung, D.; Song, Y.G.; Kang, I.M.; Choi, W.; Park, C.; Song, Y. Development and characterization of clarithromycin-smectite hybrid for burst release at high pH condition. *Mater. Res. Express* **2018**, *5*, 115408. [CrossRef]
51. Ijaz, Q.A.; Latif, S.; Rashid, M.; Arshad, M.S.; Hussain, A.; Bukhari, N.I.; Riaz, S.; Abbas, N. Preparation and Characterization of pH-Independent Sustained-Release Tablets Containing Hot Melt Extruded Solid Dispersions of Clarithromycin. *AAPS PharmSciTech* **2021**, *22*, 1–12. [CrossRef] [PubMed]
52. Balanč, B.; Kalušević, A.; Drvenica, I.; Coelho, M.T.; Djordjević, V.; Alves, V.D.; Sousa, I.; Moldão-Martins, M.; Rakić, V.; Nedović, V.; et al. Calcium–alginate–inulin microbeads as carriers for aqueous carqueja extract. *J. Food Sci.* **2016**, *81*, E65–E75. [CrossRef] [PubMed]
53. Patel, N.; Lalwani, D.; Gollmer, S.; Injeti, E.; Sari, Y.; Nesamony, J. Development and evaluation of a calcium alginate based oral ceftriaxone sodium formulation. *Prog. Biomater.* **2016**, *5*, 117–133. [CrossRef]
54. Tripathi, G.; Singh, S. Formulation and In Vitro Evaluation of pH-sensitive oil-entrapped buoyant beads of Clarithromycin. *Trop. J. Pharm. Res.* **2010**, *9*, 533–539. [CrossRef]

microorganisms

MDPI

Article

Biomimetic Synthesis of Silver Nanoparticles Using Ethyl Acetate Extract of *Urtica diocia* Leaves; Characterizations and Emerging Antimicrobial Activity

Mohammed Binsalah [1], Sandhanasamy Devanesan [2,*], Mohamad S. AlSalhi [2,*], Abdullrahman Nooh [3], Osama Alghamdi [1] and Nasser Nooh [1]

1 Department of Oral and Maxillofacial Surgery, College of Dentistry, King Saud University, Riyadh 11545, Saudi Arabia; masalah@ksu.edu.sa (M.B.); oghamdi@ksu.edu.sa (O.A.); nnooh@ksu.edu.sa (N.N.)

2 Research Chair in Laser Diagnosis of Cancers, Department of Physics and Astronomy, College of Science, King Saud University, Riyadh 11451, Saudi Arabia

3 Department of Internal Medicine, Prince Mohammed Bin Abdulaziz Hospital, Riyadh 7333, Saudi Arabia; nooh29@gmail.com

* Correspondence: dsandhanasamy@ksu.edu.sa (S.D.); malsalhi@ksu.edu.sa (M.S.A.); Tel.: +966-114676257 (S.D.)

Abstract: The current work reports the biosynthesis of silver nanoparticles (AgNPs) using the antimicrobial activities of ethyl acetate extract of *Urtica diocia* (UD) leaves as a reducing and capping agent. The synthesized UD-AgNPs were characterized using UV–visible spectroscopy, scanning electron microscopy (SEM), transmission electron microscopy (TEM), energy-dispersive X-ray analysis (EDAX), Fourier transform infrared (FTIR) spectroscopy, X-ray diffraction (XRD), and dynamic light scattering (DLS). The UD-AgNPs were evaluated against Gram-positive and Gram-negative bacteria, and their size, shape, and distribution were recorded. The average size of an NP was 19.401 nm. The zone of inhibition (ZOI) for 75 μL of UD-AgNPs against *Pseudomonas aeruginosa* (*P. aeruginosa*) was 21 ± 0.4 mm more than that of the control drug Ciprofloxacin (16 ± 10 mm). The minimum inhibitory concentration (MIC) was the lowest against *Escherichia coli* (*E. coli*) (36 ± 3 μg/mL) and *Staphylococcus epidermidis (S. epidermidis)* (38 ± 3 μg/mL). Moreover, the minimum bactericidal concentration (MBC) was the lowest against *E. coli* (75 ± 00 μg/mL) and *Enterococcus faecalis (E. faecalis* (83 ± 16 μg/mL). Thus, the UD-AgNPs synthesized using the ethyl acetate extract of UD can be used as a new antimicrobial drug.

Keywords: UD-AgNPs; characterization; green synthesis; antimicrobial effects

check for updates

Citation: Binsalah, M.; Devanesan, S.; AlSalhi, M.S.; Nooh, A.; Alghamdi, O.; Nooh, N. Biomimetic Synthesis of Silver Nanoparticles Using Ethyl Acetate Extract of *Urtica diocia* Leaves; Characterizations and Emerging Antimicrobial Activity. *Microorganisms* **2022**, *10*, 789. https://doi.org/10.3390/microorganisms10040789

Academic Editor: Kamel A. Abd-Elsalam

Received: 21 February 2022
Accepted: 5 April 2022
Published: 8 April 2022

1. Introduction

The benefits of medicinal plants have been intensively investigated for many centuries, as their natural components contain biologically active organic products, have few adverse effects, and are relatively low cost. It is estimated that plant products have provided the models for 50% of Western drugs [1]. *Urtica diocia* L. (UD), also known as stinging nettle, is an important medicinal plant that has been used since ancient times, and recent studies have reported that its leaves are a good food additive rich in medicinal value [2,3]. In addition, the crude extract of nettle plant leaves has been used to reduce inflammation [4] and maintain blood sugar levels in static conditions [5]. Multiple studies have also reported that the medicinal properties of UD include bleeding control and wound and burn healing. It also acts as a diuretic [6–8]. Moreover, crude UD extract has been reported as an antimicrobial agent against Gram-positive and Gram-negative microorganisms [9].

Antibiotic overdose as a result of an infectious disease treatment can lead to the emergence of multidrug-resistant (MDR) bacteria [10]. It has been documented that the Gram-positive bacterium *S. epidermidis* is MDR, which poses a healthcare management challenge [11]. This issue became a significant area of research into the formation of alternative neutral antibacterial agents with unique characteristics [12]. Advances in nanotechnology have allowed for the production of nanoparticles that show low toxicity against human cells and are reported to have promising bactericidal activity [13]. Phyto-nanotechnology, or plant-based nanomedicine, is used extensively in numerous fields in the form of nanoparticles (NPs) that enhance the activity of materials [14]. Recently, researchers have focused their attention on plant product-inspired metal NPs and their biomedical importance [15].

Among the different metal-based NPs, biosynthesized silver NPs (AgNPs) are well documented [16,17], biocompatible, and have few adverse effects [18]. Even a minimal amount of AgNPs has been shown to demonstrate antimicrobial [19], anticancer [20], antidiabetic [21], antiviral [22], and antioxidant properties [23]. Importantly, the synthesis of plant extract-based NPs is affordable, eco-friendly, and effective for formulating antimicrobial drugs [24]. The phytochemical screening of AgNPs containing Phoenix *dactylifera L.* leaf extract indicated polyphenols, flavonoids, and condensed tannins. These compounds play a vital role as a reducing and capping agent to inhibit particle growth and aggregation [25]. In addition, AgNPs from aqueous oregano plant leaf extract have been reported as non-infectious for sterile antimicrobial contact lenses [26].

The AgNPs synthesized with UD extract have been shown to protect the liver of cirrhotic rats [27]; similarly, AgNPs synthesized with *Urtica urens* are an effective nematicide against the root-knot nematode Meloidogyne [28]. A few studies have reported the antioxidant and antimicrobial effects of UD-AgNP aqueous extract [29,30] and methanol extract [31], indicating various effects such as the amount of $AgNO_3$; however, reports on the size of the AgNPs and the concentration of UD-AgNPs in biological activities were absent. The current study aimed to conduct a green synthesis of AgNPs with ethyl acetate leaf extract of UD. The synthesized UD-AgNPs were characterized using UV–Vis, XRD, FTIR, SEM, TEM, EDX, and DLS analysis, and they were investigated for antimicrobial activity against Gram-positive and Gram-negative bacterial strains.

2. Materials and Methods

2.1. Collection and Preparation of Plant leaves

Fresh UD leaves, collected from the Riyadh region of Saudi Arabia, were rinsed with tap water and then with Milli Q water (ultrapure and deionized water purified by a Milli-Q water system, Merk, Darmstadt, Germany) to remove surface contaminants and other dust particles. The cleaned leaves were dried entirely at room temperature, after which 100 g of the dried leaves were milled into a fine powder [29].

2.2. UD Leaf Extraction

Ethyl acetate was used as the extraction solvent since it has a medium polarity with minimal cell toxicity. In brief, 10 mL of fine blended powder of UD leaves was poured into a Soxhlet extractor chamber. Then, 250 mL of ethyl acetate was placed in a round-bottomed flask and boiled at 77.1 °C [32]. This process was carried out in 24 h, and the ethyl acetate extracted compounds were retained.

2.3. Preparation of UD-AgNPs

Briefly, 1 mM of silver nitrate ($AgNO_3$) dissolved in 250 mL of Milli Q water was added to 10 mL of ethyl acetate UD extract. The mixture was stirred continuously in a shaking incubator for 24 h and then monitored for interaction between the $AgNO_3$ and the UD extract. The reaction (color changes) of the mixture started from 8 h onwards and turned completely dark brown within 24 h. The colored solution was placed in a rotary evaporator (Rotavapor R-215, Marshall Scientific, USA) at 45° C at 75 rpm to remove the solvent. The solution was centrifuged at 15,000 rpm for 15 min three times. The supernatant was discarded, and the NPs sedimented at the bottom of the centrifuge tube. The NP sediment tube was allowed to dry in an incubator at 40 °C for 2 h before collection [33,34]. The collected semisolid UD-AgNPs were kept at 4 °C for characterization and biological study [35].

2.4. Characterization

The absorption spectrum of the reaction mixture ($AgNO_3$ + plant extract) after the stabilization of color was characterized by a UV–Vis spectrophotometer (LS 55; Perkin-Elmer, Rodgau, Germany). The size and shape of the UD-AgNPs were recorded using SEM (JEOL, Tokyo, Japan), TEM (JEOL), EDAX, XRD, and FTIR, according to the literature [14,15]. The size of the UD-AgNPs was determined by a Zetasizer DLS instrument (Malvern Instruments, Malvern, U.K).

2.5. Antimicrobial Evaluation

2.5.1. Inhibition Study

The antibacterial activity of UD-AgNPs was evaluated against the following Gram-positive pathogens: *S. aureus* (ATCC®29213), *E. faecalis* (ATCC®29212), and *S. epidermidis* (ATCC®12228), as well as the following Gram-negative pathogens: *Klebsiella pneumoniae (K. pneumoniae)* (ATCC®700603), *P. aeruginosa* (ATCC®27853), and *E. coli* (ATCC®25922). In addition, their antimicrobial susceptibility was determined using Mueller–Hinton agar (MHA) plates [11,12]. Briefly, 19 g of MHA powder was mixed with 500 mL of distilled water and then sterilized in an autoclave at 121 °C for 15 min. Next, approximately 25 mL of the medium was poured into each plate, which had been inoculated with an adjusted bacterial concentration (10^8 CFU/mL, 0.5 McFarland's standard). Then, 40 mg of UD-AgNPs was dissolved in 20 mL of dimethyl sulfoxide (DMSO). Three different concentrations of UD-AgNPs in DMSO (25, 50, and 75 µL) were added to the plates. Ciprofloxacin (10 µL) was used as a positive control drug. The negative controls were DMSO, ethyl acetate UD extract, and $AgNO_3$ (75 µL). The plates were incubated at 37 °C for 24 h. After incubation, the ZOI around the well, indicating the antimicrobial activity of the UD-AgNPs, was measured in mm [36].

2.5.2. MIC

The MIC was determined by a resazurin-based assay using UD-AgNP concentrations ranging from 15 to 50 µg/mL with an adjusted bacterial concentration (10^8 CFU/mL, 0.5 McFarland's standard). Three 96-well plates were used for the MIC evaluation. In the first plate, the first six columns were filled with 190 µL of Mueller–Hinton Broth (MHB) and 10 µL of each type of pathogen as a positive control. In comparison, the other six columns of the first plate and the second plate were treated with serial dilutions of UD-AgNPs, with concentrations ranging from 15 to 50 µg/mL (15, 20, 25, 30, 35, 40, 45, and 50 µg/mL) and 10 µL of each type of pathogen. In the third plate, three columns were filled with 150 µL of MHB and used as a negative control. After incubation for 24 h, 15 µL of 0.02% resazurin solution (5 mg in 25 mL of distilled water) was added to each well of the microtiter plates, and the plates were further incubated at 37 °C for 4 h. Color changes were recorded, and the MIC values were determined. The MIC was considered the lowest concentration of the UD-AgNPs that inhibited bacterial growth [37].

2.5.3. MBC

The MBC was determined according to the protocol of the Clinical and Laboratory Standards Institute, and MHA plates were used for the MBC experiments. Briefly, 10 µL of pathogen was inoculated onto each plate, and then different amounts of UD-AgNPs (25, 50, 75, 100, 150, 200, 250, and 500 µL) of a mixture of 40 mg of UD-AgNPs dissolved in 20 mL of DMSO were applied to the agar plates. The plates were incubated at 37 °C for 24 h. The plates started with no visible bacterial growth in specific concentrations considered as MBC. The MBC value of the UD-AgNPs for each tested microorganism was determined [38].

2.6. Statistical Analysis

The experimental data were analyzed using one-way ANOVA. The data were represented as the mean ± standard deviation (SD). The experiments were performed with a level of significance of $p \leq 0.05$.

3. Results

3.1. Characterizations

3.1.1. Visual Observation

The combination of the ethyl acetate extract of UD leaves and the $AgNO_3$ solution produced UD-AgNPs after 8 h. Initially, the $AgNO_3$ solution was colorless, and the mixture of $AgNO_3$ and the plant extract was yellow, as shown in Figure 1a,b. After stirring for 8 h, the mixture of the ethyl acetate extract of UD leaves and $AgNO_3$ solution turned dark brown, indicating the formation of UD-AgNPs (Figure 1c). The change in color was due to the interaction between the UD leaf extract and the formation of capping and reducing agents (Ag^+ to Ag).

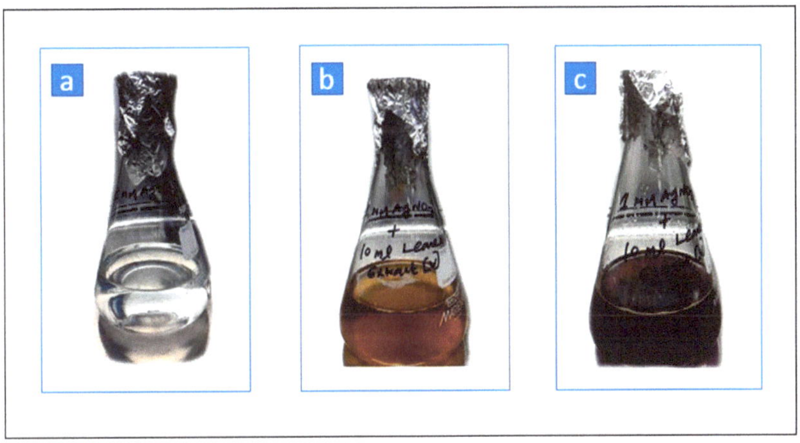

Figure 1. Ethyl acetate extract of UD leaves mediated the preparation of AgNPs. (**a**) Silver nitrate solution; (**b**) mixture of plant extract and silver nitrate; (**c**) formation of AgNPs.

3.1.2. UV–vis Spectral Study

The UV–vis spectral images at different time points (8, 16, and 24 h) showed the progression of NP formation. An absorbance peak was noted at 420 nm (Figure 2). Surface plasmon resonance experiments indicated that the solution of UD-AgNPs was stable for 24 h.

3.1.3. FTIR Spectroscopy

The FTIR spectrum indicated the ethyl acetate leaf extract alone (Figure 3a) and the synthesized UD-AgNPs (Figure 3b). The FTIR absorption spectra of acetate leaf extract alone peaked at 3433, 2925, 1730, 1624, 1166, and 611 cm^{-1}. In contrast, the synthesized UD-AgNPs absorption peaks were shifted slightly at 3430, 2924, 1626, 1018, and 564 cm^{-1}. The strong absorption peaks at 3430 cm^{-1} corresponded to the N–H stretch (primary amine). In addition, a long, narrow band was seen at 2924 cm^{-1}, indicating the presence of C–H stretching (alkane). Strong bands were also noted at 1626 cm^{-1}, representing C=C stretching (alkene). Moreover, bands were formed at 1018 cm^{-1}, corresponding to C–N stretching (amines). Another peak at 564 cm^{-1} belonged to the C–I or C–Br group of the compounds. The similar spectral features for acetate leaf extract alone and the synthesizeVirulence determinants of uropathogenic Escherichia coli in fecal strains from intestinal infections and healthy individuals. UD-AgNPs may have indicated the presence of the identified phytochemicals in the synthesized UD-AgNPs.

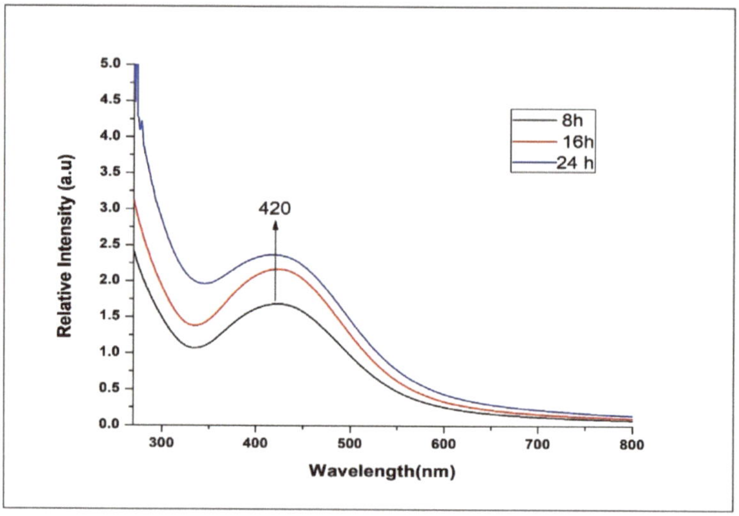

Figure 2. UV–vis spectrum of the UD-AgNPs recorded at different time points.

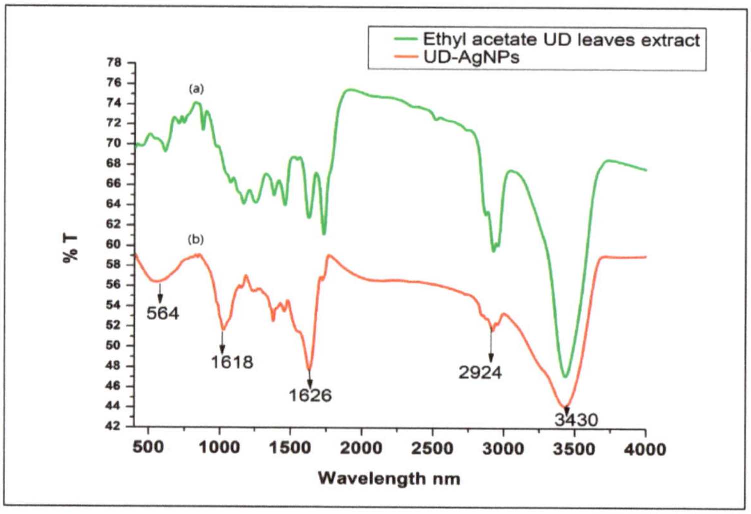

Figure 3. FTIR spectrum. (**a**) Ethyl acetate UD leaves extract; (**b**) UD-AgNPs with different functional groups.

3.1.4. XRD Pattern

The UD-AgNPs were studied using XRD, and the results are shown in Figure 4. The XRD pattern exhibited peaks at the two theta values of 38.12, 46.26, 57.40, and 76.98°, corresponding to the *hkl* plane Bragg reflections 111, 200, 220, and 311, respectively. The XRD spectrum confirmed that the synthesized UD-AgNPs had a face-centered cubic structure. The unassigned peak (*), observed at 55°, could have been caused by bioorganic compounds in the synthesized NPs.

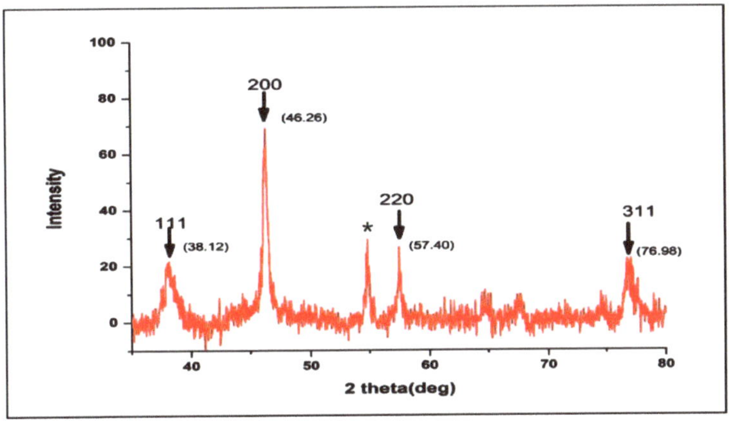

Figure 4. XRD pattern of the UD-AgNPs with peak values.

3.1.5. DLS Analysis

The DLS Zetasizer was used to determine the size distribution and the surface intensity of the synthesized UD-AgNPs (Figure 5a–b). The size distribution was measured based on the intensity fluctuation yields of the rate of Brownian motion (Figure 5a). The DLS result exhibited the average peak intensity values in the hydrodynamic diameter distribution with relatively homogenous distribution in the polydispersity index (PDI: 0.47 ± 0.31).

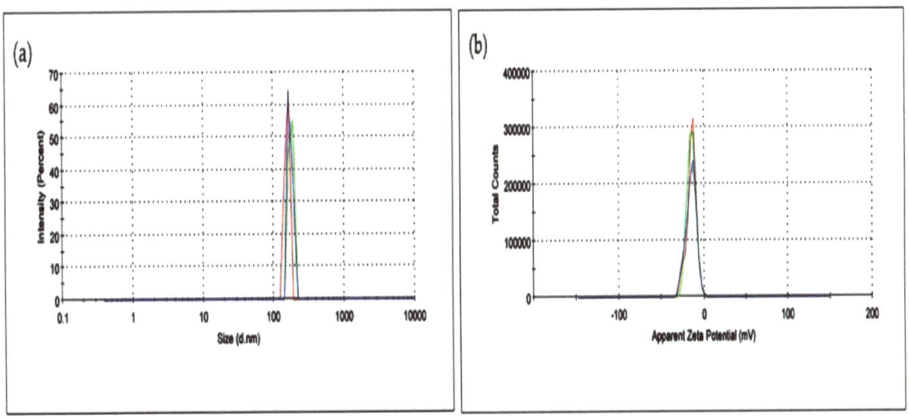

Figure 5. DLS Zetasizer analysis (**a**) Size distribution of the synthesized UD-AgNPs (**b**) Zeta potential of the synthesized UD-AgNPs.

The zeta potential result indicated that the distribution of NPs was equal, without any aggregation of UD-AgNPs (Figure 5b). The negative value (−15.5 ± 0.6 mV) of the zeta potential confirmed the size distribution and low aggregation of the synthesized UD-AgNPs, which was important for understanding their stability.

3.1.6. EDAX Analysis

As depicted in Figure 6, the EDAX for the elemental distribution in the synthesized UD-AgNPs shows that the elemental composition of Ag–L (48.53% mass) was higher than that of the other components: C–K (24.12% mass), O–K (17.80% mass), and Cl–K (9.55% mass). These results demonstrated that the AgNPs synthesized with UD-based NPs were pure.

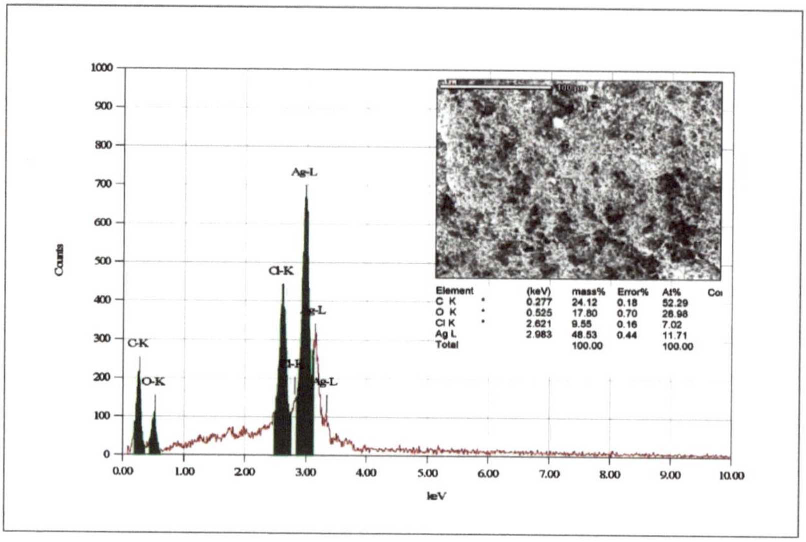

Figure 6. EADX spectrum observed elemental distribution Ag–L, C–K, O–K and Cl–K from synthesized UD-AgNPs.

3.1.7. SEM and TEM Images

The SEM images of the UD-AgNPs shown in Figure 7a,b reveal that the UD-AgNPs were spherical and did not aggregate. The images confirmed the formation of AgNPs. The TEM images of the UD-AgNPs depicted in Figure 7c,d showed that the UD-AgNPs had a spherical shape ranging from 8.2 to 56.1 nm, with an average size of 19.4 nm.

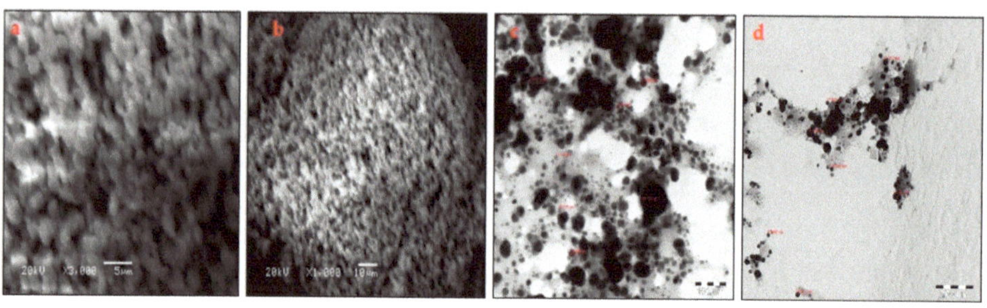

Figure 7. (**a–d**) SEM with different magnifications (**a**) 5 µm and (**b**) 10 µm; TEM images with different magnifications (**c**) 100 nm and (**d**) 200 nm exhibiting the formation of UD-AgNPs.

3.2. Antimicrobial Susceptibility

The antimicrobial susceptibility of the UD-AgNPs against Gram-positive and Gram-negative bacteria was tested at 25, 50, and 75 µL. Ciprofloxacin (10 µL) was used as a positive control, and the negative control test was performed using DMSO, ethyl acetate UD leaf extract, and $AgNO_3$ (75 µL). No record was found related to ZOI in the tested microorganisms (Figure 8a). The ZOI results of Gram-positive bacteria *S. aureus*, *E. faecalis*, and *S. epidermidis*, as well as those of the Gram-negative bacteria *E. coli*, *K. pneumoniae*, and *P. aeruginosa* are given in Figure 8b. The ZOI values are shown in Table 1. Compared to the other Gram-positive bacteria, the maximum ZOI was recorded for *S. epidermidis* (19 ± 1 mm). Among the Gram-negative bacteria, *P. aeruginosa* exhibited the highest ZOI value (21.6 ± 0.4 mm) (Table 1).

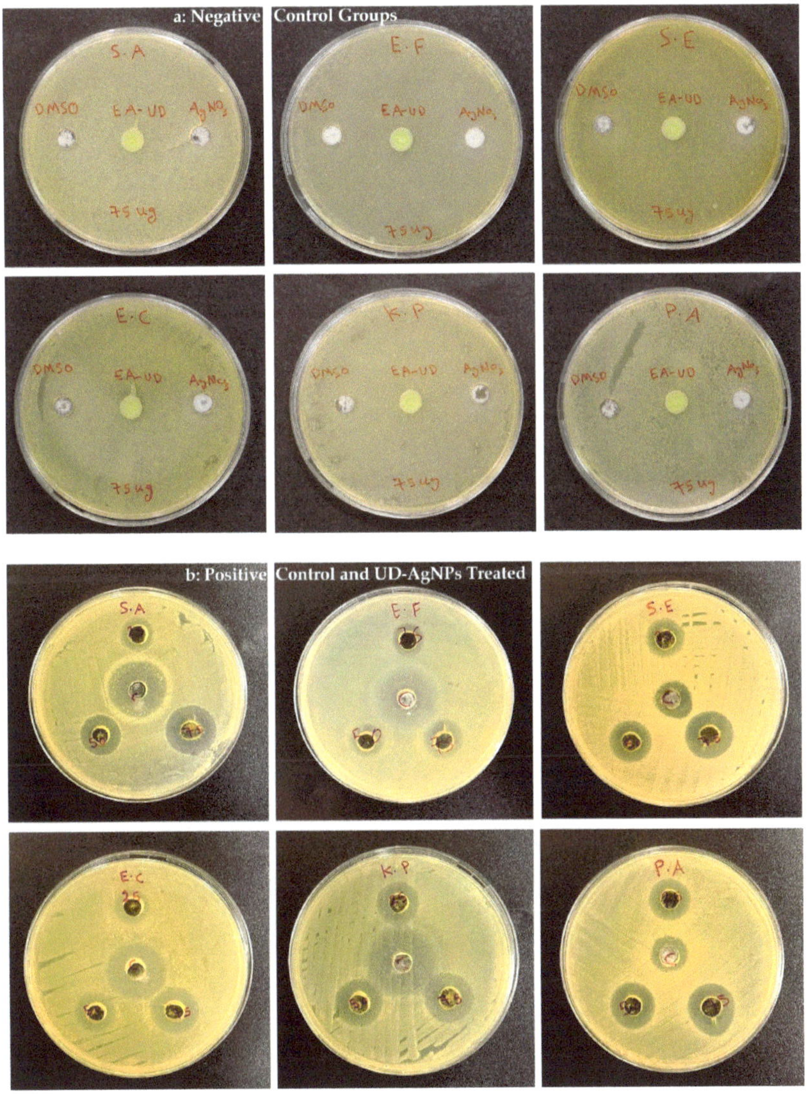

Figure 8. ZOI (**a**) control groups (DMSO, UD extract, and $AgNO_3$) and (**b**) synthesized UD-AgNPs against *S. aureus*, *E. faecalis*, *S. epidermidis*, *E. coli*, *K. pneumoniae*, and *P. aeruginosa*.

Table 1. The zone of inhibition observed for different bacterial isolates and different amounts of UD-AgNPs.

	Zone of Inhibition (mm)			
Bacteria	**25 uL (Mean ± SD)**	**50 uL (Mean ± SD)**	**75 uL (Mean± SD)**	**Control**
S. aureus	13.3 ± 0.7	16.6 ± 0.4	21.3 ± 0.7	24 ± 1
E. faecalis	12.6 ± 0.4	14 ± 1	18.6 ± 0.4	27 ± 1
S.epidermidis	14.3 ± 0.7	17.6 ± 0.6	19 ± 1	15.6 ± 0.6
K. pneumoniae	16.3 ± 0.7	17.3 ± 0.7	19.6 ± 0.6	26 ± 1
P. aeruginosa	16 ± 1	18.6 ± 0.4	21.6 ± 0.4	16 ± 1
E. coli	13.6 ± 0.6	17 ± 1	19.6 ± 0.6	24.3 ± 0.7

3.2.1. MIC Analysis

The MIC values of the UD-AgNPs against the Gram-positive bacteria *S. aureus*, *E. faecalis*, and *S. epidermidis*, as well as the Gram-negative bacteria *E. coli*, *K. pneumoniae*, and *P. aeruginosa*, were determined by assaying various concentrations of the UD-AgNPs (15, 20, 25, 30, 35, 40, 45, and 50 μg/mL). Three 96-well plates were used to assay the six microorganisms with the positive control (10 μL of each type of pathogen) and treatment (UD-AgNPs) incubated for 24 h at 37 °C. Resazurin dye (negative control) was used as a precursor to discriminate the actual MIC value of each tested organism. In this assay, the MIC values for each organism were based on color changes (Figure 9). The MIC values of the synthesized UD-AgNPs against the pathogens tested are shown in Table 2

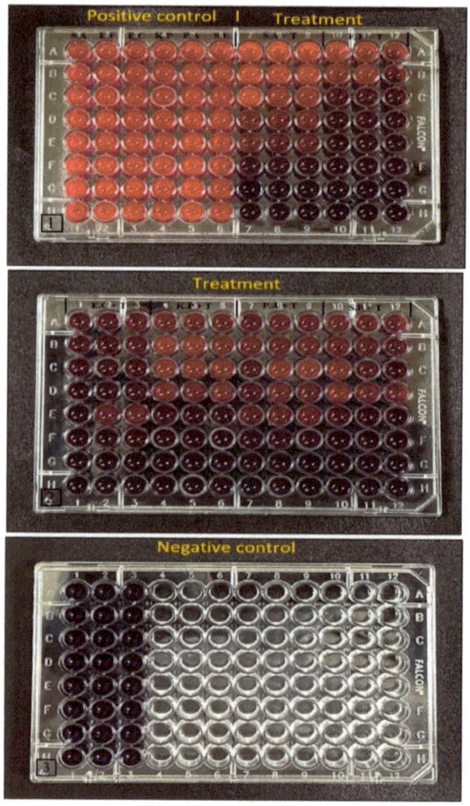

Figure 9. Resazurin Plate assay showing the MIC values of the UD-AgNPs against *S. aureus*, *E. faecalis*, *S. epidermidis*, *E. coli*, *K. pneumoniae*, and *P. aeruginosa*.

Table 2. Minimum inhibitory concentrations (MICs) and minimum bactericidal concentration (MBC) of UD-AgNPs against Gram-positive and Gram-negative bacteria.

Bacteria	MIC (μg/mL) (Mean ± SD)	MBC (μg/mL) (Mean ± SD)
S. aureus	40 ± 0	92 ± 17
S. epidermidis	38 ± 3	233 ± 33
E. faecalis	42 ± 8	83 ± 17
K. pneumoniae	40 ± 0	333 ± 33
P. aeruginosa	40 ± 0	133.± 33
E. coli	37± 3	75 ± 00

3.2.2. MBC Analysis

The MBC values for both Gram-positive bacteria and Gram-negative bacteria are shown in Table 2. The results indicated that the UD-AgNPs indeed inhibited the growth of the tested pathogens.

4. Discussion

The current research focused on synthesizing AgNPs by using the medicinal plant UD, the leaf extract of which acted as a capping and reducing agent ($AgNO_3$ into AgNPs) and presented numerous phytochemical compounds [25,26]. The formation of AgNPs was confirmed by changes in the color of the mixture [39] and the UV–vis spectral peak at 420–460 nm [40].

The FTIR spectrum showed strong absorption peaks corresponding to N–H and C–N; these amino groups played a role in the conversion process of Ag^+ to Ag. The presence of C=C and C–H bonds revealed that the alkenyl groups were involved in the strengthening of AgNPs. The alkenyl group strengthened the AgNPs as a capping agent, which could selectively bind to different types of small plane surfaces to change their specific surface-free energies. The capping agents were involved in numerous important functions, such as preventing the agglomeration of nanoparticles and reducing toxicity. Moreover, alkenyl group molecules enhanced AgNPs in bacterial activities [41,42]. Another peak at 564 cm^{-1} indicated the presence of a C–Br group, the secondary structure of proteins that interacted with the NPs [29,43]. The reported wave numbers indicating signal stretching and vibrational bending of the peaks may have been derived from phytoconstituents such as flavonoids, terpenoids, alkaloids, and soluble proteins present in plant extracts. These compounds may have been responsible for chelating and capping in the bioreduction process [44,45]. The peaks indexed to the 111, 200, and 220 planes corresponded to the cubic structure of silver, and the Debye–Scherrer equation confirmed the small and crystalline NPs [46]. Because the synthesized NPs were small, their diffusion was very fast, as reported previously [47]. According to the DLS (PDI: 0.47 $\pm$ 031) and ZP results (-15.5 ± 0.6 mV), the size and surface charge of the NPs confirmed their therapeutic potential. These two traits influenced their kinetics and interaction with cellular and biological membranes, resulting in their efficacy [45].

Synthesized UD-AgNPs showed Ag–L (48.53% mass) was more abundant than other components such as C–K, O–K, and Cl–K by EDAX, as previously reported [48]. The presence of a trace amount of oxygen (17.80% mass) in the UD-AgNPs confirmed the roles played by the UD phytochemicals in the reduction in metal ions, capping agents, and NP stability. Based on these characterizations, it was clear that the synthesized UD-AgNPs were small, spherical, and not agglomerated. Size-related properties influenced the NP mode of action [49–51].

In the present study, the antimicrobial potency of the synthesized UD-AgNPs was studied against both Gram-positive and Gram-negative pathogens. Of the Gram-positive bacteria tested, *S. epidermidis* showed the greatest ZOI, indicating its high susceptibility to the AgNPs. The earlier study showed that the ZOI was greater for *S. epidermidis*, which was more sensitive to AgNPs [52]. In the case of the other pathogens (*S. aureus* and *E. faecalis*), they had a decreased uptake of UD-AgNPs, increased efflux pump activity, and increased drug-degrading enzymatic activity, which may be the reasons for their lower susceptibility to the UD-AgNPs.

Among the three Gram-negative pathogens tested, *P. aeruginosa* exhibited the largest ZOI, followed by *E. coli* and *K. pneumoniae*, indicating sensitivity to the UD-AgNPs. The internalization of the AgNPs in *P. aeruginosa* was greater than that of the other pathogens. Previous researchers have reported that *P. aeruginosa* is highly drug-resistant, which limits the use of many antibiotics. However, it is very interesting to note that this pathogen is highly sensitive to UD-AgNPs, as reported for a few other biosynthesized AgNPs [53,54]. The Gram-negative bacteria *E. coli* and *K. pneumoniae* have been found to be resistant to many drugs, but they were sensitive to the synthesized AgNPs. It has been reported that AgNPs induce a triclosan-like bactericidal mechanism to inhibit type II fatty-acid biosynthesis and generates oxidative stress in *K. pneumoniae*, which has shown antibacterial activity in multidrug-resistant *K. pneumoniae* [37]. Like *K. pneumoniae* with extended-spectrum beta-lactamase, *E. coli* is very sensitive to NPs, and the NPs inhibit the bacterium by making the cell shrink and impairing its membrane integrity [55]. In the present study, sensitivity studies showed that the Gram-negative bacteria were more susceptible to the action of AgNPs. It has been reported that the difference in cell wall components facilitates the entry of released ions from AgNPs more into the cells of Gram-negative than Gram-positive bacteria [56].

The MIC and MBC values reflected the susceptibility of the microbes to UD-AgNPs. The high MIC values against *E. coli* and *S. epidermidis* and MBC against *E. coli* and *E. faecalis* were due to the action of lipopolysaccharides against the trapped AgNPs, as reported previously [57]. This finding reflected the strong toxicity of UD-AgNPs against the microorganism [58]. Therefore, AgNPs can be used as nanobactericides and drug carriers, as reported [57]. Hence, the outcome of this study confirmed that the ethyl acetate-mediated synthesized UD-AgNPs acted as nanobactericides, meaning that they can penetrate bacterial envelopes to damage the cell membrane structure causing oxidative stress [59–61].

5. Conclusions

UD is a medicinal plant, and its leaves possess numerous medicinal properties, including controlling cardiovascular disorders and glucose homeostasis. Using the ethyl acetate extracts of UD leaves, AgNPs were synthesized in this study. The characterization of the synthesized NPs showed that the particles were spherical and small, and since small NPs have favorable therapeutic actions, the currently synthesized NPs have promising prospects for development as drugs. This characterization study further confirmed the presence of bioactive compounds in the leaves and that they became incorporated into the AgNPs. Such morphometric changes and functional adaptations promote the attachment of the AgNPs to the cell wall of the pathogen, which then internalizes the NPs, which in turn alter the internal environment of the bacterial cell and induce oxidative stress, causing changes in the DNA and the cellular enzymes that lead to bacterial cell death or functional inhibition. The ZOI, MIC, and MBC results showed that the synthesized NPs had bactericidal action against both Gram-positive and Gram-negative bacteria. Thus, future research should focus on the activity of the newly synthesized UD-AgNPs towards multidrug resistance and extended-spectrum beta-lactamase-containing and methicillin-resistant bacteria.

Author Contributions: S.D., M.S.A. and M.B., conceptualization, methodology; M.B. and S.D., experiments and data curations, original manuscript draft preparation; N.N., A.N. and O.A., review and editing. All authors have read and agreed to the published version of the manuscript.

Funding: This research received no external funding.

Institutional Review Board Statement: Not applicable.

Informed Consent Statement: Not applicable.

Data Availability Statement: The data presented in this study are available on request from the corresponding author.

Acknowledgments: The authors are grateful to the Deanship of Scientific Research, King Saud University, for funding through the Vice Deanship of Scientific Research Chairs.

Conflicts of Interest: The authors declare that no conflict of interest.

References

1. Dar, S.A.; Ganai, F.A.; Yousuf, A.R.; Balkhi, M.-u.-H.; Bhat, T.M.; Sharma, P. Pharmacological and toxicological evaluation of *Urtica dioica*. *Pharm. Biol.* **2013**, *51*, 170–180. [CrossRef] [PubMed]
2. Shonte, T.T.; Duodu, K.G.; de Kock, H.L. Effect of drying methods on chemical composition and antioxidant activity of underutilized stinging nettle leaves. *Heliyon* **2020**, *6*, e03938. [CrossRef] [PubMed]

3. Nallan Chakravartula, S.S.; Moscetti, R.; Farinon, B.; Vinciguerra, V.; Merendino, N.; Bedini, G.; Neri, L.; Pittia, P.; Massantini, R. Stinging Nettles as *Potential* Food Additive: Effect of Drying Processes on Quality Characteristics of Leaf Powders. *Foods* **2021**, *10*, 1152. [CrossRef] [PubMed]
4. Grauso, L.; de Falco, B.; Lanzotti, V.; Motti, R. Stinging nettle, *Urtica dioica* L.: Botanical, phytochemical and pharmacological overview. *Phytochem. Rev.* **2020**, *19*, 1341–1377. [CrossRef]
5. Ziaei, R.; Foshati, S.; Hadi, A.; Kermani, M.A.H.; Ghavami, A.; Clark, C.C.; Tarrahi, M.J. The effect of nettle (*Urtica dioica*) supplementation on the glycemic control of patients with type 2 diabetes mellitus: A systematic review and meta-analysis. *Phytother. Res.* **2020**, *34*, 282–294. [CrossRef]
6. Singh, R.; Dar, S.; Sharma, P. Antibacterial activity and toxicological evaluation of semi purified hexane extract of *Urtica dioica* leaves. *Res. J. Med. Plants* **2012**, *6*, 123–135.
7. Zehraw, H.M.; Taleb, Z.M.M.; Mukhlif, B.; Al-Jabbar, S.A. Extraction and evaluation the activity of *Urtica dioica* as bleeding stop material. *Iraqi J. Biotechnol.* **2019**, *18*, 77–87.
8. Zouari Bouassida, K.; Bardaa, S.; Khimiri, M.; Rebaii, T.; Tounsi, S.; Jlaiel, L.; Trigui, M. Exploring the *Urtica dioica* leaves hemostatic and wound-healing potential. *BioMed Res. Int.* **2017**, *2017*, 1047523. [CrossRef]
9. Modarresi-Chahardehi, A.; Ibrahim, D.; Fariza-Sulaiman, S.; Mousavi, L. Screening antimicrobial activity of various extracts of *Urtica dioica*. *Rev. Biol. Trop.* **2012**, *60*, 1567–1576. [CrossRef]
10. Holmes, A.; Moore, L.; Sundsfjord, A.; Steinba, M.; Regmi, S.; Karey, A.; Guerin, P.J.; Piddoc, L.J.V. Understanding the mechanisms and drivers of antimicrobial resistance. *Lancet* **2016**, *387*, 176–187. [CrossRef]
11. Chabi, R.; Momtaz, H. Virulence factors and antibiotic resistance properties of the *Staphylococcus epidermidis* strains isolated from hospital infections in Ahvaz, Iran. *Trop. Med. Health* **2019**, *47*, 1–9. [CrossRef] [PubMed]
12. Jain, D.; Daima, H.K.; Kachhwaha, S.; Kothari, S. Synthesis of plant-mediated silver nanoparticles using papaya fruit extract and evaluation of their anti-microbial activities. *Dig. J. Nanomater. Biostruct.* **2009**, *4*, 557–563.
13. Abdel-Raouf, N.; Al-Enazi, N.M.; Ibraheem, I.B. Green biosynthesis of gold nanoparticles using *Galaxaura elongata* and characterization of their antibacterial activity. *Arab. J. Chem.* **2017**, *10*, S3029–S3039. [CrossRef]
14. Dikshit, P.K.; Kumar, J.; Das, A.K.; Sadhu, S.; Sharma, S.; Singh, S.; Gupta, P.K.; Kim, B.S. Green Synthesis of Metallic Nanoparticles: Applications and Limitations. *Catalysts* **2021**, *11*, 902. [CrossRef]
15. Devanesan, S.; AlSalhi, M.S. Green Synthesis of Silver Nanoparticles Using the Flower Extract of *Abelmoschus esculentus* for Cytotoxicity and Antimicrobial Studies. *Int. J. Nanomed.* **2021**, *16*, 3343. [CrossRef]
16. AlSalhi, M.S.; Devanesan, S.; Alfuraydi, A.A.; Vishnubalaji, R.; Munusamy, M.A.; Murugan, K.; Nicoletti, M.; Benelli, G. Green synthesis of silver nanoparticles using *Pimpinella anisum* seeds: Antimicrobial activity and cytotoxicity on human neonatal skin stromal cells and colon cancer cells. *Int. J. Nanomed.* **2016**, *11*, 4439. [CrossRef]
17. Devanesan, S.; Ponmurugan, K.; AlSalhi, M.S.; Al-Dhabi, N.A. Cytotoxic and antimicrobial efficacy of silver nanoparticles synthesized using a traditional phytoproduct, asafoetida gum. *Int. J. Nanomed.* **2020**, *15*, 4351. [CrossRef]
18. Dakal, T.C.; Kumar, A.; Majumdar, R.S.; Yadav, V. Mechanistic basis of antimicrobial actions of silver nanoparticles. *Front. Microbiol.* **2016**, *7*, 1831. [CrossRef]
19. Gurunathan, S.; Han, J.W.; Kwon, D.-N.; Kim, J.-H. Enhanced antibacterial and anti-biofilm activities of silver nanoparticles against Gram-negative and Gram-positive bacteria. *Nanoscale Res. Lett.* **2014**, *9*, 1–17. [CrossRef]
20. Wang, Y.; Zhang, X.; Bai, Y.; Li, W.; Li, X.; Xing, X.; Wang, C.; Gao, L.; Yogi, M.; Swamy, M.K. Anticancer and Antibacterial Activities of Silver Nanoparticles (AgNPs) Synthesized from *Cucumis melo* L. *J. Nanosci. Nanotechnol.* **2020**, *20*, 4143–4151. [CrossRef]
21. Das, G.; Patra, J.K.; Nagaraj Basavegowda, C.N.V.; Shin, H.-S. Comparative study on antidiabetic, cytotoxicity, antioxidant and antibacterial properties of biosynthesized silver nanoparticles using outer peels of two varieties of *Ipomoea batatas* (L.) Lam. *Int. J. Nanomed.* **2019**, *14*, 4741. [CrossRef] [PubMed]
22. Salleh, A.; Naomi, R.; Utami, N.D.; Mohammad, A.W.; Mahmoudi, E.; Mustafa, N.; Fauzi, M.B. The potential of silver nanoparticles for antiviral and antibacterial applications: A mechanism of action. *Nanomaterials* **2020**, *10*, 1566. [CrossRef] [PubMed]
23. Jadczak, P.; Kulpa, D.; Drozd, R.; Przewodowski, W.; Przewodowska, A. Effect of AuNPs and AgNPs on the Antioxidant System and Antioxidant Activity of Lavender (*Lavandula angustifolia* Mill.) from In Vitro Cultures. *Molecules* **2020**, *25*, 5511. [CrossRef] [PubMed]
24. Vanlalveni, C.; Lallianrawna, S.; Biswas, A.; Selvaraj, M.; Changmai, B.; Rokhum, S.L. Green synthesis of silver nanoparticles using plant extracts and their antimicrobial activities: A review of recent literature. *RSC Adv.* **2021**, *11*, 2804–2837. [CrossRef]
25. Laouini, S.E.; Bouafia, A.; Soldatov, A.V.; Algarni, H.; Tedjani, M.L.; Ali, G.A.; Barhoum, A. Green synthesized of Ag/Ag_2O nanoparticles using aqueous leaves extracts of *phoenix dactylifera* L. And their azo dye photodegradation. *Membranes* **2021**, *11*, 468. [CrossRef]
26. Meretoudi, A.; Banti, C.N.; Raptis, P.K.; Papachristodoulou, C.; Kourkoumelis, N.; Ikiades, A.A.; Zoumpoulakis, P.; Mavromoustakos, T.; Hadjikakou, S.K. Silver Nanoparticles from Oregano Leaves' Extracts as Antimicrobial Components for Non-Infected Hydrogel Contact Lenses. *Int. J. Mol. Sci.* **2021**, *22*, 3539. [CrossRef]
27. Kadhim, M.; Aldujaili, A.; Homady, M. Assessment of hepatoprotective role of phenolic extract of *Urtica dioica* and silver nanoparticles in male rat induced by carbon tetra-chloride. *Rasayan J. Chem.* **2017**, *10*, 305–312.

28. Nassar, A.M. Research Article Effectiveness of Silver Nano-particles of Extracts of *Urtica urens* (Urticaceae) Against Root-knot Nematode Meloidogyne incognita. *Asian J. Nematol.* **2016**, *5*, 14–19. [CrossRef]
29. Jyoti, K.; Baunthiyal, M.; Singh, A. Characterization of silver nanoparticles synthesized using *Urtica dioica* Linn. leaves and their synergistic effects with antibiotics. *J. Radiat. Res. Appl. Sci.* **2016**, *9*, 217–227. [CrossRef]
30. Joshi, N.C.; Kumar, V.; Singh, A.; Singh, R. Characterisations, Antimicrobial activities and Biological synthesis of silver (Ag) nanoparticles using the leaf extract of *Urtica dioica*. *Res. J. Pharm. Technol.* **2019**, *12*, 4429–4433. [CrossRef]
31. Flieger, J.; Franus, W.; Panek, R.; Szymańska-Chargot, M.; Flieger, W.; Flieger, M.; Kołodziej, P. Green Synthesis of Silver Nanoparticles Using Natural Extracts with Proven Antioxidant Activity. *Molecules* **2021**, *26*, 4986. [CrossRef] [PubMed]
32. Khanam, R.; Prasuna, R.G. Comparison of extraction methods and solvents for total phenolics from dairy waste. *Asian J. Dairy Food Res.* **2017**, *36*, 251–255. [CrossRef]
33. Joshi, D.R.; Adhikari, N. An overview on common organic solvents and their toxicity. *J. Pharm. Res. Int.* **2019**, *28*, 1–18. [CrossRef]
34. Thouri, A.; Chahdoura, H.; El Arem, A.; Omri Hichri, A.; Ben Hassin, R.; Achour, L. Effect of solvents extraction on phytochemical components and biological activities of Tunisian date seeds (var. *Korkobbi* and *Arechti*). *BMC Complementary Altern. Med.* **2017**, *17*, 248. [CrossRef] [PubMed]
35. Jadou, A.; Al-Shahwany, A.W. Biogenic Synthesis and Characterization of Silver Nanoparticles Using Some Medical Plants and Evaluation of Their Antibacterial and Toxicity Potential. *J. AOAC Int.* **2018**, *101*, 1905–1912. [CrossRef] [PubMed]
36. Khan, M.; Tang, H.; Lyles, J.; Pineau, R.; Mashwani, Z.R.; Quave, C.L. Antibacterial properties of medicinal plants from Pakistan against multidrug-resistant ESKAPE pathogens. *Front. Pharmacol.* **2018**, *9*, 815. [CrossRef]
37. Loo, Y.Y.; Rukayadi, Y.; Nor-Khaizura, M.-A.-R.; Kuan, C.H.; Chieng, B.W.; Nishibuchi, M.; Radu, S. In vitro antimicrobial activity of green synthesized silver nanoparticles against selected gram-negative foodborne pathogens. *Front. Microbiol.* **2018**, *9*, 1555. [CrossRef]
38. Alfuraydi, A.A.; Devanesan, S.; Al-Ansari, M.; AlSalhi, M.S.; Ranjitsingh, A.J. Eco-friendly green synthesis of silver nanoparticles from the sesame oil cake and its potential anticancer and antimicrobial activities. *J. Photochem. Photobiol. B Biol.* **2019**, *192*, 83–89. [CrossRef]
39. Shi, X.; Zou, J.; Chen, X.; Zheng, H.; Jin, Z.; Li, F.; Piao, J.-G. The Effect of Size on the Surface Enhanced Raman Scattering Property of SiO_2@PDA@AgNP Core-Shell-Satellite Nanocomposite. *Chem. Lett.* **2020**, *49*, 534–537. [CrossRef]
40. Haider, A.; Ijaz, M.; Ali, S.; Haider, J.; Imran, M.; Majeed, H.; Shahzadi, I.; Ali, M.M.; Khan, J.A.; Ikram, M. Green synthesized phytochemically (*Zingiber officinale* and *Allium sativum*) reduced nickel oxide nanoparticles confirmed bactericidal and catalytic potential. *Nanoscale Res. Lett.* **2020**, *15*, 1–11. [CrossRef]
41. Mikhailova, E.O. Silver nanoparticles: Mechanism of action and probable bio-application. *J. Funct. Biomater.* **2020**, *11*, 84. [CrossRef] [PubMed]
42. Rama, P.; Baldelli, A.; Vignesh, A.; Altemimi, A.B.; Lakshmanan, G.; Selvam, R.; Arunagirinathan, N.; Murugesan, K.; Pratap-Singh, A. Antimicrobial, antioxidant, and angiogenic bioactive silver nanoparticles produced using *Murraya paniculata* (L.) jack leaves. *Nanomater. Nanotechnol.* **2022**, *12*. [CrossRef]
43. Devanesan, S.; Jayamala, M.; AlSalhi, M.S.; Umamaheshwari, S.; Ranjitsingh, A.J.A. Antimicrobial and Anticancer properties of *Carica papaya* leaves derived di-methyl flubendazole mediated silver nanoparticles. *J. Infect. Public Health* **2021**, *14*, 577–587. [CrossRef] [PubMed]
44. Sallehudin, T.A.T.; Seman, M.N.A.; Chik, S.M.S.T. Preparation and Characterization Silver Nanoparticle Embedded Polyamide Nanofiltration (NF) Membrane. *ATEC Web Conf.* **2018**, *150*, 02003.
45. Carvalho, P.M.; Felício, M.R.; Santos, N.C.; Gonçalves, S.; Domingues, M.M. Application of light scattering techniques to nanoparticle characterization and development. *Front. Chem.* **2018**, *6*, 237. [CrossRef]
46. Rahdar, A.; Amini, N.; Askari, F.; Susan, M.; Hasan, A.B. Dynamic light scattering: A useful technique to characterize nanoparticles. *J. Nanoanal.* **2019**, *6*, 80–89.
47. Wong, A.; Liu, Q.; Griffin, S.; Nicholls, A.; Regalbuto, J. Synthesis of ultrasmall, homogeneously alloyed, bimetallic nanoparticles on silica supports. *Science* **2017**, *358*, 1427–1430. [CrossRef]
48. Lomelí-Rosales, D.A.; Zamudio-Ojeda, A.; Reyes-Maldonado, O.K.; López-Reyes, M.E.; Basulto-Padilla, G.C.; Lopez-Naranjo, E.J.; Zuñiga-Mayo, V.M.; Velázquez-Juárez, G. Green Synthesis of Gold and Silver Nanoparticles Using Leaf Extract of *Capsicum chinense* Plant. *Molecules* **2022**, *27*, 1692. [CrossRef]
49. Jahan, S.T.; Sadat, S.; Walliser, M.; Haddadi, A. Targeted therapeutic nanoparticles: An immense promise to fight against cancer. *J. Drug Deliv.* **2017**, *2017*, 9090325. [CrossRef]
50. Jurj, A.; Braicu, C.; Pop, L.-A.; Tomuleasa, C.; Gherman, C.D.; Berindan-Neagoe, I. The new era of nanotechnology, an alternative to change cancer treatment. *Drug Des. Dev. Ther.* **2017**, *11*, 2871. [CrossRef]
51. Scimeca, M.; Bischetti, S.; Lamsira, H.K.; Bonfiglio, R.; Bonanno, E. Energy Dispersive X-ray (EDX) microanalysis: A powerful tool in biomedical research and diagnosis. *Eur. J. Histochem. EJH* **2018**, *62*, 2841. [CrossRef]
52. Arshad, H.; Sami, M.A.; Sadaf, S.; Hassan, U. Salvadora persica mediated synthesis of silver nanoparticles and their antimicrobial efficacy. *Sci. Rep.* **2021**, *11*, 5996. [CrossRef] [PubMed]
53. Pareek, V.; Devineau, S.; Sivasankaran, S.K.; Bhargava, A.; Panwar, J.; Srikumar, S.; Fanning, S. Silver nanoparticles induce a triclosan-like antibacterial action mechanism in multi-drug resistant *Klebsiella pneumoniae*. *Front. Microbiol.* **2021**, *12*, 183. [CrossRef] [PubMed]

54. Hameed, A.S.H.; Karthikeyan, C.; Ahamed, A.P.; Thajuddin, N.; Alharbi, N.S.; Alharbi, S.A.; Ravi, G. In vitro antibacterial activity of ZnO and Nd doped ZnO nanoparticles against ESBL producing *Escherichia coli* and *Klebsiella pneumoniae*. *Sci. Rep.* **2016**, *6*, 24312. [CrossRef] [PubMed]
55. Baptista, P.V.; McCusker, M.P.; Carvalho, A.; Ferreira, D.A.; Mohan, N.M.; Martins, M.; Fernandes, A.R. Nano-strategies to fight multidrug resistant bacteria—"A Battle of the Titans". *Front. Microbiol.* **2018**, *9*, 1441. [CrossRef]
56. Lee, J.; Monk, I.; Goncalves da Silva, A.; Seemann, T.; Chua, K.Y.L.; Kearns, A.; Hill, R.; Woodford, N.; Bartels, M.D.; Strommenger, B.; et al. Global spread of three multidrug-resistant lineages of *Staphylococcus epidermidis*. *Nat. Microbiol.* **2018**, *3*, 1175–1185. [CrossRef]
57. Brayner, R.; Ferrari-Iliou, R.; Brivois, N.; Djediat, S.; Benedetti, M.F.; Fiévet, F. Toxicological impact studies based on *Escherichia coli* bacteria in ultrafine ZnO nanoparticles colloidal medium. *Nano Lett.* **2006**, *6*, 866–870. [CrossRef]
58. Panpaliya, N.P.; Dahake, P.T.; Kale, Y.J.; Dadpe, M.V.; Kendre, S.B.; Siddiqi, A.G.; Maggavi, U.R. In vitro evaluation of antimicrobial property of silver nanoparticles and chlorhexidine against five different oral pathogenic bacteria. *Saudi Dent. J.* **2019**, *31*, 76–83. [CrossRef]
59. Dizaj, S.M.; Lotfipour, F.; Barzegar-Jalali, M.; Zarrintan, M.H.; Adibkia, K. Antimicrobial activity of the metals and metal oxide nanoparticles. *Mater. Sci. Eng. C* **2014**, *44*, 278–284. [CrossRef]
60. Yun'an Qing, L.C.; Li, R.; Liu, G.; Zhang, Y.; Tang, X.; Wang, J.; Liu, H.; Qin, Y. Potential antibacterial mechanism of silver nanoparticles and the optimization of orthopedic implants by advanced modification technologies. *Int. J. Nanomed.* **2018**, *13*, 3311. [CrossRef]
61. Yang, W.; Shen, C.; Ji, Q.; An, H.; Wang, J.; Liu, Q.; Zhang, Z. Food storage material silver nanoparticles interfere with DNA replication fidelity and bind with DNA. *Nanotechnology* **2009**, *20*, 085102. [CrossRef]

microorganisms

MDPI

Article

Trichogenic Silver-Based Nanoparticles for Suppression of Fungi Involved in Damping-Off of Cotton Seedlings

Shimaa A. Zaki [1,2], Salama A. Ouf [1], Kamel A. Abd-Elsalam [2], Amal A. Asran [2], Mohamed M. Hassan [3], Anu Kalia [4] and Fawziah M. Albarakaty [5,*]

1 Botany and Microbiology Department, Faculty of Science, Cairo University, Giza 12613, Egypt; shim.shimshim@yahoo.com (S.A.Z.); salama@sci.cu.edu.eg (S.A.O.)
2 Plant Pathology Research Institute, Agricultural Research Centre, Giza 12619, Egypt; kamelabdelsalam@gmail.com (K.A.A.-E.); asran.amal@gmail.com (A.A.A.)
3 Department of Biology, College of Science, Taif University, Taif 21944, Saudi Arabia; m.khyate@tu.edu.sa
4 Electron Microscopy and Nanoscience Laboratory, Department of Soil Science, College of Agriculture, Punjab Agricultural University, Ludhiana 141004, Punjab, India; kaliaanu@pau.edu
5 Department of Biology, Faculty of Applied Science, Umm Al-Qura University, Makkah Al Mukarramah 21955, Saudi Arabia
* Correspondence: fmbarakati@uqu.edu.sa

Abstract: Mycogenic silver nanoparticles (AgNPs) produced by some biocontrol agents have shown the ability to inhibit the growth of numerous plant pathogenic fungi, which may be a unique method of disease management. This study describes the extracellular production of AgNPs by *Trichoderma harzianum*. The size, shape, charge, and composition of the AgNPs were subsequently studied by UV-visible spectroscopy, DLS, zeta potential, TEM, SEM, and EDX, among other methods. The AgNPs had sizes ranging from 6 to 15 nm. The antifungal activities of bio-synthesized AgNPs and two commercial fungicides (Moncut and Maxim XL) were tested against three soil-borne diseases (*Fusarium fujikuroi*, *Rhizoctonia solani*, and *Macrophomina phaseolina*). Cotton seedling illnesses were significantly reduced under greenhouse settings after significant in vitro antifungal activity was documented for the control of plant pathogenic fungi. The use of biocontrol agents such as *T. harzianum*, for example, may be a safe strategy for synthesizing AgNPs and using them to combat fungus in Egyptian cotton.

Keywords: microbial synthesis; silver nanoparticles; antifungal activity; Egyptian cotton; soil-borne pathogenic fungi

check for updates

Citation: Zaki, S.A.; Ouf, S.A.; Abd-Elsalam, K.A.; Asran, A.A.; Hassan, M.M.; Kalia, A.; Albarakaty, F.M. Trichogenic Silver-Based Nanoparticles for Suppression of Fungi Involved in Damping-Off of Cotton Seedlings. *Microorganisms* **2022**, *10*, 344. https://doi.org/10.3390/microorganisms10020344

Academic Editor: Mireille Fouillaud

Received: 9 December 2021
Accepted: 26 January 2022
Published: 2 February 2022
Corrected: 10 February 2025

1. Introduction

Cotton (*Gossypium barbadense* L.), which belongs to the family Malvaceae, is considered one of Egypt's main cash crops. Cotton seedling diseases occur due to infection caused by a complex of soil-borne organisms, including *Rhizoctonia solani* Kühn (Teleomorph: one of the most primordial Basidiomycetes is Thanatephorus cucumeris (Frank) Donk). R. solani can be found in practically all cultivated fields in vegetative form [1]. In the preponderance of Egypt's cotton-growing zones, *R. solani* was discovered to be the most important cause of cotton damping-off. A pathogenicity assay of *R. solani* belonging to different Anastomosis groups (AGs), i.e., Ag2-2, AG-4, AG-5, and AG-7, under greenhouse conditions revealed that most of the virulent isolates caused pre-emergence damping-off [2,3]. All of these isolates were extremely pathogenic, causing 100 % death in the pre-emergence stage [2–5]. *Fusarium* species are often isolated from diseased cotton roots and are recorded as a common pathogen of cotton seedling roots [6–8]. *Macrophomina phaseolina* (Tassi) Goid is a seed-borne and soil-borne disease and exhibits a wide host range that causes charcoal rot in cotton. Although early *M. phaseolina* infections in cotton occur in the seedling stage, they often remain until plant maturity [9]. When *M. phaseolina* penetrates cotton roots or stems, internal tissues are colonized quickly, and the plant dies. Dry rot is observed on diseased portions, with the occurrence of black sclerotia scattered throughout the wood and softer tissue [1].

The fungus-mediated control of pathogenic fungi is an environment friendly, cost-effective, and biocompatible approach to managing the phytopathogens [10]. The use of fungal-mediated green chemistry techniques for the synthesis of NPs can improve yield potential while lowering the input costs [11]. As a result of their high processing capacity, substantial surface area recovery, and excellent mycelial growth, fungi are an efficient and easy biosynthetic scaling agents [12]. Mycogenic nanoparticles (NPs) synthesized by beneficial fungi is a potential environmentally acceptable approach for large-scale manufacturing of various nanoparticles. Various fungal species have been used for the biosynthesis of silver nanoparticles, including *Trichoderma koningii, Aspergillus flavus, Fusarium oxysporum, Penicillium citrinum*, and *P. fellutanum* [13]. One of the most significant plant species in agriculture is *T. harzianum*, which serves as a biological control agent against phytopathogens. Hyaloid development and the release of hydrolytic enzymes breakdown the fungus's cell wall as its primary mode of action. Enzymes that are critical to mycoparasitism include chitinases, N-acetyl-β-D-glucosamine deacetyltransferase, and proteases [14–16]. In the fields of biotechnology and nanotechnology, *T. harzianum* has been extensively studied, offering new possibilities for the development of new products and uses, due to its excellent qualities for the management of phytopathogens and the simplicity with which it may be handled [17,18].

T. harzianum is employed for stabilization in the green manufacture of biogenic silver nanoparticles. Secondary metabolites released by biocontrol agents (*T. harzianum*) operate as capping and reducing agents, which can help to produce uniformly sized green and sustainable nanomaterials [18–20]. *T. harzianum* cell filtrate has been used to produce biogenic AgNPs in a simple, green, and eco-friendly way, without the need for any toxic reducing, capping, or dispersion agents [21]. The generated nanoparticles characterized using physicochemical methods exhibited face-centered cubic symmetry [21,22]. Because this fungus has a variety of qualities that aid with agricultural commodities, *Trichoderma*-based products contribute to more than half of the overall worldwide biopesticide industry employed for improving plant growth, nutrient use efficiency, and physiological response to both biotic and abiotic stresses [23]. The use of green AgNPs as an antifungal agent is seen as an environmentally benign resource, an alternative to fungicides, and a cost-effective method [24,25]. Based on updated literature, the primary goals of this investigation were to: (1) determine the ability of a cell-free extract from a *T. harzianum* natural biocontrol strain to act as a reducing agent and stabilizer in the production of AgNPs; (2) undertake physicochemical characterization to determine the structure, size, and morphology of the generated AgNPs; and (3) test the antifungal activity of AgNPs in vitro and in vivo against three fungal pathogens, namely, *R. solani, F. fujikuroi*, and *M. phaseolina*.

2. Materials and Methods

2.1. *Trichoderma* Isolates

Soil samples were taken from healthy cotton root rhizosphere soils and stored at 4 °C until used. Each soil sample was diluted five times in sterile distilled water before being applied to the surface of potato dextrose agar (PDA) (M/s HiMedia Pvt. Ltd., Mumbai, Maharashtra, India) with 0.5% of the diluted sample [26]. The single spore isolation technique was used to isolate putative *Trichoderma* colonies [27]. Petri plates inoculated with fungal strains were incubated at 28 ± 2 °C for 4 days. At the Assiut University Mycology Center (AUMC) in Assuit, Egypt, morphological approaches were used to identify *Trichoderma* species. Eppendorf tubes with sterile deionized water were used to maintain the isolates. The fungus was re-grown at 25 °C for 5 days in the dark on the freshly prepared PDA.

2.2. Preparation of Fungal Filtrate and Tricoh-Synthesis of Silver Nanoparticles

To prepare biomass for biosynthetic studies, *Trichoderma* isolates were inoculated in 250 mL conical flasks containing modified TLS liquid broth in addition to yeast extract (0.6 g/L). *T. harzianum* biomass (20 g) was mixed with 100 mL sterile deionized water in a flask. To purify *T. harzianum* filtrate, the biomass was agitated and filtered with Whatman filter paper no. 1 after incubation [28].

For three days, 50 g of fungal cell filtrate cultures were agitated at 150 rpm with 50 mL of 1 mM $AgNO_3$ solution. The fungal biomass filtrate without the silver nitrate solution served as a positive control, whereas the silver nitrate solution without the cell-free filtrate served as a negative control. Centrifugation (12,000 rpm for 20 min) was used to separate the nanoparticles, which were then washed twice with double distilled water and lyophilized for 24 h. Silver nanoparticles were generated by converting silver ions to metallic silver. The NPs were then kept in polypropylene tubes in the dark at room temperature. The generated AgNPs were subjected to several analytical methods for physicochemical characterization [19].

2.3. AgNPs Characterization

2.3.1. UV-Vis Spectrophotometer Analysis

The essential approach and the easiest way to establish the synthesis of nanoparticles in colloidal solution is via UV-Vis spectroscopy (T80 UV-Vis spectrophotometer, PG Instruments Ltd, Leicestershire, LE17 5BH, United Kingdom). After 2, 4, and 7 days of incubation, periodic sampling of 1 mL of the solution was performed, and the samples were scanned for a wavelength range of 0 to 1100 nm on a UV-Vis spectrometer. The visual shift in the color of the solution from yellow to brown was used to observe the reduction in metal ions. The distilled water was used as a blank to set the reference.

2.3.2. Transmission Electron Microscopy (TEM)

TEM was performed to ascertain the shape and measure the size of the biogenic AgNPs. The sample was made by depositing an aliquot of an aqueous AgNPs suspension onto a carbon-coated copper grid and allowing the contents to air dry in a vacuum desiccator. The dried grids were viewed in a TEM (Leo 912 AB OMEGA, Carl Zeiss, Oberkochen, Germany) operating at an accelerating voltage of 80 kV. The particle size distribution of the biogenic AgNPs was obtained through image analysis of the acquired TEM micrographs by Image J software (version 1.45, NIH, USA).

2.3.3. Dynamic Light Scattering and Zeta Potential assays

The particle size distribution of the biosynthesized AgNPs was also investigated through dynamic light scattering spectroscopy (Malvern Instruments, Southborough, MA, USA). In the data processing mode, the multi-modal resolution was set to high. The synthesized AgNPs solution was diluted (20, 40, 60, 80, and 100%), and the diluted sample was either filtered or left unfiltered using a 0.22-µm syringe-driven filter unit. All measurements were taken in triplicate at 25 °C with a 1-min temperature equilibration delay.

To determine the stability and surface charge of AgNPs obtained from *T. harzianum*, zeta potential analysis was undertaken and the average hydrodynamic diameter of the nanoparticles were also obtained from the measurements (polarity, viscosity, conductivity, charge, pH, etc.). The zeta potential of aqueous AgNPs suspension was determined on a Mansetting nano (Malvern Instruments, Southborough, MA, USA) zeta sizer equipped with Zetatrac and using Microtrac FLEX Operating Software. The sample was produced by dissolving AgNP powder in deionized water and sonicating it for 10 min in a bath sonicator (Wise WiseClean WUC A22H, Witeg Labortechnik GmbH, Wertheim, Germany).

2.3.4. Scanning Electron Microscope (SEM)

SEM investigation was undertaken to identify the size of agglomerates of the biogenic AgNPs. Initially, the dried specimen was suspended in ethanol. On a carbon-coated copper grid, a very thin coating of the sample was dropped, and a thin film of the sample was generated. The sample was then examined on a scanning electron microscope (Vega-3 SBU, Tescan, 623 00 Brno-Kohoutovice, Czech Republic) operated at a 20 kV accelerating voltage.

2.3.5. Energy Dispersive X-ray (EDX)

The EDX analysis was performed on the generated AgNPs on an EDS system (JEOL JEM-1230, Tokyo, Japan) attached to a scanning electron microscope to evaluate the presence of elemental Ag. The gold coating was performed using a gold sputtering unit (Polaron E5100, Quorum Technologies Ltd., Laughton, East Sussex England) after the dried AgNPs were mounted on the copper mesh.

2.4. *In Vitro Antifungal Activity of AgNPs*

The in vitro antifungal potential of biosynthesized AgNPs was evaluated through the agar well diffusion method. *R. solani* (RS9), *Fusarium fujikuroi* (FF10), and *M. phaseolina* isolate (MP4) were cultured for 7 days on a PDA medium at 35 °C. A cork borer was used to cut 5 mm discs of fungal inoculum and inoculate a 9 cm diameter Petri dish, which was incubated at 27 °C for 5–7 days, and the freshly generated PDA containing varying concentrations of generated AgNPs (20, 40, and 100 µg/mL) was allowed to harden; three replicates were undertaken for each concentration. Controls were nanoparticle-free PDA plates cultured under the same conditions. The fungal mycelia radial growth diameter was measured after incubation.

2.5. *Antifungal Activity under Greenhouse Conditions*

The experiments were carried out in pots using the complete randomized block design (CRBD) according to the method developed by the authors [18]. The effects of the synthesized AgNPs against three soil-borne pathogens on the cotton types Giza 90 and Giza 94 were tested using the RS9, FF10, and MP4 pathogens. Pots of autoclaved soil were infected with two-week-old cultures of the fungal-sorghum *R. solani* (RS9), *Fusarium fujikuroi* (FF10), and *M. phaseolina* (MP4) at rates of 1, 50, and 50 g/kg soil, for each pathogen, respectively, to determine the effectiveness of the treatments (Table 1). For two minutes, NaOCl was used to sterilize cotton seeds from the Giza 90 and Giza 94 kinds before they were washed four times with sterile distilled water. The sterilized seeds were then immersed for 12 h under static conditions in AgNP suspensions at concentrations of 100 and 200 µg/mL. Moncut (2 g/kg seeds) and Maxim XL (2 mL/L) were added as a coating film for cotton seeds of Giza 90 and Giza 94. The infected soil was placed into 15 cm pots with 10 seeds per container. Only in the control condition, the sterilized sorghum grains were thoroughly combined with soil. Infected controls received 1 g/kg of *R. solani* soil (RS9) and 50 g/kg of *F. fujikuroi* (FF10) and *M. phaseolina* (MP4) with no treatment. The pots were randomly sprinkled on the greenhouse bench and repeated 3 times to obtain accurate data. There were 3 duplicates (pots) for each process. Plant height, dry weight, and survival rate of plants were calculated 45 days after sowing [19,29].

Table 1. Fungicides and trichogenic AgNPs were used in the greenhouse to reduce damping-off of cotton seedlings.

Treatment	Rate of Treatments
1—autoclaved soil	*R. solani* (1 g), both of *F. fujikuroi* and *M. phaseolina* (50 g) for sterilized sorghum/kg soil
2—infested soil	*R. solani* (1 g), both of *F. fujikuroi* and *M. phaseolina* (50 g) for sterilized sorghum/kg soil
3—Moncut	2 g/kg seeds
4—Maxim XL	2 mL/L
5—AgNPs	100 µg mL^{-1}
6—AgNPs	200 µg mL^{-1}

2.6. Statistical Analysis

The experimental design of laboratory and greenhouse experiments was a randomized complete block in a factorial arrangement with three replicates (blocks). The least significant difference (LSD) test was used to examine the mean differences at $p \leq 0.05$. The MSTAT-C software was used to carry out the analysis of variance (ANOVA) of the data.

3. Results

3.1. Characterization of Synthesized Silver Nanoparticles

3.1.1. UV-Visible Spectral Analysis

A visual examination of the culture flasks revealed a color shift after the addition of $AgNO_3$ to the mycelia free cell filtrate. After 5 min, the color of the $AgNO_3$-containing mycelial free cell filtrate altered from colorless to light brown, followed by the emergence of dark brown (Figure 1).

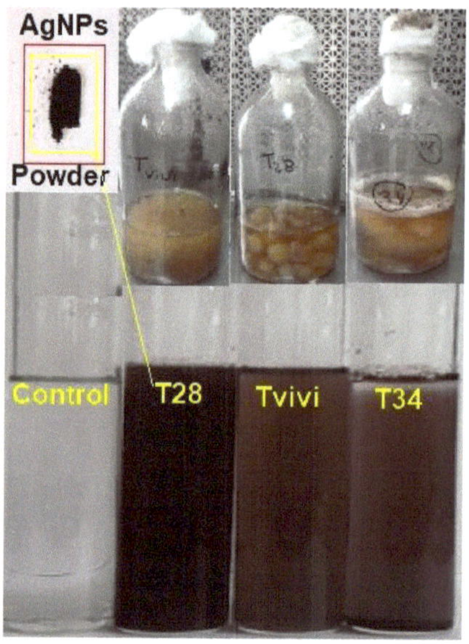

Figure 1. *Trichoderma* mycelium was produced within the liquid media of three isolates, namely, T28, Tvivi, and T34, in a 1 mM silver nitrate aqueous solution. The amount of brown hue in the sample shows the presence of silver nanoparticles (AgNPs).

The UV-Vis spectra of the mycelia-free cell filtrate incubated with silver nitrate salt of three *Trichoderma* isolates (Tvivi, T34, and T28) was obtained (Figure 2). The UV-Vis spectra of AgNPs derived from Tvivi, T34, and T28 revealed absorbance maxima of 500, 500, and 450 nm, respectively. AgNPs produced by Tvivi's strain were chosen for further investigation.

3.1.2. Dynamic Light Scattering (DLS) and Zeta Potential Analysis

The particle size distribution of AgNPs solution as determined by DLS (Figure 3A) indicated an average particle size of 52.34 nm, PdI: 0.359, and a zeta potential of −25.1 mV (Figure 3B).

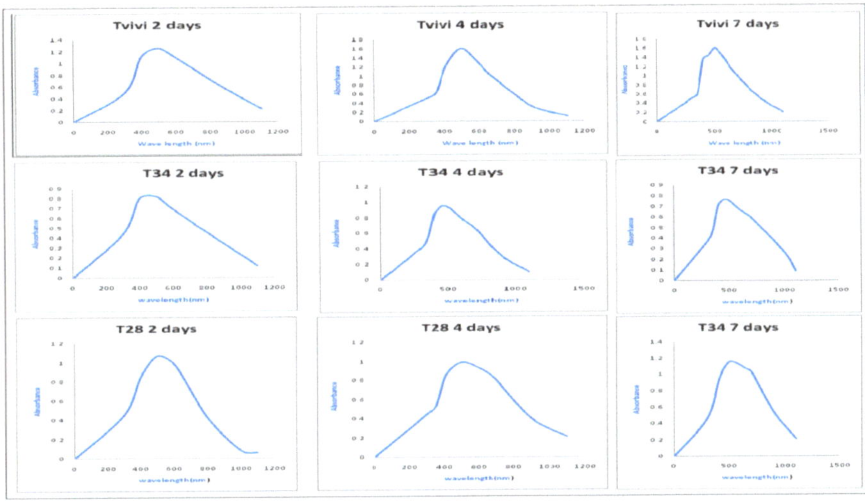

Figure 2. The UV-Vis spectrum of AgNPs generated by Tvivi, T34, and T28 was explored after 2, 4, and 7 days of synthesis.

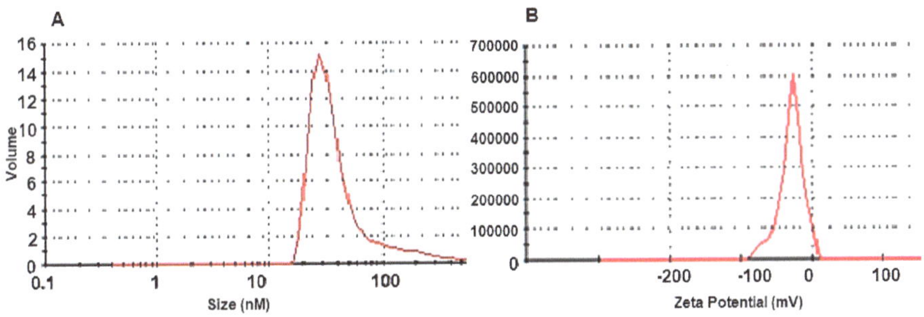

Figure 3. (**A**). Nanoparticle size distribution of silver nanoparticles (AgNPs) solution synthesized by Tvivi, (**B**) analysis of the zeta potential of synthesized AgNPs.

3.1.3. Transmission Electron Microscope (TEM) Analysis

The TEM images revealed spherical shaped AgNPs (Figure 4) with a particle size range of 6 to 15 nm, and the biogenic AgNPs were observed to be surrounded by a thin layer of organic material.

3.1.4. Scanning Electron Microscope Analysis (SEM)

The SEM micrographs clearly showed nanoparticles with almost spherical shapes. Individual nanoparticles are grouped into clusters (Figure 5).

3.1.5. Energy Dispersive X-ray Spectroscopy (EDX) Analysis

An EDX detector attached to a scanning electron microscope was used to determine the elemental composition of the powdered sample. The EDX investigation showed a strong signal for a silver element at around 3 keV (Figure 6A). The elemental analysis revealed atom (%) information as 68.60% silver, 20.11% oxygen, 9.44% carbon, and 1.86% nitrogen.

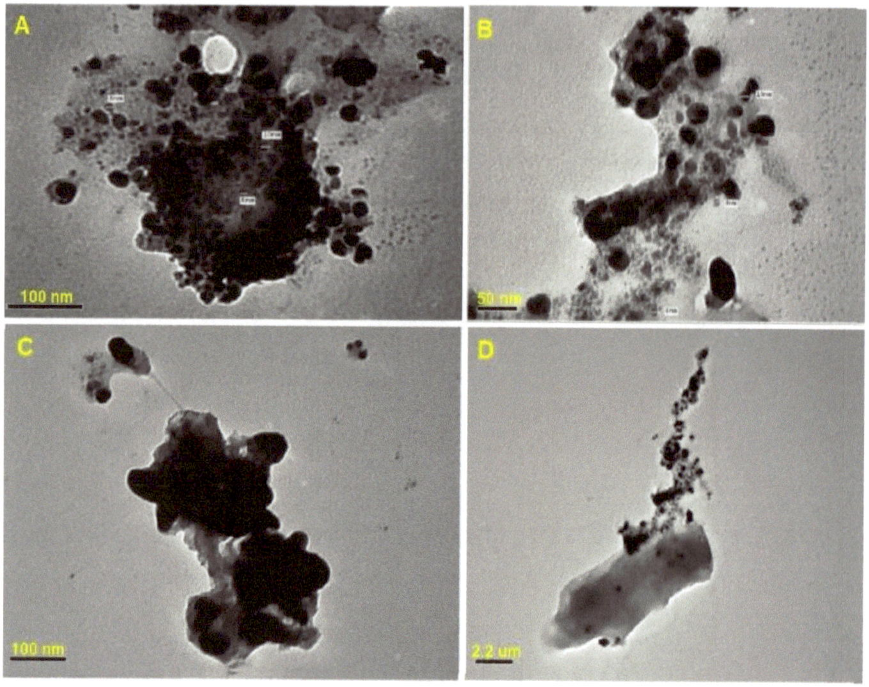

Figure 4. The TEM micrographs of the Tvivi produced silver nanoparticles (AgNPs) show the corresponding particle size distribution and shape. The electron micrographs were acquired at different magnifications **A** (×200 K), **B** (×360 K), **C** (×400 K), and **D** (×100 K).

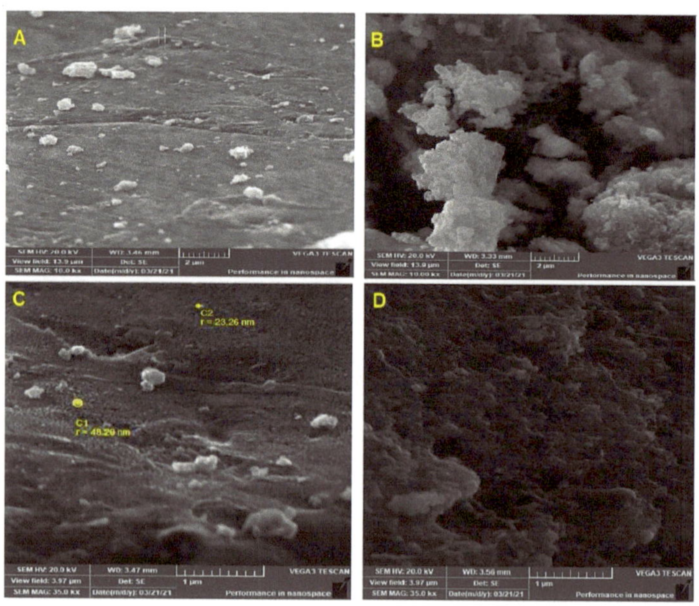

Figure 5. Scanning electron micrographs of the Tvivi synthesized silver nanoparticles (AgNPs) at 10.0 kX (**A**,**B**) and 35.0 kX (**C**,**D**) magnifications.

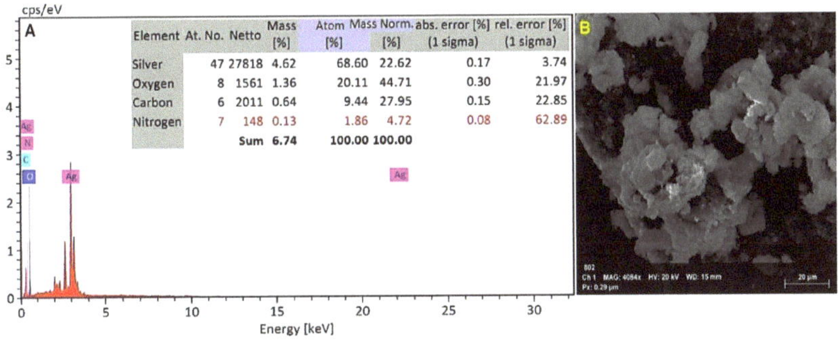

Figure 6. (**A**). The EDX spectrum and elemental analysis of the EDX spectrum of synthesized silver nanoparticles. (**B**). Screened area for the EDX spectrum of synthesized silver nanoparticles.

3.1.6. Antifungal Activity of AgNPs under In Vitro Conditions

The effects of the three concentrations of the AgNPs evaluated, and their interactions, were highly significant causes of variation ($p = 0.00$) for the fungal growth of all the studied fungi, FF10, RS9, and MP4, according to the ANOVA in Table 2.

Table 2. Analysis of variance of the antifungal effects of different concentrations of AgNPs and against the radial growth of *Fusarium fujikuroi* (FF10), *R. solani* (RS9), and *M. phaseolina* (MP4).

Growth Variables and Sources of Variation	D.F	Mean Square	F. Value	$p \geq$ F
Concentration of AgNPs(C)	3	4.606	79.302	0.000
Fungus(F)	2	48.113	828.388	0.000
F × C	6	5.622	96.798	0.000
Error	22	0.058		

Because the concentration of AgNPs × fungus interaction had a significant influence on the radial growth of fungi, to analyze the radial growth averages of treatments within each fungus in Table 3, the LSD was calculated. These studies revealed that the variations in radial growth between concentrations and controls were not the same for each fungus, indicating that fungi reacted differently to concentrations. Thus, all AgNP concentrations were effective in decreasing the linear growth of all fungi compared to the control. In addition, the AgNP concentration of 100 μg/mL was the most effective in decreasing the linear growth, as it decreased linear growth to 4.000, 2.250, and 4.167 cm for FF10, RS9, and MP4, respectively, compared to the control (9 cm); see Figure 7.

Table 3. Antifungal effects of different concentrations of silver NPs on the radial growth of *Fusarium fujikuroi* (FF10), *R. solani* (RS9), and *M. phaseolina* (MP4).

	Radial Growth (cm)			
Treatment	FF10	RS9	MP4	Mean
AgNPs 20 μg/mL	6.000	9.000	6.750	7.250
AgNPs 40 μg/mL	5.333	9.000	6.333	6.889
AgNPs 100 μg/mL	4.000	2.250	4.167	3.472
Control	9.000	9.000	9.000	9.000
Mean	6.083	7.313	6.563	6.653

LSD of fungus × concentration = 0.393.

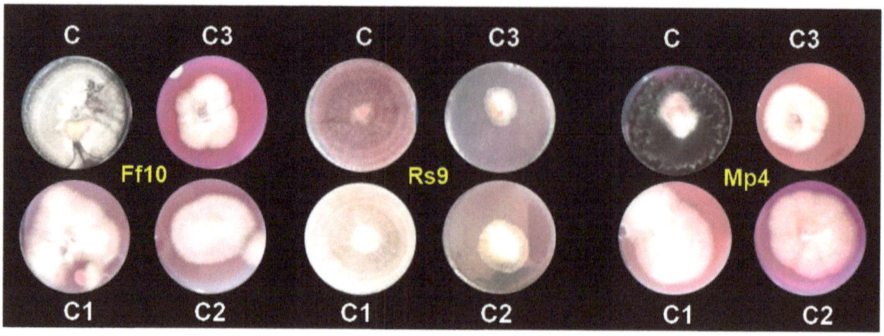

Figure 7. In vitro effect of AgNPs against damping-off pathogens caused by FF10, RS9, and MP4 isolates; (C) control, (C1) 20, (C2) 40 and (C3) 100 μg/mL after 7 days.

3.1.7. Antifungal Activity of AgNPs under Greenhouse Pot Conditions

The number of surviving seedlings, plant height, and dry weight were improved when treatments were applied to infected soil (Table 4). Treatment was a highly significant cause of variation ($p = 0.001$) of all the examined variables, according to the ANOVA shown in Table 3. Only in the case of plant height was the fungus × treatment interaction a highly significant source of variance ($p = 0.00$).

Table 4. ANOVA effects of fungus, treatments, and their interactions on several growth factors of Giza 90 cotton seedlings cultivated in contaminated soil under greenhouse conditions.

Growth Variables and Sources of Variation	D.F	Mean Square	F. Value	$p \geq$ F
Survival				
Fungi (F)	2	62.93	0.28	0.76
Treatments (T)	5	3805.20	16.89	0.00
F × T	10	71.46	0.32	0.97
Error	34	225.32		
Plant height				
Fungi (F)	2	21.65	1.72	0.19
Treatments (T)	5	212.09	16.84	0.00
F × T	10	52.94	4.20	0.00
Error	34	12.60		
Dry weight				
Fungi (F)	2	0.75	3.51	0.04
Treatments (T)	5	3.94	18.32	0.00
F × T	10	0.34	1.59	0.15
Error	34	0.22		

Because there was no fungus treatment interaction on survival, the general mean was used to compare treatment means (Table 5). Treatment with AgNPs (100 μg/mL) was the least effective treatment in controlling the condition, whereas other treatments were similarly beneficial. The difference between the examined fungi's general means was non-significant.

Table 5. The effect of a pathogenic fungus, treatments, and their interactions on the survival percentage of Giza90 cotton seedlings cultivated on infected soil in a greenhouse.

Treatments	Silver/Survival/Giza90			
	FF10 (%)	**RS9 (%)**	**MP4 (%)**	**Mean (%)**
AgNPs (100 µg/mL)	60.000	70.000	76.667	68.889
AgNPs (200 µg/mL)	90.000	86.667	93.333	90.000
Maxim ×L (2 mL)	86.667	86.667	90.000	87.778
Moncut (2 g)	86.667	83.333	83.333	84.444
Infested soil	30.000	16.667	26.667	24.445
Autoclaved soil	90.000	96.667	93.333	93.333
Mean	73.889	73.333	77.222	74.815

Before calculating the analysis of variance, the percentage values were converted into arcsine angles to achieve nearly constant variance. LSD ($p \leq 0.05$) (transformed data) for treatments was 14.01. The LSD ($p \leq 0.05$) for fungus is non-significant.

Because the interaction between cultivar treatment and plant height was significant (Figure 8), the LSD value of interaction was used to compare treatment averages within each tested fungus. In the case of *F. fujikuroi*, AgNPs (200 µg/mL) and Moncut (2 g) demonstrated the greatest efficacy in disease control (FF10). In the instance of *R. solani*, all treatments were equally effective in controlling the disease (Rs10). In the case of *M. phaseolina*, all treatments were successful in managing the disease; however, treatment with Maxim xl (2 mL) was the least effective, whereas other treatments were similarly beneficial (MP4). Due to the lack of a fungus treatment interaction on dry weight (Figure 9), the general mean was utilized to compare treatment means. All treatments were equally successful in bringing the condition under control. The difference in general means of the investigated fungus was not statistically significant.

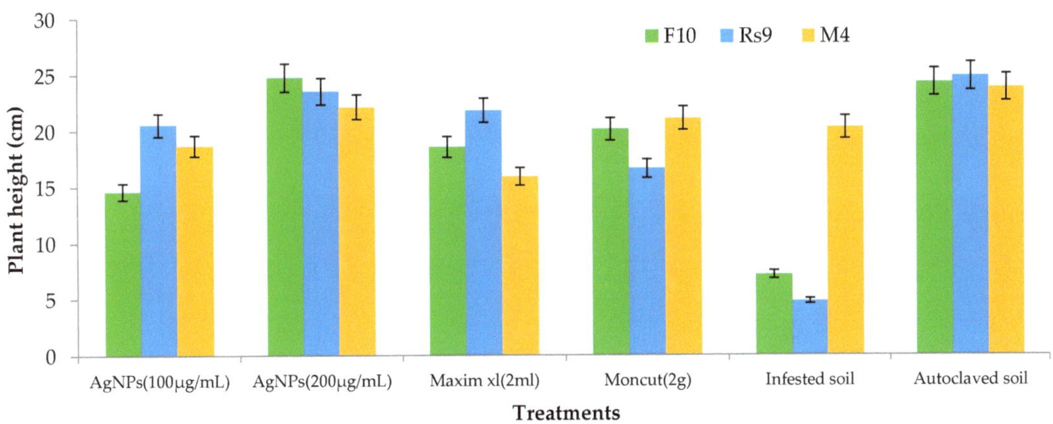

Figure 8. Effects of fungus, treatments, and their interactions on the plant height of Giza90 cotton seedlings cultivated in contaminated soil in a greenhouse. LSD ($p \leq 0.05$) for fungus × treatment = 5.74.

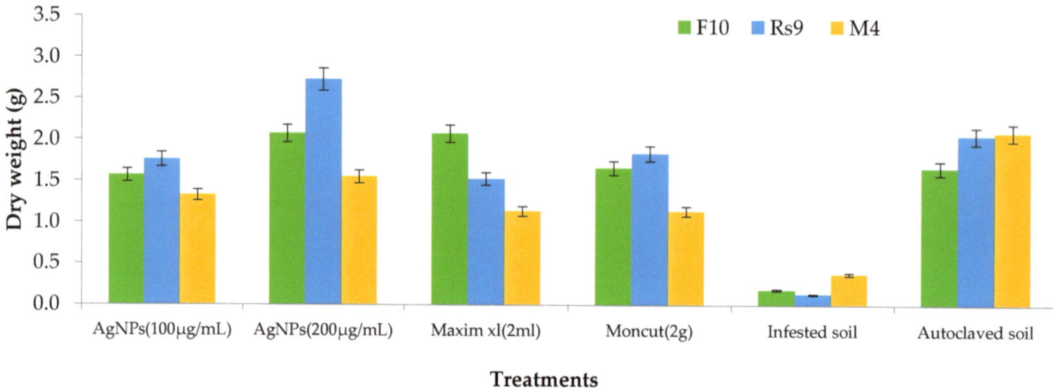

Figure 9. The effect of specific fungi, treatments, and their interactions on the dry weight of Giza 90 cotton seedlings grown in a greenhouse on infected soil. LSD for treatments = 0.43 ($p \leq 0.05$). The LSD ($p = 0.05$) for fungus was non-significant.

ANOVA revealed that treatment was a very significant source of variation ($p = 0.00$) for all examined variables (Table 6). Fungus and the fungus × treatment interaction were not significant contributors to the variance in any of the variables evaluated. Because there was no fungal treatment interaction on survival, the overall mean was used to compare treatment means (Table 7). Although all treatments were successful in managing the condition, treatment with AgNPs (100 μg/mL) was the least effective, whereas the other treatments were all equally effective. The difference between the examined fungi's overall means was not significant.

Table 6. ANOVA of the effects of specific fungi, treatments, and their interactions on some growth factors of Giza 94 cotton seedlings grown in a greenhouse on infected soil.

Growth Variables and Sources of Variation	D.F	Mean Square	F. Value	$p \geq$ F
Survival				
Fungi (F)	2	93.24	0.37	0.69
Treatments (T)	5	4579.66	18.25	0.00
F × T	10	408.13	1.63	0.14
Error	34	250.94		
Plant height				
Fungi (F)	2	7.89	0.24	0.79
Treatments (T)	5	272.93	8.17	0.00
F × T	10	40.05	1.20	0.33
Error	34	33.39		
Dry weight				
Fungi (F)	2	0.19	0.56	0.58
Treatments (T)	5	5.38	15.71	0.00
F × T	10	0.35	1.00	0.46
Error	34	0.34		

Table 7. The effects of various fungi, treatments, and their interactions on the survival percentage of Giza94 cotton seedlings grown in a greenhouse on infected soil.

Silver/Survival/Giza94								
Treatment	FF10		RS9		MP4		Mean	
	%	Transformed [a]	%	Transformed [a]	%	Transformed [a]	%	Transformed [a]
AgNPs (100 µg/mL)	40.000	38.853	16.667	15.000	80.000	63.930	45.556	39.261
AgNPs (200 µg/mL)	90.000	75.000	83.333	70.763	93.333	77.707	88.889	74.490
Maxim XL (2 mL)	80.000	67.860	76.667	65.840	76.667	61.910	77.778	65.203
Moncut (2 g)	83.333	70.077	93.333	77.707	90.000	75.000	88.889	74.261
Infested soil	16.667	19.223	30.000	28.077	6.667	12.293	17.778	19.864
Autoclaved soil	86.667	72.783	86.667	72.293	83.333	66.147	85.556	70.408
Mean	66.111	57.299	64.444	54.947	71.667	59.498	67.407	57.248

[a] Before calculating the analysis of variance, the percentage values were converted into arcsine angles to achieve nearly constant variance. LSD ($p \leq 0.05$) (transformed data) for treatments was 14.79. The LSD ($p \leq 0.05$) for fungus was non-significant.

Because there was no interaction between fungus and treatment on plant height and dry weight, the general mean was used to evaluate the means of the different treatments (Figures 10 and 11). All treatments were effective in controlling the disease for plant height. However, the therapy with AgNPs (100 µg/mL) was the least effective, whereas the other treatments were all equally effective. In terms of dry weight, all treatments were efficient in managing the disease, apart from AgNPs (100 µg/mL) (Figure 12), which was ineffective, and AgNPs (200 µg/mL), which was the most effective treatment in terms of dry weight (2.208 g). There was no statistically significant difference in the general means of the fungi that were tested.

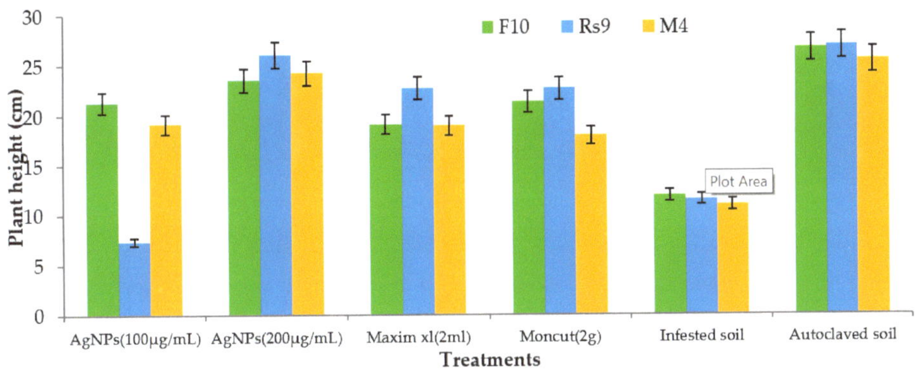

Figure 10. The effect of fungus, treatments, and interactions on the plant height of Giza 94 cotton seedlings cultivated on infected soil under greenhouse conditions. Treatment LSD ($p \leq 0.05$) = 5.39. The LSD ($p \leq 0.05$) for fungus was non-significant.

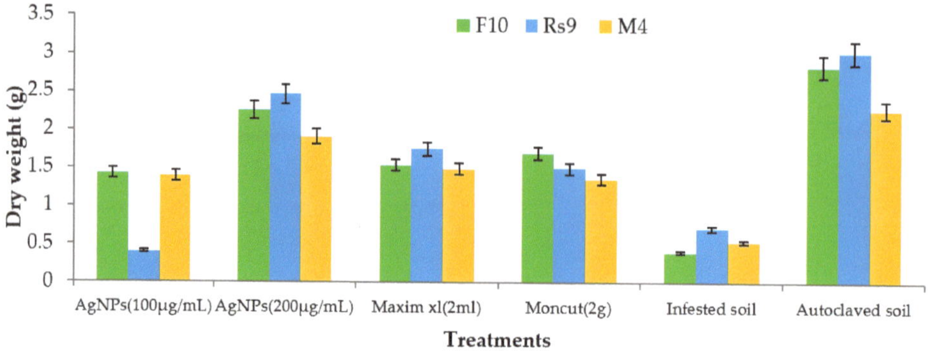

Figure 11. The effect of fungus, treatments, and interactions on the dry weight of Giza94 cotton seedlings cultivated on infected soil under greenhouse conditions is yet to be determined. LSD ($p \leq 0.05$) for treatments = 0.55. The LSD ($p \leq 0.05$) for fungus was non-significant.

Figure 12. Uncoated cotton seeds sown in sterilized soil infested with three fungal pathogens, namely, *F. fujikuroi*, *R. solani*, and *M. phaseolina* as a negative control, uncoated cotton seeds sowed in sterilized soil as a positive control, treated with two fungicides (Maxim XL and Moncut) sowed in infested soil and coated with AgNPs (100, 200 µg/mL) in infested soil. The photo was obtained after 45 days of typical growth in a greenhouse.

4. Discussion

Trichoderma is the most extensively used biocontrol agent, which exhibits characteristic features such as pathogen antagonism, stimulation of systemic resistance in the host, nutrition competition, and overall plant growth enhancement [21]. Biogenic AgNPs from *T. harzianum* extracellular filtrate have an advantage over physical and chemical processes that need high pressure, energy, and chemical precursors [19].

The ability of three *Trichoderma* species (Tvivi, T34, and T28) to generate stable AgNPs was examined. Filtrates from each fungal isolate were grown in the dark at 28 °C for 72 h with $AgNO_3$ solution. After the incubation time, the color changed from pale yellow to dark brown. The emergence of the brown color indicated that AgNPs were forming in the media [28]. The UV-visible spectrum of biosynthesized silver nanoparticles utilizing (Tvivi, T34, and T28) displayed a peak at 500, 500, and 450 nm, respectively, on all tested days, matching silver nanoparticle plasmon absorbance. The findings were consistent with previously published reports on the UV absorbance peak of generated AgNPs [29–31].

The produced particles were monodispersive. The particle size estimated by TEM was smaller than the particle size reported by DLS measurements. This is because the hydrated capping agents, proteins, and solvation effects considerably increase the particle size obtained by DLS. The findings were in line with the expectations of some previous research reports [21,28,32,33].

In the present work, the zeta potential was found to be −25.1 mV. The negative value of the zeta potential indicates particle repulsion, which increases particle stability. The findings agreed with those of Ebrahimzadeh et al. [33], who obtained biosynthesized silver nanoparticles with a zeta potential of –19.6 mV. Moreover, Ahluwalia et al. [21] obtained a zeta potential of –17.19 mV.

The morphology of silver nanoparticles produced was visualized through TEM analysis. Individual and aggregated monodispersed spherical silver nanoparticles with sizes ranging from 6 to 15 nm were observed. The NPs were coated with a thin layer of organic material within the aggregates, indicating that the NPs were stabilized by a capping agent [21,34]. Elamawi et al. [32] found that silver nanoparticles biosynthesized by *Trichoderma longibrachiatum* had monodispersed approximately spherical forms with varying diameters and a particle size distribution ranging from 1 to 25 nm. According to Nayak et al. [35], AgNPs biosynthesized by *Dillenia indica* extract were spherical, with the majority of them clustered and a few dispersed with varying sizes, as was described under TEM.

SEM images showed AgNPs having a spherical form. Clusters of AgNPs were seen, which may be the result of nanoparticle aggregation during sample preparation. These factors may have had a role in the variance in particle size. The results corroborated those of Ahluwalia et al. [21], who reported that silver nanoparticles biosynthesized by *T. harzianum* exhibited spherical AgNPs with a range of diameters within the standard error of the mean. According to Tomah et al. [36], SEM micrographs revealed that the exterior surfaces of AgNPs generated from cell-free *T. virens* HZA14 filtrate were spherical with a size range of 5 to 50 nm. Hirpara et al. [24] demonstrated that silver nanoparticles biosynthesized by *T. interfusant* had spherical and pseudospherical forms with agglomeration traces and average size distribution of 59.66 ± 4.18 nm, which was under the standard error of the mean.

The existence of AgNPs produced from the filtrate was verified by EDX analysis. This study revealed a high concentration of AgNPs (68.60 percent), followed by oxygen, carbon, and nitrogen, indicating that extracellular organic moieties from mycelial free cell filtrate were adsorbed on the nanoparticles' surface [21]. Due to surface plasmon resonance, metallic AgNPs typically exhibit an optical absorption peak at 3 keV. The results corroborated those of Konappa et al. [20], who found the existence of AgNPs (58.75 percent) at 3 keV, followed by carbon, oxygen, and chlorine. According to Tomah et al. [36], EDS examination of AgNPs found that pure silver (28.85 percent) at 3 KeV was the second most abundant component element after oxygen, nitrogen, carbon, and sulfur. Additionally, Manikandaselvi et al. [37] revealed that significant signals were observed from the Ag atoms in the nanoparticles, in addition to signals from Al, carbon, and oxygen. These signals were most likely caused by X-ray emission from the fungus's existing proteins.

In vitro, the antifungal activity of the produced AgNPs against *R. solani* (RS9), *F. fujikuroi* (FF10), and *M. phaseolina* (MP4) was determined. It was discovered that as the quantity of AgNPs increased, the diameter of the mycelium decreased. All concentrations of AgNPs were effective at reducing radial growth of all fungi when compared to the control, but AgNPs at 100 μg/mL was the most effective at reducing linear growth, as it reduced linear growth by 55.6, 75%, and 53.7% for the three pathogenic fungi FF10, RS9, and MP4, respectively.

Regardless of fungus, all treatments were successful in suppressing disease and improving life on both cotton cultivars. However, the treatment with AgNPs (100 µg/mL) was the least effective, whereas the other treatments were similarly beneficial. Commercial chemical fungicides (Maxim XL and Moncut) had the same impact as AgNPs (200 µg/mL). However, AgNPs were utilized at relatively low doses compared to other commercial chemical fungicides. Furthermore, it was discovered that using AgNPs at 200 µg/mL was more successful than using 100 µg/mL in suppressing disease and boosting survival, independent of fungus.

Several papers [38–42] emphasize the potential approach of AgNPs for the synthesis of new antibacterial agents. Our findings, which agree with those of other researchers, demonstrate the feasibility of achieving a better inhibition rate against a wide spectrum of harmful bacteria, even at lower AgNP concentrations [43–45]. This may be due to AgNPs' great propensity for attacking and adhering to fungal mycelium, destroying membrane integrity, and thereby suppressing the plant pathogen [46].

M. phaseolina fungal growth was suppressed by synthesized AgNPs with a zone of inhibition of 13 mm at 1000 µg/mL according to Vijayabharathi et al. [47]. AgNPs may disrupt the lipid bilayer, in addition to membrane permeability and electrical potential, resulting in ion and other chemical leakage and cell death [48]. The antifungal activity of biosynthesized AgNPs was tested against eight *Fusarium* species that cause cotton seedling damping-off. On Czapek Dox agar and potato dextrose agar plates, in vitro treatments with varying amounts of AgNPs were produced. From 25 to 200 ppm of AgNP interaction, fungal growth was significantly slowed [49]. Silver nanoparticles were tested for their ability to inhibit *R. solani*-induced cucumber damping-off. The fresh and dry weights of roots and shoots were considerably greater in treatments containing 30, 50, and 70 g/L of nanoparticles than in control plants. It was discovered that nanosilver harmed *R. solani* when compared to the control.

The capacity of AgNPs to permeate the cytoplasmic membrane and cell wall, RNA and DNA damage, and cell respiration all contribute to their lethal effects [50]. The NADH co-enzyme, in addition to NADH-dependent enzymes such as nitrate reductase, are found in *Trichoderma* genus strains and are vital in the creation of both nanoparticles and cappings that provide higher stability [51,52]. *T. harzianum* generates enzymes and metabolites that are involved in breaking strong ionic connections between silver and nitrate ions for its own life, perhaps through the action of hydrolytic/nitrate reductase enzymes. The enzymatic activity of extracellular fungal secondary metabolites converts hazardous Ag^+ ions into innocuous biosilver NPs. The existence of suitable functional groups in *T. harzianum*'s extracellular filtrate, which is more effective than that of other fungi and non-toxic to humans [53], provides the basis for silver bioreduction [22].

Secondary metabolites are secreted by *T. harzianum* and act as capping and reducing agents, aiding consistency and contributing to biological activity. Puerarin, genistein, isotalatizidine, and ginsenoside were identified using LC-MS/MS [20]. AgNPs can interact with the cell wall and membrane, resulting in (1) changes in permeability; (2) phosphate management disruption; (3) plasma membrane disintegration; (4) proton motive force failure; (5) ATP production suppression; (6) an impact on amino acids and enzymes that can result in bonding with amino acids and inhibition of enzyme activity due to attachment to the active core of the enzyme; (7) energy-use impediments that can lead to electron mobility in the respiratory chain and cytochrome inhibition; (8) an effect on DNA and RNA that can cause hydrogen bonding to break, nitrogen base synthesis to be inhibited, DNA and RNA synthesis to be disrupted, ribosomes to be denatured, and protein formation to be inhibited; and (9) production of reactive oxygen species (ROS) [54]. The efficiency of silver nanoparticles is determined by factors such as particle size, shape, exposure period, compound type, and target. However, the size of silver is very important because the smaller diameter provides a larger surface area and hence superior antibacterial efficacy. The amount of silver in the solution must be high enough to inhibit the growth of bacteria [55]. The AgNPs get adsorbed on the surface of the Gram-negative bacterial cell wall resulting in breakdown of the cell wall as the Ag^+ ions released from the AgNPs exhibit a strong antibacterial action [56,57].

5. Conclusions

In this study, the biocontrol agent *T. harzianum* was utilized as a reducing and stabilizing agent, resulting in the successful production of silver nanoparticles using an eco-friendly, affordable, and fast approach. UV-visible spectroscopy, DLS, zeta potential, TEM, SEM, and EDX were used to confirm the synthesis and NP structure, and to examine features such as size distribution, zeta potential, and morphology. Furthermore, in vitro and greenhouse antifungal activity against soil-borne pathogens *R. solani* (RS9), *F. fujikuroi* (FF10), and *M. phaseolina* (MP4) was demonstrated. AgNPs significantly inhibited hyphal development in the three fungal pathogens. *T. harzianum* isolates have been demonstrated to be capable of synthesizing a wide range of proteins and enzymes without the need for chemical reducers and stabilizers. Biosynthesized AgNPs have shown significant potential in protecting cotton plants from the fungal invasion caused by damping-off. In order to have more control over the size and polydispersity of AgNPs, more research is needed to discover the essential bioreducing and biotemplating molecules present in the hyphal extract. There is a need for more research on AgNPs in agroecosystems, and for more threat assessment research.

Author Contributions: Conceptualization: K.A.A.-E.; S.A.O. and S.A.Z.; methodology, S.A.Z.; investigation, S.A.Z. and A.K.; resources, F.M.A.; writing—original draft preparation, writing—review and editing, A.A.A., S.A.O. and K.A.A.-E.; visualization, S.A.Z. and A.A.A.; supervision, S.A.O. and K.A.A.-E.; funding acquisition, K.A.A.-E., M.M.H. and F.M.A. All authors have read and agreed to the published version of the manuscript.

Funding: The present study was supported by Taif University's Researchers Supporting Project (TURSP—2020/59), Taif, Saudi Arabia.

Institutional Review Board Statement: Not applicable.

Informed Consent Statement: Not applicable.

Data Availability Statement: Not applicable.

Acknowledgments: We owe thanks to Aly. A. Aly (PPathRI, ARC) for the relevant suggestions in improving the present work. The authors extend their appreciation to Taif University for supporting the current work by Taif University Researchers Supporting Project number (TURSP - 2020/59), Taif University, Taif, Saudi Arabia.

Conflicts of Interest: The authors declare no conflict of interest.

References

1. Watkins, G.M. *Compendium of Cotton Diseases*; The American Phytopathological Society: St. Paul, MN, USA, 1981; p. 87.
2. Moustafa-Mahmoud, S.M.; Ragab, M.M.; Sumner, D.R.; Ragab, M.M. Biological Control of *Rhizoctonia solani* (AG-4) in Cotton Seedlings. *Egypt. J. Agric. Res.* **1995**, *73*, 561–573.
3. El-Akkad-Salwa, A.F. Studies on Anastomosis Groups of *Rhizoctonia solani*. Ph.D. Thesis, Cairo University, Cairo, Egypt, 1997.
4. Asran-Amal, A.; Abd-Elsalam, K.A.; Omar, M.R.; Aly, A.A. Antagonistic Potential of *Trichoderma* spp. against *Rhizoctonia solani* and Use of M13 Microsatellite-Primed PCR to Evaluate the Antagonist Genetic Variation. *J. Plant Dis. Prot.* **2005**, *112*, 550–561. [CrossRef]
5. Abd-Elsalam, K.A.; Omar, M.R., Aly, A.A. First Report of *Rhizoctonia solani* AG-7 on Cotton in Egypt. *J. Phytopathol.* **2010**, *158*, 307–309. [CrossRef]
6. Roy, K.W. Epidemiological and Mycofloral Relationships in Cotton Seedling Disease in Mississippi. *Phytopathology* **1982**, *72*, 868–872. [CrossRef]
7. Colyer, P.D. Frequency and Pathogenicity of *Fusarium* spp. Associated with Seedling Diseases of Cotton in Louisiana. *Plant Dis.* **1988**, *72*, 400–402. [CrossRef]
8. Abd-Elsalam, K.A.; Asran-Amal, A.M.; Omar, M.R.; Aly, A.A.; Abd-Elsalam, K.A.; Asran-Amal, A.M.; Omar, M.R.; Aly, A.A. Frequency and Diversity of *Fusarium* spp. Colonizing Roots of Egyptian Cottons: Häufigkeit Und Viefalt von *Fusarium*-Pilfen an Wurzeln Der Ägyptischen Baumwolle. *Arch. Phytopathol. Plant Prot.* **2006**, *39*, 165–177. [CrossRef]

9. Dhingra, O.D.; Sinclair, J.B. *Biology and Pathology of Macrophomina phaseolina*; Imprensia Universidade Federal de Vicosa: Viçosa, MG, Brazil, 1978; 166p.
10. Abd-Elsalam, K.A.; Periakaruppan, R.; Rajeshkumar, S. *Agri-Waste and Microbes for Production of Sustainable Nanomaterials*; Elsevier: Amsterdam, The Netherlands, 2021; 772p. [CrossRef]
11. Singh, P.; Kim, Y.-J.; Zhang, D.; Yang, D.-C. Biological Synthesis of Nanoparticles from Plants and Microorganisms. *Trends Biotechnol.* **2016**, *34*, 588–599. [CrossRef] [PubMed]
12. Balakumaran, M.D.; Ramachandran, R.; Balashanmugam, P.; Mukeshkumar, D.J.; Kalaichelvan, P.T. Mycosynthesis of Silver and Gold Nanoparticles: Optimization, Characterization and Antimicrobial Activity against Human Pathogens. *Microbiol. Res.* **2016**, *182*, 8–20. [CrossRef] [PubMed]
13. Taha, Z.K.; Hawar, S.N.; Sulaiman, G.M. Extracellular Biosynthesis of Silver Nanoparticles from *Penicillium italicum* and Its Antioxidant, Antimicrobial and Cytotoxicity Activities. *Biotechnol. Lett.* **2019**, *41*, 899–914. [CrossRef] [PubMed]
14. Kubicek, C.P.; Mach, R.L.; Peterbauer, C.K.; Lorito, M. Trichoderma: From Genes to Biocontrol. *J. Plant Pathol.* **2011**, *83*, 11–23.
15. Almeida, F.B.d.R.; Cerqueira, F.M.; Silva, R.d.N.; Ulhoa, C.J.; Lima, A.L. Mycoparasitism Studies of *Trichoderma harzianum* Strains against *Rhizoctonia solani*: Evaluation of Coiling and Hydrolytic Enzyme Production. *Biotechnol. Lett.* **2007**, *29*, 1189–1193. [CrossRef] [PubMed]
16. Qualhato, T.F.; Lopes, F.A.C.; Steindorff, A.S.; Brandão, R.S.; Jesuino, R.S.A.; Ulhoa, C.J. Mycoparasitism Studies of *Trichoderma* species against Three Phytopathogenic Fungi: Evaluation of Antagonism and Hydrolytic Enzyme Production. *Biotechnol. Lett.* **2013**, *35*, 1461–1468. [CrossRef] [PubMed]
17. Fraceto, L.F.; Maruyama, C.R.; Guilger, M.; Mishra, S.; Keswani, C.; Singh, H.B.; de Lima, R. *Trichoderma harzianum*-Based Novel Formulations: Potential Applications for Management of Next-Gen Agricultural Challenges: Applications of *Trichoderma harzianum*-Based Novel Formulations. *J. Chem. Technol. Biotechnol.* **2018**, *93*, 2056–2063. [CrossRef]
18. Zaki, S.A.; Ouf, S.A.; Albarakaty, F.M.; Habeb, M.M.; Aly, A.A.; Abd-Elsalam, K.A. *Trichoderma harzianum*-Mediated ZnO Nanoparticles: A Green Tool for Controlling Soil-Borne Pathogens in Cotton. *J. Fungi* **2021**, *7*, 952. [CrossRef] [PubMed]
19. Guilger-Casagrande, M.; Germano-Costa, T.; Pasquoto-Stigliani, T.; Fraceto, L.F.; de Lima, R. Biosynthesis of Silver Nanoparticles Employing *Trichoderma harzianum* with Enzymatic Stimulation for the Control of *Sclerotinia sclerotiorum*. *Sci. Rep.* **2019**, *9*, 14351. [CrossRef] [PubMed]
20. Konappa, N.; Udayashankar, A.C.; Dhamodaran, N.; Krishnamurthy, S.; Jagannath, S.; Uzma, F.; Pradeep, C.K.; De Britto, S.; Chowdappa, S.; Jogaiah, S. Ameliorated Antibacterial and Antioxidant Properties by *Trichoderma harzianum* Mediated Green Synthesis of Silver Nanoparticles. *Biomolecules* **2021**, *11*, 535. [CrossRef]
21. Ahluwalia, V.; Kumar, J.; Sisodia, R.; Shakil, N.A.; Walia, S. Green Synthesis of Silver Nanoparticles by *Trichoderma harzianum* and Their Bio-Efficacy Evaluation against *Staphylococcus aureus* and *Klebsiella pneumonia*. *Ind. Crop. Prod.* **2014**, *55*, 202–206. [CrossRef]
22. Sundaravadivelan, C.; Padmanabhan, M.N. Effect of Mycosynthesized Silver Nanoparticles from Filtrate of *Trichoderma Harzianum* against *Larvae* and Pupa of Dengue Vector *Aedes egypti* L. *Environ. Sci. Pollut. Res. Int.* **2014**, *21*, 4624–4633. [CrossRef]
23. Meher, J.; Rajput, R.S.; Bajpai, R.; Teli, B.; Sarma, B.K. Trichoderma: A Globally Dominant Commercial Biofungicide. In *Soil Biology*; Springer: Cham, Switzerland, 2020; pp. 195–208. [CrossRef]
24. Hirpara, D.G.; Gajera, H.P. Green Synthesis and Antifungal Mechanism of Silver Nanoparticles Derived from Chitin-induced Exometabolites of *Trichoderma interfusant*. *Appl. Organomet. Chem.* **2020**, *34*, e5407. [CrossRef]
25. Elamawi, R.; El-Shafey, R.A. Inhibition Effects of Silver Nanoparticles against Rice Blast Disease Caused by *Magnaporthe grisea*. *Egypt. J. Agric. Res.* **2013**, *91*, 1271–1283. [CrossRef]
26. Zhou, C.; Guo, R.; Ji, S.; Fan, H.; Wang, J.; Wang, Y.; Liu, Z. Isolation of Trichoderma from Forestry Model Base and the Antifungal Properties of Isolate TpsT17 toward *Fusarium oxysporum*. *Microbiol. Res.* **2020**, *231*, 126371. [CrossRef]
27. Dou, K.; Gao, J.; Zhang, C.; Yang, H.; Jiang, X.; Li, J.; Li, Y.; Wang, W.; Xian, H.; Li, S.; et al. Trichoderma biodiversity in major ecological systems of China. *J. Microbiol.* **2019**, *57*, 668–675. [CrossRef] [PubMed]
28. Elamawi, R.M.; Al-Harbi, R.E.; Hendi, A.A. Biosynthesis and Characterization of Silver Nanoparticles Using *Trichoderma longibrachiatum* and Their Effect on Phytopathogenic Fungi. *Egypt J. Biol. Pest Control* **2018**, *28*, 28. [CrossRef]
29. Ahmadi, S.; Fazilati, M.; Mousavi, S.M.; Nazem, H. Anti-Bacterial/Fungal and Anti-Cancer Performance of Green Synthesized Ag Nanoparticles Using Summer Savory Extract. *J. Exp. Nanosci.* **2020**, *15*, 363–380. [CrossRef]
30. Göl, F.; Aygün, A.; Seyrankaya, A.; Gür, T.; Yenikaya, C.; Şen, F. Green Synthesis and Characterization of *Camellia sinensis* Mediated Silver Nanoparticles for Antibacterial Ceramic Applications. *Mater. Chem. Phys.* **2020**, *250*, 123037. [CrossRef]
31. Elgorban, A.M.; Al-Rahmah, A.N.; Sayed, S.R.; Hirad, A.; Mostafa, A.A.F.; Bahkali, A.H. Antimicrobial Activity and Green Synthesis of Silver Nanoparticles Using *Trichoderma viride*. *Biotechnol. Biotechnol. Equip.* **2016**, *30*, 299–304. [CrossRef]
32. Mukherjee, P.; Roy, M.; Mandal, B.P.; Dey, G.K.; Mukherjee, P.K.; Ghatak, J.; Tyagi, A.K.; Kale, S.P. Green Synthesis of Highly Stabilized Nanocrystalline Silver Particles by a Non-Pathogenic and Agriculturally Important Fungus *T. Asperellum Nanotechnol.* **2008**, *19*, 075103. [CrossRef]
33. Ebrahimzadeh, M.A.; Naghizadeh, A.; Amiri, O.; Shirzadi-Ahodashti, M.; Mortazavi-Derazkola, S. Green and Facile Synthesis of Ag Nanoparticles Using *Crataegus pentagyna* Fruit Extract (CP-AgNPs) for Organic Pollution Dyes Degradation and Antibacterial Application. *Bioorg. Chem.* **2020**, *94*, 103425. [CrossRef]
34. Jagtap, U.B.; Bapat, V.A. Green Synthesis of Silver Nanoparticles Using Artocar-Pus *Heterophyllus Lam.* Seed Extract and Its Antibacterial Activity. *Ind. Crop. Prod.* **2013**, *46*, 132–137. [CrossRef]

35. Nayak, S.; Bhat, M.P.; Udayashankar, A.C.; Lakshmeesha, T.R.; Geetha, N.; Jogaiah, S. Biosynthesis and Characterization of *Dillenia indica*-mediated Silver Nanoparticles and Their Biological Activity. *Appl. Organomet. Chem.* **2020**, *34*, 5567. [CrossRef]
36. Tomah, A.A.; Alamer, I.S.A.; Li, B.; Zhang, J.-Z. Mycosynthesis of Silver Nanoparticles Using Screened Trichoderma Isolates and Their Antifungal Activity against *Sclerotinia sclerotiorum*. *Nanomaterials* **2020**, *10*, 1955. [CrossRef] [PubMed]
37. Manikandaselvi, S.; Sathya, V.; Vadivel, V.; Sampath, N.; Brindha, P. Evaluation of Bio Control Potential of AgNPs Synthesized from *Trichoderma viride*. *Adv. Nat. Sci. Nanosci. Nanotechnol.* **2020**, *11*, 035004. [CrossRef]
38. Kalia, A.; Manchanda, P.; Bhardwaj, S.; Singh, G. Biosynthesized silver nanoparticles from aqueous extracts of sweet lime fruit and callus tissues possess variable antioxidant and antimicrobial potentials. *Inorganic Nanometal Chem.* **2020**, *50*, 1053–1062. [CrossRef]
39. Le Ouay, B.; Stellacci, F. Antibacterial Activity of Silver Nanoparticles: A Surface Science Insight. *Nano Today* **2015**, *10*, 339–354. [CrossRef]
40. Dobrucka, R.; Długaszewska, J. Biosynthesis and Antibacterial Activity of ZnO Nanoparticles Using *Trifolium pretense* Flower Extract. *Saudi J. Biol. Sci.* **2016**, *23*, 517–523. [CrossRef]
41. Salomoni, R.; Léo, P.; Montemor, A.F.; Rinaldi, B.G.; Rodrigues, M. Antibacterial Effect of Silver Nanoparticles in *Pseudomonas aeruginosa*. *Nanotechnol. Sci. Appl.* **2017**, *10*, 115–121. [CrossRef]
42. Youssef, M.S.; Elamawi, R.M. Evaluation of Phytotoxicity, Cytotoxicity, and Genotoxicity of ZnO Nanoparticles in *Vicia faba*. *Environ. Sci. Pollut. Res. Int.* **2020**, *27*, 18972–18984. [CrossRef]
43. Auffan, M.; Rose, J.; Bottero, J.-Y.; Lowry, G.V.; Jolivet, J.-P.; Wiesner, M.R. Towards a Definition of Inorganic Nanoparticles from an Environmental, Health and Safety Perspective. *Nat. Nanotechnol.* **2009**, *4*, 634–641. [CrossRef]
44. Ahmed, S.; Saifullah, A.M.; Swami, B.L.; Ikram, S. Green Synthesis of Silver Nanoparticles Using *Azadirachta indica* aqueous Leaf Extract. *J. Radiat. Res. Appl. Sci.* **2016**, *9*, 1–7. [CrossRef]
45. Qasim, M.; Udomluck, N.; Chang, J.; Park, H.; Kim, K. Antimicrobial Activity of Silver Nanoparticles Encapsulated in Poly-N-Isopropylacrylamide-Based Polymeric Nanoparticles. *Int. J. Nanomed.* **2018**, *13*, 235–249. [CrossRef]
46. Silver, S. Bacterial Silver Resistance: Molecular Biology and Uses and Misuses of Silver Compounds. *FEMS Microbiol. Rev.* **2003**, *27*, 341–353. [CrossRef]
47. Vijayabharathi, R.; Sathya, A.; Gopalakrishnan, S. Extracellular Biosynthesis of Silver Nanoparticles Using *Streptomyces griseoplanus* SAI-25 and Its Antifungal Activity against *Macrophomina phaseolina*, the Charcoal Rot Pathogen of Sorghum. *Biocatal. Agric. Biotechnol.* **2018**, *14*, 166–171. [CrossRef]
48. Kim, K.-J.; Sung, W.S.; Suh, B.K.; Moon, S.-K.; Choi, J.-S.; Kim, J.G.; Lee, D.G. Antifungal Activity and Mode of Action of Silver Nano-Particles on *Candida albicans*. *Biometals* **2009**, *22*, 235–242. [CrossRef] [PubMed]
49. Dawoud, T.M.; Yassin, M.A.; El-Samawaty, A.R.M.; Elgorban, A.M. Silver Nanoparticles Synthesized by *Nigrospora oryzae* Showed Antifungal Activity. *Saudi J. Biol. Sci.* **2021**, *28*, 1847–1852. [CrossRef]
50. Klasen, H.J. A Historical Review of the Use of Silver in the Treatment of Burns. II. Renewed Interest for Silver. *Burns* **2000**, *26*, 131–138. [CrossRef]
51. Kumar, S.A.; Abyaneh, M.K.; Gosavi, S.W.; Kulkarni, S.K.; Pasricha, R.; Ahmad, A.; Khan, M.I. Nitrate Reductase-Mediated Synthesis of Silver Nanoparticles from $AgNO_3$. *Biotechnol. Lett.* **2007**, *29*, 439–445. [CrossRef]
52. Vahabi, K.; Mansoori, G.A.; Karimi, S. Biosynthesis of Silver Nanoparticles by Fungus *Trichoderma reesei*. *Insci. J.* **2011**, *1*, 65–79. [CrossRef]
53. Benítez, T.; Rincón, A.M.; Limón, M.C.; Codón, A.C. Biocontrol Mechanisms of *Trichoderma* Strains. *Int. Microbiol.* **2004**, *7*, 249–260.
54. Chaloupka, K.; Malam, Y.; Seifalian, A.M. Nanosilver as a New Generation of Nanoproduct in Biomedical Applications. *Trends Biotechnol.* **2010**, *28*, 580–588. [CrossRef]
55. Guzman, M.; Dille, J.; Godet, S. Synthesis and Antibacterial Activity of Silver Nanoparticles against Gram-Positive and Gram-Negative Bacteria. *Nanomedicine* **2012**, *8*, 37–45. [CrossRef]
56. Li, G.; He, D.; Qian, Y.; Guan, B.; Gao, S.; Cui, Y.; Yokoyama, K.; Wang, L. Fungus-Mediated Green Synthesis of Silver Nanoparticles Using *Aspergillus terreus*. *Int. J. Mol. Sci.* **2012**, *13*, 466–476. [CrossRef] [PubMed]
57. Kim, S.H.; Lee, H.S.; Ryu, D.S.; Choi, S.J.; Lee, D.S. Antibacterial Activity of Silver-Nanoparticles against *Staphylococcus aureus* and *Escherichia coli*. *Microbiol. Biotechnol. Lett.* **2011**, *39*, 77–85.

MDPI AG
Grosspeteranlage 5
4052 Basel
Switzerland
Tel.: +41 61 683 77 34

Microorganisms Editorial Office
E-mail: microorganisms@mdpi.com
www.mdpi.com/journal/microorganisms